CRC Handbook
of
Biological Effects
of
Electromagnetic
Fields

Editors

Charles Polk, Ph.D.
Department of Electrical Engineering
University of Rhode Island
Kingston, Rhode Island

and

Elliot Postow, Ph.D.
Electromagnetic Radiation Program Manager
Naval Medical Research and Development Command
National Naval Medical Center
Bethesda, Maryland

CRC Press, Inc.
Boca Raton, Florida

Library of Congress Cataloging in Publication Data
Main entry under title:

CRC handbook of biological effects of electromagnetic
    fields.

    Includes bibliographies and index.
    1. Electromagnetism — Physiological effect — Handbooks,
manuals, etc.   I. Polk, Charles.   II. Postow,
Elliot.   III. Title: Handbook of biological effects
of electromagnetic fields.
QP82.2.E43C73   1986   574.19'17   85-9629
ISBN 0-8493-3265-6

    Direct all inquiries to CRC Press, Inc., 2000 Corporate Blvd., N.W., Boca Raton, Florida, 33431.

© 1986 by CRC Press, Inc.
Second Printing, 1987
Third Printing, 1988

International Standard Book Number 0-8493-3265-6
Library of Congress Card Number 85-9629
Printed in the United States

FOREWORD

The objective of this book is to present in a concise manner what is actually known at the present time about the biological effects of time invariant, low frequency and radio frequency (including microwave) electric and magnetic fields. In reviewing the vast amount of experimental data which have been obtained in recent years, the authors tried to select those results that are, in their opinion, of major importance and of lasting value. In discussing mechanisms of interaction of electromagnetic fields with living matter they have tried to differentiate between what is clearly established, what is suggested by available evidence without being convincingly proven, and what is conjecture at the present time.

No quantitative discussion of the interaction of electric and magnetic fields with living matter, and in fact no reasonable qualitative discussion, is possible without knowledge of the macroscopic electric and magnetic properties of living tissue. Therefore in Part I of the book theoretical and experimental aspects of dielectric permittivity and electrical conductivity of biological materials are treated in some detail. Description and analysis of the variation of these properties with frequency lead also to a discussion of microscopic level phenomena, although some additional modecular level processes are discussed in subsequent chapters.

Part II of the book is then concerned with phenomena that do not involve radiation, while Part III deals with electromagnetic energy which mostly enters living tissue in the form of radiated electromagnetic waves. The introduction is concerned primarily with two topics: (1) fundamental considerations important for evaluating how much of the extremely low frequency (ELF) to radio frequency (RF) electromagnetic energy present in the environment of an organism can reach its interior and (2) the difference between ''ionizing radiation'' and ''nonionizing radiation''.

Parts I, II, and III of this book are relatively independent of each other, so that they can be consulted directly by the reader whose interest is restricted to a particular, specialized topic. On the other hand the sequence of chapters is such that the book can also be used as a text for a one semester or one year course on biological effects of electromagnetic fields. One of the editors (Charles Polk) has used the manuscript in this manner with a group of graduate students at the University of Wisconsin-Madison drawn from Electrical Engineering, Medical Physics, Medicine, and Biology. Prerequisites for enrollment in the course were a good intermediate level undergraduate course in Electricity and Magnetism and at least an introductory course in Biology. The reader with such background should find this book useful as an introduction to the topics listed in the Table of Contents, as a review of recent research and as a guide to the specialized literature.

THE EDITORS

Dr. Charles Polk is Professor of Electrical Engineering at the University of Rhode Island. He was chairman of the Electrical Engineering Department at the University of Rhode Island from 1959 to 1979 and during leaves, Visiting Professor at Stanford University, Stanford, Calif. (1968/1969) and the University of Wisconsin/Madison (1983/1984). From 1975 to 1977 he was Head of Electrical Sciences and Analysis and Acting Director of the Engineering Division of the National Science Foundation in Washington, D.C. A native of Vienna, Austria, Dr. Polk has studied at the University of Paris (Sorbonne) and is a graduate of Washington University in St. Louis (B.S.E.E.) and the University of Pennsylvania in Philadelphia (M.S., Physics; Ph.D., Electrical Engineering). He worked on electrical devices, antennas, and radio propagation at R.C.A. in Camden and Princeton, N.J. and was a contributor to several books and an author of papers on electromagnetic wave propagation, antennas, electromagnetic noise of natural origin, and interaction of electromagnetic fields with ions in living systems. He was elected Fellow by the Institute of Electrical and Electronic Engineers "for contributions to understanding earth-ionosphere cavity resonances, and for leadership in engineering education". He is also a member of the Bioelectromagnetics Society, the American Geophysical Union, the American Society for Engineering Education, the New York Academy of Sciences, and AAAS.

Dr. Elliot Postow is Associate Scientific Director and Electromagnetic Radiation Program Manager at the Naval Medical Research and Development Command Bethesda, Md. From 1972 to 1978 he was Assistant Director of the Electromagnetic Radiation Project Office at the Navy Bureau of Medicine and Surgery. He was a Scientific Officer with the Office of Naval Research from 1968 to 1972. Dr. Postow is a graduate of City College in New York (B.S. in physics) and Michigan State University (Ph.D. in biophysics). He was Editor of *Bioelectromagnetics* from 1980 to 1984. Along with Shiro Takashima he edited "The Interaction of Acoustical and Electromagnetic Fields with Biological Systems" (volume 86 in Progress in Clinical and Biological Research published by Alan R. Liss, Inc., New York). In 1984 he was co-organizer of a symposium in Florence, Italy on the Interaction of Electromagnetic Fields with Biological Systems that was sponsored by the International Union of Radio Sciences (URSI). He is President of the Bioelectromagnetics Society form 1985 to 1986.

CONTRIBUTORS

Eleanor R. Adair
Associate Fellow
John B. Pierce Foundation
and
Senior Research Associate in Psychology
Yale University
New Haven, Connecticut

Frank S. Barnes
Professor
Department of Electrical and Computer
  Engineering
University of Colorado
Boulder, Colorado

Kenneth R. Foster
Associate Professor
Department of Bioengineering
University of Pennsylvania
Philadelphia, Pennsylvania

Richard B. Frankel
Senior Research Scientist
F. Bitter National Magnet Laboratory
Massachusetts Institute of Technology
Cambridge, Massachusetts

James C. Lin
Professor and Head
Department of Bioengineering
University of Illinois
Chicago, Illinois

Sol M. Michaelson
Professor
Department of Radiation Biology and
  Biophysics
University of Rochester
Rochester, New York

Morton W. Miller
Associate Professor
Radiation Biology and Biophysics
  Department
University of Rochester
Rochester, New York

Charles Polk
Professor
Department of Electrical Engineering
University of Rhode Island
Kingston, Rhode Island

Elliot Postow
Electromagnetic Radiation Program
  Manager
Naval Medical Research and
  Development Command
National Naval Medical Center
Bethesda, Maryland

Herman P. Schwan
Alfred Fitler Moore Professor Emeritus
Department of Bioengineering
University of Pennsylvania
Philadelphia, Pennsylvania

Maria A. Stuchly
Research Scientist
Radiation Protection Bureau
Health and Welfare Canada
Ottawa, Ontario, Canada

Stanislaw S. Stuchly
Professor of Electrical Engineering
Department Electrical Engineering
University of Ottawa
Ottawa, Ontario, Canada

Mays L. Swicord
Center for Devices and Radiological
  Health
Food and Drug Administration
Rockville, Maryland

T. S. Tenforde
Senior Scientist
Biology and Medicine Division
Lawrence Berkeley Laboratory
University of California
Berkeley, California

TABLE OF CONTENTS

INTRODUCTION

Charles Polk

TABLE OF CONTENTS

I. NEAR FIELDS AND RADIATION FIELDS

In recent years it has become, unfortunately, a fairly common practice, particularly in nontechnical literature, to refer to the entire subject of interaction of electric (E) and magnetic (H) fields with organic matter as biological effects of nonionizing radiation, although fields that do not vary with time and — for most practical purposes — slowly time-varying fields do not involve radiation at all. The terminology had its origin in an effort to differentiate between relatively low energy microwave radiation and high energy radiation, such as UV and X-rays, capable of imparting enough energy to a molecule or atom to disrupt its structure by removing one or more electrons. However, when applied to direct current (DC) or extremely low frequency (ELF) the terminology "nonionizing radiation" is inappropriate and misleading.

A structure is capable of efficiently radiating electromagentic waves only when its dimensions are significant in comparison with the wavelength λ. But in free space $\lambda = c/f$, where c = velocity of light in vacuum ($3 \cdot 10^8$ m/sec) and f = frequency in hertz (c/sec); therefore the wavelength at the power distribution frequency of 60 Hz, e.g., is 5000 km, guaranteeing that most available man-made structures are much smaller than one wavelength.

The poor radiation efficiency of electrically small structures (i.e., structures whose largest linear dimension $L \ll \lambda$) can be illustrated easily for linear antennae. In free space the radiation resistance R_r of a current element, i.e., an electrically short wire of length ℓ carrying uniform current along its length,[1] is

$$R_r = 80\pi^2 \left(\frac{\ell}{\lambda}\right)^2 \tag{1}$$

while the R_r of an actual center-fed radiator of total length ℓ with current going to zero at its ends, as illustrated in Figure 1 is

$$R_r' = 20\pi^2 \left(\frac{\ell}{\lambda}\right)^2 \tag{2}$$

Thus, the R_r of a 0.01-λ antenna — 50 km long at 60 Hz — would be 0.0197 Ω. Since the radiated power $P_r = I^2 R_r$ where I is the antenna terminal current, while the power dissipated as heat in the antenna wire is $I^2 R_d$ when I is uniform,* the P_r will be very much less than the power used to heat the antenna, given that the ohmic resistance R_d of any practical wire at room temperature will be very much larger than R_r (e.g., the resistance of a 50-km long, $^1/_2$-in. diameter solid copper wire would be 6.65 Ω). At DC, of course, no radiation of any sort takes place, since acceleration of charges is a condition for radiation of electromagnetic waves.

The second set of circumstances which guarantees that any object subjected to low frequency E and H fields usually does not experience effects of radiation is that any configuration that carries electric currents sets up E- and H-field components which store energy without contributing to radiation. A short, linear antenna in free space (short electric dipole) generates in addition to the radiation field E_r, an electrostatic field E_s, and an induction field E_i. Neither E_s nor E_i contribute to the P_r.[2,3] While E_r varies as $1/r$ where r is the distance from the antenna, E_i varies as $1/r^2$ and E_s as $1/r^3$. At a distance from the antenna of approximately one sixth wavelength ($r = \dfrac{\lambda}{2\pi}$) the E_i is equal to the E_r, and when $r \ll \lambda/6$ the E_r quickly

---

* For nonuniform current $I = I_0(1 - \dfrac{2|x|}{\ell})$ the dissipated power becomes $I_0^2 R_d/3$.

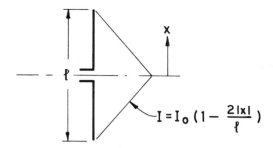

FIGURE 1.   Current distribution on short, thin, center-
fed antenna.

becomes negligible in comparison with E_i and E_s. Similar results are obtained for other antenna configurations.[4] At 60 Hz the distance $\lambda/2\pi$ corresponds to about 800 km and objects at distances of a few kilometers or less from a 60-Hz system are exposed to nonradiating field components which are orders of magnitude larger than the part of the field that contributes to radiation.

A living organism exposed to a static (DC) field or to a nonradiating near field may extract energy from it, but the quantitative description of the mechanism by which this extraction takes place is very different than at higher frequencies, where energy is transferred by radiation:

1. In the near field the relative magnitudes of E and H are a function of the current or charge configuration and the distance from the electric system. The E field may be much larger than the H field or vice versa (see Figure 2).
2. In the radiation field the ratio of the E to H is fixed and equal to 377 in free space (when E is given in volt per meter and H in ampere per meter).
3. In the vicinity of most presently available man-made devices or systems carrying static electric charges, DCs, or low frequency ($<$ 1000 Hz) currents, the E and H fields will only under very exceptional circumstances be large enough to produce heating effects inside a living object, as illustrated by Figure 3. (This statement assumes that the living object does not form part of a conducting path that permits direct entrance of current from a wire or conducting ground.) However, nonthermal effects are possible; thus an E field of sufficient magnitude may orient dipoles, or translate ions or polarizable neutral particles (see Part II, Chapters 1 and 2).
4. With radiated power it is relatively easy to produce heating effects in living objects with presently available man-made devices (see Part III). This does not imply, of course, that all biological effects of radiated radio frequency (RF) power necessarily arise from temperature changes.

Results of experiments involving exposure of organic materials and entire living organisms to static E and ELF E fields are described in the first three chapters of Part II. There, various mechanisms for the interaction of such fields with living tissue are also discussed. In the present Introduction we shall only point out that one salient feature of static (DC) and ELF E-field interaction with living organisms is that the external or applied E field is always larger by several orders of magnitude than the resultant average internal E field.[5,6] This is a direct consequence of boundary conditions derived from Maxwell's equations.[1-3]

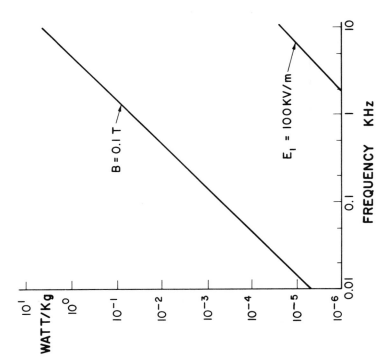

FIGURE 3. (a) Eddy current loss produced in cylinder by sinusoidally time-varying axial H field. Cylinder parameters are conductivity σ = 0.1 S/m, radius 0.1 m, density D = 1100 kg/m³, RMS magnetic flux density 0.1 T = 1000 G. Watt per kilogram = $\sigma B^2 r^2 \omega^2/8D$; see Equation 15 and use power per volume = J^2/σ; (b) loss produced by 60 Hz E field in Watt per kilogram = $\sigma E_{int}^2/D$ where external field E_1 is related to E_{int} by Equation 9 with $\epsilon_2 = \epsilon_0 \cdot 10^5$ at 1 kHz and $\epsilon_0 8 \cdot 10^4$ at 10 kHz.

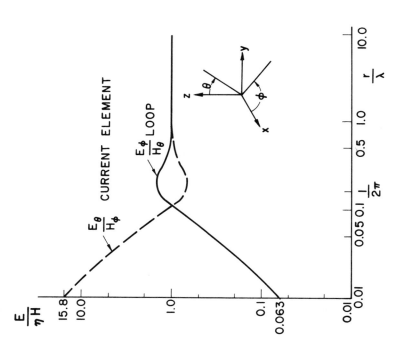

FIGURE 2. Ratio of E to H field (divided by wave impedance of free space η = 377Ω) at θ = 90°: for electric current element at origin along z-axis and for electrically small loop centered at the origin in the x-y plane.

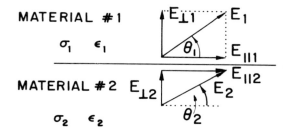

FIGURE 4.   Symbols used in description of boundary
conditions for E-field components.

II. PENETRATION OF DC AND LOW FREQUENCY
ELECTRIC FIELDS INTO TISSUE

Assuming that the two materials illustrated schematically in Figure 4 are characterized, respectively, by conductivities σ_1 and σ_2 and dielectric permittivities ϵ_1 and ϵ_2, we write E field components parallel to the boundary as $E_\parallel$ and components perpendicular to the boundary as $E_\perp$. For both static and time-varying fields

$$E_{\parallel 1} = E_{\parallel 2} \tag{3}$$

and for static (DC) fields

$$\sigma_1 E_{\perp 1} = \sigma_2 E_{\perp 2} \tag{4}$$

as a consequence of the continuity of current (or conservation of charge). The orientations of the total E fields in Media 1 and 2 can be represented by the tangents of the angles between the total fields and the boundary line

$$\tan \theta_1 = \frac{E_{\perp 1}}{E_{\parallel 1}} \quad \tan \theta_2 = \frac{E_{\perp 2}}{E_{\parallel 2}} \tag{5}$$

From these equations it follows that

$$\tan \theta_1 = \frac{\sigma_2}{\sigma_1} \frac{E_{\perp 2}}{E_{\parallel 1}} = \frac{\sigma_2}{\sigma_1} \frac{E_{\perp 2}}{E_{\parallel 2}} = \frac{\sigma_2}{\sigma_1} \tan \theta_2 \tag{6}$$

If Material 1 is air with conductivity[7] $\sigma_1 = 10^{-13}$ S/m and Material 2 a typical living tissue with $\sigma_2 \approx 10^{-1}$ S/m (compare Part I), $\tan \theta_1 = 10^{12}$, and therefore even if the field in Material 2 (the inside field) is almost parallel to the boundary so that $\theta_2 \approx 0.5^0$ or $\tan \theta_2 \approx (1/100)$, $\tan \theta_1 = 10^{10}$ or $\theta_1 = \frac{\pi}{2} - 10^{-10}$ radians. Thus an electrostatic field in air, at the boundary between air and living tissue, must be practically perpendicular to the boundary. The situation is virtually the same at ELF although Equation 4 must be replaced by

$$\sigma_1 E_{\perp 1} - \sigma_2 E_{\perp 2} = -j\omega \rho_s \tag{7}$$

and

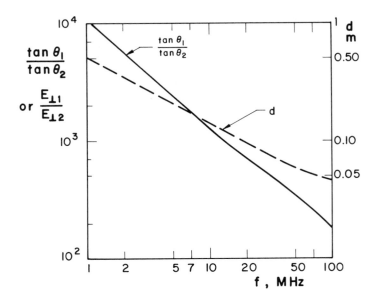

FIGURE 5.   Orientation of E-field components at air-muscle boundary (or ratio of fields perpendicular to boundary); depth (d) at which field component parallel to boundary surface decreases by approximately 50% (d = 0.693δ).

$$\epsilon_1 E_{\perp 1} - \epsilon_2 E_{\perp 2} = \rho_s \tag{8}$$

where $j = \sqrt{-1}$, ω is the radian frequency ($= 2\pi \times$ frequency), and ρ_s is the surface charge density. In Part I it is shown that at ELF the relative dielectric permittivity of living tissue may be as high as 10^6 so that $\epsilon_2 = 10^6 \epsilon_0$, where ϵ_0 is the dielectric permittivity of free space $(1/36\pi)10^{-9}$ F/m; however, it is still valid to assume that $\epsilon_2 \leq 10^{-5}$. Then from Equations 7 and 8

$$E_{\perp 1} = \frac{\sigma_2 + j\omega\epsilon_2}{\sigma_1 + j\omega\epsilon_1} E_{\perp 2} \tag{9}$$

which gives at 60 Hz with $\sigma_2 = 10^{-1}$ S/m, $\sigma_1 = 10^{-13}$ S/m, $\epsilon_2 \approx 10^{-5}$ F/m, and $\epsilon_1 \approx 10^{-11}$ F/m

$$E_{\perp 1} = \frac{10^{-1} + j4(10^{-3})}{10^{-13} + j4(10^{-9})} E_{\perp 2} \approx \frac{\sigma_2}{j\omega\epsilon_1} E_{\perp 2} = -j(2.5)10^7 E_{\perp 2} \tag{10}$$

This result, together with Equations 3 and 5, shows that for the given material properties, the field in air must still be practically perpendicular to the boundary of a living organism: $\tan \theta_1 \approx 2.5(10^7) \tan \theta_2$.

Knowing now that the living organism will distort the E field in its vicinity in such a way that the external field will be nearly perpendicular to the boundary surface, we can calculate the internal field by substituting the total field for the perpendicular field in Equation 4 (DC) and Equation 9 (ELF). For the assumed typical material parameters we find that in the static (DC) case

$$\frac{E \text{ internal}}{E \text{ external}} \approx 10^{-12} \tag{11}$$

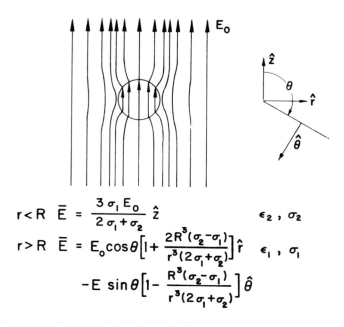

FIGURE 6. E field when sphere of radius R, conductivity σ_2, and dielectric permittivity ϵ_2 is placed into an initially uniform static field ($E = \hat{z} E_0$) within a medium with conductivity σ_1 and permittivity ϵ_1. The surface charge density is

$$\rho_f = \frac{3(\sigma_2\epsilon_1 - \sigma_1\epsilon_2) E_0}{2\sigma_1 + \sigma_2} \cos\theta \quad C/m^2$$

and for 60 Hz

$$\frac{E\ \text{internal}}{E\ \text{external}} \approx 4(10^{-8}) \tag{12}$$

Thus, a 60-Hz external field of 100 kV/m will produce an average internal E field of the order of 4 mV/m.

If the boundary between air and the organic material consists of curved surfaces instead of infinite planes, the results will be modified only slightly. Thus, for a finite sphere (with ϵ and σ as assumed here) embedded in air, the ratios of the internal field to the undisturbed external field will vary with the angle θ and distance r indicated in Figure 6, but will not deviate from the results indicated by Equations 7 and 8 by more than a factor of three.[3,8]

III. DC AND LOW FREQUENCY MAGNETIC FIELDS

DC H fields are considered in Part II, Chapter 4, and ELF H fields in Part II, Chapter 5. Since the magnetic permeability μ of most biological materials is practically equal to the magnetic permeability μ_0 of free space, $4\pi(10^{-7})$ H/m, the DC or ELF H field "inside" will be practically equal to the H field "outside". The only exception is organisms such as the magnetotactic bacteria described in Part II, Chapter 4, which synthesize ferromagnetic material. The known and suggested mechanisms of interaction of DC H fields with living matter discussed in Part II, Chapter 4 are

1.   Orientation of ferromagnetic particles, including biologically synthesized particles of magnetite.
2.   Orientation of diamagnetically or paramagnetically anisotropic molecules and/or cellular elements.[9]
3.   Generation of potential differences at right angles to a stream of moving ions (Hall effect) as a result of the magnetic force $F_m = qvB\sin\theta$, where q = the electric charge, v = velocity of the charge, B = magnetic flux density, and $\sin\theta$ = sine of the angle θ between the directions of v and B. One well-documented result of this mechanism is a "spike" in the electrocardiogram of vertebrates subjected to large DC H fields.[10,11]
4.   Changes in intermediate products or structural arrangements in the course of light-induced chemical (electron transfer) reactions, brought about by Zeeman splitting of molecular energy levels or effects upon hyperfine structure. (The Zeeman effect is the splitting of spectral lines, characteristic of electronic transitions, under the influence of an external H field; hyperfine splitting of electronic transition lines in the absence of an external H field is due to the magnetic moment of the nucleus; such hyperfine splitting can be modified by an externally applied H field.) The magnetic flux densities involved depend upon the particular system and can be as high as 2000 G but also < 100 G. Bacterial photosynthesis and effects upon the visual system are prime candidates for this mechanism.[12,13]
5.   Induction of E fields with resulting electrical potential differences and currents within an organism by rapid motion through a large static H field. Some magnetic phosphenes are due to such motion.[14]

Time-varying H fields, which are discussed in Part II, Chapter 5, may interact with living organisms through the same mechanisms that can be triggered by static H fields, provided the variation with time is slow enough to allow particles of finite size and mass, located in a viscous medium, to change orientation or position where required (Mechanisms 1 and 2), and provided the field intensity is sufficient to produce the particular effect. However, time-varying H fields, including ELF H fields, can also *induce* electric currents into stationary conducting objects. Thus, all modes of interaction of time-varying E fields with living matter may be triggered by time-varying, but not by static, H fields.

In view of Faraday's law, a time-varying magnetic flux will induce E fields with resulting electrical potential differences and "eddy" currents through available conducting paths. Since very large external ELF E fields are required (as indicated by Equations 9 to 12) to generate even small internal E fields, many man-made devices and systems generating both ELF E and H fields are more likely to produce physiologically significant internal *E* fields through the mechanism of *magnetic* induction.

The induced voltage V around some closed path is given by

$$V = \oint \mathbf{E} \cdot d\ell = -\iint \frac{\partial \mathbf{B}}{\partial t} \cdot ds \qquad (13)$$

where **E** is the induced E field. The integration $\oint \mathbf{E} \cdot d\ell$ is over the appropriate conducting path, $\partial \mathbf{B}/\partial t$ is the time derivative of the magnetic flux density, and the "dot" product with the surface element, ds, indicates that only the component of $\partial \mathbf{B}/\partial t$ perpendicular to the surface (i.e., parallel to the direction of the vector ds) enclosed by the conducting path, induces an E field. To obtain an order-of-magnitude indication of the induced current that can be expected as a result of an ELF H field, we consider the circular path of radius r, illustrated by Figure 7. Equation 13 then gives the magnitude of the E field as

$$E = \frac{\omega Br}{2} \qquad (14)$$

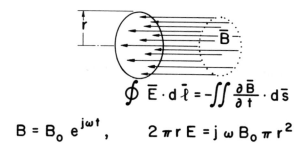

$$\oint \bar{E} \cdot d\bar{\ell} = -\iint \frac{\partial \bar{B}}{\partial t} \cdot d\bar{s}$$

$$B = B_0 \, e^{j\omega t}, \qquad 2\pi r E = j\omega B_0 \pi r^2$$

FIGURE 7. Circular path (loop) of radius r enclosing uniform magnetic flux density perpendicular to the plane of the loop. For sinusoidal time variation $B = B_0 e^{j\omega t}$.

where $\omega = 2\pi f$ and f = frequency. The magnitude of the resulting electric current density J in ampere per square meter is*

$$J = \sigma E = \frac{\sigma \omega B r}{2} \tag{15}$$

where σ is the conductivity along the path in Siemans per meter. In SI (Systeme Internationale) units used throughout this book, B is measured in tesla (1 T = 10^4 G) and r in meters. Choosing for illustration a circular path of 0.1-m radius, a frequency of 60 Hz, and a conductivity of 0.1 S/m, Equations 14 and 15 give E = 18.85 B and J = 1.885 B. The magnetic flux density required to obtain a potentially physiologically significant current density (see Section II.2) of 1 mA/m² is 0.53 mT or about 5 G. The E field induced by that flux density along the circular path is 10 mV/m. To produce this same 10-mV/m internal E field by an external 60-Hz *E* field would require, by Equation 12, a field intensity of 250 kV/m!

Since the induced voltage is proportional to the time rate of change of the H field (Equation 13), implying a linear increase with frequency (Equation 14), one would expect that the ability of a time-varying H field to induce currents deep inside a conducting object would increase indefinitely as the frequency increases; or conversely, that the magnetic flux density required to induce a specified E field would decrease linearly with frequency, as indicated in Figure 8. This is not true however, because the displacement current density $\partial \mathbf{D}/\partial t$, where $\mathbf{D} = \epsilon \mathbf{E}$, must also be considered as the frequency increases. This leads to the wave behavior discussed in Part III, implying that at sufficiently high frequencies the effects of both external E *and* H fields are limited by reflection losses (Figures 9 and 10) as well as by skin effect[15] (i.e., limited depth of penetration; d in Figure 5).

IV. RF FIELDS

At frequencies well below those where most animals and many field-generating systems have dimensions of the order of one free space wavelength — e.g., at 10 MHz where λ =

---

* Equation 15 neglects the H field generated by the induced eddy currents. If this field is taken into account, it can be shown that the induced current density in a cylindrical shell of radius r and thickness Δ is given by $[-jrH_0/\delta^2]/[1 + j\Delta r/\delta^2]$, where $H_0 = B_0/\mu_0$ and δ is the skin depth defined by Equation 17 in this Introduction. However, for conductivities of biological materials ($\sigma < 5$ S/m) one obtains at audio frequencies $\delta > 1$ m, and since for most dimensions of interest $\Delta r < 0.01$ m², the term $j\Delta r/\delta^2$ becomes negligible. The result — $-jrH_0/\delta^2$ — is then identical with Equation 15.

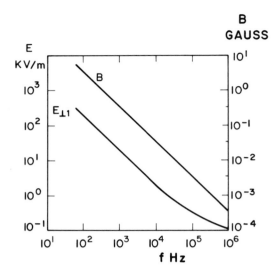

FIGURE 8.   External E and H field required to obtain an internal E field of 10 mV/m (conductivity and dielectric permittivity for skeletal muscle from Part I, Figure 10 — H-field calculation assumes a circular path of 0.1-m radius perpendicular to magnetic flux).

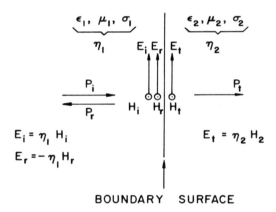

FIGURE 9.   Reflection and transmission of an electromagnetic wave at the boundary between two different media: perpendicular incidence (P_i = incident power, P_r = reflected power, P_t = transmitted power).

30 m — the skin effect limits penetration of the external field. This phenomenon is fundamentally different from the small ratio of internal to external E fields described in Equation 4 (applicable to DC) and Equation 9.

Equation 9 expresses a "boundary condition" applicable at all frequencies, but as the radian frequency ω increases (and in view of the rapid decrease with frequency of the dielectric permittivity ϵ_2 in biological materials — see Part I), the ratio of the normal component of the external to the internal E field at the boundary decreases with increasing frequency. This is illustrated by Figure 5 where $\tan \theta_1/\tan \theta_2$ is also equal to $E_{\perp 1}/E_{\perp 2}$ in

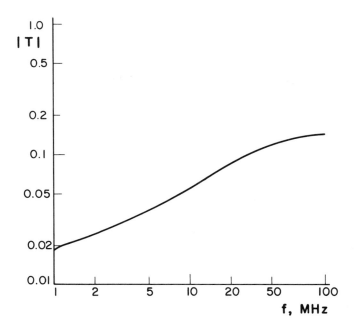

FIGURE 10.   Magnitude of transmission coefficient T for incident E field parallel to boundary surface. $T = E_t/E_i$; reflection coefficient $\Gamma = E_r/E_i$ $= T - 1$. Γ and T are complex numbers (ϵ_r and σ for skeletal muscle from Part I).

view of Equations 3, 5, and 9. However, at low frequencies the total field inside the boundary can be somewhat larger than the perpendicular field at the boundary and any field variation with distance from the boundary is not primarily due to energy dissipation, but in a homogeneous body is a consequence of shape. At RFs, on the other hand, the E *and* H fields of the incoming electromagnetic wave, after reflection at the boundary, are further decreased due to energy dissipation. Both E and H fields decrease exponentially with distance from the boundary

$$g(z) = Ae^{-\frac{z}{\delta}} \tag{16}$$

where g(z) is the field at the distance z and A the magnitude of the field just *inside* the boundary.

As defined by Equation 16 the skin depth δ is the distance over which the field decreases to 1/e ($= 0.368$) of its value just *inside* the boundary. (Due to reflection, the field A just inside the boundary can already be very much smaller than the incident external field; see Figures 9 and 10.)

Expressions for δ given below were derived[2,3,15,16] for plane boundaries between infinite media. They are reasonably accurate for cylindrical structures if the radius of curvature to skin depth (r_0/δ) is larger than about five.[15] For a good conductor

$$\delta = \frac{1}{\sqrt{\pi f \mu \sigma}} \tag{17}$$

where a good conductor is one for which the ratio p of conduction current, $J = \sigma E$, to displacement current, $\partial D/\partial t = \epsilon(\partial E/\partial t) = j\omega\epsilon E$ is large:

$$p = \frac{\sigma}{\omega\epsilon} \gg 1 \qquad (18)$$

Since for most biological materials p is of the order of one $(0.1 < p < 10)$ over a very wide frequency range (see Part I), it is frequently necessary to use the more general expression[15]

$$\delta = \frac{1}{\omega\left[\frac{\mu\epsilon}{2}\left(\sqrt{1 + p^2} - 1\right)\right]^{1/2}} \qquad (19)$$

The decrease of field intensity with distance from the boundary surface indicated by Equation 16 becomes significant for many biological objects at frequencies where $r_0/\delta \geqslant 5$ is not satisfied. However, the error resulting from the use of Equations 16 and 17 or 19 with curved objects is less when $z < \delta$; thus at $z = 0.693\ \delta$, where $g(z) = 0.5\ A$ from Equations 16 and 17, the correct values of $g(z)$, obtained by solving the wave equation in cylindrical coordinates, differs only by 20% (it is 0.6 A) even when r_0/δ is as small as 2.39.[16] Therefore, on Figure 5 the distance $d = 0.693\ \delta$, at which the field decreases to half of its value just inside the boundary surface, rather than δ is plotted for the frequency range 1 to 100 MHz, using Equation 19 with typical values for σ and ϵ for muscle from Figure 11 of Part I. It is apparent that the skin effect becomes significant for humans and larger vertebrates at frequencies > 10 MHz.

Directly related to skin depth, which is defined for fields varying sinusoidally with time, is the fact that a rapid transient variation of an applied magnetic flux density constitutes an exception to the statement that the DC H field inside the boundary is equal to the H field outside. Thus, from one viewpoint one may consider the rapid application or removal of a DC H field as equivalent to applying a high frequency field during the switching period, with the highest frequencies present being of the order of $1/\tau$, where τ is the rise time of the applied step function. Thus, if $\tau < 10^{-8}$ sec, the skin effect will be important during the transient period, since d in Figure 5 is < 5 cm above 100 MHz. It is also possible to calculate directly the magnetic flux density inside a conducting cylinder as a function of radial position r and time t when a magnetic pulse is applied in the axial direction.[17,18] Assuming zero rise time of the applied field B_0, i.e., a true step function, one finds that the field inside a cylinder of radius a is

$$B = B_0\left[1 - \sum_{k=1}^{\infty} J_0\left(r\frac{\nu_k}{a}\right) e^{-1/T_k}\right] \qquad (20)$$

where $J_0(r\ \nu_k/a)$ is the zero order Bessel function of argument $r\ \nu_k/a$ and the summation is over the nulls of J_0 designated ν_k (the first four values of ν_k are 2.405, 5.520, 8.654, and 11.792).* T_k is the rise time of the k'th term in the series and is given by

$$T_k = \frac{\mu_0\sigma a^2}{\nu_k} \qquad (21)$$

As ν_k increases, the rise time decreases and therefore the longest delay is due to the first term in the summation with $k = 1$:

---

*   This result is based upon the solution of $\partial\mathbf{B}/\partial t = (1/\mu\sigma)\nabla^2\mathbf{B}$ which is a consequence of Ampere's and Faraday's laws when the displacement current is disregarded. Equations 20 to 22 are therefore only correct when $p \gg 1$.

$$T_1 = \frac{\mu_0 \sigma a^2}{2.405} \qquad (22)$$

For a cylinder with 0.1-m radius and a conductivity $\sigma \approx 1$ S/m, which is a typical value for muscle between 100 and 1000 MHz, Equation 22 gives $T_1 = 2.6(10^{-8})$ sec. This finite rise time (or decay time in case of field removal) of the internal H field may be of some importance when pulsed H fields are used therapeutically.[19] It might also be used to measure noninvasively the conductivity of biological substances in vivo through determination of the final decay rate of the voltage induced into a probe coil by the slowly decaying internal field after the applied field is removed.[18]

Properties of biological substances in the intermediate frequency range, above ELF (> 300 Hz), and below the higher RFs, where wave behavior and skin effect begin to be important (~ 20 MHz), are discussed in Part I. However, while Part II is concerned with biological effects at DC and frequencies below a few kilohertz, Part III deals primarily with the higher RFs, > 50 MHz. One reason for this limited treatment of the intermediate frequency range is that very little animal data are available for this spectral region in comparison with the large number of experiments performed at ELF and microwave frequencies in recent years. Another reason is that most electrical processes known to occur naturally in biological systems — action potentials, EKG, EEG, ERG, etc. — are limited to DC and ELF frequencies. Thus one might expect some physiological effects from external fields of appropriate intensity in the same frequency range, even if the magnitude of such fields is not large enough to produce thermal effects. As illustrated by Figures 3 and 8, most E fields below 100 kHz set up by presently used man-made devices, and most except the very largest H fields below 10 kHz, are incapable of producing thermal effects in living organisms. (This excludes, of course, fields accompanying currents directly introduced into the organism via electrodes.) Thus, the frequencies between about 10 and 100 kHz have been of relatively little interest because they are not very likely to produce thermal or other biological effects. On the other hand, the higher RFs are frequently generated at power levels where enough energy may be introduced into living organisms to produce local or general heating. In addition, despite skin effect and the reflection loss to be discussed in more detail below, microwaves modulated at an ELF rate may serve as a vehicle for introducing ELF fields into a living organism of at least the same order of magnitude as would be introduced by direct exposure to ELF. Any effect of such ELF-modulated microwaves would, of course, require the existence of some amplitude-dependent demodulation mechanism to extract the ELF from the microwave carrier.

Part III is divided into five parts. Chapters 1 and 2 give the information necessary for establishing the magnitude of the fields present in biological objects: (1) experimental techniques and (2) analytical methods for predicting field intensities without construction of physical models made with "phantom" materials (i.e., dielectric materials with properties similar to those of living objects which are to be exposed). Since thermal effects at microwave frequencies are certainly important — although they may not be the only biological effects of this part of the spectrum — and since some (but not all) thermal effects occur at levels where the thermoregulatory system of animals is activated, thermoregulation in the presence of microwave fields is discussed in Part III, Chapter 3. Not only are therapeutic applications of microwaves based upon their thermal effects, but experimental establishment of possible nonthermal effects at the threshold of large scale tissue heating in particular living systems also requires thorough understanding of thermoregulatory mechanisms.

The vast amount of experimental data obtained on animal systems exposed to microwaves is discussed in Part III, Chapters 4 and 5. Part III, Chapter 4 emphasizes primarily application of nonmodulated fields or of modulated fields, where the type of modulation had no apparent effect other than modification of the average power level, while Part III, Chapter 5 considers

Table 1

f (MHz)	σ	ϵ_r	$\dfrac{\sigma}{\omega\epsilon_0\epsilon_r} = p$
1	0.40	2000	3.6
10	0.63	160	7.1
100	0.89	72	2.2
10^3	1.65	50	0.59
10^4	10.3	40	0.46
10^5	80	6	2.4

in some detail modulated fields and experimental evidence for nonlinear effects characterized by ''amplitude windows'' or ''frequency windows''.

At the higher radio frequencies, the external E field is not necessarily perpendicular to the boundary of biological materials (see Figures 4 and 5), and the ratio of the *total* external E field to the *total* internal field is not given by Equation 9. However, the skin effect (Equations 16 to 19) and reflection losses still reduce the E field within any biological object below the value of the external field. As pointed out in Part I, dielectric permittivity and electrical conductivity of organic substances both vary with frequency. At RFs, most biological substances are neither very good electrical conductors nor very good insulators — with the exception of cell membranes, which are good dielectrics at RF (at ELF cell membranes can act as conductors or dielectrics, but are ion selective[20-22]). The ratio p (Equation 18) is neither much smaller, nor very much larger than one as shown for typical muscle tissue[23,24] in Table 1.

Reflection loss at the surface of an organism is a consequence of the difference between its electrical properties and those of air. Whenever an electromagnetic wave travels from one material into another with different electrical properties, the boundary conditions (Equations 3 and 8) and similar relations for the H field require the existence of a reflected wave. The expressions for the reflection coefficient

$$\Gamma = \frac{E_r}{E_i} \tag{23}$$

and the transmission coefficient

$$T = \frac{E_t}{E_i} \tag{24}$$

become rather simple for loss-free dielectrics ($p \ll 1$) and for good conductors ($p \gg 1$). Since biological substances are neither, the most general expressions for Γ and T, applicable at plane boundaries, are needed.[3,15] For perpendicular incidence, illustrated by Figure 9,

$$\Gamma = \frac{\eta_2 - \eta_1}{\eta_2 + \eta_1} \tag{25}$$

$$T = \frac{2\,\eta_2}{\eta_2 + \eta_1} = 1 + \Gamma \tag{26}$$

where η_1 and η_2 are the wave impedances, respectively, of Medium 1 and 2. The wave impedance of a medium is the ratio of the E to the H field in a plane wave traveling through that medium; it is given by[15]

$$\eta = \left(\frac{j\omega\mu}{\sigma + j\omega\epsilon}\right)^{1/2} \qquad (27)$$

Clearly Γ and T are in general complex numbers, even when Medium 1 is air for which Equation 27 reduces to the real quantity $\eta_0 = \sqrt{\mu_0/\epsilon_0}$, because Medium 2, which here is living matter, usually has a complex wave impedance at RFs.

The incident, reflected, and transmitted powers are given by[15]

$$P_i = Rl|E_i|^2 \frac{1}{\eta_1^*} = \frac{|E_i|^2}{|\eta_1|^2} R_1 \qquad (28)$$

$$P_r = Rl|E_r|^2 \frac{1}{\eta_1^*} = \frac{|E_r|^2}{|\eta_1|^2} R_1 \qquad (29)$$

$$P_t = Rl|E_t|^2 \frac{1}{\eta_2^*} = \frac{|E_t|^2}{|\eta_2|^2} R_2 \qquad (30)$$

where the E fields are effective values ($E_{eff} = E_{peak}/\sqrt{2}$) of sinusoidal quantities, Rl signifies "real part of", η^* is the complex conjugate of η, and R_1 and R_2 are the real parts of η_1 and η_2. If Medium 1 is air, $\eta_1 = R_1 = 377\ \Omega$. It follows from Equations 23, 24, and 28 to 30 and conservation of energy that the ratio of the transmitted to the incident real power is given by

$$\frac{P_t}{P_i} = |T|^2 \frac{\eta_1\eta_2^* + \eta_1^*\eta_2}{2|\eta_2|^2} = 1 - \frac{P_r}{P_i} = 1 - |\Gamma|^2 \qquad (31)$$

The magnitude of the transmission coefficient T for the air-muscle interface over the 1- to 100-MHz frequency range is plotted in Figure 10, which shows that the magnitude of the transmitted E field in muscle tissue is considerably smaller than the E field in air. The fraction of the total incident power that is transmitted (Equation 31) is shown in Figure 11, indicating clearly that reflection loss at the interface decreases with frequency. However, for deeper lying tissue this effect is offset by the fact that the skin depth δ (Equation 19) also decreases with frequency (Figure 12) so that the total power penetrating beyond the surface decreases rapidly.

In addition to reflection at the air-tissue boundary, further reflections take place at each boundary between dissimilar materials. For example, the magnitude of the reflection coefficient at the boundary surface between muscle and organic materials with low water content, such as fat or bone, is shown in Table 2.

The situation is actually more complicated than indicated by Figures 10 and 11, because the wavefront of the incident electromagentic wave may not be parallel to the air-tissue boundary. Two situations are possible: the incident E field may be polarized perpendicular to the plane of incidence defined on Figure 13 (perpendicular polarization — Figure 13A) or parallel to the plane of incidence (parallel polarization — Figure 13B). The transmission and reflection coefficients[8] are different for the two types of polarization and also become functions of the angle of incidence α_1:

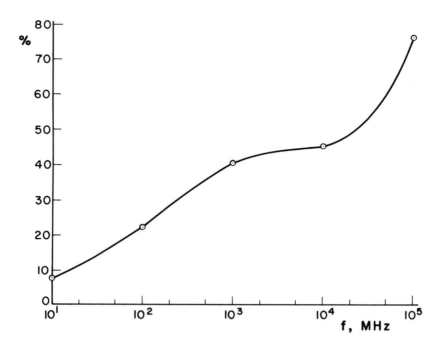

FIGURE 11.   Ratio of transmitted to incident power expressed as percent of incident power. Air-muscle interface, perpendicular incidence (Equation 31, Table 1).

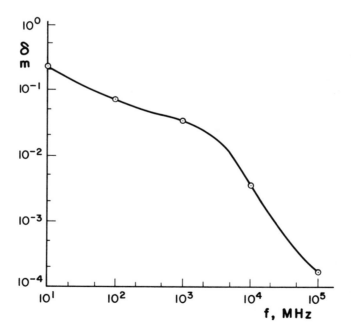

FIGURE 12.   Electromagnetic skin depth in muscle tissue from plane wave expression (Equation 19, Table 1).

Table 2

| f (MHz) | Fat or bone | | Muscle[a]—fat |
	σ (S/m)	ϵ_r	Γ
10^2	0.048	7.5	0.65
10^3	0.101	5.6	0.52
10^4	0.437	4.5	0.52

[a] σ and ϵ_r for muscle from Table 1.

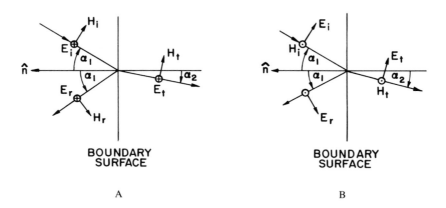

FIGURE 13.   Oblique incidence of an electromagnetic wave at the boundary between two different media. (A) Perpendicular polarization (E vector perpendicular to plane of incidence); (B) parallel polarization (E vector parallel to plane of incidence). The plane of incidence is the plane formed by the surface normal (unit vector n̂ and the direction of the incident wave); ⊗ indicates a vector into the plane of the paper; ⊙ indicates a vector out of the plane of the paper. The orientation of the field vectors in the transmitted field is shown for loss-free dielectrics. For illustration of the transmitted wave into a medium with finite conductivity, where the wave impedance η_2 becomes a complex number, see Reference 3.

Perpendicular polarization

$$T_{\perp} = \frac{2\,\eta_2\,\cos\,\alpha_1}{\eta_2\,\cos\,\alpha_1\,+\,\eta_1\,\cos\,\alpha_2} \qquad (32)$$

$$\Gamma_{\perp} = \frac{\eta_2\,\cos\,\alpha_1\,-\,\eta_1\,\cos\,\alpha_2}{\eta_2\,\cos\,\alpha_1\,+\,\eta_1\,\cos\,\alpha_2} \qquad (33)$$

Parallel polarization

$$T_{\parallel} = \frac{2\,\eta_2\,\cos\,\alpha_1}{\eta_2\,\cos\,\alpha_2\,+\,\eta_1\,\cos\,\alpha_1} \qquad (34)$$

$$\Gamma_{\parallel} = \frac{\eta_1\,\cos\,\alpha_1\,-\,\eta_2\,\cos\,\alpha_2}{\eta_2\,\cos\,\alpha_2\,+\,\eta_1\,\cos\,\alpha_1} \qquad (35)$$

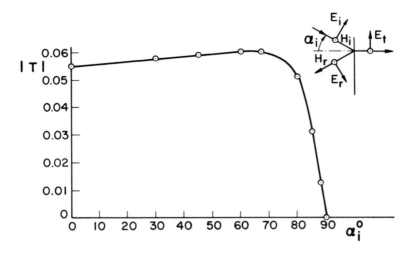

FIGURE 14.   Magnitude of complex transmission coefficient for parallel polarization vs. angle of incidence α_1 at 10 MHz (E field in plane of incidence, H field parallel to boundary plane; $\sigma_2 = 0.7$ S/m, $\epsilon_{r2} = 150$, $T = E_t/E_i$).

where α_2 is given by the generalized Snell's law (when both media have the magnetic permeability of free space) by

$$\sin \alpha_2 = \frac{\sqrt{\epsilon_1}}{\sqrt{\epsilon_2 - j\dfrac{\sigma_2}{\omega}}} \tag{36}$$

so that $\cos \alpha_2 = \sqrt{1 - \sin^2 \alpha_2}$ is a complex number unless $p_2 = (\sigma_2/\omega\epsilon_2) \ll 1$.

As illustration, the variation with angle of incidence of the transmission coefficient for parallel polarization at the air-muscle interface at 10 MHz, is shown in Figure 14. It is apparent that the transmitted field is not necessarily maximized by perpendicular incidence in the case of parallel polarization. Furthermore, whenever $p \approx 1$ or $p > 1$ (see Table 1 above) α_2 is complex, which causes the waves entering the tissue to be inhomogeneous — they are not simple plane waves, but waves where surfaces of constant phase and constant amplitude do not coincide;[3,25] only the planes of constant amplitude are parallel to the boundary surface.

Analytical solutions for nonplanar structures taking into account size and shape of entire animals have been given[26] and are also described in Part III, Chapter 2.

V. IONIZATION, IONIZING RADIATION, CHEMICAL BONDS, AND EXCITATION

Since RF fields can be characterized as nonionizing radiation, it is desirable to review the differences between ionizing and nonionizing radiation, to explain ionization phenomena and also to discuss related excitation phenomena which require less energy than ionization. We note first that the energy of electromagnetic waves is quantized with the quantum of energy (in joules) being equal to Planck's constant (see Appendix I) times the frequency. This energy can also be expressed in electronvolts, i.e., in multiples of the kinetic energy acquired by an electron accelerated through a potential difference of 1 V: 1 eV $\approx 1.6 (10^{-19})$ J. Energy quanta for a few frequencies are listed in Table 3.

Table 3

Name of radiation or application	Frequency (Hz)	Wavelength (m)	Energy of 1 quantum of radiation (eV)
UHF TV	$7\ (10^8)$	0.43	$2.88\ (10^{-6})$
Microwave radar	10^{10}	$3\ (10^{-2})$	$4.12\ (10^{-5})$
Millimeter wave	$3\ (10^{11})$	$1\ (10^{-3})$	$1.24\ (10^{-3})$
Visible light	$6\ (10^{14})$	$5\ (10^{-7})$	2.47
Ionizing UV	10^{16}	$3\ (10^{-8})$	41.2
Soft X-ray	10^{18}	$3\ (10^{-10})$	4120
Penetrating X-ray	10^{20}	$3\ (10^{-12})$	$4.12\ (10^5)$

Quantized energy can "excite" molecules; appropriate frequenceis can couple to vibrational and rotational oscillations, and if the incident energy quantum has sufficient magnitude it can tear an electron away from one of the constituent atoms. The energy required to remove one electron from the highest energy orbit of a particular chemical element is called its "ionization potential". Typical ionization potentials are of the order 10 eV — e.g., for the hydrogen atom it is 13.6 eV and for gaseous sodium it is 5.1 eV. Since chemical binding forces are essentially electrostatic, ionization implies profound chemical changes. Therefore ionization by any outside agent of the complex compounds that make up a living system leads to profound and often irreversible changes in the operation of that system.

Table 3 shows that even the highest RFs (millimeter waves) have quantum energies well below the ionization potential of any known substance; thus one speaks of nonionizing radiation when referring to electromagnetic waves below UV light frequencies. UV and higher frequency waves (X-rays, γ-rays) constitute ionizing radiation.

This explanation of the difference between ionizing and nonionizing radiation should not imply that nonionizing electromagnetic radiation cannot have profound effects upon inorganic and organic substances. Since excitation of coherent vibrational and rotational modes requires considerably less energy than ionization, it could occur at RFs; this will be discussed further on. In addition, many other possible biological effects require energies well below the level of ionizing potentials. Examples are tissue heating, dielectrophoresis, depolarization of cell membranes, mechanical stress due to piezoelectric transduction, or dielectric saturation resulting in the orientation of the polar sidechains of macromolecules leading to the breaking of hydrogen bonds. These and other mechanisms will be discussed by the authors, who will also give estimates of rates at which energy must be delivered to produce particular effects.

Returning to the discussion of ionization, it is important to note that ionization of a chemical element can be brought about not only by absorption of electromagnetic energy, but also by collision either with foreign (injected) elementary particles of the requisite energy, or by sufficiently violent collision among its own atoms. The latter process constitutes ionization by heating, or thermal breakdown, of a substance, which will occur when the kinetic energy of the colliding particles exceeds the ionization potential. Since the kinetic energy of elementary particles is related to temperature[27] by $W = kT$ where k is Boltzman's constant (see Appendix I), we find that the required temperature is

$$1.38\ (10^{-23})T \approx 5\ eV \approx (5)\ 1.6\ (10^{-19})\ J$$

$$T \approx 5\ (10^4)\ K$$

which is about twice the temperature inside a lightning stroke[28] and orders of magnitude higher than any temperature obtainable from electromagnetic waves traveling through air. Actually, initiation of lightning stroke is an example of ionization by collision with injected energetic elementary particles. A few free electrons and ions always present in the air due to ionization by cosmic rays are accelerated by the E fields generated within clouds to velocities corresponding to the required ionization energy. Only when the field is large enough to impart this energy over distances shorter than the mean free path of the free electrons or ions at atmospheric pressure can an avalanche process take place: each accelerated electron separates a low-energy electron from the molecule with which it collides, and in the process loses most of its own energy; thus, one high-energy free electron is exchanged for two free low-energy electrons and one positive ion. Both electrons are in turn accelerated again by the field, giving them high kinetic energy before they collide with neutral molecules; their collision produces four free electrons, etc. The breakdown field strength for air at atmospheric pressure is approximately $3(10^6)$ V/m, implying a mean free path of electrons

$$\Delta \ell \approx [5 \text{ eV}/3(10^6) \text{ V/m}] \approx 10^{-6} \text{ m}$$

This model is however not entirely accurate since the actual mean free path corresponds to energies of the order of 0.1 eV, which is only sufficient to excite vibrational modes in the target molecule. Apparently such excitation is sufficient to cause ionization if the collision process lasts long enough.[29]

Except for some laboratory conditions where a sufficiently high potential difference can be applied directly across a biological membrane to bring about its destruction, collisional ionization is generally not a factor in the interaction of electromagnetic waves with tissue: the potential difference required for membrane destruction[30] is between 100 and 300 mV, corresponding to a field strength of the order of $2(10^7)$ V/m (assuming a membrane thickness $d = 100$ Å; $E = V/d$). However, there is a third mechanism of ionization that is particularly important in biological systems. When a chemical compound of the type wherein positive and negative ions are held together by their electrostatic attraction, such as the ionic crystal NaCl, is placed into a suitable solvent, such as H_2O, it is separated into its ionic components. The resulting solution becomes an electrolyte, i.e., an electrically conducting medium in which the only charge carriers are ions.

In this process of chemical ionization the Na^+ cations and Cl^- anions are separated from the original NaCl crystal lattice and individually surrounded by a sheet of solvent molecules, the "hydration sheath". If the solvent is H_2O, this process is called "hydration", or more generally, for any solvent, "solvation".

A dilute solution of NaCl crystals in H_2O is slightly cooler than the original constituents were before the solvation process, indicating that internal energy of the system was consumed. Actually energy is consumed in breaking up the original NaCl bonds and some, but less, energy is liberated in the formation of the solvation sheath. The process is due to the interaction between the dipole moment of the solvent molecule (H_2O in our example) and the electric charges on the ions. Thus, solvents with higher relative dielectric constant ϵ_r indicating higher inherent electric dipole moment per unit volume (P), solvate ions more strongly ($\epsilon_r = 1 + P/[\epsilon_0 E]$ where E is the electric field applied during the measurement of ϵ_r). For example, H_2O with $\epsilon_r \approx 80$ solvates more strongly than methanol with $\epsilon_r \approx 33$. For biological applications it is worth noting that solvation may affect not only ionic substances, but also polar groups (i.e., molecular components which have an inherent dipole moment) such as $-C=O$, $-NH$, or $-NO_2$. Details of the process are discussed in texts on electrochemistry.[31,32]

In biological processes not only chemical ionization and solvation of ionic compounds but, of course, all kinds of chemical reactions, take place. One of the central questions in the study of biological effects of E and H fields is therefore not only whether they can cause or influence ionization, but whether they can affect — speed up, slow down or modify — any naturally occurring biologically important chemical reaction.

Table 4

Type of bond	Change in free energy (binding energy)	
	kcal/mol	eV/molecule
Covalent	50—100	2.2—4.8
Van der Waals	1—2	0.04—0.08
Hydrogen	3—7	0.13—0.30
Ionic[a]	5	0.2
Av thermal energy at 310 K	0.62	0.027

Note: For conversion from kcal/mol to eV/molecule see Appendix I.

[a] For ionic groups of organic molecules such as COO^-, NH_3^+ in aqueous solution.

In Table 4 typical energies for various types of chemical bonds are listed. For comparison the thermal energy per elementary particle at 310 K is also shown.

Complementing the numbers in Table 4 one should also point out that

1. The large spread in the energies of thermal motion guarantees that at physiological temperatures some molecules always have sufficient energy to break the strongest weak bonds.[33]
2. The average lifetime of a weak bond is only a fraction of a second.
3. The weak binding forces are effective only between surfaces in close proximity and usually require complementary structures such as a (microscopic) plug and hole (as are thought to exist between antigen and antibody).[33]
4. Most molecules in aqueous solution form secondary bonds.
5. The metabolism of biological systems continuously transforms molecules and therefore also changes the secondary bonds that are formed.

Comparison of the last columns in Tables 3 and 4 shows that millimeter waves have quantum energies which are only about one order of magnitude below typical Van der Waals energies (waves at a frequency of 10^{12} Hz with a quantum energy of 0.004 eV have a wavelength of 0.3 mm and can still be classified as millimeter waves). One might expect therefore that such waves could initiate chemically important events (i.e., configurational changes), e.g., by multiple transitions between closely spaced vibrational states at successively higher energy levels.[47]

Energies associated with transition from one to another mode of rotation of a diatomic molecule are given by $W = \ell(\ell + 1)A$,[27,34] where $\ell = 0,1,2,3...$ and $A = 6(10^{-5})$ eV; thus already an electromagnetic wave with a frequency of 29 GHz — still in the microwave region — can excite a rotational mode. *Vibrational* modes of diatomic molecules[27,34] correspond to energies of the order of 0.04 eV, requiring excitation in the IR region. Vibrational frequencies in a typical H-bonded system[35] are of the order of 3000 GHz; however, attenuation at this frequency by omnipresent free H_2O may prevent any substantial effect.[35]

Kohli et al.[36] predict that longitudinal and torsional modes of double helical DNA should not be critically damped at frequencies > 1 GHz (although relaxation times are of the order of picoseconds) and Kondepudi[37] suggests the possibility of an influence of millimeter waves at approximately $5(10^{11})$ Hz upon oxygen affinity of hemoglobin due to resonant excitation of heme plane oscillations. Although Furia et al.[38] did not find resonance absorption at millimeter waves in yeast, such was reported by Grundler et al.[39,48] The latter experiment has been interpreted[40] as supporting Fröhlich's theory of cooperative phenomena in biological systems. That theory postulates "electric polarization waves" in biological membranes which are polarized by strong biologically generated[20] fields ($\sim 10^7$ V/m). Fröhlich[41-43] suggests that metabolically supplied energy initiates mechanical vibrations of cell membranes. The frequency of such vibrations is determined by the dimensions and the elastic constants of the membranes: based on an estimate of the sound velocity in the membrane of 10^3 m/sec and a membrane thickness of 100 Å (equal to one half wavelength) one obtains a frequency of $5(10^{10})$ Hz. Individual molecules within and outside the membrane may also oscillate, and frequency estimates vary between 10^9 Hz for helical RNA[44] and $5(10^{13})$ Hz for hydrogen-bonded amide structures.[45] Since the membranes and molecules involved are strongly polarized, the mechanically oscillating dipoles produce oscillating electromagnetic fields that are then able to transmit energy, at least in some situations, over distances much larger than the distance to the next adjacent molecule.

Electromagnetic coupling of this type may produce long range cooperative phenomena. In particular, Fröhlich[46] has shown that two molecular systems may exert strong forces upon each other when their respective oscillation frequencies are nearly equal, provided the dielectric permittivity of the medium between them is strongly dispersive or excitation is supplied by pumping (i.e., by excitation at the correct frequency from an external source). The mechanism is nonlinear in the sense that it displays a step-like dependence on excitation intensity. Possible long range effects may be, for example, attraction between enzyme and substrate.[43] These and related topics have been discussed in detail by Illinger[35] and are reviewed in the present volume in Part III, Chapter 5.

ACKNOWLEDGMENTS

Preparation of this Introduction and my part of the editorial work for this book were done mostly while I was a visitor at the University of Wisconsin in Madison. I wish to express my thanks to the University of Rhode Island for granting the sabbatical leave which made this visit possible and to the University of Wisconsin for providing the opportunity for stimulating discussions with faculty and students on the subject matter of this book. I also want to sincerely thank Ms. Alyce Hegge and Ms. Kay Ewers for first-rate secretarial assistance.

REFERENCES

1. **Jordan, E. C.,** *Electromagnetic Waves and Radiating Systems,* Prentice-Hall, Englewood Cliffs, N.J., 1950.
2. **Schelkunoff, S. A.,** *Electromagnetic Waves,* D Van Nostrand, New York, 1943, 133.
3. **Stratton, J. A.,** *Electromagnetic Theory,* McGraw-Hill, New York, 1941, 435.
4. **Van Bladel, J.,** *Electromagnetic Fields,* McGraw-Hill, New York, 1964, 274.
5. **Kaune, W. T. and Gillis, M. F.,** General properties of the interaction between animals and ELF electric fields, *Bioelectromagnetics,* 2, 1, 1981.
6. **Bridges, J. E. and Preache, M.,** Biological influences of power frequency electric fields — a tutorial review from a physical and experimental viewpoint, *Proc. IEEE,* 69, 1092, 1981.
7. **Iribarne, J. V. and Cho, H. R.,** *Atmospheric Physics,* D. Reidel, Boston, 1980, 134.
8. **Zahn, M.,** *Electromagnetic Field Theory, A Problem Solving Approach,* John Wiley & Sons, New York, 1979.

9. **Raybourn, M. S.,** The effects of direct-current magnetic fields on turtle retina in vitro, *Science,* 220, 715, 1983.

10. **Gaffey, C. T. and Tenforde, T. S.,** Alterations in the rat electrocardiogram induced by stationary magnetic fields, *Bioelectromagnetics,* 2, 357, 1981.

11. **Tenforde, T. S., Gaffey, C. T., Moyer, B. R., and Budinger, T. F.,** Cardiovascular alterations in *Macaca* monkeys exposed to stationary magnetic fields: experimental observations and theoretical analysis, *Bioelectromagnetics,* 4, 1, 1983.

12. **Schulten, K.,** Magnetic field effects in chemistry and biology, *Festkörperprobleme/Advances in Solid State Physics,* Vol. 22, Heyden, Philadelphia, 1982, 61.

13. **Blankenship, R. E., Schaafsma, T. J., and Parson, W. W.,** Magnetic field effects on radical pair intermediates in bacterial photosynthesis, *Biochim. Biophys. Acta,* 461, 297, 1977.

14. **Sheppard, A. R.,** Magnetic field interactions in man and other mammals: an overview, in *Magnetic Field Effect on Biological Systems,* Tenforde, T. S., Ed., Plenum Press, New York, 1979, 33.

15. **Jordan, E. C.,** *Electromagnetic Waves and Radiating Systems,* Prentice-Hall, Englewood Cliffs, N.J., 1950, 132.

16. **Ramo, S., Whinnery, J. R., and Van Duzer, T.,** *Fields and Waves in Communication Electronics,* John Wiley & Sons, New York, 1965, 293.

17. **Smyth, C. P.,** *Static and Dynamic Electricity,* McGraw-Hill, New York, 1939.

18. **Bean, C. P., DeBlois, R. W., and Nesbitt, L. B.,** Eddy-current method for measuring the resistivity of metals, *J. Appl. Phys.,* 30(12), 1959, 1976.

19. **Bassett, C. A. L., Pawluk, R. J., and Pilla, A. A.,** Augmentation of bone repair by inductively coupled electromagnetic fields, *Science,* 184, 575, 1974.

20. **Plonsey, R. and Fleming, D.,** *Bioelectric Phenomena,* McGraw-Hill, New York, 1969, 115.

21. **Houslay, M. D. and Stanley, K. K.,** *Dynamics of Biological Membranes,* John Wiley & Sons, New York, 1982, 296.

22. **Wilson, D. F.,** Energy transduction in biological membranes, in *Membrane Structure and Function,* Bittar, E. D., Ed., John Wiley & Sons, New York, 1980, 182.

23. **Johnson, C. C. and Guy, A. W.,** Nonionizing electromagnetic wave effects in biological materials and systems, *Proc. IEEE,* 60, 692, 1972.

24. **Schwan, H. P.,** Field interaction with biological matter, *Ann. N.Y. Acad. Sci.,* 303, 198, 1977.

25. **Kraichman, M. B.,** *Handbook of Electromagnetic Propagation in Conducting Media,* NAVMAT P-2302, U.S. Superintendent of Documents, U.S. Government Printing Office, Washington, D.C., 1970.

26. **Massoudi, H., Durney, C. H., Barber, P. W., and Iskander, M. F.,** Postresonance electromagnetic absorption by man and animals, *Bioelectromagnetics,* 3, 333, 1982.

27. **Sears, F. W., Zemansky, M. W., and Young, H. D.,** *University Physics,* 5th ed., Addison-Wesley, Reading, Mass., 1976, 360.

28. **Uman, M. A.,** *Lighting,* McGraw-Hill, New York, 1969, 162.

29. **Coelho, R.,** *Physics of Dielectrics for the Engineer,* Elsevier, Amsterdam, 1979, 155.

30. **Schwan, H. P.,** Dielectric properties of biological tissue and biophysical mechanisms of electromagnetic field interaction, in *Biological Effects of Nonionizing Radiation,* Illinger, K. H., Ed., ACS Symposium Series 157, American Chemical Society, Washington, D.C., 1981, 121.

31. **Koryta, J.,** *Ions, Electrodes and Membranes,* John Wiley & Sons, New York, 1982.

32. **Rosenbaum, E. J.,** *Physical Chemistry,* Appleton-Century-Crofts, Educ. Division, Meredith Corp., New York, 1970, 595.

33. **Watson, J. D.,** *Molecular Biology of the Gene,* W. A. Benjamin, Menlo Park, Calif., 1976, 91.

34. **Rosenbaum, E. J.,** *Physical Chemistry,* Appleton-Century-Crofts, Educ. Division, Meredith Corp., New York, 1970, 285.

35. **Illinger, K. H.,** Electromagnetic-field interaction with biological systems in the microwave and far-infrared region, in *Biological Effects of Nonionizing Radiation,* Illinger, K. H., ACS Symposium Series 157, American Chemical Society, Washington, D.C., 1981, 1.

36. **Kohli, M., Mei, W. N., Van Zandt, L. L., and Prohofsky, E. W.,** Calculated microwave absorption by double-helical DNA, in *Biological Effects of Nonionizing Radiation,* Illinger, K. H., Ed., ACS Symposium Series 157, American Chemical Society, Washington, D.C., 1981, 101.

37. **Kondepudi, D. K.,** Possible effects of 10^{11} Hz radiation on the oxygen affinity of hemoglobin, *Bioelectromagnetics,* 3, 349, 1982.

38. **Furia, L., Gandhi, O. P., and Hill, D. W.,** Further investigations on resonant effects of mm-waves on yeast, Abstr. 5th Annu. Sci. Session, Bioelectromagnetics Society, University of Colorado, Boulder, June 12 to 17, 1983, 13.

39. **Grundler, W., Keilman, F., and Fröhlich, H.,** Resonant growth rate response of yeast cells irradiated by weak microwaves, *Phys. Lett.,* 62A, 463, 1977.

40. **Fröhlich, H.,** Coherent processes in biological systems, in *Biological Effects of Nonionizing Radiation,* Illinger, K. H., Ed., ACS Symposium Series 157, American Chemical Society, Washington, D.C., 1981, 213.

41. **Fröhlich, H.,** What are non-thermal electric biological effects?, *Bioelectromagnetics,* 3, 45, 1982.
42. **Fröhlich, H.,** Coherent electric vibrations in biological systems and the cancer problem, *IEEE Trans. Microwave Theory Tech.,* 26, 613, 1978.
43. **Fröhlich, H.,** The biological effects of microwaves and related questions, in *Advances in Electronics and Electron Physics,* Marton, L. and Marton, C., Eds., Academic Press, New York, 1980, 85.
44. **Prohofsky, E. W. and Eyster, J. M.,** Prediction of giant breathing and rocking modes in double helical RNA, *Phys. Lett.,* 50A, 329, 1974.
45. **Careri, J.,** Search for cooperative phenomena in hydrogen-bonded amide structures, in *Cooperative Phenomena,* Haken, H. and Wagner, W., Eds., Springer-Verlag, Basel, 1973, 391.
46. **Fröhlich, H.,** Selective long range dispersion forces between large systems, *Phys. Lett.,* 39A, 153.
47. **Barnes, F. S. and Hu, C.-L. J.,** Nonlinear interactions of electromagnetic waves with biological materials, in *Nonlinear Electromagnetics,* Uslenghi, P. L. E., Ed., Academic Press, New York, 1980, 391.
48. **Grundler, W., Keilmann, F., Putterlik, V., Santo, L., Strube, D., and Zimmermann, I.,** Nonthermal resonant effects of 42 GHz microwaves on the growth of yeast cultures, in *Coherent Excitations of Biological Systems,* Fröhlich, H. and Kremer, F., Eds., Springer-Verlag, Basel, 1983, 21.

Part I — Dielectric Permittivity and Electrical Conductivity of Biological Materials

DIELECTRIC PROPERTIES OF TISSUES

Kenneth R. Foster and Herman P. Schwan

TABLE OF CONTENTS

I. INTRODUCTION

The bulk electrical properties of biological materials have been of interest for many reasons for over a century. The dielectric properties of tissues are needed for the calculation of the internal electric fields resulting from exposure to nonionizing electromagnetic (EM) fields, and are thus important in the development of diagnostic and therapeutic medical applications of this energy and studies of possible hazards of EM fields. Dielectric properties of biological materials other than tissues are important in developing applications of EM fields in a variety of areas, e.g., food processing, other agricultural purposes, and the drying of various products. On a more fundamental level, study of these properties gives important information about possible mechanisms by which external fields can produce effects in an organism.

It has been 26 years since the publication of a comprehensive review of the bulk electrical properties of tissues and cell suspensions by either Schwan or myself.[1] The subsequent years have seen the development of computer-controlled instrumentation for precise and rapid measurements of dielectric properties, and a greater understanding of dielectric phenomena in tissues and other complex materials. Books dealing with dielectric properties of biological materials have been published by Cole,[2] Pethig,[3] Grant et al.,[4] and Schanne and P.-Ceretti.[5] To fully explain the advances of the past 3 decades and to summarize previously published work would greatly exceed the presently available space.

This chapter will complement the previous review[1] while providing sufficient introductory material to be self-contained. The emphasis will be on the mechanisms that are responsible for the observed properties. We believe that sufficient dielectric data are presently available for most routine engineering applications, and that the mechanisms for the dielectric properties of tissues at submegahertz through low gigahertz frequencies are largely understood even though much remains to be done. These data, and an understanding of the mechanisms, are the starting materials for a critical analysis of claims of biological effects of nonionizing radiation, and proposed biomedical applications of this energy.

The specific aims of this chapter are

1. To summarize some of the very early work on this subject and establish a historical perspective.
2. To review the basic concepts of dielectric phenomena along with the formalism that is used to describe these phenomena in biological materials. The most important relaxation mechanisms will be summarized as they pertain to tissues. This material has been reviewed often before, but is important to an understanding of the phenomena at hand. Considerable advances have been made on the theory of dielectric properties of heterogeneous systems since the earlier review.
3. To review the dielectric properties of several tissues at frequencies from audio through microwave ranges. We wish to discuss in some detail more recent data and the underlying mechanisms. The new data and their interpretation support conclusions arrived at before, but in some areas considerable progress has been made, e.g., in determining the relative roles of bulk-like and motionally restricted water in the microwave absorption properties of tissues. Also, the new data extend to higher frequencies, and a far better picture of the dielectric properties of low water content tissues such as bone is now available.
4. To discuss effects of cellular structure on the dielectric properties of tissues. The coupling between externally applied EM fields and intracellular compartments, including membranes, can be understood as a logical extension of this subject.
5. To examine the limits of dielectric relaxation spectroscopy, in particular, to explore the limits to which the distribution of relaxation times can be inferred from dielectric data. These limits were formerly suggested,[1] but a more general discussion has recently become possible. We hope that this will help resolve several ongoing disputes involving different interpretations of rather similar dielectric data from tissues and other biological materials.
6. To review the dielectric relaxation properties of tissue water and their relation to other transport properties of biological systems. Many new data are available and the contribution of water to the total absorption of microwave energy can be discussed with more assurance than previously possible.

Several extended summaries of the dielectric properties of tissues and cell suspensions have already been published in addition to the volumes cited above. Data up to 1938 are well covered by Rajewsky.[6] The previous review[1] presented data over a range from 10 Hz to almost 10 GHz, which have been frequently quoted and used in many calculations of the average and spatial distributions of absorbed power in irradiated subjects, considered in other chapters in this volume. An exhaustive tabulation of these and more recent data is presented by Stuchly and Stuchly.[8] The mechanisms of interaction of EM fields with biological materials have been reviewed by Schwan and Foster[9] and Stuchly.[10] The Stuchly papers complement our own and are highly recommended.

A. Historical Introduction

Interest in the electrical properties of biological materials began about 100 years ago, as instruments for the measurement of electrical resistance, and eventually capacitance, became available. This interest gradually developed into a major theme in electrophysiological research and yielded, even from the first, important insights into pertinent mechanisms. To provide a historical introduction to the field, we will informally review some of the early work. Detailed references are found in recent books by Schanne and P.-Ceretti[5] and Cole.[2] Cole, in particular, presents an excellent discussion of work performed in the 1920s through 1940s, in which he and his colleagues played such an important role. Rajewsky[6] and Gildemeister[7] extensively review earlier literature.

Interest was initially centered on the bulk electrical properties of tissues and blood. Almost immediately, it became apparent that such properties are much more complex than those of simple electrolyte solutions. For example, it was noted that upon the application of a direct current (DC) potential step, the current changes with time, beginning with a charging spike; hysteresis phenomena also were observed. The reasons for these effects were not entirely clear for some time, and in retrospect the possibility that electrode polarization contributed to these observations cannot be ruled out. There emerged rather early from this work the concept of a tissue "polarization" which was thought to be largely responsible for these phenomena. These early attempts were described by the eminent neurophysiologist du Bois-Reymond, in a book entitled *Investigations about Animal Electricity,* published in Berlin in 1849.[11]

Some of the early developments in understanding the electrical properties of tissues can be summarized as follows: Peltier[12] discovered in 1834 that "animal bodies" behave as "polarization cells" if exposed to DC currents, i.e., they are able to store electricity and release it after the termination of the current. Hermann[13] reported in 1872 that the anisotropy in the DC resistance of muscle is a factor of 4 to 9 in freshly excised tissue, but the anisotropy slowly disappears after death. He correctly interpreted this result as arising from the microscopic structure of the tissue, but proposed incorrectly that the mechanism is different ionic permeabilities in directions perpendicular and parallel to the fibers. Hermann was probably also the first to measure the resistance of tissues to "alternating" currents (AC), using an apparatus of his own construction that alternately applied current pulses of opposite sign to the tissue.

Several authors noted the complex frequency dependence of the electrical properties. Du Bois[14] found in 1898 that skin behaves like a capacitor if subjected to DC currents. Galler[15] in 1913 reported the first true AC measurements on a tissue (frog skin) and found that resistance at 1 kHz is several times lower than at DC. Gildemeister,[16] in 1919, and Einthoven,[17] in 1923, further examined the frequency dependence of tissue resistance. Gildemeister and others observed the current peak that occurs when skin is exposed to a voltage step and interpreted this as an initial small resistance which increases with time; they considered this initial resistance as the "true" resistance, and the larger final resistance as "pretended". In these experiments, the observed time constants were of the order of 1 msec, and the apparent increase in resistance with time was thought to result from developing counter potentials of a "polarization" kind. Whether this early "time domain spectroscopy" principally detected the alpha dispersion or its combination with electrode polarization effects is difficult to decide.

Early studies on the impedance properties of blood led to important insights that could be applied to more complex tissues. It was almost immediately recognized that at low frequencies erythrocytes behave as nonconducting particles in a conducting medium. In his famous treatise, Maxwell[18] derived the mixture theory that could be applied to this tissue. Such attempts were met with only limited success, however, in part because erythrocytes are not spherical as Maxwell's theory demanded, and in part because of aggregation of the erythrocytes. A number of dielectric studies on blood appeared, including physiologically oriented work such as by Stewart,[19] who measured the low frequency resistance of blood while developing indicator dilution measurements of blood flow. But it was particularly the work of Höber[20,21] that clarified the underlying reason for the strong frequency dependence of the resistance of this tissue. Since continuous-wave, high frequency generators were not yet available, Höber indirectly measured the high frequency resistance of erythrocyte suspensions by observing the damping of high frequency oscillations in a resonant circuit produced by a spark discharge generator. By substituting salt solutions of appropriate ionic strength for the blood sample to produce the same effect, he determined the apparent ionic strength of the blood sample at high frequencies, and showed it was significantly higher than the corresponding quantity at low frequencies. These studies were performed only a few years after the final statement of Bernstein's hypothesis in 1902 of a cell membrane. Höber speculated that the membrane excluded low frequency currents, and passed high frequency currents. In short, he discovered and correctly identified the mechanism of what is now called the "beta dispersion". His interpretation was hotly debated for many years, but eventually prevailed.

Simultaneously with these developments, there emerged during the late 1880s much interest in the effects of high frequency electric currents on tissues. The idea of using electricity to produce biological and medical effects had been highly popular during the previous century (among both charlatans and reputable physicians), and as techniques were developed for producing high frequency currents these too began to be applied. An excellent history of the field of electrotherapy is provided by Licht[22] (which includes a bibliography listing 923 dissertations and books on the subject, most from the 19th century!). D'Arsonval was the first in 1982 to conduct experiments in this regard. Tesla followed using ACs of an unspecified frequency, which was probably in the range of 10 to 100 kHz. His explanation of the unexpectedly small effect of such currents was that tissues are "condensors", i.e., they offer low resistance at high frequencies.[23] The famous physiologist, Nernst, recognized that electrical currents are progressively less able to excite tissue with increasing frequency. He proposed in 1908 that the threshold strength of ACs should increase in proportion to the square root of the frequency of the applied currents.[24] Subsequent investigators observed an increase with frequency,[25] although significant deviations were noted from single square root behavior, and his theory was eventually extended. It is of interest to note that a number of outstanding physicists, physiologists, and engineers of the time contributed to these developments, including Wien, Nernst, and Tesla.

The understanding gained before World War I can thus be summarized as follows. It was recognized that tissue conducts electricity, that its resistance varies with frequency, and that the mechanism for the conduction is the movement of ions. Cellular structure and the presence of membranes were recognized as influencing the electrical properties of tissues, and the anisotropy in the electrical properties of muscle and nerve tissues was also recognized, if not fully understood. The concept of tissue "polarization" was established, which reflects the capacitive properties of tissues as we know them today. Clearly, this was an extremely active period and much insight was gained. But the origin of the tissue "polarization" remained elusive and the complex events that occur when tissue is electrically excited could not be well separated from linear processes such as the capacitive charging of the cell membrane.

After World War I, dielectric studies encompassed a steadily increasing range of frequencies and variety of materials, and more sophisticated attempts were made to understand the underlying mechanisms. Philippson[26] extended data on many tissues over a wide frequency range, which demonstrated that the electrical properties of all tissues are frequency dependent. McClendon[27] reported the resistance and capacitance of blood from 266 Hz to 2 MHz, and was perhaps the first to measure both components of the complex admittance of tissues over this extended range. By now all sorts of equivalent circuits had been proposed to describe the observed frequency dependence of the impedance, and there remained little doubt that cellular structure and cell membranes were largely responsible for these effects.

During the 1920s and 1930s a number of cellular systems were successfully characterzied. Osterhout[28] published an influential study of the conductance of the marine kelp, in which he interpreted the variations in tissue conductance after death as arising from changes in the membrane permeability. Fricke, Curtis, and Cole entered the field, and developed advanced instrumentation that permitted the accurate measurement of both conductivity and permittivity over a wide frequency range. In a famous series of experiments, they studied the dielectric properties of various cell suspensions and successfully applied the complex extension of the

Maxwell mixture theory in analyzing these data. Dänzer[29] developed a rigorous theory for the beta dispersion in cell suspensions, and Rajewsky and colleagues Dänzer, Osken, Schaefer, Wachter, and Schwan measured the conductivity of a large number of systems, including many tissues and biological solutions from a few kilohertz to 100 MHz, using bridges at audio and radio frequencies, and transmission lines at very high frequencies. Wachter was probably the first to measure the absorption of water, blood, milk, ascites, and other biological fluids at 1 to 10 GHz using damped oscillations. He obtained values 45 years ago that are close to those accepted today. These efforts and others not mentioned produced extensive conductivity and permittivity data from many systems, together with a quantitative understanding of the mechanisms responsible for the beta dispersion for cell suspensions. The work of these investigators, in particular Fricke, Curtis, and Cole, is noteworthy for application of sophisticated physical concepts for the solution of important problems in biology.

To account in detail for the dielectric properties of tissues was a much more difficult problem. While a dispersion due to the charging of cell membranes clearly existed in tissues as well as cell suspensions, tissues exhibit a much broader range of relaxation time constants than is evident in suspensions of individual cells. It was suggested that the cell membrane exhibited a frequency independent impedance phase angle analogous to that of the polarization impedance of the interface of an electrolyte and a metal surface. This concept had originated much earlier; Hermann had speculated about issue "polarization" and Gildemeister had introduced a circuit composed of a resistance in series with a parallel RC combination which can be readily transformed into the circuit modeling the alpha or beta dispersion. Philippson used such equivalent circuits to model the data, and tried to associate their elements with biological structures such as membranes. The concept of a membrane impedance phase angle that is independent of frequency thus appears as a continuation of a process that started in the previous century with speculations about tissue "polarization". It is interesting to note that this concept was seriously entertained as recently as 1972 by Cole.[2]

World War II subsequently stimulated rapid developments in instrumentation. Oscillators were developed that operated at frequencies up to 10 GHz, and transmission line and waveguide methods became available to measure dielectric properties through the microwave frequency range. Rajewsky and Schwan[30] were the first in 1948 to report complex permittivity data on blood at frequencies up to nearly 1 GHz, which were soon confirmed and extended to still higher frequencies by Cook[31] in England. A number of British investigators, including Cook, Buchanan, Grant, and England, became interested in the dielectric properties of biological systems at microwave frequencies, and made significant contributions that are still being continued to the present day. Schwan established a laboratory at the University of Pennsylvania in 1950, and in the 1950s and 1960s he and his colleagues reported data from tissues covering the three principal relaxation ranges that, together with results obtained from 1 to 8.5 GHz by Herrick,[132] have been published in several reviews and widely quoted by still other investigators. Schwan and colleagues including Li, Carstensen, Schwarz, Takashima, Maczuk, Foster, Pauly, and others have investigated the dielectric properties of many other materials including protein and DNA solutions, vesicles and organelles, tissues of many sorts, nonbiological materials, and electrodes, at frequencies that have ranged from roughly 1 mHz to 18 GHz. These investigators identified several polarization mechanisms in addition to the previously established cell membrane (beta) dispersion, including counterion polarization and dipolar relaxation of water associated with proteins.

In addition to the work at the University of Pennsylvania, excellent dielectric studies on biological materials have been carried out at many other places, most extensively by Grant and his group in London, Carstensen and colleagues at Rochester, Hanai and Irimajiri in Japan, and Pauly in Erlangen. Extensions of the theory, including numerical studies, have been made by Hanai, Irimajiri, Carstensen, and Grant, as well as many others.

FIGURE 1. Equivalent circuit of an idealized parallel plate capacitor filled with material of relative permittivity ϵ^1 and conductivity σ. The plate area is A and the distance between plates is d. In a real circuit, a small additional shunt capacitance would occur from the stray fields in the air outside of the material.

With the introduction of modern network analyzer and time domain measurement systems, the determination of the dielectric properties of biological samples is easier, which should encourage increased efforts in this field. Many opportunities exist, and a number of problems still have to be resolved. In principle it is now possible to rapidly assess cellular parameters by electrical means, in either single cells or populations of cells. More work needs to be done to resolve the various contributions to the alpha dispersion, to clarify the origin of the apparent "constant phase angle" behavior of tissues, and to study in more detail the dielectric properties of water in tissues and their relation to other transport properties. The manipulation of cells using alternating electric fields has in recent years been of great interest with practical application in electric field-induced cell fusion. In order to fully understand and optimize these techniques, the electrical properties of the cells themselves must be more precisely known.

About 10 years ago many believed that the field was saturated and no longer had much to offer. Today work on dielectric properties of biological systems and their response to electric fields is being conducted at an increasing number of places. This work will surely provide new insights into how electrical fields can influence cells and tissues, and will lead to beneficial applications of EM energy.

B. Definitions and Basic Concepts

The dielectric permittivity ϵ and conductivity σ of a material are, respectively, the charge and current densities induced in response to an applied electric field of unit amplitude. These definitions are illustrated by examples of an idealized parallel plate capacitor, of plate area A and separation d, which contains the material under investigation (Figure 1). Initially let the region between the plates be a vacuum. A constant voltage difference V between the plates will induce a charge density D given by

$$D = \epsilon_0 \, V/d \qquad (1)$$

where ϵ_0 is the permittivity of free space (8.85×10^{-12} F/m). The capacitance C is the ratio of the total induced charge on the plates to the applied voltage (ignoring the contribution from the fringing fields outside of the region between the plates)

$$C = \epsilon_0 \, A/d \qquad (2)$$

If a material is now introduced between the plates, an additional charge density P_s will be induced on the plates from the polarization of charges within the material. For sufficiently low voltages this charge density is proportional to E and can be written

$$D = \epsilon_0 \, E + P_s$$

$$= \epsilon_s \, \epsilon_0 \, E \qquad (3)$$

and the capacitance is now

$$C = \epsilon_s \epsilon_0 A/d \qquad (4)$$

where ϵ_s is the static permittivity of the material. The circuit might also include a conductance G in parallel with the capacitance

$$G = \sigma_s A/d \qquad (5)$$

where σ_s is the (DC) conductivity of the material. In response to alternating fields, the dielectric properties will vary with frequency if dielectric relaxation processes are present.

For sinusoidally oscillating fields, the electrical properties of the circuit of Figure 1, and hence the dielectric properties of the material, can be specified in several equivalent ways:

1.   At circular frequency ω the complex admittance Y* can be written

$$Y* = G + j\omega C$$

$$= (A/d)(\sigma + j\omega\epsilon_0\epsilon) \qquad (6a)$$

from which is defined the complex conductivity $\sigma*$ of the material

$$\sigma* \equiv \sigma + j\omega\epsilon_0\epsilon \qquad (6b)$$

2.   Equivalently, the complex capacitance C* can be defined as

$$C* = Y*/j\omega \qquad (7a)$$

which leads to the definition of the complex relative permittivity $\epsilon*$

$$\epsilon* \equiv \epsilon - j\sigma/\omega\epsilon_0 \equiv \epsilon' - j\epsilon'' \qquad (7b)$$

where ϵ' is obviously the same as ϵ and ϵ'' equal to $\sigma/\omega\epsilon_0$. The complex conductivity and permittivity are related by

$$\sigma* = j\omega\epsilon*\epsilon_0 \qquad (7c)$$

We will often refer to the real part of the complex permittivity as ϵ when no relaxation is present, and otherwise as ϵ'. The term "dielectric constant" applies to ϵ or ϵ'. (This term is principally used in the chemical literature with reference to the permittivity of pure liquids at audio or radio frequencies, which is typically independent of frequency in this range.)

3.   Finally, the series equivalent impedance Z* can be written

$$Z* = 1/Y*$$

$$= R + jX$$

$$= \frac{G - j\omega C}{G^2 + (\omega C)^2} \qquad (8a)$$

from which the complex specific impedance of the material z* is defined

$$z^* = 1/\sigma^*$$

$$= \frac{\sigma - j\omega\epsilon_0\epsilon}{\sigma^2 + (\omega\epsilon_0\epsilon)^2} \tag{8b}$$

Conventionally, the resistivity ρ is defined as the inverse of the conductance, or $1/\sigma$, and is not in general equal to the real part of z^*.

Obviously, dielectric data can be represented in several equivalent ways to best suit the purposes at hand. Frequently in the older physiologically oriented literature, tissue dielectric properties were often quoted as series equivalent impedances rather than as complex permittivity values. However, the fundamental processes, dielectric polarization and electrical conduction, occur in parallel, and the electrical properties are more logically presented in terms of a parallel combination of a capacitive and a conductive element. For example, aqueous electrolytes have conductivity and permittivity values that are practically independent of frequency up to the microwave range, while the corresponding specific impedance would be highly frequency dependent as seen from Equation 8b. We shall express data in terms of conductivity and permittivity unless there are good reasons for doing otherwise.

The illustration of an ideal parallel plate capacitor is only meaningful at frequencies at which the free space wavelength is much larger than the dimensions of the capacitor. A more general definition of the relative permittivity ϵ and conductivity σ is by the relations

$$D = \epsilon\epsilon_0 E$$

$$J = \sigma E$$

where D and J are, respectively, the displacement and the conduction current densities. Moreover, the measured electrical properties of parallel plate capacitors are quite different from ideally predicted values, because of a variety of parasitic effects that can be modeled as series or shunt impedance elements in the circuit. An extensive review of practical measurement techniques is found in Schwan.[32]

C. General Relaxation Theory

The dielectric polarization arises from physical displacement of charge and requires time to develop. Therefore, the response of a material to a voltage step is a more or less complicated relaxation process that depends on the kinetics of the charge displacement. We review in this section the mathematical relation between these relaxation processes and the frequency dependence of the dielectric properties; in Section II we will summarize the major dielectric relaxation mechanisms that occur in tissues and in Section IV survey the corresponding dielectric properties of several tissues.

In the simplest case, the polarization of a sample will relax towards the steady state as a first order process characterized by a relaxation time τ. This time-constant can range from picoseconds to microseconds for partial orientation of molecular dipoles, and to seconds for other relaxation processes. An additional polarization arises from the distortion of electron clouds, but the time-constant is so short ($\ll$ 1 psec) that it can at present be considered instantaneous. The dielectric response of a first order system to a step in the applied field of magnitude E can then be written

$$\frac{D(t)}{\epsilon_0 E} = \epsilon_\infty + (\epsilon_s - \epsilon_\infty)(1 - e^{-t/\tau}) \tag{9}$$

where ϵ_∞ arises from the electronic polarizibility. (The subscript ∞ refers to the polarizability that would be measured at infinite frequency, i.e., that observed in the limit of zero time after the application of a step change in electric field.) The dielectric response of this system to a sinusoidal field Eexp (jωt) is obtained by Laplace transformation of Equation 9 and yields the Debye equation:

$$\epsilon^* = \frac{D(j\omega)}{\epsilon_0 E(j\omega)}$$

$$= \epsilon_\infty + \frac{(\epsilon_s - \epsilon_\infty)}{1 + j\omega\tau} \tag{10}$$

This equation can be separated into real and imaginary parts:

$$\epsilon^* \equiv \epsilon' - j\epsilon'' \tag{11a}$$

$$\epsilon' = \epsilon_\infty + \frac{\epsilon_s - \epsilon_\infty}{1 + (\omega\tau)^2} \tag{11b}$$

$$\epsilon'' = \frac{(\epsilon_s - \epsilon_\infty)\,\omega\tau}{1 + (\omega\tau)^2} + \frac{\sigma_s}{\omega\epsilon_0} \tag{11c}$$

$$\sigma = \sigma_s + \frac{(\sigma_\infty - \sigma_s)(\omega\tau)^2}{1 + (\omega\tau)^2} \tag{11d}$$

in which the limit values ϵ_s, ϵ_∞, σ_s, and σ_∞ are interrelated by

$$\frac{(\epsilon_s - \epsilon_\infty)\,\epsilon_0}{\tau} = \sigma_\infty - \sigma_s \tag{11e}$$

The term σ_s has been included to allow for a frequency independent ionic conductance. Since the permittivity and conductivity cannot vary independently, in principle the frequency dependence of one quantity will determine the frequency dependence of the other (see Section I.E).

In the absence of any physical justification, one might just as well look at the time dependence of the current density J

$$J/E = \sigma_\infty + (\sigma_s - \sigma_\infty)(1 - e^{-t/\tau}) \tag{12}$$

This transforms into the admittance equivalent of Debye Equation 10

$$\sigma^* = \sigma_\infty + \frac{(\sigma_s - \sigma_\infty)}{1 + j\omega\tau} + \omega\epsilon_\infty\epsilon_0 \tag{13}$$

where $\omega\epsilon_\infty\epsilon_0$ has been included to account for a frequency independent addition to the relaxation effect. Since

$$J = \frac{dD(t)}{dt} + \sigma_s E$$

the equations for D and J readily yield the coupling Equation 11e.

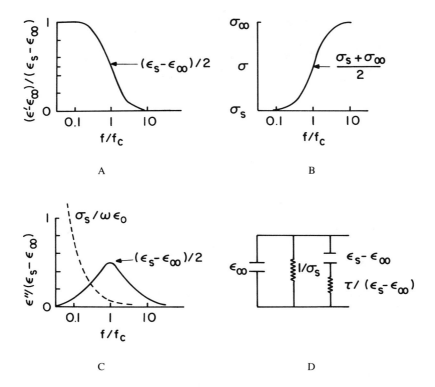

FIGURE 2. Single, time-constant relaxation. (A) Relative permittivity vs. frequency; (B) conductivity vs. frequency; (C) loss vs. frequency showing (broken line) a contribution from a DC conductivity; (D) an equivalent circuit that models the Debye equations. In this single, time-constant situation, the relaxation frequencies defined from (A) and (B) would be identical; in the presence of a distribution of time-constants, the relaxation frequency from the conductivity plot (B) would in general exceed that from the permittivity plot (A).

The relaxation (or characteristic) frequency $f_c = (2\pi\tau)^{-1}$ is the frequency at which the permittivity is halfway between its low and high frequency values. The variation in the permittivity ϵ', loss ϵ'', and conductivity $\sigma(= \omega\epsilon''\epsilon_0)$ are shown in Figure 2 for this single, time-constant process. In this case, the dispersion extends over roughly a decade in frequency from $f_c/3$ to $3\ f_c$. As Hasted[33] observed, Equations 10 and 11 represent perhaps the broadest absorption line in all physics, the great breadth of which is due to the process of relaxation, not resonance, as is the case with other spectroscopic phenomena. (Conventionally, the term ''line'' would be applied only to resonance-type phenomena.) No unequivocal evidence has been found for the presence of resonance-type dielectric polarization in biological materials in the frequency range of interest here.

Dielectric relaxation behavior can be best represented by Argand diagrams on the complex plane. Corresponding to Equations 6 to 8, this can be accomplished in various ways, each emphasizing different aspects on the relaxation behavior.

1. In the normalized complex admittance plane $Y*\frac{d}{A}$, or equivalently, the complex conductivity plane (see Equation 6)

$$\left[\sigma - \frac{\sigma_s + \sigma_\infty}{2}\right]^2 + [\omega\epsilon_0(\epsilon' - \epsilon_\infty)]^2 = \left[\frac{(\epsilon_s - \epsilon_\infty)\epsilon_0}{2\tau}\right]^2$$

$$= \left[\frac{\sigma_\infty - \sigma_s}{2}\right]^2 \qquad (14)$$

Thus a locus of $\sigma'' - \omega\epsilon_0\epsilon_\infty$ vs. σ' is a circle centered at $\sigma' = (\sigma_s + \sigma_\infty)/2$, of which only the half corresponding to positive values of σ' and $\sigma'' - \omega\epsilon_0\epsilon_\infty$ has physical meaning. At the relaxation frequency when $\omega = 2\pi f_c = 1/\tau$, the ordinate becomes $(\sigma_\infty - \sigma_s)/2$, which corresponds to the apex of the semicircle.

2.   In the complex permittivity plane

$$\left[\epsilon' - \frac{\epsilon_s + \epsilon_\infty}{2}\right]^2 + \left[\epsilon'' - \frac{\sigma_s}{\omega\epsilon_0}\right]^2 = \left[\frac{\epsilon_s - \epsilon_\infty}{2}\right]^2$$

Thus a plot of $\epsilon'' - \sigma_s/\omega\epsilon_0$ vs. ϵ' will lead to a semicircle for single, time-constant relaxation, whose center is on the real axis at the point $\epsilon_\infty + (\epsilon_s - \epsilon_\infty)/2$ and whose apex corresponds once again to f_c. In the presence of a distribution of relaxation times, the relaxation frequencies obtained from the peaks of plots in the complex permittivity and complex admittance planes differ, in general, with that obtained from the complex admittance plane of the higher frequency.

3.   In the complex impedance plane the magnitude of the material impedance follows a semicircular orbit for constant ϵ and σ as can be seen from Equation 8. The plot of the imaginary part against the real part of Z^* follows the equation

$$\left(r - \frac{1}{2\sigma}\right)^2 + x^2 = \left(\frac{1}{2\sigma}\right)^2$$

which intercepts the real axis at $r = 0$ in the limit $\omega \to \infty$ and $r = 1/\sigma$ as $\omega \to 0$. The peak of the locus at $r = 1/2\sigma$ occurs at the frequency $f = \sigma/2\pi\epsilon_0\epsilon$ assuming constant ϵ and σ. More generally, a circle in the admittance plane with finite limit values must transform into circle in the impedance plane as required by the properties of the conformal mapping $z^* = 1/y^*$, where z^* and y^* represent the complex impedance and admittance, respectively. Thus, a semicircular locus for dielectric data in the impedance plane does not imply the presence of a dielectric relaxation process in the conductivity or permittivity, although one might nevertheless be present.

D. Distribution of Relaxation Times

For many reasons, most real materials do not exhibit single, time-constant relaxation behavior. For example, multiple relaxation processes might occur in parallel, each with a different relaxation time, or perhaps relaxation processes might occur that have kinetics that are not of first order. Moreover, in concentrated systems the electrical interactions between the relaxing species will usually lead to a distribution of macroscopic relaxation times. The frequent difficulty in experimentally distinguishing among these various possibilities is a persistent problem in dielectric spectroscopy.

Assuming linear response, the dielectric response to a step change E in field strength can be represented as the superposition of first order processes

$$\frac{D(t)}{\epsilon_0 E} = \epsilon_\infty + \Delta\epsilon_1(1 - e^{-t/\tau_1}) + \Delta\epsilon_2(1 - e^{-t/\tau_2}) + \dots$$

$$\text{or} \quad \epsilon^*(\omega) = \epsilon_\infty + \frac{\Delta\epsilon_1}{1 + j\omega\tau_1} + \frac{\Delta\epsilon_2}{1 + j\omega\tau_2} + \dots \tag{16}$$

In the limit where

$$\tau_1 \ll \tau_2 \ll \tau_3 \dots$$

a plot of the permittivity and conductivity vs. frequency will exhibit a series of clearly resolved dispersions, separated by plateaus at intermediate frequencies. (This will, in general, require an order-of-magnitude separation between the relaxation times.) A plot of $\epsilon'' - \sigma_s/\omega\epsilon_0$ vs. ϵ' in the complex permittivity plane will then produce a series of semicircles, each with a center on the real (permittivity) axis, which intersect the real axis at points

$$\epsilon_\infty, \quad \epsilon_\infty + \Delta\epsilon_1, \quad \epsilon_\infty + \Delta\epsilon_1 + \Delta\epsilon_2, \quad \dots$$

Often the relaxation times $\tau_1, \tau_2, \dots$ are not so well separated, and the material will exhibit a broad, perhaps featureless dispersion extending from roughly one third the lowest to three times the highest relaxation frequency that is present. If so, Equation 16 could be replaced by an integral

$$\frac{D(t)}{\epsilon_0 E} = \epsilon_\infty + (\epsilon_s - \epsilon_\infty) \int_0^\infty p(\tau)(1 - e^{-t/\tau})d\tau \tag{17}$$

where

$$\int_0^\infty p(\tau)d\tau = 1$$

In the frequency domain the complex dielectric permittivity becomes

$$\epsilon^* = \epsilon_\infty + (\epsilon_s - \epsilon_\infty) \int_0^\infty \frac{p(\tau)d\tau}{1 + (j\omega\tau)} - j\sigma_s/\omega\epsilon_0 \tag{18}$$

In principle, the distribution function $p(\tau)$ can be obtained numerically from the data; in practice this is a very uncertain procedure (see Section V.A below). It should be emphasized that the decomposition represented by Equations 16 or 18 is a formal process only, and its success does not demonstrate that a multiplicity of real physical processes is actually present.

A variety of empirical relaxation functions have been proposed which can fit dielectric data from many experiments.[33] In the absence of theoretical justification, they chiefly serve to parameterize the data without necessarily leading to a deeper understanding of the underlying mechanisms involved. One of the more useful functions was first proposed by Cole and Cole.[34,35]

$$\epsilon^* = \epsilon_\infty + \frac{(\epsilon_s - \epsilon_\infty)}{1 + (jf/f_c)^{1-\alpha}} - \frac{j\sigma_s}{\omega\epsilon_0} \tag{19}$$

which separates into real and imaginary parts

$$\frac{\epsilon' - \epsilon_\infty}{\epsilon_s - \epsilon_\infty} = \frac{1 + (f/f_c)^{1-\alpha} \sin(\alpha\pi/2)}{1 + (f/f_c)^{2(1-\alpha)} + 2(f/f_c)^{1-\alpha} \sin(\alpha\pi/2)} \tag{20}$$

$$\frac{\epsilon'' - \sigma_s/\omega\epsilon_0}{\epsilon_s - \epsilon_\infty} = \frac{(f/f_c)^{1-\alpha} \cos(\alpha\pi/2)}{1 + (f/f_c)^{2(1-\alpha)} + 2(f/f_c)^{1-\alpha} \sin(\alpha\pi/2)} \tag{21}$$

At frequencies $\gg f_c$, the dielectric relaxation approaches a fractional power law function. As the frequency is varied, a plot of $\epsilon'' - \sigma_s/\omega\epsilon_0$ against ϵ' in the complex permittivity plane describes a circular arc given by

$$(x - x_1)^2 + [y + (r - y_1)]^2 = r^2 \tag{22}$$

where $\quad x = (\epsilon' - \epsilon_\infty)/(\epsilon_s - \epsilon_\infty)$

$$y = (\epsilon'' - \sigma_s/\omega\epsilon_0)/(\epsilon_s - \epsilon_\infty)$$

$$x_1 = 1/2$$

$$y_1 = \frac{\cos(\alpha\pi/2)}{2[1 + \sin(\alpha\pi/2)]}$$

and $\quad r = [2 \cos(\alpha\pi/2)]^{-1}$

Again, the apex corresponds to the mean relaxation frequency f_c. The angle at the center of the circle between a line parallel to the Y axis and the radial line, to the intersection of the arc with the X axis, is $(1 - \alpha)\pi/2$ radians (Figure 3).

The distribution function $p(\tau)$ that corresponds to the Cole-Cole function is

$$p(\tau) = \frac{1}{2\pi} \frac{\sin(\alpha\pi)}{\cosh[(1 - \alpha) \ln(\tau/\tau_0)] - \cos(\alpha\pi)} \tag{23}$$

where $\tau_0 = (2\pi f_c)^{-1}$ is the mean relaxation time.

The current density induced by a unit step in the electric field is[35]

$$\frac{J(t)}{\epsilon_0} = \frac{(\epsilon_s - \epsilon_\infty)}{\tau_0} (1 - \alpha) \left(\frac{t}{\tau_0}\right)^{-(2-\alpha)} \sum_{n=1}^{\infty}$$

$$\frac{(-1)^{n-1} n}{\Gamma[(1 - n(1 - \alpha)]} \left(\frac{t}{\tau_0}\right)^{-(n-1)(1-\alpha)} \tag{24}$$

which approaches the limiting forms

$$\frac{J(t)}{\epsilon_0} = \begin{cases} \dfrac{(\epsilon_s - \epsilon_\infty)}{\tau_0} \dfrac{1}{\Gamma(1 - \alpha)} \left(\dfrac{t}{\tau_0}\right)^{-\alpha} & t \ll \tau_0 \qquad (25) \\[3em] \dfrac{(\epsilon_s - \epsilon_\infty)}{\tau_0} \dfrac{(1 - \alpha)}{\Gamma(\alpha)} \left(\dfrac{t}{\tau_0}\right)^{-(2-\alpha)} & t \gg \tau_0 \qquad (26) \end{cases}$$

A comparable equation can be written for the dielectric properties expressed in the complex conductivity form

$$\sigma^* = \sigma_\infty + \frac{\sigma_s - \sigma_\infty}{1 + (-jf/f_c)^{1-\alpha}} + j\omega\epsilon_0\epsilon_\infty \tag{27}$$

and for the complex impedance

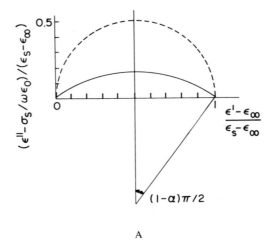

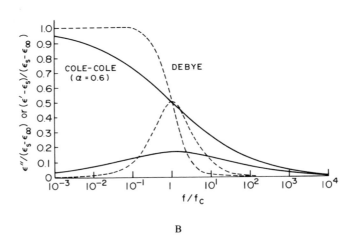

FIGURE 3. (A) The complex permittivity plot of the Cole-Cole equation (Equation 19), showing the depressed semicircular behavior that is characteristic of many biological materials. The parameters in this figure refer to Equation 19; (B) the corresponding permittivity and conductivity plot. The corresponding Debye relaxation is shown by the broken lines. The distribution parameter α is 0.6, corresponding to a broad range of relaxation times.

$$z^* = R_\infty + \frac{R_s - R_\infty}{1 + (jf/f_c)^{1-\alpha}} \tag{28}$$

These equations are formally similar but are not equivalent; in practice each could be used to empirically summarize the dielectric properties of tissues over wide frequency ranges.

Circles in the admittance and impedance planes can be related as follows. The admittance circle can be written

$$y^* = \sigma_\infty + \frac{\sigma_s - \sigma_\infty}{1 + (j\omega\tau)^{1-\alpha}} \tag{29}$$

which can be changed to

$$\frac{\sigma_s - y^*}{y^* - \sigma_\infty} = (j\omega\tau)^{1-\alpha} \tag{30}$$

Equation 28 may be written

$$\frac{Z^* - \rho_s}{\rho_\infty - Z^*} = \frac{\sigma_\infty}{\sigma_s} (j\omega\tau)^{1-\alpha} \tag{31}$$

and compared with the impedance circle

$$\frac{Z^* - \rho_s}{\rho_\infty - Z^*} = (j\omega\tau)^{1-\alpha} \tag{32}$$

with limit values ρ_s and ρ_∞ of the resistivity at low and high frequencies. The time-constants τ and τ' of the relaxations in the admittance and impedance planes are different; in fact

$$\tau^{1-\alpha} \frac{\sigma_\infty}{\sigma_s} = (\tau')^{1-\alpha} \tag{33a}$$

or

$$\left(\frac{\tau}{\tau'}\right)^{1-\alpha} = \frac{\sigma_s}{\sigma_\infty} \tag{33b}$$

It is clear that the peak frequency of the impedance circle will always be smaller than that of the admittance circle. The difference can be very pronounced. For example, the beta dispersion of muscle (see Section IV.A.2) presents a conductivity ratio of 4:5, with an α value of 0.3. Thus, $\tau'/\tau \approx 7$ to 9.

The biophysical significance of the apparent fractional power law relaxation in tissue has long been of interest. An electrical circuit to model one of Equations 19 to 28 would include an admittance or impedance element with constant phase angle,[34] which is the principal rationale for the supposed constant impedance phase angle of the cell membrane (see Section I.A). There seems to be little independent experimental evidence for such a property, however, and in some cases other explanations are possible. A superposition of several different relaxation processes with widely different relaxation times could lead to an overall behavior that would be difficult to distinguish experimentally from constant phase angle behavior. Moreover, relaxation mechanisms can be imagined that are not of the single, time-constant type. For example, diffusion-controlled processes show transient responses that typically vary as $t^{-1/2}$, consistent with an α parameter of 0.5 in Equation 25. In later sections we will consider one tissue (liver) in which the relaxation at radio frequencies (RFs) can be modeled by the Cole-Cole equation in which the appearance of constant phase angle behavior arises from the structural heterogeneity, and another tissue (bone) in which such behavior appears to arise from a diffusion-controlled or charge-hopping process.

Characteristically, dielectric relaxation in solids more typically exhibits fractional power dependence expected from Equations 20 and 21 in the region $f > f_c$, rather than single, time-constant behavior represented by the Debye equations. Hill and Jonscher[36] have proposed that such fractional power law dependence is an inherent feature of dielectric relaxation in solids and arises from cooperative relaxation processes. This analysis, however, is rather formal and has yet to be applied in detail to dielectric relaxation in biological materials. A full explanation for the apparently broad spectrum of relaxation times in tissues is consequently a difficult and still unsolved problem.

E. Kramers-Kronig Relations

In order for the complex dielectric permittivity to represent a system that is physically realizable, it must be consistent with the Kramers-Kronig relations[37]

$$\epsilon'(f) - \epsilon_\infty = \frac{2}{\pi} \int_0^\infty \frac{x \, \epsilon''(x)}{x^2 - f^2} \, dx \tag{34}$$

$$\epsilon''(f) \quad = \frac{-2f}{\pi} \int_0^\infty \frac{\epsilon'(x) - \epsilon_\infty}{x^2 - f^2} \, dx \tag{35}$$

where x is a variable of integration. For given values of ϵ_s and ϵ_∞, the total area under a loss peak when ϵ'' is plotted against $\ln(f)$ must be independent of the distribution of relaxation times, since letting $f \to 0$, in Equation 34 gives

$$\epsilon_s - \epsilon_\infty = \frac{2}{\pi} \int_0^\infty \frac{\epsilon''(x)dx}{x}$$

$$= \frac{2}{\pi} \int_{-\infty}^\infty \epsilon''(\ln f)d(\ln f) \tag{36}$$

The Kramers-Kronig relations follow from very general assumptions, the principal ones being linearity of response and causality (i.e., no effect preceding a stimulus). They apply to dielectric materials with any arbitrary distribution of relaxation times, and even hold in the presence of resonance-type behavior such as is found in optical spectra. The principal limitation is that the applied field must be of sufficiently low amplitude that the response is linear — which is practically always the case in dielectric measurements on biological materials. A second limitation — that the properties of the system do not vary with time — is not strictly true for most biological preparations, but is nearly always an excellent approximation for the timescale in which dielectric measurements are carried out.

II. DIELECTRIC RELAXATION MECHANISMS

The above discussion was entirely general in that no specific dielectric polarization mechanisms were considered. We review briefly three dominant relaxation processes that occur in tissues: interfacial polarization, dipolar orientation, and ionic diffusion.

A. Interfacial Polarization (Maxwell-Wagner Effect)

If a material is electrically heterogeneous, a dielectric relaxation occurs in the bulk properties from the charging of the interfaces within the material. This relaxation does not arise within the bulk phases of the material, but is a consequence of the boundary conditions that the internal electric fields must satisfy at the interfaces between phases. Such effects often principally determine the dielectric properties of colloids and emulsions. Recent comprehensive reviews include Hanai[38] and Dukhin.[39,40] We consider several examples for illustration.

1. Two Slabs in Series

Consider a situation in which two slabs of thickness d_1 and d_2 are joined together with their interface oriented perpendicular to the direction of the external field. If the permittivity and conductivity of the two materials are, respectively, ϵ_1, σ_1 and ϵ_2, σ_2, the boundary condition on the electric field component normal to the interface is

$$\epsilon_1 E_1 = \epsilon_2 E_2 \tag{37}$$

if the interface is free of charge, i.e., the displacement vectors $D_1 = \epsilon_1 E_2$ and $D_2 = \epsilon_2 E_2$ are equal. The current densities at the interface are therefore related by

$$j_1/j_2 = \sigma_1 E_1/\sigma_2 E_2$$

$$= \sigma_1 \epsilon_2/\sigma_2 \epsilon_1 \tag{38}$$

Thus, if $\sigma_1 \epsilon_2 \neq \sigma_2 \epsilon_1$, the interface will become charged at a rate that is proportional to the difference between j_1 and j_2. Equivalently, the permittivity of the composite can be found by considering the material to consist of two capacitors in series. The bulk permittivity can be cast into the form of the Debye equations, where

$$\tau = \epsilon_0 \frac{\epsilon_1 d_2 + \epsilon_2 d_1}{\sigma_1 d_2 + \sigma_2 d_1} \tag{39}$$

is the relaxation time and

$$\epsilon_s = \frac{(\epsilon_2 \sigma_1 - \epsilon_1 \sigma_2)^2 (d_1 + d_2)(d_1 d_2)}{(\epsilon_1 d_2 + \epsilon_2 d_1)(\sigma_1 d_2 + \sigma_2 d_1)^2} + \epsilon_\infty \tag{40}$$

$$\sigma_s = (d_1 + d_2) \frac{\sigma_1 \sigma_2}{\sigma_1 d_2 + \sigma_2 d_1} \tag{41}$$

$$\epsilon_\infty = \frac{(d_1 + d_2) \epsilon_1 \epsilon_2}{(\epsilon_1 d_2 + \epsilon_2 d_1)} \tag{42}$$

$$\sigma_\infty = \sigma_s + \frac{(\epsilon_2 \sigma_1 - \epsilon_1 \sigma_2)^2 (d_1 + d_2) d_1 d_2}{(\epsilon_1 d_2 + \epsilon_2 d_1)^2 (\sigma_1 d_2 + \sigma_2 d_1)} \tag{43}$$

$$\text{where } \sigma_\infty \equiv \lim_{\omega \to \infty} (\epsilon'' \omega \epsilon_0). \tag{43}$$

2. Dilute Suspension of Spherical Particles

The corresponding equations for a dilute suspension of spherical particles were originally developed for the DC case by Maxwell[18] and extended to AC by Wagner.[41] If (ϵ_i*) and (ϵ_a*) are the relative complex dielectric permittivities of the suspended and continuous phases, and p is the volume fraction of the suspension (which is assumed to be small), a solution of Laplace's equation yields an expression for the complex permittivity of the mixture ϵ_m*

$$\epsilon_m^* = (\epsilon_a^*) \frac{2\epsilon_a^* + \epsilon_i^* - 2p(\epsilon_a^* - \epsilon_i^*)}{2\epsilon_a^* + \epsilon_i^* + p(\epsilon_a^* - \epsilon_i^*)} \tag{44}$$

which can also be cast into the form of the Debye equations with

$$\epsilon_\infty = \epsilon_a \frac{2\epsilon_a + \epsilon_i - 2p(\epsilon_a - \epsilon_i)}{2\epsilon_a + \epsilon_i + p(\epsilon_a - \epsilon_i)} \tag{45}$$

$$\epsilon_s - \epsilon_\infty = \frac{9(\epsilon_a \sigma_i - \epsilon_i \sigma_a)^2 \, p(1 - p)}{[2\epsilon_a + \epsilon_i + p(\epsilon_a - \epsilon_i)][2\sigma_a + \sigma_i + p(\sigma_a - \sigma_i)]^2} \tag{46}$$

$$\sigma_s = \sigma_a \frac{2\sigma_a + \sigma_i - 2p(\sigma_a - \sigma_i)}{2\sigma_a + \sigma_i + p(\sigma_a - \sigma_i)} \tag{47}$$

$$\sigma_\infty - \sigma_s = \frac{9(\sigma_a \epsilon_i - \sigma_i \epsilon_a)^2 \, p(1 - p)}{[2\sigma_a + \sigma_i + p(\sigma_a - \sigma_i)][2\epsilon_a + \epsilon_i + p(\epsilon_a - \epsilon_i)]^2} \tag{48}$$

$$\tau = \frac{2\epsilon_a + \epsilon_i + p(\epsilon_a - \epsilon_i)}{2\sigma_a + \sigma_i + p(\sigma_a - \sigma_i)} \epsilon_0 \tag{49}$$

If $\epsilon_i \ll \epsilon_a$ and $\sigma_i \ll \sigma_a$, Equation 44 can be expanded to first order in ϵ_i/ϵ_a and σ_i/σ_a to yield

$$\epsilon \cong \left[\frac{1 - p}{1 + p/2}\right] \epsilon_a + \left[\frac{9p}{(2 + p)^2}\right] \epsilon_i \tag{50}$$

$$\sigma \cong \left[\frac{1 - p}{1 + p/2}\right] \sigma_a + \left[\frac{9p}{(2 + p)^2}\right] \sigma_i \tag{51}$$

In this approximation, the mixture properties are simply weighted averages of the permittivity and conductivity values of the constituent phases. No dispersion in the bulk sample is expected from the approximate Equations 50 and 51; in contrast, a small dispersion is predicted by the full Equations 45 to 49 with time-constant $\tau \approx \epsilon_a \epsilon_0/\sigma_a$. In the opposite case ($\epsilon_i \gg \epsilon_a$, $\sigma_i \gg \sigma_a$) a very large dispersion will occur.

Equation 44 can be written in the form

$$\frac{\sigma_m^* - \sigma_a^*}{\sigma_m^* + \gamma\sigma_a^*} = p \frac{\sigma_i^* - \sigma_a^*}{\sigma_i^* + \gamma\sigma_a^*}$$

with $\quad \sigma^* = \sigma + j\omega\epsilon_0\epsilon'$

and $\quad \gamma = 2 \tag{52}$

This calculation was extended by Fricke[42,43] to the case where the suspended particles are prolate or oblate spheroids, by allowing the shape factor γ to vary in the above equations. In particular, for the case of cylindrical particles oriented normal to the field, $\gamma = 1$. Velick and Gorin[44] further extended this analysis to allow for ellipsoids with three different axes. Equation 52 will be referred to as the "Maxwell-Fricke equation" in the subsequent discussion.

Equation 52 can also be written in the form 53a below, for a sphere of radius R inside a larger sphere of radius R + d since $[R/(R + d)]^3$ is the volume fraction occupied by the smaller sphere. A rigorous derivation for this was given by Maxwell for the DC case and can be readily extended to the AC case.

3. Dilute Suspension of Membrane-Covered Spheres

The Maxwell-Wagner theory can be extended to a suspension of spheres, each of which is a composite of a sphere of dielectric properties (ϵ_i and σ_i) surrounded by a thin shell of properties (ϵ_{sh} and σ_{sh}). One first calculates the equivalent complex permittivity of the membrane-covered sphere, then the complex permittivity of the suspension by two consecutive applications of Equation 52. If R and d are the radius of the sphere and thickness of the shell, the equivalent complex conductivity σ^+ of the sphere is found from

$$\frac{\sigma^* - \sigma_{sh}^*}{\sigma^* + 2\sigma_{sh}^*} = \left(\frac{R}{R + d}\right)^3 \frac{\sigma_i^* - \sigma_{sh}^*}{\sigma_i^* + 2\sigma_{sh}^*} \tag{53a}$$

If $d \ll R$ this reduces to

$$\sigma^* = \frac{\sigma_i^* + \left(\dfrac{2d}{R}\right)(\sigma_i^* - \sigma_{sh}^*)}{1 + \left(\dfrac{d}{R}\right)\dfrac{\sigma_i^* - \sigma_{sh}^*}{\sigma_{sh}^*}} \tag{53b}$$

These results can be applied to a suspension of membrane-covered cells of radius R and unit membrane capacitance C_m and conductance G_m given by

$$C_m = \epsilon_{sh}\epsilon_0/d \qquad \text{(F/m}^2\text{)}$$

$$G_m = \sigma_{sh}/d \qquad \text{(S/m}^2\text{)}$$

$$\text{with} \quad \sigma_{sh}^+ = G_m d + j\omega C_m d$$

The full set of equations that results is quite involved, but simplifying assumptions can be made that are appropriate for a biological cell suspension.[1,45] For materials of conductivity and permittivity that would pertain to cells at RF frequencies and below

$$\sigma_i \gg \omega\epsilon_0\epsilon_i$$

$$\sigma_a \gg \omega\epsilon_0\epsilon_a$$

$$\text{and} \quad |\sigma_i^*| > |\sigma_{sh}^*|$$

The resulting simplified expression can be expanded to first order in p and cast into the form of the Debye equations with

$$\epsilon_s - \epsilon_\infty \cong \frac{9pRC_m}{4\epsilon_0\left[1 + RG_m\left(\dfrac{1}{\sigma_i} + \dfrac{1}{2\sigma_a}\right)\right]^2} \tag{54}$$

$$\rightarrow \frac{9pRC_m}{4\epsilon_0}$$

$$\sigma_s \cong \sigma_a\left[1 - \frac{3p}{2}\frac{1 + RG_m(1/\sigma_i - 1/\sigma_a)}{1 + RG_m\left(\dfrac{1}{\sigma_i} + \dfrac{1}{2\sigma_a}\right)}\right] \tag{55}$$

$$\rightarrow \sigma_a\left(1 - \frac{3p}{2}\right)$$

$$\tau \cong RC_m\frac{\sigma_i + 2\sigma_a}{2\sigma_i\sigma_a + RG_m(\sigma_i + 2\sigma_a)} \tag{56}$$

$$\rightarrow RC_m\left(\frac{1}{2\sigma_a} + \frac{1}{\sigma_i}\right)$$

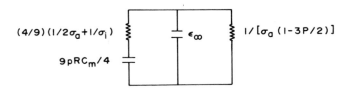

FIGURE 4. Equivalent circuit for the Maxwell-Wagner dispersion due to RC charging of the cell membranes. The parameters are σ_i = conductivity of the cytoplasm (S/m), σ_a = conductivity of the extracellular fluid, C_m = specific membrane capacitance (F/m^2), R = cell radius (M), p = volume fraction of cells in suspension.

$$\epsilon_\infty \cong \epsilon_a \left(1 - 3p \, \frac{\epsilon_a - \epsilon_i}{2\epsilon_a + \epsilon_i} \right) \tag{57}$$

The corresponding relation for σ_∞ is given by Equation 57 with all permittivity values replaced by the corresponding conductivities. Since $\sigma_i \simeq \sigma_a \sim 10^{-1}$ S/m, while G_m often is in the range 10^{-1} to 10^{-3} S/m^2 for biological membranes, the product RG_m is negligible compared to σ_i or σ_a for cellular dimensions, and Equations 54 to 56 assume the simple limiting forms given above. A physical interpretation of these results is suggested by an equivalent circuit (Figure 4). At low frequencies, the current flows mostly through the extracellular spaces. At high frequencies, the cell membranes become short-circuited and the current flows through the cells as well. The relaxation frequency $f_c \equiv 1/2\pi\tau$ corresponds to the transition between these two limiting cases, and is determined by the charging time-constant of the cell membrane capacitance through the combined resistance of the extra-cellular and intracellular fluids.

It should be noted that the assumption of low volume fraction p was made twice to derive the equations for cell suspensions: the first to obtain the Maxwell-Wagner result for particle suspensions and the second to obtain the simple closed form expressions, Equations 54 to 57. Therefore, these equations must be considered as approximations to a rather more complex situation. Closed form expressions can be obtained from Equation 52, valid (within approximations inherent in Equation 52) for any value of p, but they are quite involved and of slight advantage since Equation 52 is itself only valid for small p.

4. Dilute Suspensions of Spheres Surrounded by Multiple Membranes

The Maxwell-Wagner theory was extended to such cases by Fricke[46] and more recently by Irimajiri.[47] The procedure consists of repeatedly calculating the equivalent homogeneous specific admittance of the sphere, including successive shells in turn. The resulting equations yield multiple dielectric dispersions, corresponding to the number of interfaces lying between successive shell phases. This theory was used to analyze the dielectric properties of cultured lymphoma cells, in which the contribution of the cell nucleus to the permittivity of the suspension is pronounced at high frequencies.[48] This analysis lends itself to a general discussion of the coupling of EM fields with intracellular membranes (see Section V.C).

5. More Concentrated Suspensions

The results presented in examples 2 to 4 above pertain only to dilute suspensions in which the electrical interactions among the suspended particles are negligible. The extension of the mixture theory to higher volume fractions of suspension is a difficult problem, for which no exact solution has been found even for "simple" suspensions of identical spheres. Dukhin[39] presents an excellent summary of the difficulties involved as well as the various approaches taken to extend the theory. One method is to mathematically replace the medium surrounding a particle by an equivalent homogeneous medium with a suitably defined "effective" permittivity, and calculate the incremental change in the permittivity of the suspension resulting from the successive additions of incremental volume fraction of particles. Following this approach, Hanai[38] calculated the complex permittivity ϵ_m^* of a suspension of spherical particles of volume fraction p

$$\left(\frac{\epsilon_m^* - \epsilon_i^*}{\epsilon_a^* - \epsilon_i^*}\right)\left(\frac{\epsilon_a^*}{\epsilon_m^*}\right)^{1/3} = 1 - p \tag{58}$$

where ϵ_i^* and ϵ_a^* are the complex permittivities of the suspended and continuous phases, respectively. This theory was recently extended to ellipsoidal particles by Boned and Peyrelasse.[49]

The Hanai theory predicts an interfacial polarization dispersion in the suspension that has a similar underlying mechanism, as considered by the Maxwell-Wagner theory, but with considerably more complicated expressions. The following limiting expressions can be obtained from Equation 58

$$\left(\frac{\epsilon_\infty - \epsilon_i}{\epsilon_a - \epsilon_i}\right)\left(\frac{\epsilon_a}{\epsilon_\infty}\right)^{1/3} = 1 - p \tag{59}$$

$$\epsilon_s\left(\frac{3}{\sigma_s - \sigma_i} - \frac{1}{\sigma_s}\right) = 3\left(\frac{\epsilon_a - \epsilon_i}{\sigma_a - \sigma_i} + \frac{\epsilon_i}{\sigma_s - \sigma_i}\right) - \frac{\epsilon_a}{\sigma_a} \tag{60}$$

$$\left(\frac{\sigma_s - \sigma_i}{\sigma_a - \sigma_i}\right)\left(\frac{\sigma_a}{\sigma_s}\right)^{1/3} = 1 - p \tag{61}$$

$$\sigma_\infty\left[\left(\frac{3}{\epsilon_\infty - \epsilon_i}\right) - \frac{1}{\epsilon_\infty}\right] = 3\left(\frac{\sigma_a - \sigma_i}{\epsilon_a - \epsilon_i} + \frac{\sigma_i}{\epsilon_\infty - \epsilon_i}\right) - \frac{\sigma_a}{\epsilon_a} \tag{62}$$

In these expressions, σ_s, σ_∞, and ϵ_∞ are the values of the conductivity or permittivity of the suspension in the zero or infinite frequency limits. In general, the relaxation is not of the single, time-constant (Debye) type, because the local field at each particle is in part determined by neighboring particles. The Hanai theory was tested by dielectric measurements on water-in-oil suspensions which exhibit large interfacial polarization effects, and shown to yield significantly better correspondence with the data than the simpler Maxwell-Wagner theory.[50] It should be noted that Equations 59 to 62 assume that the dielectric properties of the suspended and continuous phases (ϵ_i, σ_i, ϵ_a, and σ_a) are independent of frequency. If a dispersion occurs in one of these phases, Equation 58 could be solved numerically, to predict the dielectric properties of the bulk suspension.

The Hanai theory can be extended to the case of suspensions of membrane-covered spheres by calculating an "effective" permittivity of the sphere using Equation 53, and then solving Equation 58 numerically, to obtain the dielectric properties of the suspension. Such an approach was taken by Hanai[51] to predict the dielectric properties of concentrated erythrocyte suspensions. It was found that the resulting complex permittivity plots deviated significantly from semicircular behavior for suspensions of very high volume fraction (p > 0.8). For suspensions of lower concentration (p ~ 0.3 to 0.5), the complex permittivity plots resemble the depressed circular arcs produced by the Cole-Cole equation, while at still lower concentrations the dispersion closely approximates the single, time-constant behavior expected from the Maxwell-Wagner theory (Figure 5).

Some comments are necessary about the limitations of the Maxwell mixture theory. The Maxwell results can be shown to be exact only to first order in p.[52] A more rigorous analysis

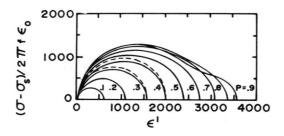

FIGURE 5. Complex permittivity plots of the dielectric properties of suspensions of shell-spheres calculated from Equations 53 and 58 (solid curves). The broken lines are semicircles representing single, time-constant relaxation. The parameters that were used are $\epsilon_a = 80$, $\sigma_a = \sigma_i = 0.25$ S/m, $\epsilon_i = 50$, $\epsilon_s = 6.5$, $\sigma_s =$ S/m, d = 5 nm, D = 3.8 μm. (From Hanai, T., Asami, K., and Koizumi, N., *Bull. Inst. Chem. Res. Kyoto Univ.*, 57, 297, 1979. With permission.)

would calculate the dielectric properties of the mixture as a power series in p, with terms to $0(p^n)$ representing n-wise interactions among the particles. An exact calculation of higher order terms is a formidable task, that requires more detailed information about the structure of the dispersion than is usually available. However, it can be shown that the contributions from higher order terms depend very much on the dielectric properties of the suspended vs. continuous phases. Such contributions are generally small if the complex permittivity (or conductivity) of the suspended phase is much smaller than that of the suspending medium, but can be large in the opposite case.[52] The predictions of the Hanai and Maxwell equations do not differ much for the case of particles of low permittivity or conductivity suspended in a medium of much higher conductivity or permittivity, as expected.

The above considerations help explain the perhaps unexpected success of the Maxwell theory. In view of the low DC conductance of most cell membranes, a cell suspension can be modeled for purposes of calculating the bulk conductivity as a collection of nonconducting particles in an electrolyte solution. This model was applied quite literally by early investigators who used sand as an electrical model for blood at low frequencies.[43] More recently, Cole found that the Maxwell theory could accurately predict the low frequency conductivity of suspensions of sea urchin eggs of volume fractions as high as 80%,[2] and of electrical analogues of tissues of volume fractions up to 90%.[53] In these cases, contributions from higher order effects would be expected to be small.

On the other hand, the effective permittivity of cells at low frequencies is much higher than that of extracellular fluids, and the Maxwell theory can be expected to be less satisfactory in calculating the dielectric dispersion of concentrated cell suspensions. (This was shown numerically by the work of Hanai and colleagues; see Figure 5.) However, the Maxwell theory does not fail drastically. At moderate volume fractions, it appears from the Hanai theory that nearest-neighbor effects give rise to a spread of relaxation times but do not greatly shift the mean relaxation frequency. The extracted parameters for the cell (such as C_m and σ_i) depend on the limiting permittivity at low frequencies and average relaxation frequency, and are not greatly affected. Indeed, it has been our experience that such parameters calculated using the Maxwell theory from dielectric measurements do not depend strongly on p, and thus are rather insensitive to interaction effects. We conclude that the Maxwell theory can be effectively used in spite of its limitations. Since it leads to simple algebraic results it is attractive for first order analyses even in situations in which interparticle interactions might be important. However, it is by no means exact and must be used with discretion.

It is interesting to note that the same mixture theory can be applied to other transport properties as well (thermal conductivity, magnetic susceptibility, and self-diffusion),[54] and comparative studies can be used to great advantage. This will be discussed again (see Section V.B).

6. Very Small Cell Size

Equation 53b applies when the membrane thickness d is small compared with the cell radius R. This is usually the case, since d is about 10^{-8} m and the cell size is in the micrometer range. An extension of the theory for particles in size down to 10^{-7} m has been given by Schwan and Morowitz,[55] and a general treatment for any size is given by Pauly and Schwan.[45] No simple, closed-form expressions for the parameters of the dispersion could be obtained, but is was found that the suspension exhibits two dispersions. One dispersion occurs in the RF range and is predicted by the cellular dispersion equations quoted above. The second is of small magnitude and occurs at much higher frequencies for suspensions of cells of typical dimensions. If the cell diameter is smaller than 0.1 μm, the two dispersion overlap in frequency, and numerical techniques must be employed to separate them.[51,57]

B. Dipolar Relaxation Mechanisms

A second polarization mechanism is the partial orientation of permanent dipoles. Excellent reviews of dipolar relaxation theory[58] and its applications to protein solutions[59,60] are available. We present the results of the simple Debye theory to suggest the magnitude of the effects that are involved.

Consider an ensemble of independent dipoles, each with dipole moment μ. An externally imposed electric field of intensity F at each dipole will exert a torque of magnitude μF sin θ, where θ is the angle between the dipole moment and the field. The tendency of this torque to orient the dipole will be opposed by random thermal agitation, and the ensemble will approach an equilibrium orientation ⟨cos θ⟩. Evaluation of the thermodynamic partition function[58] yields

$$<\cos \theta> = \coth \frac{\mu F}{kT} - \frac{kT}{\mu F} \qquad (63a)$$

where k is Boltzmann's constant and T the absolute temperature. For reasonable field strengths, the maximum orientational energy of a dipole (μF) is much less than its mean kinetic energy (which is of the order of kT) and Equation 63a reduces to

$$<\cos \theta> \sim \mu F/3kT \qquad (63b)$$

The net polarization of the sample, p, is given by

$$p = N \mu <\cos \theta> \qquad (64)$$

where N is the number of molecules per unit volume, or approximately

$$p \cong N \frac{\mu^2 F}{3kT} \qquad (65)$$

The local field intensity F at each molecule can be related to the external field strength E by the Lorentz expression

$$F = \frac{2 + \epsilon}{3} E \qquad (66)$$

Table 1
DIELECTRIC PROPERTIES OF VARIOUS PROTEINS AT 25°C[59]

Protein	M.W. ($\times 10^3$)	$\Delta\epsilon/g/\ell$	μ (D.U.)	$\tau \times 10^8$ (sec)	a/b
Ovalbumin	44	0.10	250	18; 4.7	5
Horse serum albumin	70	0.17	380	36; 7.5	6
Horse carboxyhemoglobin	67	0.33	480	8.4	1.6
Horse serum pseudoglobulin	142	1.08	1100	250; 28	9
β — Lactoglobulin (in 0.25 M glycine)	40	1.51	730	15; 5.1	4
Myoglobin	17	0.15	170	2.9	

Note: M.W. = molecular weight (daltons); μ = dipole moment (Debye units); τ = rotational relaxation times; a/b = axial ratio; $\Delta\epsilon/g/\ell$ = total increase in permittivity per gram of protein per gram of solution, measured at low frequencies above the permittivity of pure water.

which leads to an expression for the static relative permittivity

$$(\epsilon_s - 1)/(\epsilon_s + 2) = N_0\mu^2/9kTV\epsilon_0 \tag{67}$$

where N_0 is Avogadro's number and V is the molar volume equaling N_0/N.

After application of an electric field, the ensemble can be assumed to approach thermal equilibrium with first order kinetics leading to dielectric relaxation behavior of the Debye type, with rotational relaxation time-constant τ estimated from the Stokes law[58]

$$\tau = \frac{4\pi\eta a^3}{kT} \tag{68}$$

where a is the radius of the dipole and η the viscosity of the medium. (The macroscopic relaxation time, however, will be larger by a small factor because of local field effects.) This time-constant ranges from microseconds for large globular proteins, to picoseconds for smaller polar molecules, e.g., water. Consequently, the center frequency of the dispersion will be in the megahertz to gigahertz range. While the Debye theory is entirely macroscopic, it is not drastically in error for nonassociated molecular solutions, and the Stokes law calculation is surprisingly good even for a strongly associated liquid such as water.[61] (Equation 67 will fail in this case, however.)

In tissues, several dipolar relaxation effects can be anticipated. Globular proteins typically exhibit total dielectric increments (defined as the maximum increase in permittivity above that of the water divided by the concentration) in the range of 1 to 10/(g of protein per 100 g of solution) with relaxation frequencies $\leq$ 10 MHz[59,60] (Table 1), and are expected to be a small contribution to the permittivity of tissue at RFs. Partial orientation of polar sidechains has been suggested as a mechanism contributing to a small relaxation that is observed between 0.1 to 1 GHz in protein solutions.[1,62] Above some tens of megahertz, the major contribution to the dielectric permittivity is the water that constitutes about 80% of most soft tissues. Pure liquid water exhibits nearly single, time-constant dipolar relaxation with a characteristic frequency of 20 GHz at room temperature, or 25 GHz at 37°C.[63,64] It appears that the hydration water associated with proteins in solution exhibits a dipolar relaxation frequency substantially below that of the pure liquid.[65-67] Consequently, tissue water can be expected to exhibit a major relaxation at 20 to 25 GHz due to free water, with an additional component at relaxation frequencies near or perhaps below 1 GHz.[66,67] Tissue dielectric data (see Section IV) do indeed show these effects. The dielectric relaxation properties of water in tissues and other complex solutions will be considered in Section V.B.

Table 2
ELECTRICAL PROPERTIES OF
SUSPENSIONS OF POLYSTYRENE
SPHERES OF VOLUME FRACTION
CLOSE TO 30%

Particle radius (μm)	ϵ_s	f_c (kHz)
0.59	10,000	0.6
0.28	3,000	1.8
0.094	2,450	15
0.044	540	80

Adapted from Schwarz, G., *J. Phys. Chem.*, 66, 2636, 1962.

In summary, orientation of permanent dipoles makes only a small contribution to the permittivity of tissues below 100 MHz (where it is often masked by much larger interfacial and counterion polarization effects) but becomes more noticeable above 10 MHz and dominant above 1 GHz.

C. Counterion Diffusion Polarization

A third major class of polarization mechanisms arises from ionic diffusion in the electrical double layers adjacent to charged surfaces. These effects must be clearly distinguished from the Maxwell-Wagner polarization due to the charging of interfaces within heterogeneous materials, which depends on the differences in bulk dielectric properties of the constituent materials and is a macroscopic phenomenon. Counterion polarization effects, in contrast, are surface phenomena. In counterion polarization, the time-constant is of the form L^2/D, where L is a governing distance for diffusion (i.e., the size of the particle) and D a diffusion coefficient. This is in contrast to Maxwell-Wagner mechanisms, for which the time-constant is of the form RC or $\epsilon\epsilon_0/\sigma$ where R (or $1/\sigma$) is a resistance and C (or $\epsilon\epsilon_0$) represents a capacitance.

Counterion polarization is pronounced at audio and radio frequencies in emulsions,[38-40] suspensions of charged polystyrene spheres,[68] microorganisms,[69] and long chain macromolecular polyions such as DNA.[70] To illustrate the large magnitude of such effects, suspensions of submicron-sized polystyrene spheres exhibit permittivity values approaching 10^4 with relaxation frequencies in the kilohertz range (Table 2).

A rigorous analysis of counterion polarization is difficult since it must take into account coupled electrical and hydrodynamic phenomena in the double layers surrounding the particle, which are described by nonlinear equations. Nevertheless, a theory by Schwarz[71,72] successfully predicts the order-of-magnitude of the phenomena in spite of rather severe approximations. Schwarz considered the case of a macroscopic sphere of radius a with counterion surface charge density δ_0 (in units of m^{-2}), in which the thickness of the electrical double layer is much smaller than the particle diameter. The frequency dependent surface conductivity was found by solving the electrodiffusion equation to obtain the effective permittivity ϵ_p^* of the particle

$$\epsilon_p^* = \epsilon_b + \frac{1}{1 + j\omega\tau} \frac{e_0^2\delta_0}{\epsilon_0 kT} \tag{69}$$

where ϵ_b is the permittivity of the bulk material of the particle, and e_0 is the charge of the counterion. The relaxation time τ is proportional to the square of the radius

$$\tau = \epsilon_0 a^2 / 2ukT \qquad (70)$$

where u is the surface mobility of the counterions (m^2/Vsec). (The bulk permittivity of the suspension is then calculated using the Maxwell mixture theory.) This model was extended by Einolf and Carstensen[69] to account for the dielectric properties of bacteria, allowing for the diffusion of ions both inside and outside the cell wall. It is curious that the Schwarz theory, if applied to globular proteins, leads to dielectric increments and relaxation times that are in the same range as those expected from the permanent dipole mechanism.[70] This theory, however, does not apply for such small particles, which are comparable in size to the thickness of the electrical double layer. Nevertheless, it appears that counterion processes might contribute noticeably to dispersion at RFs in globular proteins.

When applied to rod-like molecules, the theory predicts dielectric increments that are proportional to the length, and relaxation times that vary as the square of the length.[72] (In contrast, the rotational, diffusional relaxation time varies as the cube of the length — see Equation 68.) Normally, biological macromolecules are highly charged, and the attraction between the counterions and the fixed surface ions becomes the dominating process. For linear biopolymers, the time-constant becomes

$$\tau = \frac{\epsilon_{eff} a^2}{2u\bar{z}e_0^2}$$

$$\text{which holds if} \quad \frac{\bar{z}e_0^2}{\epsilon_{eff}} \ll kT \qquad (71)$$

where ϵ_{eff} is the effective absolute permittivity of the molecule within the ionic atmosphere, and $\bar{z}$ is the mean number of counterions per unit of length.[70] Counterion polarization is clearly responsible for the large dielectric increments of DNA solutions at low frequencies,[70,73] and is presumably also important in the dielectric properties of other long chain biopolymers at low frequencies. It is reasonable to assume that such processes contribute to the alpha dispersion observed in tissues at low frequencies (see Section III), although exactly how much of this dispersion arises from ionic diffusion processes, as opposed to membrane charging effects, is difficult to decide.

The dipolar and counterion relaxation theories summarized above pertain to linear responses to weak fields. In the case of dipolar relaxation (Equation 63a), field strengths typically of hundreds of kilovolts per meter are needed before the interaction energy μF becomes comparable to kT, at least for molecular dipoles. However, counterion polarization can be produced in macroscopic objects, and then much smaller fields are expected to be sufficient to produce a nonlinear response. The approximate energy gained by an ion of charge e_0 in diffusing across a sphere of radius R in field E is e_0ER. This is comparable to kT for a 100-μm sphere in a field strength of about 250 V/m. Thus, significant displacements of the counterion atmosphere may be produced, in sufficiently large particles, at moderate field strengths.[74] The problem is that the response is then very slow — with a relaxation frequency of about 0.02 Hz for the 100-μm particle, for example. In smaller particles the response would be faster (cf. Equation 70) but the thresholds correspondingly higher, e.g., 2.5 kV/m in a 10-μm sphere.

The Schwarz theory has been successful in accounting for the observed effects, but it has been criticized for its *ad hoc* nature, in particular for its neglect of ionic diffusion in the electrolyte surrounding the charged surfaces. Quite different models have recently been developed, that can account for the dielectric polarization in the polystyrene spheres[68] as well or better than that of Schwarz. Notable work has been done in this regard by Dukhin,[39,40] Fixman,[134,135] and Chew and Sen.[136-138]

The calculations by these authors, particularly by Chew and Sen, are complex and the reader is referred to their papers for detailed discussion. We describe briefly the important differences from the Schwarz theory.

These authors abandon the model of a tightly bound layer of charges assumed by Schwarz, in favor of the Guoy-Chapman model of a diffuse double layer in which the ionic charge distribution is given by the Boltzmann distribution in terms of the potential.[137] The potential is given in terms of the charge distribution by Poisson's equation. In some cases hydrodynamic effects (electrically induced fluid flow) are considered as well. The coupled fluid mechanical and electrical problem is then solved using various approximations.

The principal finding is that counterion polarization is influenced by diffusion of charges in the bulk fluid surrounding the particle. The electrical currents that are set up in the double layer by the applied field discharge partly through a diffusion cloud that extends a distance of roughly $\sqrt{2D/\omega}$ from the particle, where D is the self-diffusion coefficient of the ion in the bulk fluid and ω is the frequency, as well as by conduction processes in the fluid. The result is that the surface currents have a component that is out of phase with the applied field, leading to a large effective permittivity of the particle. Moreover, the kinetics of the diffusion process lead to a complex permittivity that is not of the single, time-constant (Debye) type. In a recent paper Chew[138] calculated the dielectric dispersion of a spherical particle that, if fitted to the Cole-Cole equation, would give a distribution parameter α of 0.4. The mean relaxation time is not very different from that predicted by the Schwarz theory and still scales as a^2.

So far, the theories of Dukhin, Fixman, and Chew and Sen have not been applied in detail to biological materials. Given the complexity of the situation (both biological and theoretical) it is difficult to see exactly how they could be applied in any compelling way to tissues. But the work is noteworthy in that it leads to a physical explanation of the far-broader-than-Debye relaxation behavior, even in the simple case of a dielectric sphere in an electrolyte.

III. DIELECTRIC DISPERSION IN TISSUES

A. Conductivity

The rate of heat generation by an electric field of R.M.S. amplitude E in a material of conductivity σ is σE^2. In tissues, the dominant contributions to the conductivity at low frequencies are ionic conduction through the extracellular and intracellular fluids, while above 100 MHz, conductivity resulting from dielectric loss becomes increasingly significant. While the changes in conductivity with frequency are coupled to changes in permittivity by the Kramers-Kronig relations, it is instructive to consider the conductivity separately, with reference to the dielectric mixture theory. Several frequency regions will be considered; in each range mixture theory can be applied with a different value of the volume fraction that is effectively excluded from the current as it flows through the suspension. A later section (Section V.B) will apply this theory to examine the dielectric properties of tissue water.

1. Low Frequencies (< 0.1 MHz)

For times longer than the RC charging time τ of the cell membranes (Equation 56), the effective conductivity of a cell can be considered to be negligible and the excluded volume of the suspension will equal the total volume of the cells themselves. For AC, the corresponding frequency range is below the beta relaxation frequency $1/2\pi\tau$.

This simple idea is the basis for interpreting the low frequency conductivity of blood whose variation with hematocrit and other factors was first studied almost a century ago. (An excellent review, including many references to early work, has recently been prepared by Trautman and Newbower.[75]) The low frequency conductivity of blood can be fitted to the Maxwell-Fricke equation (Equation 52), but the required shape factor γ varied from 0.65

to 1.39 in different studies, depending on the species of animal and measurement conditions. Since the red cell of most species is nonspherical and is subject to changes in aggregation and orientation with flow, these variations in the form factor are understandable and important, and certainly contributed to the difficulties in applying the simple Maxwell theory to data from blood.

Conductivity data from solid tissues are even more difficult to predict because of the great structural complexity of such tissue. However, a simple analysis based on the mixture theory is surprisingly accurate. Assuming a typical fraction of the extracellular fluid in tissue to be 0.1 and a typical conductivity σ_a of 2 S/m for normal biological fluids and effectively nonconducting cells, Equation 51 predicts a DC conductivity of 0.14 S/m, which is in the range of observed values for soft, high water content tissues at low frequencies. An extensive compilation of resistivity values from many tissues at low frequencies is given by Geddes and Baker.[76]

2. Frequencies between 30 to 300 MHz

At frequencies above the beta relaxation frequency, the cell membranes are short-circuited, and the mixture theory can be applied with reasonable success, assuming the excluded volume to consist of poorly conducting protein plus its water of hydration. In this range, the conductivity of most tissues is nearly independent of frequency; it increases again above 100 MHz where contributions from relaxation processes become significant with respect to the (frequency independent) ionic conductivity.

3. UHF and Microwave Frequencies (> 100 MHz)

At these frequencies, the conductivity reflects increasingly large contributions from relaxational effects. Three possible mechanisms can be identified.

1. A Maxwell-Wagner process due to interfacial polarization of the electrolyte and relatively nonconducting protein molecules (cf. Equations 44 to 49). Assuming appropriate values for the electrical properties of the electrolyte and protein in a typical tissue, Equations 44 to 49 predict increases in conductivity of a few hundredths of a millisiemens per meter with a mean relaxation frequency of about 300 MHz. This was suggested as a minor contribution to the dielectric relaxation in barnacle muscle at frequencies between 0.1 to 1 GHz.[77]

2. Dielectric loss of small polar molecules and polar sidechains on proteins. The principal relaxational range of protein molecules is in the low megahertz range; but by virtue of their smaller size, single amino acids and polar sidechains of protein molecules can undergo a dielectric relaxation at much higher frequencies. In view of the proportionality of the total increase in conductivity to the mean relaxation frequency (Equation 11e), such a relaxation process can contribute observably to the loss above 100 MHz even though the corresponding changes in permittivity might be small.

3. Dielectric relaxation of water. Pure water exhibits a dielectric relaxation that is nearly characterized by a single time process with relaxation time of 8 psec, corresponding to a relaxation frequency at 20 GHz at 25°C.[63] Consequently, its conductivity will rise approximately quadratically with frequency below the relaxation frequency (cf. Equation 11d). For typical high water content tissues, this increase in conductivity becomes comparable to the ionic conductivity at 3 to 5 GHz. Moreover, the conductivity of tissues and protein solutions will include a contribution from dipolar relaxation of the water of hydration, which exhibits a relaxation frequency an order-of-magnitude or more below that of bulk water. It appears that in barnacle muscle tissue, the contribution to the total conductivity from dipolar relaxation of motionally restricted water, ionic conductivity, and dipolar loss of bulk tissue water are all of roughly comparable

magnitude at about 3 to 5 GHz, although the uncertainties in separating the contributions from "bulk-like" and motionally altered water are quite large.[77]

The relaxational properties of tissue water can be investigated by comparing the data to those expected on the assumption of single, time-constant relaxation. Equation 11d is rewritten to represent the sum of a dipolar plus an ionic term

$$\sigma_a = \frac{2\pi f^2 (\epsilon_s - \epsilon_\infty)\epsilon_0/f_c}{1 + (f/f_c)^2} + \sigma_s \tag{72}$$

where the parameters ϵ_s, ϵ_∞, and f_c pertain to pure water and σ_s is an ionic conductivity. Assuming that the proteins are relatively nonconductive, Equation 51 can be employed

$$\sigma \rightarrow \frac{1 - p'}{1 + p'/2}\sigma_a + \frac{9p'}{(2 + p')^2}\sigma_i \tag{73}$$

where σ_i is the effective conductivity of the protein plus its water of hydration, and p' is the volume fraction of hydrated protein that is excluded from conduction. This excluded volume fraction is greater than p, the volume fraction of anhydrous protein; the difference $(p' - p)$ is by hypothesis the volume fraction of motionally altered water.

If the frequency chosen is sufficiently high, the frequency dependence of σ should reflect that of σ_a, and variations in σ_i with frequency will become insignificant. If so, the total conductivity σ will approach a linear function of $f^2/(1 + [f/f_c]^2)$ if f_c is chosen to equal the relaxation frequency of the bulk liquid. From the slope of this function, the effective excluded volume p' and consequently the fraction of hydration water can be obtained. This assumes an oversimplified model, in an attempt to separate the "free" and motionally restricted water. Our later discussion, however, shows that this approach yields entirely reasonable estimates of protein hydration and is consistent with other evidence as well.

B. Permittivity

Typically, soft tissues exhibit a continuous monotonic decrease in permittivity with frequency, together with an associated increase in conductivity. Permittivity values can exceed 10^5 to 10^6 at subaudio frequencies and can be expected to reach limiting values of 4 to 5 at frequencies approaching 100 GHz. Three major dispersion regions can be identified (Figure 6), that would correspond to separate relaxation processes with total dielectric increments, $\Delta\epsilon_\alpha$, $\Delta\epsilon_\beta$, and $\Delta\epsilon_\gamma$, respectively. These might be modeled by an equation of the form of Equation 16, although each relaxation region would itself more likely be characterized by distribution of relaxation times.

The alpha dispersion is manifested by the very large increase in permittivity at audio frequencies. In this frequency range, large permittivity values are produced by ionic diffusion processes in micron and larger sized objects (cf. Section II.C); such effects must be presumed to occur in tissues as well at these low frequencies. Other possible low frequency polarization mechanisms that are specific to individual tissues would include active membrane conductance phenomena, the charging of intracellular membrane-bound organelles that connect with the outer cell membrane, and perhaps a frequency dependence in the membrane impedance itself. A review of the current understanding and unsolved questions about the alpha dispersion has been provided by Schwan.[78] For a total dielectric increment $\Delta\epsilon$ of 10^6 and relaxation frequency of 100 Hz, the increase in conductivity is expected from the Kramers-Kronig relations to be roughly 0.005 S/m, which is negligible compared to the ionic conductivity of most biological preparations. At these low frequencies the tissue impedance is overwhelmingly resistive in spite of the tremendous permittivity values that are measured.

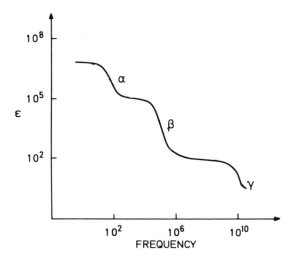

FIGURE 6. A schematic illustration of three major dispersion regions found in typical tissues, with permittivity and conductivity values typical of soft tissues.

Consequently, for engineering applications, the alpha dispersion is of little significance. Its implications for bioeffects studies remain to be established.

The beta dispersion occurs at RFs due principally to the capacitive charging of cellular membranes in tissues (Section II.A), although smaller contributions are also expected from dipolar relaxation of proteins in the tissue. Blood exhibits a total dielectric increment $\Delta\epsilon_\beta$ of 2000 and a beta relaxation frequency of 3 MHz, and consequently a total conductivity increase of roughly 0.4 S/m of which by far the largest part is due to the increase in the volume fraction of the suspension that is available to conduction, i.e., the intracellular as well as extracellular spaces. For tissues, the total change in permittivity through the beta dispersion approaches 10^4 relative to free space, and the relaxation frequency is about 500 kHz. The larger static permittivity values and longer relaxation times of tissues vs. blood are consequences of the typically larger cell sizes. The implications of the beta dispersion for bioeffects studies will be discussed later (Section V.C).

The gamma dispersion occurs with a center frequency near 25 GHz at body temperature, due to the dipolar relaxation of the water that constitutes 80% of the volume of most soft tissues. For a total dielectric increment $\Delta\epsilon_\gamma$ of 50 (typical of soft tissues containing 80% water) and relaxation frequency of 25 GHz, the total increase in conductivity is about 70 S/m.

In addition to the above three major dispersion regions, there is a smaller, rather poorly defined delta dispersion in the range 0.1 to 3 GHz, for which no single, dominant relaxation process has been identified. In tissues, the total observed changes in relative permittivity between 0.1 and 1 GHz are typically in the 10 to 20 range relative to free space, with associated increases in conductivity of 0.4 to 0.5 S/m. This is presumed to arise in part from the dipolar relaxation of water of hydration, and in part from rotational relaxation of polar sidechains (Section V.B), and possibly also from ionic effects of the Maxwell-Wagner type or counterion diffusion along small regions of charged surfaces. The lack of a single, dominant mechanism makes the analysis of this dispersion region in tissues rather difficult.

IV. DIELECTRIC PROPERTIES OF SELECTED TISSUES

A. Properties of Excised Tissues

1. Blood

As indicated earlier, the study of the dielectric properties of blood has been important in the development of the understanding of the underlying mechanisms giving rise to the bulk electrical properties of tissues. This subject continues to be important in view of the many practical applications requiring accurate conductivity or permittivity values for this tissue.

An excellent historical review of the early work by Höber, Fricke, and others is presented by Cole[2] and by Schanne and P.-Ceretti,[5] the latter of whom summarize in detail the many dielectric relaxation studies that have been conducted on this tissue since the time of Fricke.

The dielectric relaxation properties of blood and hemoglobin have been reviewed by Schwan,[79] with emphasis on the pertinent mechanisms. In whole blood, no alpha dispersion is observed; the gamma and delta dispersions appear to be identical to those of hemoglobin solutions of appropriate concentration (as would be expected). Whole blood exhibits a low frequency limiting value of the permittivity of approximately 2000 for a hematocrit of 40%, and a center beta relaxation frequency of 3 MHz. A table of the derived electrical properties of the erythrocyte, from the many previous investigations, is provided by Schanne and P.-Ceretti.[5]

The dielectric properties of concentrated erythrocyte suspensions are quite complex, due to aggregation and other effects. Moreover, the properties cannot be exactly predicted by the simple Maxwell-Fricke theory for dilute suspensions of spherical cells, but require empirical shape factors or other modifications to the theory. An excellent review of the low frequency conductivity of erythrocyte suspensions under various conditions is by Trautman and Newbower,[75] who present several empirical functions that summarize the conductivity under various experimental conditions. A recent paper by Pfützner[125] reports the permittivity of blood from several species at frequencies at 50 KHz, as a function of the hematocrit. Rather surprisingly, the permittivity reaches a peak at a hematocrit of roughly 60% and declines at higher volume fractions, an effect attributed to accidental variations in the current pathways around regions of packed cells.

2. Skeletal Muscle

The principal electrical feature of muscle is its extreme anisotropy, with at least a seven- to tenfold variation in conductivity (of dog skeletal muscle) at low frequencies, and similar pronounced anisotropies in the permittivity (Figure 7). Similar variations have been noted in skeletal and heart muscle tissue from many other species as well,[80-82] that persist at all frequencies at which measurements have been carried out. A comprehensive summary of the low frequency conductivity of various muscle tissues is prepared by Geddes and Baker.[76] We summarize briefly the dielectric properties at the three major dispersion regions described above.

a. Alpha Dispersion

It was earlier suggested that the alpha dispersion in muscle arises principally from polarization of counterions near the membrane surface;[1] however, Falk and Fatt[83,84] pointed out that this would require unusually high fixed charge densities on the membrane, and proposed instead that the alpha dispersion arises from an access impedance to the sarcotubular system, i.e., at low frequencies the current penetrates deeply into the sarcotubular system, leading to an increase in the total capacitance between the sarcoplasm and the extracellular medium. Consequently, the relaxation time-constant is determined by a combination of the sarcotubular capacitance and resistance of the fluid within these tubules. For tissues in the perpendicular orientation, the permittivity at low frequencies is proportional to the membrane

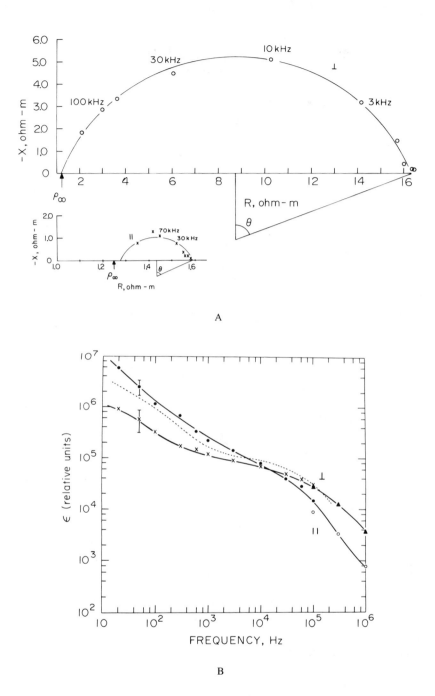

A

B

FIGURE 7. The electrical properties of dog skeletal muscle at 37°C in the perpendicular and parallel orientations. (A) Specific impedance plots. The arrow indicates the specific resistance measured at 100 MHz in nonoriented tissue, representing the limiting resistivity of the beta dispersion; (B) permittivity; (C) conductivity vs. frequency. Also shown (dotted lines) are the dielectric properties of nonoriented frog skeletal muscle at 25°C.[81] The symbols ⊥ and ∥ refer to tissues in the transverse and the parallel orientations, respectively. (From Epstein, B. R. and Foster, K. R., *Med. Biol. Eng. Comput.*, 21, 51, 1983. With permission.)

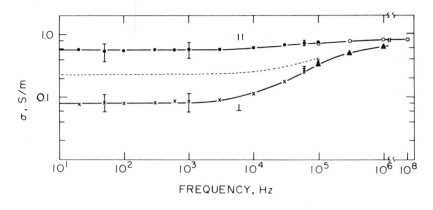

FIGURE 7C

capacitance (Equation 54). Thus, the dispersion resulting from the charging of the sarco-tubular system would be associated with an increase in the bulk permittivity of the tissue, corresponding to the increase in apparent membrane capacitance from that of the outer cell membrane to that with outer cell membrane plus sarcotubular system. The total capacitance of the sarcotubular system can exceed that of the outer cell membrane by an order-of-magnitude or more.

This is consistent with a variety of impedance studies on single muscle fibers. Schwan[81] first observed the alpha dispersion in frog skeletal muscle, which corresponded to an apparent increase in the capacitance of the membrane by a factor of 30 with a center relaxation frequency of approximately 100 Hz. Takashima studied the impedance of single frog skeletal muscle fibers in the radial direction using a vaseline gap technique,[85] and found also a dispersion in the apparent capacitance of the membrane centered at 100 Hz. The total increase in capacitance, however, was only a factor of 5. The difference between these two results might in part arise from different measurement techniques, i.e., the use of single fiber preparations vs. dielectric measurements on bulk muscle tissue. The data in Figure 7, which are also obtained from bulk tissue, are consistent with Schwan's earlier results.

In the dog skeletal muscle, the alpha dispersion is even more pronounced in the longitudinal orientation. The sarcotubular system is a complex, membrane-bound structure that runs along the length of the fiber, and it is reasonable to assume that the dielectric polarization might also be produced by fields parallel to the fiber axis, associated with charging of surfaces within each fiber. Conceivably, contributions could also arise from counterion polarization or charge-hopping along fibrils or membrane surfaces tangential to the field direction.

b. Beta Dispersion

For fibers oriented parellel to the field, a beta dispersion will be produced by the charging of the outer cell membrane, although it will occur at subaudio frequencies in view of the length of typical muscle cells. At audio frequencies, the conductivity of the tissue in the parallel orientation should be close to that of the cytoplasm.

In the perpendicular orientation, the outer cell membranes are charged with much smaller time-constants that can be calculated using the theory presented in Section II. Although the cell radii and total fraction of cells depend on the muscle type and condition, typical values for dog skeletal muscle are 50 μm (average cell radius) and 0.85 (average cell volume fraction). Equations 54 to 57, modified for the case of a cylindrical cell, then yield

$$\epsilon_s - \epsilon_\infty \cong aC_m/\epsilon_0 = 5 \times 10^4$$

$$\sigma_s \cong \sigma_a(1 - p)/(1 + p) = 0.06 \text{ S/m}$$

$$f_c \cong \sigma_i/(2\pi aC_m) = 250 \text{ kHz} \tag{74}$$

assuming values of σ_a and σ_i that are both equal to the measured conductivity of the tissue at 100 MHz, and assuming the membrane capacitance C_m to be 10^{-2} F/m^2. The data in Figure 7 exhibit, as expected, a plateau at a value roughly 10^5 relative to free space, and a relaxation frequency (at which the conductivity is halfway between its high and low frequency limits) around 250 kHz. The conductivity in the longitudinal orientation is higher and varies much less with frequency, as expected.

c. Gamma and Delta Dispersions

The dielectric properties of nonoriented dog skeletal muscle are shown in Figure 8 at RFs, together with those of barnacle muscle fibers (perpendicular orientation) at RF-microwave frequencies.[77,86] While the data suggest a plateau near 0.1 GHz, there is no distinct separation between the beta and higher frequency dispersions. The dielectric properties of barnacle muscle fibers (perpendicular orientation) at 1°C are plotted on the complex permittivity plane in Figure 9. It appears that the gamma dispersion for the barnacle muscle at 1°C can be characterized by a single, time-constant relaxation centered at 8 to 9 GHz, compared to 9 GHz for pure water at that temperature.

The magnitude $\Delta\epsilon_\gamma$ of the gamma dispersion is found in Figure 9 from the intercept of the semicircle with the real axis. Thus $\Delta\epsilon_\gamma$ is about 57 relative to free space. The excluded volume, calculated from this figure by the mixture theory, is about 0.2, approximately 25% higher than the known volume fraction of anhydrous protein, reflecting the presence of water of hydration with lower relaxation frequency. While the possibility of a small shift in relaxation frequency cannot be excluded, it is clear that the relaxation frequency of the major fraction of water in the muscle tissue is close, if not identical, to that of the pure liquid.

The presence of the delta dispersion is seen by the variation in permittivity with frequency below 3 GHz, at which the permittivity of pure liquid water has essentially reached its low frequency limiting value. This delta dispersion in the tissue is most clearly seen in a plot of the conductivity vs. the quantity $f^2/(1 + [f/f_c]^2)$ (Figure 10) where f_c is the relaxation frequency of pure liquid water. The conductivity data approach a linear function at high frequencies indicating that the dipolar loss from the free tissue water is the dominant contribution to the conductivity. At low gigahertz frequencies, however, the delta dispersion contributes significantly to the total conductivity. As Figure 8 shows, the dielectric properties of muscle and gelatin suspensions are similar between 0.1 and 10 GHz but diverge at lower frequencies due to membrane charging effects.

d. Distribution of Relaxation Times

At audio and RFs, muscle clearly exhibits a broad distribution of apparent relaxation times. It is curious that the permittivity increase ($\epsilon_s - \epsilon_\infty$) approximates a simple power function of frequency over 5 or more decades in frequency. In the complex impedance plane this corresponds to a depressed semicircular locus with a constant phase angle of about 70° (Figure 8; Reference 2). A distribution of diameters of the muscle fibers will contribute somewhat to the distribution. However, the spread in fiber diameters is at most a factor of 3 or so, which will only be noticeable near the center frequency of the beta dispersion. The presence of organelles and proteins will contribute somewhat to the spread in the relaxation at frequencies above the center relaxation frequency of the outer cell membrane. It would appear that the broad spectrum at low frequencies arises from the complex structure of the sarcotubular system.

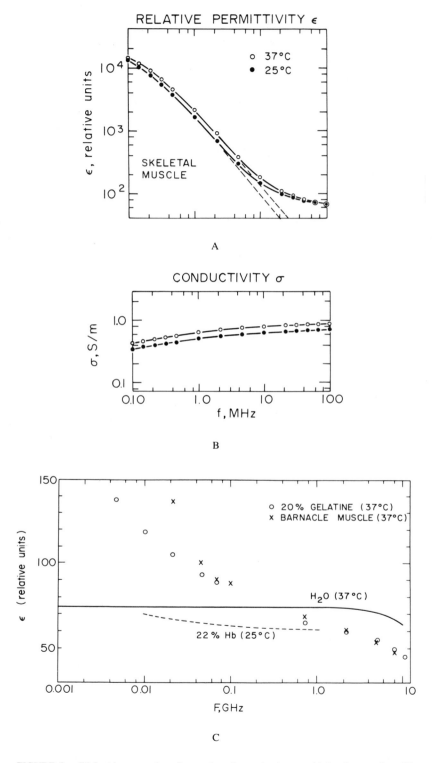

FIGURE 8.   Dielectric properties of nonoriented muscle tissue at higher frequencies. (A) Permittivity of dog skeletal muscle;[86] (B) conductivity;[86] (C) permittivity of barnacle muscle fibers (perpendicular orientation) at UHF and microwave frequencies, with the permittivity of pure water and two protein solutions of the same volume fraction shown for comparison.[77]

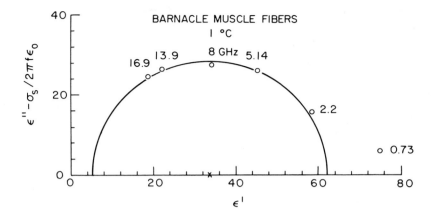

FIGURE 9. Complex permittivity plot of the dielectric properties of barnacle muscle fibers (perpendicular orientation) at 1°C, showing a single, time-constant relaxation with center frequency near 8 to 9 GHz with an additional delta dispersion evident at low gigahertz frequencies. (From Foster, K. R., Schepps, J. L., and Schwan, H. P., *Biophys. J.*, 29, 271, 1980. With permission.)

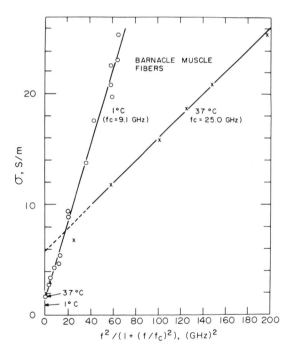

FIGURE 10. The conductivity of barnacle muscle fibers (perpendicular orientation) plotted vs. the quantity $f^2/[1 + (f/f_c)^2]$, with f_c assumed equal to the relaxation frequency of pure water at the indicated temperatures. The conductivity of the tissue at 100 MHz at these two temperatures is indicated by the arrows at left. The data show the delta dispersion that adds significantly to the conductivity of the tissue above 0.1 GHz, evident by the 4.5 S/m difference at 37°C between the zero frequency intercept of the data (6 S/m) and the conductivity at 100 MHz (1.5 S/m). (From Foster, K. R., Schepps, J. L., and Schwan, H. P., *Biophys. J.*, 29, 271, 1980. With permission.)

In summary, the dielectric relaxation properties of this tissue are quite complex. The various dispersion regimes — alpha, beta, gamma, and delta — are readily recognized but not clearly separable. The great difficulties these and similar data from other tissues presented to earlier investigators is readily apparent.

e. Temperature Coefficients

Figures 11 and 12 show the variation in dielectric properties of muscle and another tissue (brain) with temperature. As expected, the conductivity at UHF frequencies and below has the same temperature coefficient at 0.1 GHz as that of simple electrolytes, i.e., about 2 %/°C near room temperature. At frequencies below the relaxation frequency of water, the conductivity arising from dipolar absorption varies inversely as f_c (cf. Equation 72) and will thus exhibit a negative temperature coefficient of about $-2\%/°C$. Therefore, the temperature coefficient of the tissue conductivity will decrease at gigahertz frequencies to $-2\%/°C$ with a crossover point near 2 GHz, where the ionic and dipolar contributions to the conductivity are equal. The temperature coefficients in the permittivity are comparable or smaller. At all frequencies, the temperature coefficients for both the permittivity and conductivity fall roughly within the range $\pm$ 2%/°C, and are probably negligible for most applications. (A fuller discussion[8] as well as additional temperature coefficients from various tissues at 200 to 900 MHz[126] are available.)

3. Other Soft Tissues

The dielectric properties of several soft, high water content normal and tumor tissues are shown in Figures 13 and 14.[86,87] The dielectric properties of all of these tissues are quite similar and can be discussed as a group.

a. Beta Dispersion

The dielectric relaxation shown in Figure 12 clearly reflects a broad distribution of relaxation times which arises principally from the presence of membrane-bound structures of widely varying dimension. This was established by Stoy et al.[86] for liver tissue using a simple geometric model for the tissue and assuming that the contributions to the permittivity from each major tissue structure are additive. Thus,

$$\epsilon' = \Sigma_i \frac{\Delta\epsilon_i}{1 + (f/f_{ci})^2} + \epsilon_\infty \tag{75}$$

where ϵ_∞ is the contribution of tissue water and protein, and $\Delta\epsilon_i$ and f_{ci} are, respectively, the total increase in permittivity and center relaxation frequency of the ith structure. The parameter ϵ_∞ was taken to be the measured permittivity at 100 MHz; the values $\Delta\epsilon_i$ and f_{ci} were calculated using Equations 54 to 57 for several major membrane-bound structures in the tissue (Table 3). The dielectric increment and relaxation frequency of the tissue protein were assumed to be similar to those of globular proteins given in Table 1. The bulk permittivity predicted by this simple model agrees remarkably well with the measured properties above 1 MHz (Figure 15; References 86 and 88), but is somewhat too low at lower frequencies.

A few words about this approach are needed. At low frequencies the interior of the cells is shielded from the externally applied electric fields since the other cell membranes prevent current from entering the cell, and consequently the organelles do not contribute to the bulk permittivity, nor do their membranes contribute at sufficiently high frequencies since they are shorted out by their capacitance. Consequently, there exists a limited frequency range in which the charging of subcellular membranes will contribute significantly to the permittivity of the tissue, which extends roughly from the beta relaxation frequency of the entire cell to that of the organelle. The ratio of these two relaxation frequencies is, from Equation

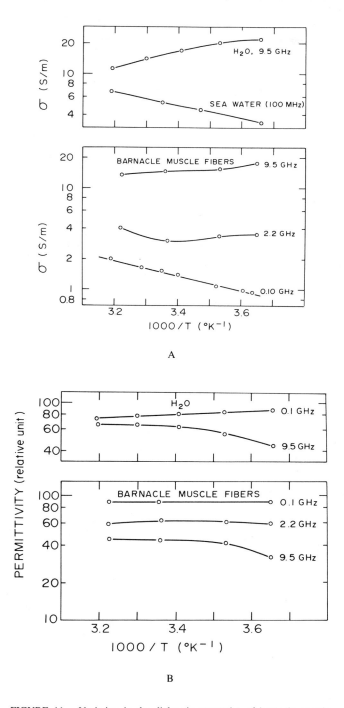

FIGURE 11. Variation in the dielectric properties of barnacle muscle fibers with reciprocal temperature. The corresponding properties of pure water or seawater are shown for comparison. At frequencies near 1 to 2 GHz, two opposite changes (increase with temperature in the ionic conductivity of tissue electrolyte and decrease in the dipolar loss of water) combine so that the temperature coefficient of the tissue is very small. (A) conductivity; (B) permittivity. (From Foster, K. R. and Schepps, J. L., *J. Microwave Power,* 16, 107, 1981. With permission.)

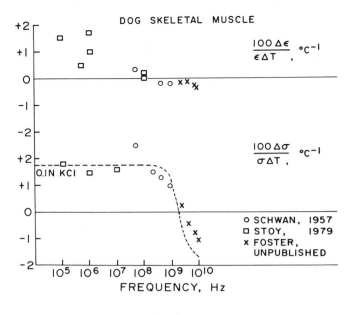

A

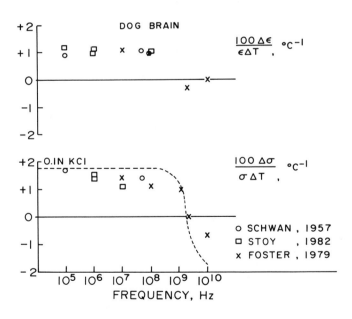

B

FIGURE 12.   Fractional change with reciprocal temperature of dielectric properties of (A) dog muscle and (B) brain near 37°C. The sources of the data are Schwan,[1] Foster and Schepps,[87] and Stoy.[86] In most cases, the temperature coefficients have been calculated from two measurements at 25 to 28 and 37°C.

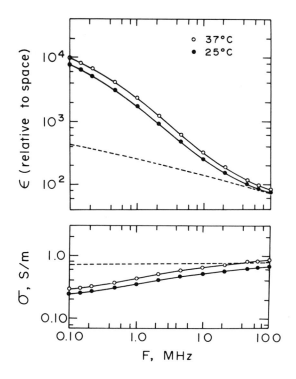

FIGURE 13. Dielectric permittivity and conductivity of dog pancreas at 25 and 37°C, together with the fitted Cole-Cole function (Equations 20 and 21). The parameters are ϵ_s = 10,480 (12,960) relative to free space, ϵ_∞ = 41.3 (36.7) relative to free space, σ_s = 0.226 (0.268) S/m, f_c = 230 (254) kHz, α = 0.232 (0.228) for the data. The figures in parentheses refer to results at 37°C; otherwise at 25°C. Also shown (dotted line) are the dielectric properties (25°C) after homogenization of the tissue, showing the elimination of most of the beta dispersion by disruption of the cells. (From Stoy, R. D., Foster, K. R., and Schwan, H. P., *Phys. Med. Biol.*, 27, 501, 1982. With permission.)

56, the ratio of the diameters of the cell and organelle. The fact that the organelles do not contribute at low frequencies to the permittivity of the tissue may result in the little ''hump'' seen in Figure 15 near 10 MHz, which corresponds to the relaxation frequency of the micron-sized mitochondria.[89]

While Stoy's analysis is obviously very approximate, it nevertheless leads to the following conclusions:

1. The dielectric dispersion in liver tissue in the range of 1 to 100 MHz represents the high frequency end of the beta dispersion of the cells together with the organelles.
2. The large range in dimensions of major membrane-bound structures within the tissue accounts for the broad spread of relaxation frequencies observed at RFs.
3. The permittivity values observed below 10^5 Hz exceed those expected due to the polarization of outer cells alone.
4. The contribution of the tissue protein to the total permittivity at RFs appears to be small compared to that of membrane-bound structures.

b. Gamma Dispersion

The contribution of the tissue water to the permittivity at microwave frequencies can be

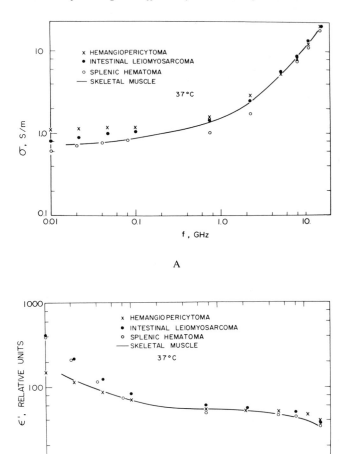

FIGURE 14.   Dielectric properties at RFs of three representative tumor tissues and dog skeletal muscle at 37°C. (A) Conductivity; (B) permittivity. (From Schepps, J. L. and Foster, K. R., *Phys. Med. Biol.*, 25, 1149, 1980. With permission.)

estimated by fitting the complex permittivity data above 5 GHz to the Debye equations (Equations 11b and c) assuming a relaxation frequency identical to that of pure liquid water,[90] and arbitrarily choosing a value of 4 for ϵ_∞ (Figure 16). The parameter that is obtained from this fit, ϵ_s^m (corresponding to ϵ_s in Equations 11b and c) is roughly equal to the contribution of the tissue water to the total permittivity at frequencies below approximately 1 GHz. The conductivity values of these tissues at 100 MHz ($\sigma_{0.1}$) closely approximate the conductivity of the cytoplasm in the absence of cell membranes (Figure 17). The fitted parameters are summarized in Table 4.

c. Delta Dispersion

The tissues exhibit a much broader relaxation at microwave frequencies than is expected from dipolar relaxation of bulk water alone. After subtracting the contribution of the free

Table 3
MORPHOLOGY OF RAT LIVER: CONTRIBUTIONS
TO PERMITTIVITY

Structure	Average radius (m)	Vol fraction	$\Delta\epsilon$	f_c (MHz)
Hepatocyte	8.9×10^{-6}	0.83	9400	0.72
Total nuclei	3.9×10^{-6}	0.05	470	1.6
Mitochondria	5.0×10^{-7a}	0.22	277	13
Endoplasmic reticulum	2.5×10^{-8b}	0.15	8	250
Protein	c	0.16	100	3

[a] The mitochondria are cylindrical in shape, with diameters of 2.4×10^{-6} m by 0.6×10^{-6} m. The average radius is taken as the radius of a sphere with the same total volume as the cylinder.

[b] The average radius of the endoplasmic reticulum is taken as half of the mean cisternal width.

[c] The total tissue protein was assumed to have a static permittivity and relaxation frequency typical of most globular proteins. While these values are uncertain for the mostly fibrous tissue proteins, the contribution of tissue protein to the total dielectric increment is probably rather small at most frequencies, compared to the contribution from membrane polarization effects.

Modified from Stoy, R. D., Foster, K. R., and Schwan, H. P., *Phys. Med. Biol.,* 27, 501, 1982.

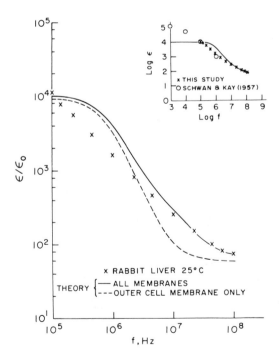

FIGURE 15.   Comparison between the measured permittivity of rabbit liver tissue (crosses and open circles) and calculated values obtained from Equation 75 (solid line). Also shown (broken line) is the calculated contribution of the outer cell membranes. (From Stoy, R. D., Foster, K. R., and Schwan, H. P., *Phys. Med. Biol.,* 27, 501, 1982. With permission.)

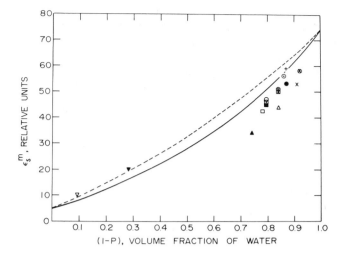

FIGURE 16.   Extrapolated permittivity ϵ_s^m obtained by fitting data from several normal and tumor tissues above 3 GHz to the Debye equation with the same relaxation time as pure water vs. (1-p) the volume fraction of solid in the tissues. The solid and broken lines are the predictions of the Maxwell-Fricke theory for suspended spheres (broken curve) and suspended ellipsoids of revolution of axial ratio 5:1 (solid curve) for comparison. Identification of the symbols is in Table 4. The parameter ϵ_s^m is obtained by fitting the data above 3 GHz to single, time-constant relaxation (Equations 11a to d) with center frequency equal to that of water at the same temperature ($\epsilon_\infty = 4$). (From Schepps, J. L. and Foster, K. R., *Phys. Med. Biol.* 25, 1149, 1980. With permission.)

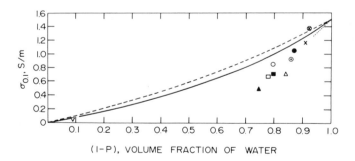

FIGURE 17.   The conductivity of several tumor and normal tissues at 100 MHz vs. the volume fraction of solids in the tissue (1-p). Identification of the other symbols is in Table 4. The solid and broken lines are the predictions of the Maxwell-Fricke mixture theory for suspended, nonconductive spheres (broken curve) and prolate ellipsoids of revolution of axial ratio 5 (full curve). The dotted line is the locus of data from various proteins at 1 kHz. (From Schepps, J. L. and Foster, K. R., *Phys. Med. Biol.*, 25, 1149, 1980. With permission.)

tissue water, the delta dispersion is revealed in the tissues extending down into the RF region, that is remarkably similar in several tissues (Figure 18).

d. Parameterization of Dielectric Properties at UHF-Microwave Frequencies

Figures 16 and 17 together with the above analysis can be used to obtain empirical equations

Table 4
SUMMARY OF DATA

	Symbol[a]	Vol fraction of water (1-P)	Extrapolated microwave permittivity (ϵ_s^m)	Conductivity at 0.1 GHz ($\sigma_{0.1}$ S/m)	Extrapolated microwave conductivity ($\sigma_{0.1}$ S/m)
Brain (gray matter)	△	0.84	44	0.70	1.13
Brain (white matter)	◀	0.74	34	0.48	0.75
Splenic hematoma	○	0.86	56	0.92	1.60
Splenic hematoma (different sample)	○	0.795	47	0.85	1.30
Skeletal muscle	■	0.795	47	0.70	2.40
Hemangiopericytoma	×	0.91	53	1.16	2.90
Hemangiopericytoma (different sample)	⊗	0.92	58	1.37	3.60
Fat	▷	0.09	10	0.005	0.10
Intestinal leiomyosarcoma	●	0.87	53	1.05	2.30
Liver	□	0.795	43	0.67	2.30
Vaginal fibroleiomyoma	+	0.87	59	—[b]	1.80
Pulmonary papillary adenosarcoma	⊕	0.84	51	—	1.40
Lipoma	▶	0.28	20	—	0.40
Renal tubular adenosarcoma	⊞	0.84	50	—	2.40
Spleen	□	0.795	46	—	1.90

[a] Symbols refer to Figures 16 and 17.
[b] The symbol (—) indicates that no value was measured.

From Schepps, J. L. and Foster, K. R., *Phys. Med. Biol.*, 25, 1149, 1980. With permission.

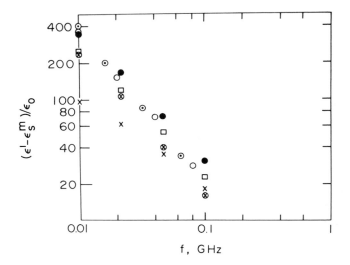

FIGURE 18. Delta dispersion of several tumor and normal tissues found by subtracting the parameter ϵ_s^m from the measured permittivity (see text). The symbols are identified in Table 4. (From Schepps, J. L. and Foster, K. R., *Phys. Med. Biol.*, 25, 1149, 1980. With permission.)

that summarize the relative permittivity and conductivity of a variety of soft, high water content tissues above 100 MHz:[90]

$$\epsilon'/\epsilon_0 = 1.71 \ f^{-1.13} + \frac{\epsilon_s^m - 4}{1 + (f/25)^2} + 4$$

$$\sigma = 1.35 \ f^{0.13} \ \sigma_{0.1} + [0.00222(\epsilon_s^m - 4)f^2]/[1 + (f/25)^2]$$

(f in GHz, σ in S/m). (76)

In these equations, the Debye terms represent the contribution of the bulk tissue water to the dielectric properties; the exponential terms represent the delta dispersion. The parameters ϵ_s^m and $\sigma_{0.1}$ would be estimated for a given tissue of known solid content from Figures 16 and 17. Since the origin of the delta dispersion is not well defined, it can be expected to vary somewhat among different tissues. Therefore, these equations provide reasonable guesses but should be used with caution in predicting tissue properties below 1 GHz.

e. Tumor vs. Normal Tissues

It is well known that tumor tissues have significantly higher water content than homologous normal tissues.[91,92] We quote from the text by Homburger and Fishman:[91]

Of the gross constituents..., the most constant difference between normal and tumor tissues is the increased water content of the latter. Most authors who have studied the chemistry of tumors have made the observation that the dry weight of tumor tissue is less than that of the corresponding normal tissue. The high water content of tumor tissue is shared by embryonic tissue, and therefore cannot be considered as uniquely characteristic of neoplasia. The significance of the increased water content is not known, but the suggestion has been made that it represents a tendency toward increased protein hydration.

For example, the water content of normal epidermis by weight is 60.9%, while that of carcinoma of the skin is 81.7%;[91] normal rat liver has a water content of 71.4% while hepatoma tissue has 81.9%.[92] Therefore, it is expected that tumor tissues will exhibit somewhat larger permittivity and conductivity values than homologous normal tissues at UHF and microwave frequencies. This is conceivably of practical significance in hyperthermia treatment of tumors, although the differences in dielectric properties between tumor and normal tissues at the frequencies of use in hyperthermia (27 MHz and above) are small. Joines[93] has pointed out that such differences might be considered in choosing an optimal frequency in the design of hyperthermia applicators, although other engineering considerations (the trade-offs possible between penetration depth and diameter of focus) will probably be dominant in the design of such applicators.

Little is known about the extent, if any, of differences between the dielectric properties of tumor vs. homologous normal tissues at lower radio and audio frequencies. Nearly 60 years ago, Fricke and Morse[94] reported that the permittivity of breast tumor tissues (at 20 kHz) was higher than that of nonmalignant breast tumor and other breast tissues, and suggested the feasibility of diagnosis based on this difference. To our knowledge, this has not been followed up. *Note added in proof:* Recent measurements show that the conductivity at audiofrequencies of an implanted liver tumor in the rabbit is 7 to 10 times higher than that of the surrounding normal tissue. This apparently arises from necrosis in the tumor nodule, a common phenomenon.[94a]

4. Bone

The dielectric properties of bone under near-normal physiological conditions have only recently been reported.[95,96] Fluid-saturated bone exhibits a broad, featureless dispersion that reaches a limiting (relative) permittivity of about 1000 at audio frequencies (Figure 19), and of about 10 to 20 at high RFs; the corresponding increase in conductivity is about 0.05 S/m between DC and 100 MHz, consistent with requirements of the Kramers-Kronig relations. These dielectric properties can be closely fitted by a Cole-Cole function with distribution parameter of 0.5 and center relaxation frequency of a few kilohertz which is, however, proportional to the conductivity of the surrounding electrolyte.[96] The mechanisms for this have not been well established but several possibilites can be discussed.

a. Maxwell-Wagner Polarization

Bone contains 15 to 20% fluid by volume, and a few percent cells by volume (the remaining tissue consisting of mineral and collagen). Consequently, Maxwell-Wagner polarization effects are possible from polarization at fluid-solid interfaces (Equations 45 to 48) or charging of cellular membranes within the tissue (Equations 54 to 57). Use of Equations 45 to 48 with reasonable values for the permittivity and conductivity of the suspended electrolyte, assuming a relative permittivity of about 20 and negligible conductivity for the matrix, leads to relaxation frequencies above 100 MHz and a total dielectric increment of order 1 relative to free space for the Maxwell-Wagner effect. It does not appear that this mechanism is the principal explanation of the relaxation that is observed.

A possibly more significant contribution is the polarization of membranes within the tissue. Equations 54 to 57, with a volume fraction of 2% of osteocytes and other cells, predict a total dispersion magnitude of some 100 relative to free space, with center frequency at roughly 0.1 MHz, depending on the dimensions of the cells. The corresponding increase in the conductivity will be of the order of 6 mS/m. Such a dispersion appears to be too small to account for the data in Figure 19, but in view of the uncertainties in this calculation it might nevertheless be a significant contribution to the bulk dielectric properties.

b. Ionic Processes

A larger contribution is expected from polarization of ions near charged surfaces in the tissue. There are a great number of interfaces in bone, e.g., between lamellae, between fibrils, and in the many small capillaries that permeate the tissue. A Cole-Cole distribution

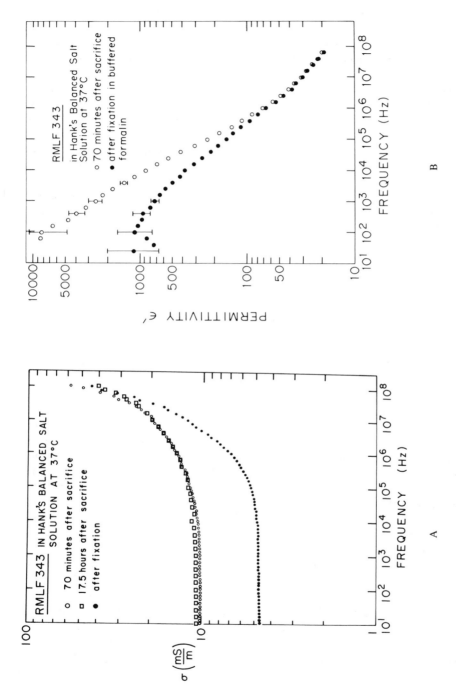

FIGURE 19.   Dielectric properties of fluid-saturated rat femur (radial direction) in freshly excised samples and after formalin fixation. (A) Conductivity; (B) relative permittivity. (From Kosterich, J. D., Foster, K. R., and Pollack, S. R., *IEEE Trans. Biomed. Eng.*, 30, 81, 1983. With permission.)

parameter of 0.5 is consistent with a diffusion-controlled polarization mechanism although a charge-hopping mechanism as considered by Hill and Jonscher[36] might also be possible. The problem remains how to quantitatively relate these mechanisms to the data.

The DC conductivity of the rat femur is proportional to that of the electrolyte surrounding the tissue and is due to simple ionic conduction through the channels that traverse the tissue.[96] There is a threefold variation in the DC conductivity with orientation in the several mammalian haversian bones that have been studied,[95] that arises from the orientation and dimensions of the major channels through the tissue. The corresponding variation in permittivity with orientation is much less well established.

B. In Vivo vs. In Vitro Properties

The dielectric properties reported above pertain to excised tissues that are far removed from a physiologically normal state. The question naturally arises whether these properties are likely to be representative of tissues *in situ*. In the past few years there have been developed methods of measuring the dielectric properties of tissues at radio and microwave frequencies *in situ* in living animals.[97-101] Measurements of the dielectric properties of various tissues at lower frequencies in living animals were reported much earlier.[102,103] A related question is whether such measurements can be used to measure physiologically important changes. The answer is yes, with important qualifications. We offer the following comments.

1. Changes in Dielectric Properties after Excision

The dielectric properties of muscle tissue have been measured by Schwan[81] as a function of time after excision. At low frequencies, pronounced changes were found within hours after excision; the permittivity gradually decreased, almost eliminating the alpha dispersion. In one sample the permittivity at audio frequencies decreased by nearly tenfold within 2 days. In contrast, changes in the dielectric properties at RFs were only apparent after several days. In our experience, the dielectric properties of tissues at RFs generally do not change observably for days after excision, provided the sample is kept under refrigeration and gross deterioration is prevented. These observations are consistent with the fact (Equations 54 to 57) that the beta dispersion is almost independent of the membrane conductance (a physiologically variable quantity) but rather depends on the membrane capacitance, which is much less variable, as well as on the resistivity of the electrolyte in the tissue. The earlier changes in the alpha dispersion would suggest that the sarcotubular system is more sensitive to disruption. Overall, the dielectric properties that have been reported at RF through microwave frequencies in living tissues *in situ* are in excellent agreement with those measured in excised tissues; any consistent difference is apparently swamped by the variability arising from the heterogeneity of the tissues.

2. Changes in Dielectric Properties with Physiological Changes In Situ

Impedance techniques at audio frequencies for measurement of blood flow or other physiological parameters are of course well established.[104] We consider the possible changes in dielectric properties at radio and microwave frequencies with physiological changes.

Several studies have investigated changes in dielectric properties of tissues at radio or microwave frequencies *in situ* immediately after the death of the animal, or with pronounced changes in blood perfusion of an organ. Stuchly et al.[105] observed no changes in the permittivity of cat brain tissue *in situ* between 0.1 and 1 GHz within the first minute after sacrifice of the animal by overdose of pentobarbital. Burdette[106] reported pronounced changes in the permittivity at 2.45 GHz of isolated kidneys with changes in blood flow into the renal artery, with less pronounced changes in the conductivity. Since kidney has an unusually high blood volume, amounting to 25% or so of the total volume of the organ,[107] some changes with blood perfusion are to be expected. However, the large magnitude of the

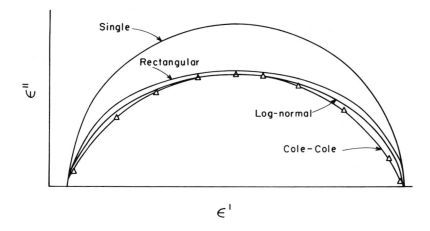

FIGURE 20. Complex permittivity plot obtained from the Cole-Cole distribution (Equation 77), log-rectangular distribution (Equation 78), and log-normal distribution (Equation 79). The triangles represent values that were calculated from the Cole-Cole distribution and used as trial data in determining the ranges of possible distribution functions.

changes (i.e., a decrease in permittivity from 63 to 45, with an increase in total organ perfusion from 0 to 60 mℓ/min) seems to be too large and in the wrong direction to be explained by a simple increase in the blood volume, and the mechanism responsible for the effects is not clear at present. Burdette[108] also reported measurable changes in the dielectric properties of kidney after perfusion with 6% DMSO (a cryoprotectant). Since DMSO has a lower permittivity than water (47 vs. 78) these changes are understandable and perhaps of practical use. Another application where in vivo dielectric properties at microwave frequencies can be related to physiological changes is the monitoring of pulmonary edema.[109]

There is, in summary, clear indication that dielectric measurements at radio and higher frequencies can yield information of physiological consequence. It would appear that the changes that would most readily be observed occur not at the cellular level but rather at the tissue level (e.g., from relatively large changes in blood volume, or pulmonary edema) or due to experimentally induced changes in electrolyte (e.g., replacement of some tissue water by DMSO). We expect the well-known problem with clinical impedance measurements at lower frequencies — that the impedance variations in a subject due to physiological changes are smaller than the variability among normal subjects — will also be encountered at higher frequencies as well.

V. OTHER TOPICS

A. Distribution of Relaxation Times

Dielectric data from biological systems commonly exhibit relaxation behavior that is not of the simple Debye type, which could arise from the presence of several relaxation processes, or the presence of relaxation processes whose kinetics are not of first order. A fundamental question is what is the distribution of relaxation time-constants that gives rise to the observed properties. A common approach is to decompose the relaxation into a sum of single, time-constant components (e.g., Equation 16) which might be identified with different physical processes. This procedure is reminiscent of the compartmental analysis used in many other fields. We consider in this section the inherent limitations in the interpretation of dielectric data to determine such distributions.

In the earlier review[1] it was noted that typical experimental data can be represented by a wide variety of distributions of macroscopic relaxation times. For example, Figure 20 com-

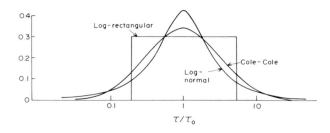

FIGURE 21. Distribution functions represented by Equations 77 to 79.

pares the predicted dielectric properties calculated from Equation 18 using the Cole-Cole and other distribution functions. These functions are (Figure 21)

Cole-Cole distribution

$$p[\ln(\tau/\tau_0)] = \frac{(\sin \alpha\pi)/2\pi}{\cosh[(1 - \alpha) \ln \tau/\tau_0] - \cos(\alpha\pi)} \quad \alpha = 0.23 \tag{77}$$

Log-rectangular distribution

$$\begin{aligned} p[\ln(\tau/\tau_0)] &= 0.3 & |\ln \tau/\tau_0| &\le 1.67 \\ &= 0 & |\ln \tau/\tau_0| &\ge 1.67 \end{aligned} \tag{78}$$

Log-normal distribution

$$p[\ln(\tau/\tau_0)] = \frac{k}{\sqrt{\pi}} \exp\left[-k^2\left(\ln \frac{\tau}{\tau_0}\right)^2 \right] \quad k = 0.60 \tag{79}$$

(The points shown in Figure 20 were calculated from the Cole-Cole function.)

Figure 21 shows that dielectric data that suggest a depressed semicircular arc in the complex permittivity or conductivity plane (i.e., the Cole-Cole distribution) might be rather well fitted using a rectangular or Gaussian distribution, possibly supporting quite different physical interpretations of the data. The question arises: what is the total range of distribution functions that are consistent with a given set of data, within specified maximum errors? The range gives an indication of the fundamental uncertainties in the interpretation of such data.

Colonomos and Gordon[110] recently showed how linear programming techniques can be used to obtain precise bounds on the cumulative distribution functions of relaxation times that can give rise to any physically meaningful set of dielectric relaxation data. The following constraints must be satisfied

$$L'(\omega_k) = \int_0^\infty \frac{p(\tau)d\tau}{1 + \omega_k^2 \tau^2} \tag{80}$$

$$L''(\omega_k) = \int_0^\infty \frac{p(\tau)\omega\tau d\tau}{1 + \omega_k^2\tau^2} \tag{81}$$

where $L'(\omega_k)$ and $L''(\omega_k)$ are related to the real and imaginary parts of the function $(\epsilon^* - \epsilon_\infty)/(\epsilon_s - \epsilon_\infty)$ at frequency ω_k and $p(\tau)$ is the strength of the distribution at relaxation time τ. The cumulative distribution function $\Sigma(\tau)$ is defined as

DISTRIBUTION OF RELAXATION TIMES

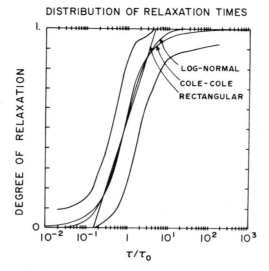

FIGURE 22. Cumulative distribution functions defined by Equation 82 for the three distribution functions (Equations 77 to 79), together with the upper and lower bounds calculated, using the linear programming technique. In these calculations, maximum relative errors of 5% were assumed for the complex permittivity values.

$$\Sigma(\tau) = \int_0^\tau p(\tau')d\tau' \tag{82}$$

These bounds are shown in Figure 22 for the cumulative distributions that would fit the data from Figure 20, within maximum errors of 5%. Obviously, for larger permissible errors, the corresponding bounds would be even less restrictive. The result is that a very large range of distribution functions can fit dielectric data of typical accuracy. It is not surprising, therefore, that the three functions in Figure 20 could satisfactorily fit the data. This point — made earlier[1] — is much more forcefully shown by the present analysis.

This calculation provides only the range of cumulative distribution functions that are consistent with the data within specific experimental errors; it does not determine the most probable function. Several theoretically determined distribution functions could be compared using chi-square or other statistical techniques, and the "best" could be chosen even though all fall within the limits provided by the linear programming results. On the other hand, it is difficult to rule out the possibility of systematic errors in real experimental data that are smaller than the scatter in the data, typically 1 to 5% relative uncertainty. Thus, the analysis of Colonomos and Gordon provides a severe warning about the dangers of curve-fitting dielectric data, unless a theoretical expression can be used that has some independent justification. We will return to this point in the next section.

B. Electrical Properties of Water in Tissues

The dielectric properties of tissues in several ways reflect the properties of the water that constitutes the major fraction of most tissues. The ionic conductivity of the cytoplasm and extracellular fluids appears to be determined principally by the conductivity of the aqueous electrolyte, together with the volume fraction of nonconducting protein that is contained within it. The dielectric dispersion at microwave frequencies (the gamma dispersion) arises from the dipolar relaxation of the bulk tissue water, while the UHF dispersion (the delta dispersion) reflects in part the dipolar relaxation of the water of hydration of the protein. Recent work has provided a clearer understanding of the relationships between these phenomena and the underlying physical properties of the tissue water, and consequently a better understanding of the mechanisms of absorption of nonionizing energy by the tissues. We consider these relations in the present section.

1. Ionic Conductivity

The ionic conductivity of protein solutions in electrolyte has been of interest for many years. We summarize studies of the ionic conductivity of protein solutions, cell suspensions, and tissues to show the underlying similarity of the behavior.

Bull and Breese[111] measured the electric conductivity of a variety of proteins in several electrolyte solutions at 1 kHz. The data were interpreted using the Maxwell-Fricke mixture theory assuming that the total nonconductive volume fraction consists of the protein plus its associated water of hydration, to yield "bound" water fractions of 0.6 water per gram protein. This value is somewhat above the conventionally accepted range of about 0.3 to 0.4 water per gram protein obtained from thermodynamic studies of water binding in proteins.[112]

In their study of the electric properties of blood at high frequencies, Pauly and Schwan[113] estimated the conductivity of the interior of the human erythrocyte to be 0.518 S/m at 25°C, compared with a value 1.45 S/m calculated from the known ionic composition of the cell. This difference in part arises from the excluded volume due to the protein content and its associated water of hydration. However, the decrease in conductivity was somewhat larger than would be expected assuming reasonable values of the protein hydration which was attributed in part to a "frictional effect" arising from hydrodynamic forces on the ions due to the presence of the protein surfaces. In both of these studies, however, the calculated fractions of "bound" water were not drastically above conventionally accepted values.

The ionic conductivity of the cytoplasm of tissues can be estimated from the conductivity near 100 MHz, which is above the beta dispersion frequency of the membrane-bound structures but below the frequency at which significant contributions to the conductivity from dipolar relaxational effects are expected.[90] The conductivity of the several tissues in Figure 17 at 100 MHz extrapolates in the limit of zero protein concentration to a value of roughly 1.5 S/m, which is about two thirds that of 0.15 M NaCl at body temperature. Since the electrolyte in mammalian tissues typically consists of 0.15 M small inorganic cations, balanced by Cl^- and large organic anions, the extrapolated conductivity is about as expected. The conductivity of the various tissues at 100 MHz can be considered to be a function of their solid content, and analyzed in a manner similar to the previously cited data from protein solutions.

Finally, the ionic conductivity of a simple polymer poly(ethylene) oxide in dilute electrolyte solution was reported recently.[114] These experiments, and their interpretation using the Maxwell mixture theory, were similar to the previous studies on the biological materials.

All of these results are summarized in Figure 23. Remarkably, the conductivity of the tissues, protein solutions, and polymer solutions show comparable decreases below the predictions of the mixture theory, due to restriction in the motional freedom of the water by hydration effects. The self-diffusion coefficients of water in these various preparations show similar trends.[114] The similar results in all of these materials suggest that the degree of motional restriction is not sensitive to the chemical nature of the suspended molecule.

2. UHF Dispersion (Delta Dispersion)

Figure 10 shows the conductivity of barnacle muscle at two temperatures, i.e., plotted vs. the quantity $f^2/[1 + (f/f_c)^2]$, showing the delta dispersion at low gigahertz frequencies. The conductivity of a solution of poly(ethylene) oxide shows similar behavior (Figure 24).[114]

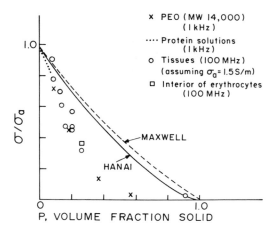

FIGURE 23.   Conductivity of a variety of biological and nonbiological suspensions divided by the conductivity of the suspending electrolyte, showing similar hydration effects. The dashed line is the locus of the conductivity of several protein solutions;[111] other data are poly(ethylene) oxide (PEO) of 14,000 mol wt in 0.1 *N* KCl[114] erythrocyte cytoplasm,[113] and tumor and normal tissues at 100 MHz.[90] The predictions of the Maxwell and Hanai relations (Equations 44 and 58) are also shown. A limiting conductivity of 1.5 S/m was assumed in representing the tissue, which was obtained from Figure 17 in the limit of zero solid content.

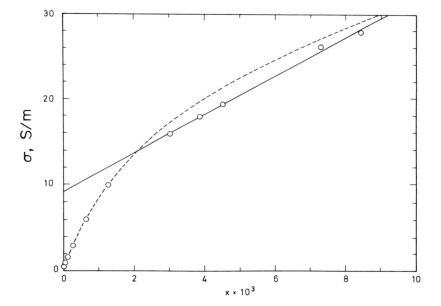

FIGURE 24.   Plot of the conductivity vs. the quantity

$$x = \frac{(2 \pi f^2/f_c)\epsilon_0}{1 + (f/f_c)^2}$$

for a 25% (volume) solution of poly(ethylene) oxide, showing the delta dispersion at low gigahertz frequencies that adds a total conductivity of ca. 9 S/m to the solution conductivity at microwave frequencies. This is to be compared with the similar plot (Figure 10) for muscle tissue, and shows the similar delta dispersion in these two materials. (From Foster, K. R., Cheever, E., Leonard, J. B., and Blum, F., *Biophys. J.*, 45, 975, 1984. With permission.)

Moreover, a similar dispersion is observed in protein solutions[115,116] and ionic and nonionic microemulsions[117,118] that lack the polarization mechanisms associated with polar sidechains or ionic processes that might occur in the tissues. This dispersion in the nonionic microe-mulsions and polymer solutions no doubt arises from the dipolar reorientation of the water of hydration, since no other suitable mechanisms are present. We conclude that this is at least one of possibly several mechanisms for the delta dispersion observed in tissues.

3. Microwave Dispersion (Gamma Dispersion)

In all of the cases described above, the dielectric relaxation at microwave frequencies approaches single, time-constant behavior with the same relaxation time as that of pure liquid water. The apparent amount of the bulk-like water equals the total water content minus the expected fraction of motionally restricted water.

These various results cited lead to a consistent picture of the dielectric properties of the water in these materials. Some fraction of water, that is equal or only somewhat larger in magnitude to the "bound" water fraction obtained by thermodynamic measurement, exhibits a rotational and translational correlation time that is increased by roughly tenfold compared to that of the bulk liquid water. The remaining water is bulk-like in its dielectric and self-diffusion properties. The precise relation of this motionally restricted water to thermodyn-amically "bound" water, and the mechanisms which produce the restriction, are largely unknown.

It has been suggested that the average dielectric relaxation time of the water in muscle tissue,[119] myoglobin solution,[120] and ocular and brain tissues[121,122] is longer than that of the pure liquid water by a small factor of 1.5 or less. The data from which these conclusions have been derived are similar to those presented here. However, the methods of analyzing the data in these other studies were quite different. Grant and colleagues[120,121] fitted the permittivity and conductivity data to various relaxation equations by computer. We, on the other hand, have determined the limiting behavior of the dielectric permittivity or conductivity at high frequencies, to extract the contribution of the bulk-like water from the total observed relaxation. Masszi's analysis was based on fitting only the conductivity data, assuming single, time-constant behavior (i.e., the Debye equations) over a limited frequency range (2 to 4 GHz) to extract a mean relaxation time.

These different studies proceeded from different assumptions, and consequently reached different conclusions. The problem, at heart, is the impossibility of determining the distri-bution function $p(\tau)$ directly from the data. One can only assume a function, and fit the data to it. But there is agreement among all groups on the following major points:

1. The dielectric relaxation of tissue and protein solutions at UHF and microwave fre-quencies exhibits a distribution of relaxation times.
2. The mean relaxation time in this distribution is shifted above that of bulk water at the same temperature.

This is expected if some fraction in the water had a longer relaxation time than bulk water.

The additional conclusion that we would add is that the longer average dielectric relaxation time is associated with a reduction in other transport properties. Dielectric measurements at microwave frequencies reflect the motional properties of water on a short (picosecond) timescale rather than time-averaged properties, as with water self-diffusion or ionic con-ductivity. They thus contain some information about the distribution of mobilities of the water, although the uncertainties in extracting this are very great. The comparision of several transport properties (ionic conductivity, NMR relaxation, self-diffusion, dielectric relaxation) in the same systems can be a powerful tool in the study of the motional properties of water in these complex systems, that can help overcome the limitations inherent in one set of measurements alone.

C. Coupling Considerations

The dielectric properties of tissues and their underlying mechanisms provide insight into the interactions of EM fields with biological systems. Much is known about the interactions at the molecular and membrane levels.[8,9] It is necessary, however, to translate fields that are typically measured outside of the exposed subject to the strength of the *in situ* fields that actually act on the cell membranes or other components. We will show that the simple theory described previously can yield coupling factors that relate external to *in situ* field strengths, and that the coupling factors are strongly frequency dependent. This discussion is based on a recent paper by Schwan.[123]

Consider a spherical body of dielectric properties (σ_1, ϵ_i) in a homogeneous medium of properties (σ_a, ϵ_a), with an external field of initial strength E_0. The field strength inside the sphere, E_i, can be found from a solution of the Laplace equation[2]

$$E_i = \frac{3E_0\,\sigma_a^*}{\sigma_i^* + 2\sigma_a^*} \tag{83}$$

where the complex conductivities (σ_i^*, σ_a^*) pertain to the dielectric properties inside and outside of the sphere, and are defined in Equation 6b. This calculation applies in the quasistatic limit, where the wavelength of the radiation is much greater than the diameter of the sphere. If the outer medium is air while the sphere has dielectric properties typical of soft tissue,

$$\sigma_a = j\omega\epsilon_0$$

$$\text{and} \quad E_i = \frac{j3E_0\epsilon_0\omega}{2\sigma_a} \tag{84a}$$

$$j_i = \frac{j3E_0\epsilon_0\omega}{2} \tag{84b}$$

where j_i is the internal current density. The factor $j = \sqrt{-1}$ indicates, as usual, that there is a 90° phase shift. For typical tissues at 10 Hz, $\sigma_i \cong 1$ S/m, and the internal field strength is seven orders-of-magnitude below that of the external field. We conclude that for air-coupled objects:

1.   At ELF frequencies, the internal field strength is far smaller than the external field strength.
2.   At microwave frequencies the two are comparable.
3.   The internal field strength is proportional to the external field strength and (since the conductivity of most tissues varies only threefold or so between ELF frequencies and 1 GHz) it scales approximately as the frequency.

These results can be extended to the case of homogeneous ellipsoids. Here too the internal field is constant throughout the interior, but its magnitude depends on the axial ratio of the ellipsoid and the orientation with respect to the external field.

We consider the application of these principles to several cases of importance in bioeffects studies.

1. Coupling into Macroscopic Objects

Equation 84 provides a simple estimate of the electric fields and current densities induced in an isolated animal or human situated in an external electric field, that applies at sufficiently low frequencies, e.g., below the resonance frequency of about 70 MHz for an erect human. It demonstrates that the internal fields at ELF frequencies in the human are far weaker than the external field.

Irregularities in the shape of the body will affect the distribution of currents. For example, in a man standing in a field perpendicular to ground, the average current density in the legs is greater than in the trunk, by a factor that is approximately equal to the ratios of trunk to leg cross-sectional areas.[124] Variations in the conductivity of the various tissues will also affect the current distribution; in muscle, at frequencies below 1 MHz, current will tend to flow in a direction parallel to the fiber axis. Current will tend to flow around relatively nonconductive tissues such as lung, fat, and bone, and through relatively conductive tissues such as blood. But these complications do not contradict the conclusions obtained from Equation 84 that the average current densities flowing in the body are at ELF frequencies very small.

2. Coupling into Cells and Organelles

Consider a sphere of radius R covered by a membrane of negligible thickness and capacitance C_m F/m². The membrane potential V_m induced by external field strength E is given by[123]

$$V_m(\theta) = \frac{1.5 \text{ ER } \cos\theta}{1 + j\omega\tau} \tag{85}$$

where τ is the time-constant for the beta dispersion (Equation 56), and θ is the angle between the radius to a point on the membrane and the direction of the external field. The maximum membrane potential is of the order of millivolts for cells of typical (10 μm) diameter in low frequency fields of 100 V/m intensity. At frequencies above the beta relaxation frequency $1/2\pi\tau$, the membrane potential is approximately $V_m(\theta)$

$$V_m(\theta) \cong \frac{1.5 \text{ E } \cos\theta}{\omega C_m[1/\sigma_i + 1/(2\sigma_a)]} \tag{86}$$

using the approximate result (Equation 56). In the high frequency limit, the current density through the membrane will correspond to that through the external medium, since the impedance of the cell membrane becomes vanishingly small. The membrane potential developed by a current density J will then be $J/\omega C_m$. At 1 GHz this would correspond to a maximum potential of a tenth of a microvolt, for a current density of 10 A/m² through the extracellular medium. (If the conductivity of the cytoplasm were different from that of the extracellular medium, a small factor of the order of 1 must be included in this calculation.)

These considerations can be easily extended to the case where the external field is a nonsinusoidal function of time, by writing Equation 85 with the Laplace variable s replacing $j\omega$, and E(s) the Laplace transform of the external field. The membrane voltage would then be found as a function of time by evaluating the inverse Laplace transform.

If only the electric field in the air surrounding the preparation is specified, then an additional coupling factor such as given by Equation 84 is necessary.

Similar considerations enter into the calculation of the field strength within the cytoplasm. The total voltage drop across the cell from Equation 85 is $3RE_0$. This is divided between the membrane and the cytoplasm; however there is a 90° phase shift because of the capacitive reactance of the membrane. Thus the cytoplasmic voltage and induced membrane potential must be added vectorially. The field strength E_i inside the cytoplasm is [123]

$$E_i = \frac{(1.5)j\omega\tau E}{1 + (j\omega\tau)} \qquad (87)$$

This suggests that the field strength in the cytoplasm in the high frequency limit will be 1.5 E. A more accurate calculation would take into account the possible difference in conductivity of the cytoplasm compared to that of the external electrolyte. Since typically, $\sigma_i \simeq \sigma_a/2$, Equation 83 would suggest a field strength of about 1.2 E in the cytoplasm in the high frequency limit.

This process can be readily extended to the case of a cell containing membrane-bound organelles. A full calculation would require the solution of the Laplace equation for the structure, as was done for a single, membrane-covered sphere.[45] This procedure, if carried out in detail, will lead to lengthy algebraic expressions. One such calculation was recently reported by Drago et al.[133] A simpler treatment, presented here, ignores factors of order-of-magnitude unity but shows more clearly the underlying principles involved. The assumption is that the organelle is sufficiently smaller than the cell that it negligibly perturbs the field within the cell as some distance from the organelle. Thus, Equations 85 and 87 can be applied several times in turn to estimate the field strengths in different compartments of the cell.

Consider, for example, a cell of radius R containing a nucleus of radius R_n with corresponding beta relaxation times τ_1 and τ_2. The frequency dependence of the field strengths in the various parts of this cell can be predicted by several applications of Equations 85 and 87

a. Induced outer membrane potential:

$$V_m = \frac{1.5\ ER\ \cos\theta}{1 + j\omega\tau_1} \qquad (88)$$

b. Cytoplasmic field strength (far from the nucleus):

$$E_i = \frac{(1.5)j\omega\tau_1 E}{1 + j\omega\tau_1} \qquad (89)$$

c. Induced nuclear membrane potential:

$$V_n = \frac{(9/4)(j\omega\tau_1)\ ER_n\ \cos\theta}{(1 + j\omega\tau_1)(1 + j\omega\tau_2)} \qquad (90)$$

d. Nuclear field strength:

$$E_n = \frac{(9/4)E(j\omega\tau_1)(j\omega\tau_2)}{(1 + j\omega\tau_1)(1 + j\omega\tau_2)} \qquad (91)$$

Since τ_2 / τ_1 is approximately equal to the ratio of the nuclear to cellular radii, the two cut-off frequencies involved might be different by a factor of 2 to 10 or more. Equations 88 and 90 suggest that the outer cell membrane and nuclear membranes will experience different induced potentials in the high frequency limit. This, however, is probably an artifact of an oversimplified calculation. If the conductivities of the cytoplasm and nuclear contents are

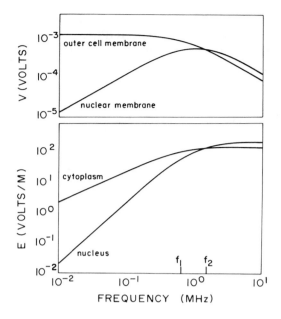

FIGURE 25. Internal field strengths and membrane potentials in a cell with a nucleus, calculated from Equations 88 to 91. The external field strength was assumed to be 100 V/m and the diameters of the cell and nucleus are, respectively, 10^{-5} and 5×10^{-6} m, with corresponding beta relaxation frequencies (f_1 and f_2) of 0.7 and 1.6 MHz as indicated on the figure. It is expected that a more precise calculation would yield nuclear field strengths and membrane potentials that approach those in the cytoplasm and outer cell membrane at high frequencies.

equal, the same current density should flow throughout the membranes, and their potentials should be equal, given equal membrane impedance. A more exact analysis would lead to comparable field strengths and membrane potentials (i.e., $V_n \approx V_m$ and $E_n \approx E_i$) in the limit of high frequencies, if the conductivity of the cytoplasm and nuclear contents were comparable.

The physical meaning of these results can be easily summarized. At very low frequencies the outer membrane shields the interior, thus preventing the nuclear membrane from being polarized. Above the beta relaxation frequency, the reactances of both the outer and nuclear membranes become very small and the voltage drop consequently falls off as the inverse of the frequency. The greatest potential induced across the nuclear membrane occurs at frequencies between the beta dispersion frequencies of the cell and nucleus, and is, very roughly, equal to the product of the external field strength times the radius of the organelle. To illustrate typical frequency ranges in liver (Table 3) the beta dispersion frequency of the outer cell membranes is calculated to be roughly 0.7 MHz and for the nuclei two times higher. Resulting field strengths and membrane potentials calculated from Equations 88 to 91 are given in Figure 25.

More complicated effects are expected for organelles which connect to the outer membrane, such as the sarcoplasmic reticulum in the muscle fiber. The organelle could be charged from the electrolyte surrounding the cell, in which case the time-constant would be that of the alpha dispersion, since the series component of the circuit in Figure 4 would be identified with an access impedance to the sarcotubular system; or, the organelle could be charged

Table 5

SUMMARY OF THE COUPLING PROPERTIES OF EXTERNAL FIELDS TO CELLULAR MEMBRANES AND COMPARTMENTS

	$f < f_c^a$ ($\sim$ 1 MHz)	$f_c < f < f_n^b$	$f > f_n$ ($\sim$ 10 MHz)
Cell			
Membranes	Polarized	Not polarized	Not polarized
Interior	Doubly shielded	Shielded	Exposed
(proteins, etc.)			
Organelles			
Membranes	Not polarized	Partially polarized	Not polarized
Interior	Doubly shielded	Shielded	Exposed
(nucleic acids)			
Connecting organelles			
Membranes	Polarized	Not polarized	Not polarized
Interior	Not exposed	Exposed	Exposed

[a]   The beta dispersion frequency of the outer cell membrane.
[b]   The beta dispersion frequency of the organelle.

Adapted from Schwan, H. P., in *Biological Effects and Dosimetry of Nonionizing Radiation*, Grandolfo, M., Michaelson, S. M., and Rindi, A., Eds., Plenum Press, New York, 1983, 213.

through the cytoplasm, as would occur in an organelle not connecting with the external medium. It is expected that the maximum induced potential across the organelle would be larger for the low frequency process, and be comparable to the potential developed across the outer cell membrane. In the higher frequency process, the maximum potential developed would be roughly equal to the field strength times a mean diameter of the organelle, a far smaller quantity.

Table 5 summarizes the results of this analysis, applied to a hypothetical cell containing one intracellular compartment. The two major conclusions are

1.   For a constant field strength in the electrolyte outside of the cell, the maximum induced membrane potentials are proportional to the diameter of the cell or subcellular structure. These induced membrane potentials fall off as the inverse of the frequency above the respective beta dispersion frequency. For outer cell membranes this decline begins at RFs; for membranes of organelles this decline begins at ELFs if the organelle connects with the outer cell membrane or at RFs if not.
2.   The interiors of cells and subcellular components are well shielded from the external field at low frequencies, and are increasingly exposed at RFs.

The above considerations suggest that the local fields inside tissues resulting from the application of external fields vary substantially with frequency. This variation is also found, to a much lesser degree, in homogeneous solutions in which the local fields that act on individual molecules are functions of the bulk dielectric properties of the medium and consequently also of frequency. The cell membranes lead to coupling factors that exhibit bandpass frequency response in which the potential developed across internal membranes within a cell exhibits a peak at a frequency that is intermediate between the beta dispersion of the outer cell membrane and that of the organelle. These peaks will be quite broad, even in homogeneous cell suspensions, and will be broader still in tissues that exhibit great structural heterogeneity, including a great diversity of membrane-bound structures.

We expect for this reason that a frequency dependence will be introduced in any biological effect arising at the level of intracellular membranes or cytoplasm, in addition to whatever time-constants characterize the response itself. Investigations of the frequency dependence of a biological effect could help clarify at what cellular level the effect is produced. Unfortunately, so far little work has been done on analyzing reported biological effects of EM fields over the extended ELF-RF range with appropriate considerations for the prevailing potentials and field strengths at the cellular and subcellular levels.

VI. DIELECTRIC PROPERTIES OF TISSUES — TABULATED

Table 6 presents selected permittivity and conductivity data from various tissues. Where available, up to three values for the tissues have been presented, including measurements performed on excised tissues of various species and in vivo. The aim is to present primarily new data; in some cases earlier data have been included to fill out the table.

Other tabulations should be noted, in particular by Schwan,[1] which extensively summarizes older data, Geddes and Baker,[76] with resistivity values for many tissues at low frequencies, and the very extensive table by Stuchly and Stuchly[8] which includes data from "phantom" tissue materials. Recent extensive data from various excised tissues at RFs are by Stoy[86] and various tissues in vitro and in vivo by Stuchly and colleagues.[100,101,105,128]

All data are reported for tissues at body temperature (37 to 38°C). Typical temperature coefficients are summarized in Figures 11 and 12, but are probably negligible for most practical applications.

ACKNOWLEDGMENTS

This work was supported in part by grant CA26046 from the National Cancer Institute, and Contract number N00014-78-C-0392 from the Office of Naval Research. We gratefully acknowledge the suggestions and helpful advice from many colleagues, in particular Drs. Charles Polk and Edwin L. Carstensen, and assistance with the numerical computations by Mr. Erik Cheever.

Table 6

Conductivity (S/m)

Frequency	A Skeletal muscle parallel (nonoriented)	B Skeletal muscle perpendicular (nonoriented)	C Liver	D Lung	E Spleen	F Kidney	G Brain white matter	H Brain grey matter	I Bone	J Whole blood	K Fat
1	0.52	0.076	0.12	0.089							
2 (10 Hz)											
3											
4	0.52	0.076	0.13	0.092					0.0126	0.60	
5 (100 Hz)											
6											
7	0.52	0.08	0.13	0.096					0.0129	0.68	0.02—0.07
8 (1 kHz)											
9											
10	0.55	0.085	0.15	0.11					0.0133	0.68	
11 (10 kHz)											
12											
13	0.65	0.40	0.15		0.62	0.24—0.25	0.12—0.15	0.17	0.0144	0.55	
14 (100 kHz)	0.56—0.59		0.16							0.68	
15	0.38—0.44										
16	0.83—0.85		0.27		0.63	0.37—0.39	0.14—0.19	0.21	0.0173	0.71	
17 (1 MHz)	0.58—0.63		0.30								
18											
19	0.86—0.87		0.47		0.84	0.64—0.68	0.21—0.28	0.35	0.0237	1.11	
20 (10 MHz)	0.92—0.96		0.46		0.55—0.53	0.50—0.57	0.30	0.38			
21	0.69—0.75		0.42—0.46				0.29—0.31	0.45—0.63			
22	0.95—0.99		0.72	0.53	1.05	0.94—1.05	0.36—0.48	0.69	0.0574	1.0	0.02—0.07
23 (100 MHz)	0.9 ± 0.08		0.70		0.73—0.76	0.66—0.72	0.45	0.7			
24	0.75—0.82		0.60—0.71		0.80 ± 0.02	0.75 ± 0.02	0.48—0.51	0.52—0.85		0.7—0.8	
25	1.38—1.45		0.98	0.73	1.2		0.89—0.94		0.05	1.4—1.6	
26 (1 GHz)	1.3		1.2		1.09—1.13	0.95—0.97	0.80	1.1		1.3	0.03—0.09
27	1.5		0.95—1.0		2.0	1.0	0.81—0.82	0.89—1.17			
28			2.0		2.5 ± 0.03		1.8—2.1				
29 (3 GHz)	2.7 ± 0.07		2.4				1.5	2.0	0.16	2.5—3.1	
30	2.8		2.8		2.7	2.3 ± 0.05					

Relative Permittivity

Frequency												
10 Hz	10^7				5×10^7	2.5×10^7	10^6				$3,800$	1.5×10^5
100 Hz	1.1×10^6				8.5×10^5	4.5×10^5	3.2×10^5				$1,000$	5×10^4
1 kHz	2.2×10^5				1.3×10^5	8.5×10^4	1.2×10^5				640	
10 kHz	8×10^4				5.5×10^4	2.5×10^4	7×10^4				280	2×10^4
100 kHz	1.5×10^4	$24,800$—$27,300$	$14,400$—$15,800$	$3,260$	$9,760$		3×10^4	$10,900$—$12,500$	$3,800$		87	
1 MHz	$2,460$—$2,530$	$1,900$—$2,150$		$1,450$	$1,970$ $1,970$		$2,390$—$2,690$	543—827	$1,250$	$2,040$	37	
10 MHz	170—190	187—204	162—181	321 352—410	338 300 251—265	35	431—499 190—204	163—209 200 190—191	352 380 237—289	200	23	
100 MHz	67—72	68 ± 2	64—70 57—59	83 71—76 81 ± 3	77 79 65—68	35	89—95 56—62 85 ± 1	57—66 65 58—64	90 90 65—80	8		67 72—74
1 GHz	58	48		54 50—51 50 $52 \, 0.6$	55 47—49 42		43 46	40—44 35 38—39 35—41	45 47—51		7.5	58—62 63—67 63
3 GHz	52.5 ± 0.7	46	40—42	46 42	42—43 34—38 37 53		47.5 ± 1	33	44		8	55—56 63
10 GHz	46	40—42	37 35	38	34—38 37		30—37	25	40		8	50—52 45

Additional column / reference values (as printed at head of columns): 8.3, 7.7, 8.8; 6.5, 10.0; 5.8—6.7, 10.0; 4.5—7.4; 10; 0.5—1.7, 9.1, 10.5; 0.3—0.4; 4.5—7.5; 4.3—7.5; 3—6; 4—7; 3.5—4.0; $2,900$; $2,810$; $4,000$, $2,740$.

Row grouping key:
1, 2, 3 } 10 Hz
4, 5, 6 } 100 Hz
7, 8, 9 } 1 kHz
10, 11, 12 } 10 kHz
13, 14, 15 } 100 kHz
16, 17, 18 } 1 MHz
19, 20, 21 } 10 MHz
22, 23, 24 } 100 MHz
25, 26, 27 } 1 GHz
28, 29, 30 } 3 GHz
31, 32, 33 } 10 GHz

Table 6 (continued)

References

Coordinates	Ref.	Tissues
1 A,B 4 A,B 7 A,B 10 A,B 13 A,B	80	Dog skeletal muscle, 37°C (av of 5 measurements, SD ~ 30%)
15 A,B 18 A,B 21 A,B 24 A,B	86	Nonoriented dog skeletal muscle, 37°C (range of 3 measurements)
14 A,B 17 A,B 20 A,B	86	Nonoriented rat skeletal muscle, 37°C (range of 2 measurements)
13 C 16 C 19 C 22 C	86	Dog liver, 37°C (single specimen)
14 C 17 C 20 C 23 C	86	Rabbit liver, 37°C (single specimen)
13 E 16 E 19 E 22 E	86	Dog spleen, 37°C (single specimen)
13 F 16 F 19 F 22 F	86	Dog kidney, 37°C (range of 2 measurements)
13 G,H 16 G,H	86	Dog brain, white and gray matter, 37°C (range of 2 measurements)

Coordinates	Ref.	Tissues
19 G,H 22 G,H		
27 E 30 E 33 E	90, 115	Dog spleen, 37°C (values at 1,3 GHz interpolated, single specimen)
27 A,B 30 A,B 33 A,B	90, 115	Dog skeletal muscle, 37°C (values at 1,3 GHz interpolated, single specimen)
1 C 4 C 7 C 10 C	103	Dog liver, *in situ* (av of 20 measurements, SD ~ 25%)
1 D 4 D 7 D 10 D	103	Dog lung, inflated, *in situ* (av of 20 measurements, SD ~ 25%)
7 K 10 K	103	Dog fat, *in situ*
19 A,B 22 A,B 25 A,B	105	Cat skeletal muscle, in vivo, 31°C (range of 3 measurements)
20 E 23 E 26 E	105	Cat spleen, in vivo, 35°C (range of 3 measurements)
20 F 23 F 26 F	105	Cat kidney, in vivo, 35°C (range of 3 measurements)
21 G,H 24 G,H 27 G,H	105	Cat brain, in vivo, 33°C (range of 3 measurements)

Code	Description	Ref.
21 C	Cat liver, in vivo, 35°C (range of 3 measurements)	105
24 C		
27 C		
28 C	Bovine liver, 37°C	100
23 D,J	Beef blood	126
22 K	Excised human tissues (deflated lung)	
26 K	27°C (measurement frequencies 0.2—0.9 GHz)	
26 D		
27 J		
23 A,B	Various cat tissues, in vivo (av ± SD, 55 measurements in 4 animals); value at 10 GHz extrapolated from 8.0 GHz; 1 GHz interpolated	128
26 A,B		
29 A,B		
32 A,B		
24 E		
25 E		
28 E		
31 E		
24 F		
27 F		
30 F		
33 F		
26 C	Dog liver, 37°C (value at 1.3 GHz extrapolated)	90, 115
29 C		
32 C		
4 I	Rat femur, 37°C, immersed in Hank's buffered saline, radial direction (single sample)	95
7 I		
10 I		
13 I		

Code	Description	Ref.
16 I	Mixed brain tissue, mouse, 37°C (value at 3 GHz interpolated)	122
19 I		
22 I		
25 G	Rabbit blood, room temperature	130
28 G		
8 J		
11 J		
14 J		
17 J		
20 J	Normal human blood, hematocrit 40%, 21°C (50 kHz)	125
13 J		
25 C	Various tissues, dog, horse, 38°C (except 25 I, 28 I, 25°C); the measurements were made at 8.6 GHz and extrapolated to 10 GHz	132
30 C		
25 K		
28 K		
25 J		
28 J		
25 I		
28 I		
31 A,B,I,K		
32 J	Human (9.4 GHz), 37°C	131
20 G,H	Dog brain, white and gray matter, 37°C	127
23 G,H		
26 G,H		
29 G,H		
32 G,H		
4 J	Sheep blood, 18°C	122
7 J		

REFERENCES

1. **Schwan, H. P.,** Electrical properties of tissue and cell suspensions, in *Advances in Biological and Medical Physics,* Vol. 5, Academic Press, New York, 1957, 147.
2. **Cole, K. S.,** *Membranes, Ions, and Impulses,* University of California Press, Berkeley, 1972.
3. **Pethig, R.,** *Dielectric and Electronic Properties of Biological Materials,* John Wiley & Sons, New York, 1979.
4. **Grant, E. H., Sheppard, R. J., and South, G. P.,** *Dielectric Behavior of Biological Molecules in Solution,* Oxford University Press, Oxford, 1978.
5. **Schanne, O. F. and P.-Ceretti, E. R.,** *Impedance Measurements in Biological Cells,* John Wiley & Sons, New York, 1978.
6. **Rajewsky, B.,** *Ultrakurzwellen, Ergebnisse der biophysikalischen Forschung,* Georg Thieme, Verlag, Stuttgart, 1938.
7. **Gildemeister, M.,** Die passiven elektrischen Erscheinungen im Tier- und Pflanzenreich, in *Handbuch d. norm. u. pathol. Physiol.,* Vol. 8, Bethe, A. Ed., 1928, 657.
8. **Stuchly, M. A. and Stuchly, S. S.,** Dielectric properties of biological substances — tabulated, *J. Microwave Power,* 15, 19, 1980.
9. **Schwan, H. P. and Foster, K. R.,** R-F field interactions with biological systems: electrical properties and biophysical mechanisms, *Proc. IEEE,* 68, 104, 1980.
10. **Stuchly, M.,** Interaction of radiofrequency and microwave radiation with living systems, *Radiat. Environ. Biophys.,* 16, 1, 1979.
11. **du Bois-Reymond, E.,** *Untersuchungen über tierische Elektricität,* G. Reimer, Berlin, 1849.
12. **Peltier, J. C. A.,** cited in **du Bois-Reymond, E.,** *Untersuchungen über tierische Elektricitat,* G. Reimer, Berlin, 1860.
13. **Hermann, L.,** Über eine Wirkung galvanischer Ströme auf Muskeln und Nerven, *Pflügers Arch. Gesamte Physiol.,* 5, 223, 1872.
14. **Du Bois, M.,** *C. R. Acad. Sci.,* 126, 1790, 1898.
15. **Galler, H.,** Über den elektrischen Leitungswiderstand des tierischen Körpers, *Pflügers Arch. Gesamte Physiol.,* 149, 156, 1913.
16. **Gildemeister, M.,** Über elektrischen Widerstand, Kapazität und Polarisation der Haut, *Pflügers Arch. Gesamte Physiol.,* 176, 84, 1919.
17. **Einthoven, W. and Bijtel, J.,** Über Stromleitung durch den menschlichen Körper, *Pflügers Arch. Gesamte Physiol.,* 198, 439, 1923.
18. **Maxwell, J. C.,** *Treatise on Electricity and Magnetism,* Oxford University Press, London, 1873.
19. **Stewart, G. N.,** The relative volume or weight of corpuscles and plasma in blood, *J. Physiol.,* 24, 356, 1899.
20. **Höber, R.,** Eine Methode, die elektrische Leitfähigkeit im Innern von Zellen zu messen, *Pflügers Arch. Gesamte Physiol.,* 133, 237, 1910.
21. **Höber, R.,** Ein zweites Verfahren die Leitfähigkeit im Innern von Zellen zu messen, *Pflügers Arch. Gesamte Physiol.,* 148, 189, 1912.
22. **Licht, S.,** History of electrotherapy, in *Therapeutic Electricity and Ultraviolet Radiation,* 2nd ed., Licht, S., Ed., Waverly Press, Baltimore, 1967.
23. **Tesla, N.,** High frequency oscillators for electro-therapeutic and other purposes, lecture at 8th Annu. Meet. Am. Electro-Therapeutic Assoc., Buffalo, N. Y., September 1898, in *Nikola Tesla, Lectures, Patents, Articles,* Nikola Tesla Museum, Belgrade, Yugoslavia, 1956.
24. **Nernst, W.,** Zur Theorie des elektrischen Reizes, *Pflügers Arch. Gesamte Physiol.,* 122, 275, 1908.
25. **Coppée, G.,** La pararesonance dans l'excitation par les courants alternatifs sinusoidaux, *Arch. Int. Physiol.,* 90, 1, 1934.
26. **Philippson, M.,** Sur la résistance électrique des cellules et des tissus, *C. R. Soc. Belge Biol.,* 83, 1399, 1920.
27. **McClendon, J. F.,** Colloidal properties of the surface of the living cell. II. Electric conductivity and capacity of blood to alternating currents of long duration and varying in frequency from 266 to 2,000,000 cycles per second, *J. Biol. Chem.,* 69, 733, 1926.
28. **Osterhout, W. J. V.,** *Injury, Recovery and Death in Relation to Conductivity and Permeability,* Lippincott, Philadelphia, 1922.
29. **Dänzer, H.,** Über das Verhalten biologischer Körper im Hochfrequenzfeld, *Ann. Phys.,* 21(5), 463, 1934.
30. **Rajewsky, B. and Schwan, H. P.,** Die Dielektrizitätskonstante und Leitfähigkeit des Blutes bei ultrahohen Frequenzen, *Naturwissenschaften,* 35, 315, 1948.
31. **Cook, H. F.,** Dielectric behavior of human blood at microwave frequencies, *Nature,* 168, 247, 1951.
32. **Schwan, H. P.,** Determination of biological impedances, in *Physical Techniques in Biological Research,* Vol. 6, Oster, G. et al., Eds., Academic Press, New York, 1963, 323.

33. **Hasted, J. B.,** *Aqueous Dielectrics,* Chapman and Hall, London, 1973.
34. **Cole, K. S. and Cole, R. H.,** Dispersion and absorption in dielectrics. I. Alternating current characteristics, *J. Chem. Phys.,* 9, 341, 1941.
35. **Cole, K. S. and Cole, R. H.,** Dispersion and absorption in dielectrics. II. Direct current characteristics, *J. Chem Phys.,* 10, 98, 1942.
36. **Hill, R. M. and Jonscher, A. K.,** The dielectric behavior of condensed matter and its many-body interpretations, *Contemp. Phys.,* 24, 75, 1983.
37. **Kronig, R. de L.,** On the theory of dispersion of X-rays, *J. Opt. Soc. Am.,* 12, 547, 1926.
38. **Hanai, T.,** in *Emulsion Science,* Sherman, P., Ed., Academic Press, New York, 1968, chap. 5.
39. **Dukhin, S. S.,** Dielectric properties of disperse systems, *Surface Colloid Sci.,* 3, 83, 1971.
40. **Dukhin, S. S. and Shilov, V. N.,** *Dielectric Phenomena and the Double Layer in Disperse Systems,* John Wiley & Sons, New York, 1974.
41. **Wagner, K. W.,** *Arch. Electrochem.,* 2, 371, 1914.
42. **Fricke, H.,** A mathematical treatment of the electric conductivity and capacity of disperse systems. I. The electric conductivity of a suspension of homogeneous spheroids, *Phys. Rev.,* 24, 575, 1924.
43. **Fricke, H.,** A mathematical treatment of the electric conductivity and capacity of disperse systems. II. The capacity of a suspension of conducting spheroids surrounded by a non-conducting membrane for a current of low frequency, *Phys. Rev.,* 26, 678, 1925.
44. **Velick, S. and Gorin, M.,** The electrical conductance of suspensions of ellipsoids and its relation to the study of avian erythrocytes, *J. Gen. Physiol.,* 23, 753, 1940.
45. **Pauly, H. and Schwan, H. P.,** Über die Impedanz einer Suspension von kugelförmigen Teilchen mit einer Schale, *Z. Naturforsch.,* 14B, 125, 1959.
46. **Fricke, H.,** The complex conductivity of a suspension of stratified particles of spherical cylindrical form, *J. Phys. Chem.,* 59, 168, 1955.
47. **Irimajiri, A., Hanai, T., and Inouye, A.,** A dielectric theory of "multi-stratified shell" model with its application to a lymphoma cell, *J. Theor. Biol.,* 78, 251, 1979.
48. **Irimajiri, A., Doida, Y., Hanai, T., and Inouye, A.,** Passive electrical properties of cultured murine lymphoblast (L5178Y) with reference to its cytoplasmic membrane, nuclear envelope, and intracellular phases, *J. Membr. Biol.,* 38, 209, 1978.
49. **Boned, C. and Peyrelasse, J.,** Etude de la permittivité complexe d'ellipsoides disperses dans un milieu continu. Analyses theorique et numerique, *Colloid Polym. Sci.,* 261, 600, 1983.
50. **Hanai, T. and Koizumi, N.,** Dielectric relaxation of W/O emulsions in particular reference to theories of interfacial polarization, *Bull, Inst. Chem. Res. Kyoto Univ.,* 53, 153, 1975.
51. **Hanai, T., Asami, K., and Koizumi, N.,** Dielectric theory of concentrated suspensions of shell-spheres in particular reference to the analysis of biological cell suspensions, *Bull. Inst. Chem. Res. Kyoto Univ.,* 57, 297, 1979.
52. **Chiew, Y. C. and Glandt, E. D.,** The effect of structure on the conductivity of a dispersion, *J. Colloid Interface Sci.,* 94, 90, 1983.
53. **Cole, K. S., Li, C., and Bak, A. F.,** Electrical analogues for tissues, *Exp. Neurol.,* 24, 459, 1969.
54. **Batchelor, G. K.,** Transport properties of two-phase materials with random structure, *Annu. Rev. Fluid Mech.,* 6, 227, 1974.
55. **Schwan, H. P. and Morowitz, H. J.,** Electrical properties of the membranes of the pleuropneumonia-like organism A 5969, *Biophys. J.,* 2, 395, 1962.
56. **Schwan, H. P., Takashima, S., Miyamoto, V. K., and Stoeckenius, W.,** Electrical properties of phospholipid vesicles, *Biophys. J.,* 10, 1102, 1970.
57. **Redwood, W. R., Takashima, S., Schwan, H. P., and Thompson, T. E.,** Dielectric studies on homogeneous phosphatidylcholine vesicles, *Biochim. Biophys. Acta,* 255, 557, 1972.
58. **Debye, P.,** *Polar Molecules,* Chemical Catalog Co., New York, 1929.
59. **Oncley, J. L.,** in *Proteins, Amino Acids, and Peptides,* Cohn, E. J. and Edsall, J. T., Eds., Reinhold, New York, 1943, 543.
60. **Takashima, S.,** Dielectric properties of proteins. I. Dielectric relaxation, in *Physical Principles and Techniques of Protein Chemistry,* Academic Press, New York, 1969.
61. **Abragam, A.,** *The Principles of Nuclear Magnetism,* Oxford University Press, London, 1961, 327.
62. **Pennock. B. E. and Schwan, H. P.,** Further observations on the electrical properties of hemoglobin bound water, *J. Phys. Chem.,* 73, 2600, 1969.
63. **Schwan, H. P., Sheppard, R. J., and Grant, E. H.,** Complex permittivity of water at 25°C, *J. Chem. Phys.,* 64, 2257, 1976.
64. **Grant, E. H., Szwarnowski, S., and Sheppard, R. J.,** Dielectric properties of water in the microwave and far-infrared regions, in *Biological Effects of Nonionizing Radiation,* (ACS Symposium Series 157), Illinger, K. H., Ed., American Chemical Society, Washington, D.C., 1981.
65. **Buchanan, T. J., Haggis, G. H., Hasted, J. B., and Robinson, B. G.,** The dielectric estimation of protein hydration, *Proc. R. Soc. London, Ser.* A, 213, 379, 1952.

66. **Schwan, H. P.,** Electrical properties of bound water, *Ann. N.Y. Acad. Sci.,* 125, 344, 1965.

67. **Grant, E. H.,** Electrical properties of bound water, *Ann. N.Y. Acad. Sci.,* 125, 418, 1965.

68. **Schwan, H. P., Schwarz, G., Maczuk, J., and Pauly, H.,** On the low frequency dielectric dispersion of colloidal particles in electrolyte solution, *J. Phys. Chem.,* 66, 2626, 1962.

69. **Einolf, C. W. and Carstensen, E. L.,** Passive electrical properties of microorganisms. V. Low frequency dielectric dispersion in bacteria, *Biophys. J.,* 13, 8, 1973.

70. **Schwarz, G.,** Dielectric relaxation of biopolymers in solution, *Adv. Mol. Relax. Processes,* 3, 281, 1972.

71. **Schwarz, G.,** A theory of the low frequency dielectric dispersion of colloidal particles in electrolyte solution, *J. Phys. Chem.,* 66, 2636, 1962.

72. **Schwarz, G.,** Über die Dispersion des Orientierungsfeldeffektes von Polyelektrolyten in hochfrequenten elektrischen Feldern, *Z. Phys. Chem. (Frankfurt),* 19, 286, 1959.

73. **Mandel, M.,** Dielectric properties of charged linear macromolecules with particular reference to DNA, *Ann. N.Y. Acad. Sci.,* 303, 74, 1977.

74. **Rabinowitz, J. R.,** The effect of electric field induced perturbation of the distribution of ions near the cell surface on migration of charged membrane components, *J. Theor. Biol.* 99, 377, 1982.

75. **Trautman, E. D. and Newbower, R. S.,** A practical analysis of the electrical conductivity of blood, *IEEE Trans. Biomed. Eng.,* 30, 141, 1983.

76. **Geddes, L. A. and Baker, L. E.,** The specific resistance of biological material — a compendium of data for the biomedical engineer and physiologist, *Med. Biol. Eng.,* 5, 271, 1967.

77. **Foster, K. R., Schepps, J. L., and Schwan, H. P.,** Microwave dielectric relaxation in tissue. A second look, *Biophys. J.,* 29, 271, 1980.

78. **Schwan, H. P.,** Electrical properties of cells: principles, some recent results and some unresolved problems, in *Biophysical Approach to Excitable Systems,* Adelman, W. S. and Goldman, D., Eds., Plenum Press, New York, 1981, 3.

79. **Schwan, H. P.,** Electrical properties of blood and its constituents: alternating current spectroscopy, *Blut,* 46, 185, 1983.

80. **Epstein, B. R. and Foster, K. R.,** Anisotropy in the dielectric properties of skeletal muscle, *Med. Biol. Eng. Comput.,* 21, 51, 1983.

81. **Schwan, H. P.,** Die elektrischen Eigenschaften von Muskelgewebe bei Niederfrequenz, *Z. Naturforsch.,* 9b, 245, 1954.

82. **Rush, S., Abildskov, J. A., and McFee, R.,** Resistivity of body tissues at low frequencies, *Circ. Res.,* 12, 40, 1963.

83. **Fatt, P.,** An analysis of the transverse electrical impedance of striated muscle, *Proc R. Soc. London, Ser. B,* 159, 606, 1964.

84. **Falk, G. and Fatt, P.,** Linear electrical properties of striated muscle fibers observed with intracellular electrodes, *Proc R. Soc. London, Ser. B,* 160, 69, 1964.

85. **Takashima, S.,** Passive electrical properties and voltage dependent membrane capacitance of single skeletal muscle fibers, *Pflügers Arch. Gesamte Physiol.,* 403, 197, 1985.

86. **Stoy, R. D., Foster, K. R., and Schwan, H. P.,** Dielectric properties of mammalian tissues from 0.1 to 100 MHz: a summary of recent data, *Phys. Med. Biol.,* 27, 501, 1982.

87. **Foster, K. R. and Schepps, J. L.,** Dielectric properties of tumor and normal tissues at radio through microwave frequencies, *J. Microwave Power,* 16, 107, 1981.

88. **Schwan, H. P. and Kay, C. F.,** Capacitive properties of living tissues, *Circ. Res.,* 5, 439, 1957.

89. **Pauly, H., Packer, L., and Schwan, H. P.,** Electrical properties of mitochondrial membranes, *J. Biophys. Biochem. Cytol.,* 7, 589, 1960.

90. **Schepps, J. L. and Foster, K. R.,** The UHF and microwave dielectric properties of normal and tumor tissues: variation in dielectric properties with tissue water content, *Phys. Med. Biol.,* 25, 1149, 1980.

91. **Homburger, F. and Fishman, W. H., Eds.,** *The Physiopathology of Cancer,* Hoeber-Harper, New York, 1953, 554.

92. **Greenstein, J. P.,** *Biochemistry of Cancer,* Academic Press, New York, 1947, 246.

93. **Joines, W. T., Jirtle, R. L., Rafal, M. D., and Schaefer, D. J.,** Microwave power absorption differences between normal and malignant tissue, *Int. J. Radiat. Oncol. Biol. Phys.,* 6, 681, 1980.

94. **Fricke, H. and Morse, S.,** The electric capacity of tumors of the breast, *J. Cancer Res.,* 10, 340, 1926.

94a. **Smith, S. R. and Foster, K. R.,** Dielectric properties of VX-2 carcinoma vs. normal liver tissue, *IEEE Trans. Biomed. Eng.,* in press.

95. **Kosterich, J. D., Foster, K. R., and Pollack, S. R.,** Dielectric permittivity and electrical conductivity of fluid saturated bone, *IEEE Trans. Biomed. Eng.,* 30, 81, 1983.

96. **Kosterich, J. D., Foster, K. R., and Pollack, S. R.,** Dielectric properties of fluid saturated bone: effect of variation in conductivity of immersion fluid, *IEEE Trans. Biomed. Eng.,* 31, 369, 1984.

97. **Burdette, E. C., Cain, F. L., and Seals, J.,** In vivo probe measurement technique for determining dielectric properties at VHF through microwave frequencies, *IEEE Trans. Microwave Theory Tech.,* 18, 414, 1980.

98. **Magin, R. and Burdette, E. C.,** Electrical properties of tissue at microwave frequencies: a new approach to detection and treatment of abnormalities, in *Non-Invasive Physiological Measurements,* Vol. 2, Rolfe, P. M., Ed., Academic Press, New York, 1983.

99. **Athey, T. W., Stuchly, M. A., and Stuchly, S. S.,** Dielectric properties of biological substances at radio frequencies. Part I. Measurement method, *IEEE Trans. Microwave Theory Tech.,* 30, 82, 1982.

100. **Brady, M. M., Symonds, S. A., and Stuchly, S. S.,** Dielectric behavior of selected animal tissues *in vitro* at frequencies from 2 to 4 GHz, *IEEE Trans. Biomed. Eng.,* 28, 305, 1981.

101. **Stuchly, M. A., Kraszewski, A., Stuchly, S. S., and Smith, A. M.,** Dielectric properties of animal tissues *in vivo* at radio and microwave frequencies: comparison between species, *Phys. Med Biol.,* 27, 927, 1982.

102. **Schwan, H. P. and Kay, C. F.,** Specific resistance of body tissues, *Circ. Res.,* 4, 664, 1956.

103. **Schwan, H. P. and Kay, C. F.,** Conductivity of living tissues, *Ann. N.Y. Acad. Sci.,* 65, 1007, 1957.

104. **Geddes, L. A. and Baker, L. E.,** *Principles of Applied Biomedical Instrumentation,* 2nd ed., John Wiley & Sons, New York, 1975.

105. **Stuchly, M. A., Athey, T. W., Stuchly, S. S., Samaras, G. M., and Taylor, G.,** Dielectric properties of animal tissues *in vivo* at frequencies 10 MHz—1 GHz, *Bioelectromagnetics,* 2, 93, 1981.

106. **Burdette, E. C., Cain, F. L., and Seals, J.,** In situ permittivity measurements: perspectives, techniques, results, in *Medical Applications of Microwave Imaging,* Krone, R., Ed., IEEE Press, New York, 1986, 13.

107. **Lippold, O. C. J. and Winton, F. R.,** *Human Physiology,* Little, Brown, Boston, 1968, 660.

108. **Burdette, E. C. and Karow, A. M.,** Kidney model for study of electromagnetic thawing, *Cryobiology,* 15, 142, 1978.

109. **Iskander, M. F. and Durney, C. H.,** Microwave methods of measuring changes in lung water, *J. Microwave Power,* 18, 265, 1983.

110. **Colonomos, P. and Gordon, R. G.,** Bounded error analysis of experimental distributions of relaxation times, *J. Chem. Phys.,* 71, 1159, 1979.

111. **Bull, H. B. and Breese, K.,** Electrical conductance of protein solutions, *J. Colloid. Interface Sci.,* 29, 492, 1969.

112. **Cook, H. F. and Kuntz, I. D.,** The properties of water in biological systems, *Annu. Rev. Biophys. Bioeng.,* 3, 95, 1974.

113. **Pauly, H. and Schwan, H. P.,** Dielectric properties and ion mobility in erythrocytes, *Biophys. J.,* 6, 621, 1966.

114. **Foster, K. R., Cheever, E., Leonard, J. B., and Blum, F.,** Transport properties of polymer solutions: a comparative approach, *Biophys. J.,* 45, 975, 1984.

115. **Schepps, J. L.,** The Measurement and Analysis of the Dielectric Properties of Normal and Tumor Tissues at UHF and Microwave Frequencies, Ph.D. dissertation, University of Pennsylvania, Philadelphia, 1980.

116. **Foster, K. R., Schepps, J. L., and Epstein, B. R.,** Microwave dielectric studies on proteins, tissues, and heterogeneous suspensions, *Bioelectromagnetics,* 3, 29, 1982.

117. **Foster, K. R., Epstein, B. R., Jenin, P. C., and Mackay, R. A.,** Dielectric studies on nonionic microemulsions, *J. Colloid Interface Sci.,* 88, 233, 1982.

118. **Epstein, B. R., Foster, K. R., and Mackay, R. A.,** Microwave dielectric properties of ionic and nonionic microemulsions, *J. Colloid Interface Sci.,* 95, 218, 1983.

119. **Masszi, G., Szuarto, A., and Grof, P.,** Investigations on the ion- and water-binding of muscle by microwave measurements, *Acta Biochim. Biophys. Acad. Sci. Hung.,* 11, 129, 1976.

120. **Grant, E. H., Nightingale, N. R. V., Sheppard, R. J., and Gough, S. R.,** Dielectric properties of water in myoglobin solution, in *Biological Effects of Nonionizing Radiation,* (ACS Symposium Series 157), Illinger, K. H., Ed., American Chemical Society, Washington, D.C., 1981.

121. **Gabriel, C., Sheppard, R. J., and Grant, E. H.,** Dielectric properties of ocular tissues at 37°C, *Phys. Med. Biol.,* 28, 43, 1983.

122. **Nightingale, N. R. V., Goodridge, V. D., Sheppard, R. J., and Christie, J. L.,** The dielectric properties of cerebellum, cerebrum, and brain stem of mouse brain at radiowave and microwave frequencies, *Phys. Med. Biol.,* 28, 897, 1983.

123. **Schwan, H. P.,** Biophysics of the interaction of electromagnetic energy with cells and membranes, in *Biological Effects and Dosimetry of Nonionizing Radiation,* Grandolfo, M., Michaelson, S. M., and Rindi, A., Eds., Plenum Press, New York, 1983, 213.

124. **Guy, A. W., Davidow, S., Yang, G.-Y., and Chou, C.-K.,** Determination of electric current distributions in animals and humans exposed to a uniform 60 Hz high intensity electric field, in *The Interaction of Electromagnetic Fields with Biological Systems,* Takashima, S. and Postow, E., Eds., A.R. Liss, New York, 1982.

125. **Pfützner, H.,** Dielectric analysis of blood by means of a raster-electrode technique, *Med. Biol. Eng. Computing,* 22, 142, 1984.

126. **Schwan, H. P. and Li, K.,** Capacity and conductivity of body tissues at ultrahigh frequencies, *Proc. IRE,* 41, 1735, 1953.
127. **Foster, K. R., Schepps, J. L., Stoy, R. D., and Schwan, H. P.,** Dielectric properties of brain tissue between 0.01 and 10 GHz, *Phys. Med. Biol.,* 24, 1177, 1979.
128. **Kraszewski, A., Stuchly, M. A., Stuchly, S. S., and Smith, A. M.,** *In vivo* and *in vitro* dielectric properties of animal tissues at radiofrequencies, *Bioelectromagnetics,* 3, 421, 1982.
129. **Schwan, H. P.,** Über die Niederfrequenzleitfähigkeit von Blut und Blutserum bei verschiedenen Temperaturen, *Z. Gesamte Exp. Med.,* 109, 531, 1941.
130. **Fricke, H. and Curtis, H. J.,** The electric impedance of hemolyzed suspensions of mammalian erythrocytes, *J. Gen. Physiol.,* 18, 821, 1935.
131. **England, T. S. and Sharples, N. A.,** Dielectric properties of the human body in the microwave region of the spectrum, *Nature,* 163, 487, 1949.
132. **Herrick, J. F., Jelatis, D. G., and Lee, G. M.,** Dielectric properties of tissues important in microwave diathermy (abstract only), *Fed. Proc. Fed. Am. Soc. Exp. Biol.,* 9, 60, 1950.
133. **Drago, G. P., Marchesi, M., and Ridella, S.,** The frequency dependence of an analytical model of an electrically stimulated biological structure, *Bioelectromagnetics,* 5, 47, 1984.
134. **Fixman, M.,** Charged macromolecules in external fields. I. The sphere, *J. Chem. Phys.,* 72, 5177, 1980.
135. **Fixman, M.,** Thin double layer approximation for electrophoresis and dielectric response, *J. Chem. Phys.,* 78, 1483, 1983.
136. **Chew, W. C. and Sen, P. N.,** Potential of a sphere in an ionic solution in thin double layer approximations, *J. Chem. Phys.,* 77, 2042, 1982.
137. **Chew, W. C. and Sen, P. N.,** Dielectric enhancement due to electrochemical double layer: thin double layer approximation, *J. Chem. Phys.,* 77, 4683, 1982.
138. **Chew, W. C.,** Dielectric enhancement and electrophoresis due to an electrochemical double layer: a uniform approximation, *J. Chem. Phys.,* 81, 4541, 1984.

Part II — Effects of DC and Low Frequency Fields

Chapter 1

INTERACTION OF DC ELECTRIC FIELDS WITH LIVING MATTER

Frank S. Barnes

TABLE OF CONTENTS

I. INTRODUCTION

The fact that electric currents can affect the behavior of biological systems has been known for more than 2000 years, and electric shocks have been used to treat a wide variety of ailments off and on ever since. However, it is surprising how incomplete our knowledge is today of the effects of direct current (DC) voltages and currents on biological systems.

Electrical signals are clearly important in the control of biological processes and in carrying information from one part of the body to another. Nerve cells propagate electrical signals from sensors of pressure, temperature, light, sound, etc. to the brain and return control signals to muscles, the heart, etc., yet if we choose to stimulate these processes with external electrical inputs, we have a relatively limited understanding of how a given electrical signal will affect various biological organs, what the safe limits of exposure are (particularly over extended periods of time), and how electrical signals are carried across cell membranes or propagated along nerves.

The purpose of this review is to bring together some of the physics which underlie the interaction between electric fields and biological materials with the objective of providing background for determining safe levels of exposure and new applications for the use of electricity in therapy. It begins by examining some of the forces that are exerted on charged particles in fluids, proceeds to some of the effects of electric fields on membranes, and looks at secondary effects of current flow due to heating. This is followed by a description of a few gross effects on whole animals. All this information is presented with the object of specifying the general level or intensity of fields, currents, and temperatures where one can expect to observe a given class of biological responses. The next section contains data on the levels of typical naturally occurring and man-made fields and information on current safety standards. This section also includes information on both electrical signals and noise levels in the body. The final sections introduce the relationships of electric fields to tissue growth and applications to therapy.

II. PHYSICS OF THE INTERACTIONS OF ELECTRIC FIELDS WITH BIOLOGICAL MATERIALS

Biological systems consist of complex physical subsystems, and in an attempt to understand them, we will need to start at the most elementary level. Perhaps the simplest level, which is already surprisingly complicated, is the effect of electric fields on biological fluids. These fluids contain a large number of components, including ions, polar molecules such as water, proteins, lipids, hormones, and colloidal particles. At low current densities conductivity is linear; however, at moderate to high current densities, nonlinearities begin.[1] The next level of complexity involves the interaction of the fields with membranes that behave like porous solids for fields applied perpendicular to their surface, and like viscous liquids for fields in the plane of the membrane.[2,3] Membranes are inhomogeneous so that different portions of them may be affected differently by the perturbing fields. Additionally, membranes are involved in active chemical reactions that change their porosity to various ions in a selective way so that both electrical potentials and chemical signals may change the membrane conductivity by orders of magnitude. The next level of complexity occurs in the interactions between the biological fluids and the membranes in the presence of electric fields. Electric fields affect the selective transport of ions or molecules through the membrane. They change the build-up of layers at the surface and change the way new molecules are incorporated into the membrane or are bound to its surface. The result of changes in the transport of molecules or ions across cell membranes is changes in the performance of the cells and in turn, of the organs of which they are a part. For example, a biasing electric current across a pacemaker cell in the heart will change its firing rate and thus the pumping rate of the heart.

The fundamental law describing forces on charged particles is

$$\vec{F} = q\,(\vec{E} + \vec{v} \times \vec{B})\tag{1}$$

where $\vec{F}$ is the force acting on a charged particle, q is its effective charge, $\vec{E}$ is the electric field, $\vec{v}$ is the velocity of the particle, $\vec{B}$ is the magnetic field, and $\rightarrow$ denotes a vector quantity. It is worth noting that $\vec{E}$ and $\vec{B}$ are coupled by the Maxwell equation so that a time-varying magnetic field generates an $\vec{E}$ field and vice versa. Additionally, if nonlinearities or time-varying impedances are present, alternating current (AC) fields can be rectified to produce DC. Here we will neglect the magnetic field since the forces are often quite small and are treated in detail in Part II, Chapters 4 and 5.

In addition to forces on charged particles, we can induce forces on polarizable neutral molecules or molecules with dipole moments in an electric field which varies in space. To first order these forces are described by

$$\vec{F}_d = (\vec{P} \cdot \nabla\vec{E})\tag{2}$$

$$F_i = \alpha V(\vec{E} \cdot \nabla E)\tag{3}$$

where $\vec{F}_d$ is the force on a molecule with a permanent dipole moment $\vec{P}$, $\nabla\vec{E}$ is the gradient of the electric field, and $\vec{F}_i$ is the force on a molecule with an induced dipole moment $\vec{P} = \alpha V\vec{E}$ where α is the tensor polarizability and V is the volume.

Since most biological materials are highly inhomogeneous, there are nearly always important local variations in the electric field constituting a significant electric field gradient. This in turn means that when we are looking at the detail of the movement of molecules, membranes, and ions, we need to consider the field gradients as well as the fields.

The electric current density within a homogeneous fluid in an electric field is given by

$$\vec{J} = \sum_j q_j\mu_jC_j\vec{E}\tag{4}$$

where q_j is the charge on the ion of type j, μ_j is its mobility, and C_j is its concentration. Typical ion mobilities are given in Table 1. The conductivity of biological fluids is in the vicinity of $\sigma = 0.4$ Sm^{-1}. Note that if the fluid channels are relatively thick, the fluids are relatively good conductors and thus tend to cause a short circuit across voltages that might otherwise appear across membranes which typically have conductivities at least a thousand times smaller.

Current flow superimposes a drift velocity on top of the much larger random thermal velocity in opposite directions for positively and negatively charged particles. It can lead to a redistribution of ions or molecules as a result of their differential mobilities and to an increase in the concentration of the appropriately charged ions at the interfaces. The average drift velocities $\vec{v}$ are given by

$$\vec{v} = \mu_i\vec{E}\tag{5}$$

This effect is known as electrophoresis and is frequently used to separate large molecules or charged colloidal particles.[4]

For spherical particles in an insulating fluid, the mobility μ_i is given by

Table 1
TYPICAL VALUES OF BIOLOGICAL COEFFICIENTS

Mobilities

Proteins	$\mu = 10^{-10}$—10^{-8} m²/V sec	Ca++	$\mu = 6.2 \times 10^{-8}$ m²/V sec
Na+	$\mu = 5.2 \times 10^{-8}$ m²/V sec	Mg++	$\mu = 5.4 \times 10^{-8}$ m²/V sec
K+	$\mu = 7.6 \times 10^{-8}$ m²/V sec	Cl−	$\mu = 7.9 \times 10^{-8}$ m²/V sec

Conductivities and Resistivities

Blood	$\sigma = 0.67$ S/m	
Lung	$\sigma = 0.05$ S/m	
Liver	$\sigma = 0.14$ S/m	
Fat	$\sigma = 0.04$ S/m	
Cell fluid	$\sigma = 0.5$ S/m	$\rho_f = 2\ \Omega m$
Membrane	$\sigma = 10^{-5}$—10^{-7} S/m	$\rho_m = 10^5$—$10^7\ \Omega m$

$$\mu_i = \frac{q}{6\pi\eta a} \tag{6}$$

where η is the viscosity of the fluid and a is the radius of the particle. In a conducting medium, counterions, or ions with a charge of opposite sign to that of the surface charge of the particle, are attracted to it. They change the effective radius of the particle and partially shield its charge.[4,5] Additionally, small counterions may flow in the opposite direction to the particle motion, exerting a viscous drag. The theory for the motion of a rigid sphere through a conducting liquid is complicated if all these effects are taken into account. Often some of the parameters, including the charge on the sphere, are not measurable. However, a relatively simple expression for the electrophoretic mobility is often used

$$\mu_1 = \frac{\epsilon_1 \zeta}{4\pi\eta} \tag{7}$$

ϵ_1 and η are the dielectric permittivity and the viscosity of the fluid and ζ is the electrical potential drop from the particle surface across the bound fluid to the interface where the liquid begins to flow under the shear stress. Stated another way, ζ is the potential at the surface where the stationary fluid and the liquid which is moving with the particle meet. It is to be noted that ζ is less than the total potential ψ across the charged double layer surrounding the charged particles.

Neutral particles with either permanent or induced dipole moments have different mobilities, depending upon their size and dipole moment. This can lead to changes in their distribution when they are placed in an electric field gradient.

In the case of an induced dipole moment for an ideal dielectric sphere with dielectric permittivity ϵ_2 and $\sigma_2 = 0$, placed in an ideal dielectric fluid with dielectric permittivity ϵ_1 and $\sigma_1 = 0$, and a nonuniform electric field prior to inserting the sphere, the force $\vec{F}_d$ acting on the sphere can be shown to be[5]

$$\vec{F}_d = \alpha V(\vec{E}_1 \cdot \nabla \vec{E}_1) = 4\pi a^3 \epsilon_1 \frac{\epsilon_2 - \epsilon_1}{\epsilon_2 + 2\epsilon_1} (\vec{E}_1 \cdot \nabla \vec{E}_1) \tag{8}$$

where $\vec{E}_1$ is the field in the fluid prior to insertion of the sphere. Written another way

$$\vec{F}_d = \frac{3}{2} (\text{volume})\ \epsilon_1 \frac{\epsilon_2 - \epsilon_1}{\epsilon_2 + 2\epsilon_1} \nabla |\vec{E}_1|^2 \tag{9}$$

If we assume that the viscous drag on a spherical particle is given by Stokes' law

$$\vec{F}_s = 6\pi a \eta \vec{v} \tag{10}$$

The mobility x_1 is given by

$$x_i = \frac{2}{3} \frac{a^2}{\eta} \epsilon_1 \frac{\epsilon_2 - \epsilon_1}{\epsilon_2 + 2\epsilon_1} \nabla E_1 \tag{11}$$

In order for particles with dipole moments to change their distribution under the influence of an electric field gradient, the force $\vec{F}_d$ has to be large enough to overcome the other forces which may be binding them. One of the fundamental forces which must frequently be overcome is due to osmotic pressure or diffusion. The osmotic pressure (Π) due to a concentration of C_i particles per unit volume is given by $\Pi = C_i kT$,[5] where k is the Boltzmann constant and T is the absolute temperature. The osmotic pressure can be thought of as the force per unit area arising from diffusion or the random motion of the particles. The average differential force on a particle is proportional to the gradient of the osmotic pressure and is given by

$$\vec{F}_{os} = -\frac{1}{C_i} \nabla \Pi = \frac{-kT\nabla C}{C_i} - k\nabla T \tag{12}$$

If we consider the case of a spherical volume with a radial concentration gradient at constant temperature, the force is given by

$$\vec{F}_{os} = -kT \frac{\Delta C_i}{C_i} \cdot \frac{r'_o}{\Delta r} \tag{13}$$

where r_0' is the unit vector, ΔC is the incremental change in concentration, and Δr is the incremental change in distance. The maximum change is

$$\frac{\Delta C_i}{C_i} = 1 \tag{14}$$

when the presence or absence of a particle occurs on the distance $\Delta r = 2a$, where a is the particle radius. In this case, we get the maximum force.

$$F_{os\,max} = -\frac{kT}{2a} \tag{15}$$

To get an idea of the size of these forces, consider a particle of fat with $a = 1$ μm in water. The maximum osmotic pressure at $T = 300$ K is $F_{os(max)} = 2 \times 10^{-15}$ N. The dielectric constant for water is approximately $\epsilon_1 = 80\epsilon_0$ and for a fat particle $\epsilon_2 = 2\epsilon_0$ where $\epsilon_0 = 8.854 \times 10^{-12}$ F/m. To get a dielectric force greater than the maximum osmotic force, we need a value of $\nabla|E^2| > 10^{12}$ V^2/m^3. This is given approximately by a voltge of 100 V across a 5-mm gap when the E field goes from zero to a peak value of 5×10^4 V/m over the same gap. For a particle with a single charge in a uniform field, we would need a field of $E = 1.3 \times 10^4$ V/m to get an equal force.

The electric current densities generated by a concentration gradient are given by

$$J_D = -qD\nabla C_i \tag{16}$$

where D is the diffusion constant and is given by

$$D = vkT \tag{17}$$

where v is the hydrodynamic mobility with the dimensions of $\dfrac{\text{velocity}}{\text{force}}$ and D has the dimensions of square meters per second.

There are four forces which may become important when considering the interaction between two particles in a fluid. These are the osmotic or diffusion force, electrostatic force, Van der Waals force, and hydration force.[6] These forces may all become important in considering the interaction between particles or bilipid membranes in an aqueous fluid. Electrostatic or coulomb forces between particles of like charge are repulsive. Because the charged particles attract free ions of the opposite sign, which produces a double layer, they are effectively shielded, or are screened by the charged ions of the opposite sign when immersed in a conducting fluid. This force decays exponentially or

$$F_c = F_0 \exp \frac{-r}{\lambda_d} \tag{18}$$

where λ_d is known as the Debye screening length,[7]

$$\lambda_d = \left(\frac{2q^2n}{\epsilon kT} \right)^{1/2} \tag{19}$$

and n is the density of the ion species doing the shielding, q is the charge, ϵ is the dielectric constant of the solution, k is the Boltzmann constant, and T is the absolute temperature. For physiological saline of approximately 0.14 *M* solution the Debye length is approximately 8.3 Å.[8] Thus the electrostatic forces are important only at very short ranges.

For like particles, the Van der Waals forces are repulsive at short distances (1 to 2 Å) and attractive at longer ranges. These forces may be thought of as being generated by transient electromagnetic fields due to fluctuations or displacements of the electric charges in the membranes or particles. These fluctuations occur because of thermal agitation or natural uncertainties in the position and momenta of the electrons and atomic nuclei. If one thinks of the local transient fluctuations in terms of the underlying contributions from oscillations at all possible frequencies, it can be shown that the strength of the contribution due to the local fluctuations at a given frequency is proportional to the absorption of light at that frequency by the material. For an individual atom these forces fall off very rapidly as $\dfrac{1}{r^7}$. However, when they are integrated over the surface of a membrane, which is thick compared to an atomic layer, they are correlated over many atoms, as the wavelengths are large compared to an atomic diameter.

A calculation of these fields has been done starting with quantum field theory.[9] The size of the forces and the rate at which they decay depends upon the distance between the membranes and on the difference of the bulk polarizability of the membrane and the aqueous gap in a complex way. All frequencies of charge fluctuations contribute to the attraction, and each gives rise to a different relationship between energy and the distance of separation.

One case of interest is for two membranes with a distance d_w across the aqueous gap between them. The thickness of the membranes is assumed to be large compared to the spacing. The force between these two membranes is approximately given by[8]

$$F_w = \frac{H}{6d_w^3} \tag{20}$$

where H is the Hamaker coefficient. In a typical situation, the distances over which the Van der Waals force is estimated to be important extend out to separations of 100 to 200 Å, which is substantially longer than a Debye length or the rate of fall-off for the electrostatic or coulomb forces.

The hydration forces are repulsive forces that rise extremely rapidly as the membrane bilayers approach a separation distance of approximately an atomic spacing. Experimentally, for neutral phosphorous lipid, these forces can be expressed in the form[8]

$$P = P_0 \exp \frac{-d_w}{\Lambda} \qquad (21)$$

where Λ is a scaling constant, and in the case of egg phosphomonoesterase, bilayers $P_0 \simeq 7 \times 10^{13}$ N/m^2 and $\Lambda = 2.56$ Å. This force may be important up to about 20 Å and is assumed to come about as a consequence of the work required to remove water from the hydrophilic surface of the membrane.

All these forces act over relatively short ranges in typical biological fluids. The ion densities are so large that charge neutrality is maintained everywhere except very close to the charged surfaces.

The effect of an electric field or an electric field gradient in a fluid is to superimpose a small drift velocity on a relatively large random thermal velocity. For example, if we apply an electric field of 10^3 V/m to a Na$^+$ ion, we would expect a drift velocity of about 5×10^{-5} m/sec as compared to a thermal velocity of about 4×10^2 m/sec. For a protein we would expect a drift velocity approximately one tenth of the speed of the Na$^+$ ion, although at higher fields. This means that if we are to transport proteins or other small charged particles over appreciable distances of a few millimeters, we can expect it to take periods of time of minutes or longer. In the case of bacteria, we have measured drift velocities of a 10^{-6} m/sec at 100 Hz in fields of about 10^4 V/m, and gradients of 5×10^6 V/m^2 or about 10% of the velocity of the Na$^+$ ions in the same field.[5,10]

III. THE EFFECTS OF ELECTRIC FIELDS ON CELL MEMBRANES

Electric fields play a very important role in the normal biological function of membranes. It would be surprising if externally applied fields did not affect them. The effects from fields in the plane of the membrane, where large molecules such as proteins are free to move as in a viscous fluid, are significantly different from the effects of fields in the transverse direction where the membrane components are bound in a layer typically 50 to 150 Å thick.

The interiors of cells are normally negatively biased with respect to the surrounding fluid by 50 to 150 mV, which leads to transverse electric fields up to tens of millions of volts per meter.[11] The effective membrane resistance per unit (R_m) area takes on values of 0.14 to 15 Ω/m^2 in the transverse direction. This corresponds to resistivities in the range of $\rho_m = 10^7$ Ωm to $\rho_m = 10^9$ Ωm. Both the surrounding fluid and the interior of a cell have resistivities (ρ_f) of about $\rho_f = 2$ Ωm. This means that the cell membrane tends to shield the interior of a cell very effectively from externally applied fields.

Consider the case of a hypothetical rectangular cell as shown in Figure 1. An external field, E, causes a current density, $J_f = E/\rho_f$, to flow in the external medium, where ρ_f is the resistivity of the fluid. The corresponding voltage drop is $V = El = J_f \rho_f l$ which we can consider to be applied to the cell. This voltage is distributed across the cell with

$$V = [\rho_m(2d) + \rho_f(1 - 2d)] J_m \qquad (22)$$

where J_m is the current density through the cell and ρ_m is the resistivity of the membranes.

FIGURE 1.   Current distribution in a hypothetical rectangular cell.

Typical cell membrane thicknesses are 6 to 10 nm and typical dimensions are 10 to 150 μm. Setting $l = 100$ μm and

$$\rho_m 2d = 10 \ \Omega m^2 \tag{23}$$

we get

$$V = (10 \ \Omega m^2 + 2 \times 10^{-4} \ \Omega m^2) J_m \tag{24}$$

This shows that essentially all of the transverse voltage drop occurs across the membrane, and the interior of the cell is almost completely shielded from external fields. The anisotropic characteristics of cells are reflected in an anisotropy of the dielectric and conductive properties of tissue which may give variations of as much as 10 to 1 in conductivity, depending on direction of measurement relative to cell orientation[12] as shown in Part I.

A number of fish, including sharks, have been shown to use very long cells to increase the voltage drop across a sensitive membrane in order to sense fields as low as 10^{-6} V/m.[13] The long cell may be thought of as an antenna that concentrates the field across a very thin, voltage-sensitive detector membrane. This membrane appears to have a built-in amplifier that allows detection of signals which are only a little above the natural electrical noise.

Membranes are not just simple linear resistors, but usually are nonlinear and in the case of nerve cells, are time-varying resistors. For a passive membrane the Nernst equation predicts a diode rectifying characteristic of the form[11]

$$I = I_0 \left[\exp \left(\frac{V_m}{\eta V_T} \right) - 1 \right] \tag{25}$$

where V_T is the voltage of equivalent temperature

$$V_T = \frac{kT}{q} = 0.026 \ V \tag{26}$$

at

$$T = 300 \ K \tag{27}$$

q is the charge on the electron, V_m is the voltage across the membrane, and I_0 and η are constants. Thus for currents flowing through a membrane in one direction, the current is nearly constant while in the other it increases exponentially with voltage. In addition to passive currents, cells also use the energy from metabolic processes for the active transport of ions against the fields established by the concentration gradients. These processes are usually modeled as current sources and described as pumps.[11] A thermodynamic approach to pumping shows that ions can be pumped if they form a compound with a material which can flow through the membrane and which is created on one side of the membrane and destroyed on the other.[14] However, the details of the pumping process are not well understood.

In the case of pacemaker cells, there are also feedback processes that lead to an oscillating membrane potential, and a membrane resistance which is a function of time. The current flow for these cells is described empirically by the Hodgkin Huxley equation.[11] An alternate approach which treats the nerve pulse like a plasma instability has been proposed by Triffet and Green[15] and Vaccaro and Green.[16] Na^+ and K^+ currents are the dominant carriers for the propagation of nerve impulses along a cell. It is generally believed the Na^+ and K^+ currents, which flow through the membrane in opposite directions, are carried through separate channels. Ca^{++} ion currents are involved in the activation of at least a portion of the K^+ currents and are voltage-gated. By activating the K^+ currents, the Ca^{++} ions shorten the length of time the cell is depolarized, and thus speed up the firing cycle.[17] A statistical approach to the formation of protein channels in the membrane by Baumann and Easton[18-20] predicts many of the observed characteristics.

During the firing of a nerve cell, the Na^+ current pulse precedes the K^+ current pulse which returns the cell to its resting potential.[11] The overall concentration balances are maintained by active ion pumps. Cl^-, Mg^{++} and possibly OH^-, H^+ ions may also be involved in the current flow across a cell membrane.

The firing of a nerve cell typically involves voltage spikes of 10^{-1} V and peak current densities of 1.5 A/m². Changes in the firing rate can be induced by the injection of charge through a microelectrode of $< 10^{-9}$ A for a few milliseconds. However, in cases where electrodes are used to stimulate muscles or to control epilepsy, the current is injected through a series of cell membrane fluid boundaries at a distance from the controlling nerve fiber. Thus typical injected currents to produce behavioral changes in cells are milliamps, and current densities are 10 A/m² or higher.

For fields parallel to the plane of the membrane, it is possible to obtain electrophoresis or a rearrangement of charged particles. This has been shown by Poo[3] in striking fashion in cultured embryonic *Xenopus* myotomal muscle cells. Receptors on the surface of the cell were labeled with a fluorescent dye and allowed to uniformly distribute themselves. Exposures to electric fields of 10^2 to 10^3 V/m were sufficient to concentrate the fluorescent-labeled receptors on the side of the anode in about 10 min. After shutting off the field, diffusion returned the dye to its uniform distribution in about 2 hr. This corresponds to an in-plane diffusion constant of about 3×10^{-12} m²/sec. The force on the receptor molecules or particles in the membrane includes not only $q\vec{E}$ but also any viscous drag which may be generated by the flow of ions of the opposite sign moving along the surface in the opposite direction. The direction of motion for a given charged particle seems to depend on whether it has a larger or smaller zeta (ζ) potential than the potential across the charged double layer at the interface between the particle and the fluid (see Equation 7).

Additional work has shown that the distribution of acetylcholine (ACh) receptors is changed by external fields.[21,22] These receptors are concentrated on the cathode-facing surface of the cell in fields of 10^3 V/m over a period of 30 min by literally rearranging channels already existing in the cell membrane. The concentration or clustering persists for at least 5 hr after the field has been turned off, indicating that the clustering is relatively stable. Single channel patch measurements show both a higher density of ACh channels in the clusters near the cathode and a longer mean duration of the pulses through the transmembrane channels. The length of the current pulse near the anode does not differ from the controls, indicating that the field itself does not have a direct effect on the channel kinetics. The lateral diffusion coefficient, D, of ACh receptors in the plasma membrane of cultured *Xenopus* embryonic muscle cells is estimated to be 2.6×10^{-6} m²/sec at 22°C.[21]

IV. THERMAL EFFECTS

One important effect of current flow due to electric fields is heating. The power input to a given volume of material can be expressed by $P' = I^2R$ where I is the total current and R is the resistance of the sample. For many calculations, a more useful expression is given by the power per unit volume or $P = \sigma E^2 = \rho J^2$ where σ is the conductivity, E is the electric field intensity, ρ is the resistivity, and J is the current density. The temperature rise resulting from this heat input is determined by the thermal capacity of the volume and the mechanisms for carrying the heat energy away. Typically these thermal loss mechanisms include a combination of conduction and convection processes. For short current pulses, the heat dissipation is usually dominated by thermal conduction. If we consider a homogeneous sphere of radius a immersed in an infinite fluid, the thermal conductive relaxation time is approximately given by[23]

$$\tau_c = \frac{a^2}{4\overline{K}} \tag{28}$$

where $\overline{K}$ is the thermal diffusivity and is measured in square meters per second. The thermal diffusivity is given by

$$\overline{K} = \frac{K'}{\rho' C_P} \tag{29}$$

where K' is the thermal conductivity in cal/m s °C, ρ' is the material density in kg/m^3, and C_P is the thermal capacity in cal/°C kg. If an applied current pulse is short compared to τ_c, the maximum temperature change is given by[23]

$$\Delta T_{max} = \left(\frac{3}{2\pi e}\right)^{3/2} \frac{\overline{H}}{\rho' C_P a^2} \tag{30}$$

where $\overline{H}$ is the total input energy in calories and e is the base of natural logarithms. For current inputs that are long compared to the thermal relaxation time τ_c, the peak temperature is determined by a balance between the input power P and the dissipation processes controlled by conduction and convection. For many situations where the volume involved is a cubic millimeter or larger, the thermal time constant is controlled by the amount of blood flowing through the volume. In these cases, temperatures are often more easily measured than calculated since a complicated thermal and electrical boundary value problem would have to be solved analytically to calculate the temperature rise. This is particularly true since the viscosity η and other thermal and electrical parameters such as ρ, C_P, $\overline{K}$, etc. are functions of temperature. For example, C_P for an artificial bilipid membrane is shown in Figure 2.[24] Another example of the importance of change in temperature is the conductivity of saline[25]

$$\sigma \approx C_1 \left[10^{\left(\frac{1}{T} + \alpha\right)\left(\frac{1}{b}\right)} \right] \times 10^{-4} \text{ S/m} \tag{31}$$

where C_1 is the concentration of NaCl in milligram equivalents per liter, T is the absolute temperature $\alpha \approx -6.23 \times 10^{-3}$, and $b \approx -1.4 \times 10^{-3}$. In the range around 37.5°C, this means that a 5°C change in temperature corresponds to a little less than 9% change in conductivity.[25]

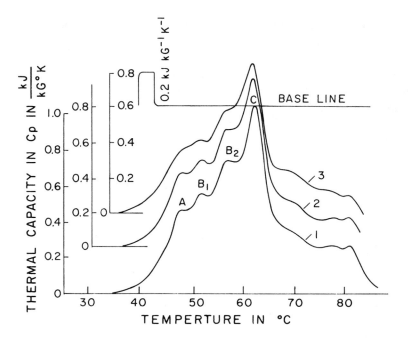

FIGURE 2. Differential changes in the heat capacity Cp erythrocyte membranes as a function of temperature in 5 mmol/ℓ sodium phosphate with pH = 7.4 and a concentration of 5 mg protein per milliliter. The changes at A, B, B_2, and C correspond to changes in the structure of the membrane with temperature and are irreversible. Curve (1) intact membranes; (2) irradiated at 330 mHz for 5 min (SAR 9 m2/g); (3) irradiated at 330 mHz for 30 min (SAR 9 mw/g).

Changes in temperature are important, not only because they change transport properties such as viscosity η, mobility μ, and diffusion coefficient D, but also because they change chemical reaction rates. Typical biochemical reactions can be described by an equation of the form[26]

$$\frac{dS}{dt} = -K'S \tag{32}$$

where S is the fraction of the material which has undergone the chemical reaction, t is the time, and K′ is the reaction rate. K′ is often given

$$K' = \frac{kT}{h} \exp\left(\frac{+\Delta H' - T\Delta S'}{R'T}\right) \tag{33}$$

where k is the Boltzmann constant, T is the absolute temperature, H′ is the free energy, S′ is the entropy, h is Planck's constant, and R′ is the gas constant. The significant feature is that the reaction rate K′ varies exponentially with temperature, and ΔH′ and TΔS′ are large numbers. Thus, very small changes in temperature can lead to big changes in chemical reaction rates. A rule of thumb which the author uses to estimate whether or not significant biological changes are likely is to see if ΔT is > 10°C for 10^{-6} sec, or 5°C for 1 sec, or 2°C for hours. If the ΔTs are larger, then they can be expected to lead to important changes in the biological system. Typical mammalian temperature regulatory systems will hold the internal body temperature constant to better than 0.5°C. (Compare Part III, Chapter 3.)

In addition to the magnitude of the temperature change, it can be shown that the rate of temperature rise, $\dfrac{dT}{dt}$, is important and can induce current to flow across membranes. Changes in the firing rate of pacemaker cells from the ganglion of *Aplysia* have been induced by total temperature changes of as little as 1/10°C when the rates of change are about 1°/sec.[27] This change of the firing rate corresponds to the injection of approximately 1 nA to the cell. By taking the time derivative of the Nernst equation, which describes the passive equilibrium potential across a membrane for a single ion, it can be shown that a current proportional to the temperature derivative is to be expected, or[28]

$$ I = -q, \, V_1' \, C_1 \left(\frac{\phi}{\Phi_T} \right) \left(\frac{\dot{\phi}}{\phi} - \frac{\dot{T}}{T} \right) \tag{34} $$

where q is the charge on the ion, V_1' is the volume of the cell, C_1 is the concentration of ions inside the cell, ϕ is the resting potential, Φ_T is the voltage equivalent temperature, $\Phi_T = \dfrac{kT}{q}$, $\dot{\phi}$ is the derivative of the membrane potential with respect to time, T is the temperature, and $\dot{T}$ is the temperature derivative with respect to time.

To get an idea of the magnitudes of both heating (as described in Equation 30) and the effect of the rate of rise (as given by Equation 34), consider the case of liver tissue with σ = 0.14 S/m and a field strength in the tissue of 2×10^3 V/m. The rate of temperature rise is approximately 13°C/sec, assuming no conduction or diffusion heat losses and the thermal capacity of water. For this high field a significant temperature rise requires about 1/2 sec. However, the rate of rise has been shown to be significant in exciting a brain slice from a mouse with pulses of 10^{-3} sec.[29]

V. NATURAL FIELDS AND MAN-MADE FIELDS

One way of approaching the effects of electric and magnetic fields on man is to examine what we are exposed to naturally, and then to compare it with man-made sources. This approach is in a sense, a global way of getting boundaries on the size of the fields to which we can be expected to have adapted by evolution, and is complementary to the detailed aproach of looking for the effects of electric and magnetic fields on each biological system or subsystem.

This approach needs to be carried out on at least two levels. First, we would like to know the approximate magnitudes of the fields that occur in nature outside man or the biological system of interest. Second, we would like to have values for the internal or physiological fields.

The natural electric fields at the surface of the earth have both DC and AC components.[30] One may think of the earth as a spherical capacitor where the surface is negatively charged with respect to an electrical conducting ionosphere which is about 50 km above the surface. This capacitor is being continuously charged by about 100 lightning strokes per second from thunderstorms worldwide. Since the atmosphere is a finite conductor, it also discharges with an RC time constant of about 18 sec. The result is an average electric field of about 130 V/m. This field is not uniform with height and typically falls off to 30 V/m at 1 km above the surface. The local values vary widely with temperature and humidity. In the Sahara during dust storms caused by winds in the dry season, a field of 1500 V/m has been measured with the polarity reversed from the normal. In thunderstorms, fields of up to 3000 V/m have been measured without lightning, and the polarity has been known to reverse in minutes. Storms as far as 50 km away have been shown to affect local fields.

The atmosphere is a relatively poor conductor, and as such will suspend a significant number of charged ions, dust particles, etc. This helps to contribute to local field variations of 20 to 50% over the course of the day, and is a normal characteristic of our environment. However, it is interesting to note that as we go up in frequency, the magnitude of the natural electric and magnetic fields decreases rapidly. At 60 Hz, the average natural electric field is reduced to about 1×10^{-3} V/m.

One result of electric discharges in the atmosphere, as well as natural ionizing radiation, is the creation of small positive and negative ions in the atmosphere. In clean country or mountain air the typical ion density is about 10^{10}/m³ with an average ion lifetime of a few minutes.[31] During a hot dry wind, positive ions created by the shearing forces can increase in concentration significantly. It has been shown that increases in the negative ion concentration reduce the amount of serotonim (5-HT) in mice and rabbits, possibly by accelerating the enzymatic oxidation process.[31] A similar result has been demonstrated in the oxidation of cytochrome c. Positive ions appear to block monoamine oxidase action, thus raising the concentration of free 5-HT.[31] Changes in 5-HT levels produce significant changes in the central nervous system, with high levels of positive ions raising the anxiety levels under stress. Other effects of increased positive ion concentration include a decrease in the survival rate of mice exposed to a measured dose of influenza, while an increase of negative ions reduced the mortality rate.[31]

The significance of these results is that it is relatively easy to change the ion concentration in air using high-voltage DC systems where a leakage current of 1 μA from a burr or other sharp point would correspond to the generation of about 10^{12} ions per second.

Relatively few high-voltage DC transmission lines are in use today for distribution of power, and because the shocks resulting from a short contact across a high DC voltage are so painful and obviously dangerous these systems are nearly always shielded. Thus one is rarely exposed to DC electric fields $> 10^3$ V/m. An additional feature of this exposure is that air is such a good insulator that the DC currents through the body in a noncontacting situation are very small, as explained in the Introduction. For example, 1000 V across a 1-cm gap would yield a current density of approximately 10^{-7} A/m² flowing across the air gap. Thus the principal hazards from DC fields occur when parts of the body make contact with a conductor.

VI. MEDICAL APPLICATIONS

The use of electricity to treat a great variety of ills dates back at least 2000 years. In early times, the electric shock was obtained from electric eels or torpedo fish and was used to treat such things as gout and hemorrhoids.[32] Until the 1920s, various kinds of electric shocks were reputed to have all the curative powers of a patent medicine and were used to treat a great variety of ills. Even Benjamin Franklin is reputed to have cured a girl of convulsions by giving her four shocks, morning and evening, for a period of about 3 months.

The first use in modern medicine seems to have been in pacemakers for defective hearts in 1960. The pacemaker electrodes are placed in contact with the heart muscle and pulses of 0.1 to 10 mA, a few milliseconds long, are used to synchronize the firing of the cells. The original pacemaker operated at a fixed rate; however, more recent devices can supply pulses on demand.[33] A second use is for defibrillation or to synchronize the firing of the heart cells by overwhelming the natural signals which have gotten out of sequence with a large current pulse. Defibrillation pulses may be as large as 20 A at 6000 V and last for a few milliseconds. Typical energies for adults are 200 to 400 J.[34,35] Still another use of external electrical signals includes the stimulation of muscles in which the control nerves communicating with the brain have been severed. Typical current pulses for stimulating muscles in the leg range from 1 to 10 mA, and in some cases, as much as 20 mA depending on location and the efficiency of coupling from the electrode to the tissue to be activated.[32] These pulses can be from 10 to 60 μsec long and the rate of firing determines the strength of the muscle response. Typical values are from 10 to 35 pulses per second. Current experiments include an attempt to use computer sequencing of muscle stimulation to permit walking by quadriplegics.

Still another example of the application of electrical stimulation has been the treatment of epilepsy.[36] In one study, rectangular millisecond pulses were coupled at a 10-Hz rate through electrodes placed over the anterior lobe of the cerebellum and automatically switched between two areas at 8-min intervals. After an exposure of 24 hr/day for 30 days at a level just below that which would induce headaches, a significant reduction in the frequency of epileptic seizures was shown in five out of six chronic patients. Seizure occurrence rates were reduced by factors of from 2 to 10 and, in one case, from 100/month to none.

The current level was not reported for these treatments, but it seems likely that the peak electrode currents were a few milliamps. A supporting study on cat brains showed the effect of the current pulses was first to stimulate and then to suppress the firing rate of cerebral neurons. Over the period of days involved in this study, it seems likely that there was a change in the level of neurotransmitter agents as a result of electrical stimulation. A related use of electricity is electroconvulsive therapy.[37] In this treatment, electrodes may be applied to either the temple or forward part of the scalp and voltages from 70 to 130 V at 60 Hz are applied for from 0.1 to 0.5 sec. The corresponding currents are between 0.2 and 0.6 A. The objective of the current pulses is to induce convulsions which in turn release chemicals that modify brain behavior. The treatment has been most successful in relieving severe depression where the potential for suicide is high. However, because of the possible side effects which may induce partial loss of memory and brain damage, this form of treatment is controversial.[38]

It is clear that we still have much to learn about how and where to apply electric signals to the brain. We have still not reached the stage of technology where we can separately contact individual nerve fibers in a bundle and supply pulses to them in a controlled fashion over extended periods of time. When we succeed, there will be many exciting applications. For example, accident victims may be enabled to regain the use of their limbs.

Another important application of electric fields is the modification of cells. A technique developed by Zimmermann et al.[39-43] utilizes dielectrophoresis, or the forces exerted on an induced dipole moment of the cell in a large electric field gradient to line up the cells into "pearl chains" to be fused. A theory worked out by Schwan[44] and also by Hu[45] shows that particles with a dielectric permittivity ϵ_1 immersed in a fluid of dielectric permittivity ϵ_2 take on the lowest energy configuration in an alternating electric field. At low frequencies this means that the cells form long chains upon the application of field strengths of 1 to 2 $\times 10^4$ V/m. Depending on the cell size and the suspending fluid, the optimum frequency is usually in the range of 10 kHz to 100 MHz; however, we have also successfully used frequencies from 60 Hz to 1 kHz to move bacteria.

If a short ($\approx$ 10 μsec) pulse of approximately 6×10^4 V/m is applied to contacting cells in the pearl chain, a channel is formed between the cells which allows intracellular material to flow back and forth between the two cells. The dynamics of the cell membrane are such that this channel grows until there is no longer a barrier between the cells. The mixing of genetic material from merged cells of different types makes it possible to form new varieties, and considerable effort is currently being expended to generate new types of plants for agriculture.

This high field fusion technique can be applied to a monolayer of cultured cells. Experiments on Swiss mouse 3T3-C2 fibroblast cells in a monolayer culture have resulted in approximately 20% fusion on exposure to fields of 10^5 V/m for 50 μsec.[46]

Both the pulse length and the strength of the applied field are critical to the successful fusion of two or more cells. If the pulse length is too long, permanent damage is done to the cell membranes and cell fusion is not obtained.

For pulses of 10 to 100 μsec applied to a single cell, transient, reversible increases in porosity occur. The membrane voltage drops and the conductivity increases. If the field required to reach breakdown is exceeded by a factor of 2 to 6, or if the pulse length is extended to milliseconds or seconds, permanent mechanical damage is done to the cell membrane. The initial breakdown field strength is inversely proportional to the pulse length and the temperature and varies with cell type. The pore openings which are created by the initial breakdown may last from a few microseconds up to a minute or so, depending on the cell type and the location of the breakdown. These temporary pores in the cell allow for the transport of material into or out of the cell to the surrounding medium.

The ability to fuse cells of a wide variety of types and to inject or extract material from the surrounding medium in a controlled way is adding a new dimension to cell research.

VII. THE EFFECTS OF ELECTRIC FIELDS ON GROWTH

It has been shown that electric fields can be used either to inhibit or to stimulate growth under appropriate conditions. This is currently a field of active research.[47,48] Studies of wound healing, for example, have shown that application of a negatively biased electrode to an ulcer suppressed both healing and infection, while a positive bias stimulated growth. Thus, a method of treatment has evolved in which the wound is first biased negatively until it can be considered clinically noninfected, and then the potential is reversed to stimulate growth. The net result for a set of 14 ulcers was about a fourfold increase in the healing rate. The current levels for these experiments were in the range of 200 μA to 1 mA.[48]

The processes for growth occur over extended periods of time and typically involve currents which are several orders of magnitude smaller than those used to stimulate nerve or muscle cells.

Naturally growing plant cells typically show an inward flowing current to the growth region of tens of picoamps, and current densities in the range 0.01 to 1 A/m². In lily pollen grain, a relatively steady transcellular current begins to flow hours before germination or the initiation of visible growth.[49] The interior of the growth area is negatively biased, which implies an inward flow of K^+ and other positive ions. In the case of barley roots the dominant current has been shown to be H^+.[50] This was shown by raising the pH from 5.5 to 7.1 and by insensitivity to changes in Na^+, K^+, and Ca^{++} and Cl^- concentrations. On top of the gross inward current are superimposed microcurrent loops through each growing root hair. These microcurrents enter the growing root hair tip with current densities of about 2×10^{-2} A/m² and correlate well with the tip growth. Nongrowing hairs did not show the inward currents. The current return path for both the macro- and microcurrents includes a pathway through the xylem, and possibly through the epidermal layer between the zones of elongation and the root hair zone. There are also apparently alternating bands of alkalinity and acidity near growing barley roots with strong steady currents entering the alkaline zones. Peak current densities are about 8×10^{-2} A/m².[51] In *Nitella* cells the elongation or growth occurs in the acidic region or regions of current exit.[51] Growth in these regions was stimulated by external acidification and inhibited by alkalinization. This pattern is contrary to the general rule of cell growth at current leaks or points of current entry.

Ca^{++} currents appear to precede visible tip growth in various plant cells. There is direct evidence with radioactive Ca^{++} 45 for leaks in both fucoid eggs and in lily pollen.[51] In developing plantlets of the water mold *Blastocladiella*[51] Ca^{++}, tip currents of 10^{-2} A/m² have been observed. These currents are blocked by La^{3+}, and Ca^{++} has been shown to be both necessary and sufficient to maintain the current. The La^{3+} also blocks spore growth.

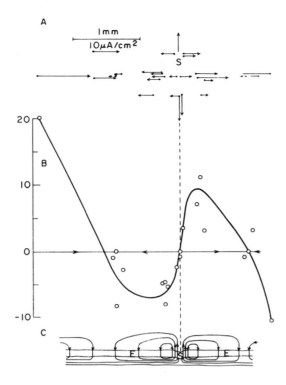

FIGURE 3.   The pattern of current that flows across the streak of the embryo, measured in a horizontal plane 0.2 mm above the epiblast. (A) Map of a scan across the middle of the streak together with all measurements made nearby. (B) Graph of the main scan. As the arrowheads on the abscissa indicate, positive values indicate current going right and negative ones show current going left. (C) Diagram of inferred current pattern in a cross-section during the streak. The thickness of the embryo is exaggerated.

In animal egg cells, between fertilization and the first cleavage, steady currents are found to enter at the animal pole and leave at the vegetal pole. In frog and sea urchin eggs, the current first enters the cell near the area where the cell eventually divides. Shortly afterwards, it reverses to flow out of the cleavage region. These currents may be accompanied by voltage drops across the interior of the cell of up to 10 mV. In the case of root cells, potential differences of up to 60 mV have been measured along the surface of the cell corresponding to currents of 0.1 to 0.4 μa and current densities of 0.1 A/m^2.

On a larger scale, strong electrical currents have been shown to flow out of the primitive streak of growing chick embryos.[52] Current densities as high as 1 A/m^2 leave the streak and return over a broader area. A current of about 10^{-4} A flows out of the streak. The component of the current flowing along the streak reverses direction about 2×10^{-4} m behind Hensen's node with current flowing away from this epicenter in all directions. Figures 3 and 4 illustrate these currents. These currents are believed to be strong enough to help control the organization of the embryo growth.

The application of external fields to growing cells can modify both the growth rate and the orientation of the growth. Patel and Poo[53] have shown that fields in the range of 10^{-3} to 10^{-1} V/m have a large effect on the growth of neurite from single, dissociated *Xenopus* neurons in culture. Upon exposure to fields of 2.5×10^{-2} V/m for 6 hr, the growth rate

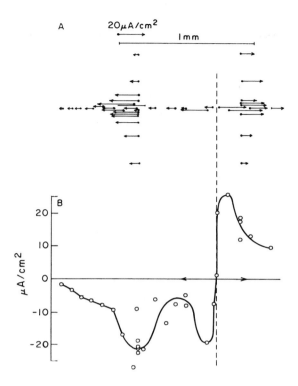

FIGURE 4.   The pattern of current that flows along the streak
of the embryo, measured in a plane 0.2 mm above the epiblast.
(A) Map of a scan along the streak together with all measurements made nearby; (B) graph of the main scan. Positive values
indicate current going right (toward the front end of the embryo)
and negative ones show current going left. The dashed line goes
through the central reversal point, or epicenter.

toward the cathode becomes approximately double that toward the anode. Additional neurites
which were initially growing at right angles to the field bend to grow toward the cathode.
These effects are reversible for exposures of up to at least 2 hr with the asymmetry in the
growth disappearing several hours after the field is turned off. No asymmetry in the growth
is seen with application of alternating fields. The asymmetry in growth seems to be associated
with the concentration of receptor for concanavalin A (Con A) under the influence of the
field toward the cathode, and can be blocked by binding tetramethylrodamine (R-Con A)
to the neurons prior to the application of the electric field. This binding of R-Con A
presumably cross-links and immobilizes the Con A receptors on the cell surface. The overall
implication is that the underlying mechanism for the field-induced orientation of neurite
growth toward the cathode is the influence of the electric field upon the distribution of
growth-controlling glycoproteins.

A similar result of directed nerve growth toward the cathode and suppression on the anode
side is reported for explants of embryonic chick dorsal root ganglia exposed to electric fields
of 50 V/m for 3 to 29 hr.[49]

The work on nerve regrowth also includes a study of enhanced spinal cord regeneration
in lamprey when electric fields of approximately 10 V/m are imposed across the neural
break.[54] With a total delivered current of about 10 μA for a period of 5 or 6 days and current
densities of 1 A/m², about 73% of the spinal cords would propagate action potentials in both

directions across the lesions. This compared with 23% for the untreated controls. Additionally, microscopic examination showed twice as many swollen tips and tips with neurites for the spinal cord breaks exposed to the electric fields as those which were left to heal on their own.

The mechanism by which the regeneration takes place is not understood but it is to be noted that the imposed current is in the opposite direction to the natural injury current. The injury current density starts out at about 5 A/m^2 and drops to 4 or 5 $\times$ 10^{-2} A/m^2 over the first 36 hr continues at about this level for at least 4 days.

In another example, Yamamoto[55] has shown that unfertilized eggs of the medaka fish can be activated by a field of 200 V/m applied for 20 to 60 sec. Activation occurs at either electrode and twice as rapidly at the anode. This field should produce a voltage drop of 360 mV across the 1.2-mm egg. It is likely that the nonlinear characteristics of the cell membrane lead to a hyperpolarization of the cell and a large influx of Ca^{++}, which is known to be a central event in the activation of these eggs.[49]

In a second example, when the sign is opposite to that for the *Xenopus* neurites, eggs from the algae *Pelvetia* normally tend to grow on the side nearest a positive electrode under the influence of fields of 30 to 60 V/m, corresponding to a voltage drop across the cell of 3 to 6 mV.[49] These cells also grow toward the highest concentration in a Ca^{++} gradient, and it is likely that externally applied fields lead to an increased flux of Ca^{++} ions.

In summary, there are strong natural ionic currents associated with cell growth and there is evidence that these currents may be driven by a separation of the leaks and pumps within the cell membranes.[47] The voltage gradients (2.5 V/m) developed by these currents in a semi-infinite surrounding medium would seem to be too small to be important in organizing the cell growth. However, when these currents are confined to narrow intercellular spaces, voltage drops from 0.1 to several millivolt may be generated over the length of a cell and these voltages have been shown to be large enough to rearrange channels in membranes and thus affect cell growth. It is also likely that there is positive feedback so that the currents grow in time up to some limiting value, and current generated by one cell contributes to inducing the current in adjacent cells. The evidence for the effects of naturally generated currents on cell growth and patterning has been reviewed in detail by Jaffe.[56,57]

Many attempts to stimulate the growth of multicelled plants and animals with electric fields have failed to produce useful results.[58] This may be, in large part, because the cells have requirements for polarization which vary in angle so that currents aiding one cell inhibit the next. However, one situation where this is clearly not the case is in the growth of bone. Bone crystallites are ordered and additionally, bone is piezoelectric.[59] The normal loading and unloading of a bone generates electric field pulses. Voltages from 10 μV to 4 mV have been generated by applying loads of between 10 and 90 N with a Bytrex load cell in bovine cartilage. The lowest voltages were produced by tracheal cartilage (10 to 400 μV) and the highest (1 to 4 mV) from articular cartilage and meniscus. Both pulsed and DC fields have been shown to stimulate bone growth. Growth occurs from a negatively charged bone fragment toward the positive electrode. In bone growth from a single electrode, typical threshold currents are 3 to 50 μA with an optimum between 10 to 25 μA. Currents in excess of 50 μA caused deleterious effects.[60]

A better approach than implanting electrodes for the regrowth of bone across a fracture seems to be the generation of electric fields by induction.[60] A pair of coils to induce fields of 0.2 to 20 V/m across fractures has been used successfully to stimulate bone growth. Pulse lengths of 0.15 msec and repetition rates of 65 Hz were used in experiments on dogs. Growth rate improvements by a factor of 2 are not uncommon, and in some clinical cases, bones which would not regrow naturally have been repaired. In these cases, typical exposures extend over several weeks to a few months (see also Section II).

VII. DISCUSSION AND SUMMARY

In this review, we have examined some of the mechanisms by which nontime-varying or slowly pulsed electric fields affect biological systems. An attempt has been made to list a few typical values of electric field strength and current densities that are known to affect transport properties so that the reader can make his or her own estimates of the probability that a given exposure will be significant. In very rough terms, both the medical and safety literature indicate threshold levels of concern for total currents leaving and entering the body above a few milliamperes or current densities of 10^{-1} A/m^2, and electric field strengths inside the body in excess of several volts per meter for short periods of time, in the order of seconds. For long exposures, on the order of days or weeks, the level of concern drops to 10^{-2} A/m^2 and about 10 V/m or less. Our understanding of the details of how changes in electric fields affect the transport of various ions and molecules across membranes and along their surfaces, and of the various feedback systems that control their biological function is still incomplete. Thus, it becomes very difficult to set safety standards. Standards which are too restrictive will result in very large costs for unnecessary insulation and limit the use of electrical instruments and appliances. Standards set too high may permit unnecessary biological damage. At the present time, our understanding is most incomplete with respect to the effects of long-term, low-level fields. This is not surprising, since the necessary experiments are the most difficult to do and the most subject to uncontrolled changes in the biological subject and the environment.

ACKNOWLEDGMENTS

The author wishes to express appreciation to Maria Stuchly, Ross Adey, Mike Marron, and Charles Polk for their many helpful comments and suggestions. He also wishes to express his appreciation to the Office of Naval Research under Contract N00014-81-K-0387 for financial support of his work in this area.

REFERENCES

1. **Barnes, F. S. and Hu, C. J.**, Nonlinear interactions of electromagnetic waves with biological materials, in *Nonlinear Electromagnetics,* Uslenghi, P. L. L., Ed., Academic Press, New York, 1980, 391.
2. **Singer and Nicolson,** The fluid mosaic model of the structure of cell membranes, *Science,* 175, 720, 1972.
3. **Poo, M.,** In situ electrophoresis of membrane components, in *Annual Review of Biophysics and Bioengineering,* Mullins, L. J., Ed., Academic Press, New York, 1981, 245.
4. **Bler, Ed.,** *Electrophoresis: Theory, Methods, and Applications,* Vol. 2, Academic Press, New York, 1967.
5. **Pohl, H. A.,** Dielectrophoresis, the behavior of neutral matter, in *Nonuniform Electric Fields,* Cambridge University Press, Cambridge, 1978.
6. **Rand, E. P.,** Interacting phospholipids bilayers: measured forces and induced structural changes, in *Annual Review of Biophysics and Bioengineering,* Vol. 10, Mullins, L. J., Ed., Academic Press, New York, 1981, 277.
7. **Pethig, R.,** *Dielectric and Electronic Properties of Biological Materials,* John Wiley & Sons, New York, 1979, 184.
8. **Parsegian, J. L.,** Long range physical forces in the biological milieu, in *Annual Review of Biophysics and Bioengineering,* Mullins, L. J., Ed., Academic Press, New York, 1973, 221.
9. **Lifshitz, E. M.,** *Zh. Eksp. Teor. Fiz.,* 29, 95, 1955; *Sov. Phys. JETP,* 2, 73, 1956.
10. **Barnes, F., Ginley, R., and Shulls, W.,** AC electric field effects on bacteria, presented at BEMS 3rd Ann. Conf., Washington, D.C., 1981.
11. **MacGregor, R. J. and Lewis, E. R.,** *Neural Modeling,* Plenum Press, New York, 1977.
12. **Epstein, B. R. and Foster, K. R.,** Anisotropy in the dielectric properties of skeletal muscle, *Med. Biol. Eng. Comp.,* 21, 1983.

13. **McCleave, J. D., Romell, A. S., and Cathcart,** Weak electric and magnetic fields in fish orientation, in *Orientation: Sensory Basis,* Adler, H. E., Ed., New York Academy of Sciences, 188, 270, 1971.

14. **Ransom, B. P. and Eyring, H.,** Membrane permeability and electrical potential, in *Ion Transport Across Membranes,* Clarke, H. J. and Nachmansohn, D., Eds., Academic Press, New York, 1954, 103.

15. **Triffet, T. and Green, H. S.,** Information and energy flow in a simple nervous system, *J. Theor. Biol.,* 86, 3, 1980.

16. **Vaccaro, S. R. and Green, H. S.,** Ionic processes in excitable membranes, *J. Theor. Biol.,* 81, 771, 1979.

17. **Eckert, R. and Ewald, D.,** Residual calcium ions depress activation of calcium-dependent current, *Science,* 216, 730, 1982.

18. **Baumann, G. and Easton, G.,** Modeling state-dependent sodium conductance data by memoryless random process, *Math Biosci.,* 60, 265, 1982.

19. **Schauf, C. L. and Baumann, G.,** Experimental evidence consistent with aggregation kinetics in sodium current of myxicola giant axons, *Biophys. J.,* 35, 707, 1981.

20. **Baumann, G. and Eastman, G. S.,** Charge immobilization linked to inactivation in the aggregation model of channel gating, *J. Theor. Biol.,* 99, 249, 1982.

21. **Poo, M.,** Rapid lateral diffusion of functional ACh receptors in embryonic muscle cell membrane, *Nature,* 295, 332, 1982.

22. **Poo, M. M., Poo, W. J., and Lam, J. W.,** Laterial electrophoresis and diffusion of concanavalin A receptors in the membrane of embryonic muscle cell, *J. Cell Biol.,* 76, 483, 1978.

23. **Hu, C. L. and Barnes, F. S.,** The thermal-chemical damage in biological materials under laser irradiation, *IEEE Trans. Biomed. Eng.,* 17, 220, 1970.

24. **Shuyrou, V. L. Zhodan, G. G., and Akorv, I. G.,** Effects of 330 mHz radiofrequency, in *Int. Biol. Physics,* USSR Academy of Science, Pushchino Moscow Region U.S.S.R., personal communication.

25. **Trautman, E. D. and Newbower, R. S.,** A practical analysis of the electrical conductivity of blood, *IEEE Trans. Biomed. Eng.,* 30, 141, 1983.

26. **Johnson, F. H., Eyring, H., and Stover, B. J.,** *The Theory of Rate Processes in Biology and Medicine,* John Wiley & Sons, New York, 1974.

27. **Chalker, R.,** The Effect of Microwave Absorption and Associated Temperature Dynamics on Nerve Cell Activity in *Aplysia,* M.S. thesis, University of Colorado, Boulder, 1982.

28. **Barnes, F. S.,** Cell membrane temperature rate sensitivity predicted from the Nernst equation, *Bioelectromagnetics,* in press.

29. **Adey, G., McNaughton, B. L., and Wachtel, H.,** A system for recording microwave effects on isolated mammalian brain slices, presented at BEMS 5th Annu. Conf., Boulder, Colo., 1983.

30. **Polk, C.,** Sources, propagation amplitude and temporal variations of extremely low frequency (0—100 Hz) electromagnetic fields, in *Biological and Clinical Effects of Low-Frequency Magnetic and Electric Fields,* Llaurado, J. G., Sances, A., Jr., and Battocletti, J. H., Eds., Charles C Thomas, Springfield, Ill., 1974, 21.

31. **Krueger, A. and Reed, E.,** Biological impact of small air ions, *Science,* 193, 1209, 1976.

32. **Peckham, P. and Martimer, J. T.,** Restoration of hand function in the quadriplegic through electrical stimulation, in *Biomedical Engineering and Instrumentation,* Vol. 3, Hambrech, F. T. and Rewick, J. R., Eds., Marcel Dekker, New York, 1977.

33. **Schaldach, M., Furman, S., Hein, F., and Thull, R., Eds.,** *Advances in Pacemaker Technology,* Springer-Verlag, Basel, 1975.

34. **Tacker, W. A., Jr. and Geddes, L. A.,** Fibrillation and defibrillation, in *Electrical Defibrillation,* CRC Press, Boca Raton, Fla., 1980, chap. 1.

35. **Tacker, W. A., Jr., Galioto, F. M., Jr., Giuliani, E., Geddes, L. A., and McNamara, D. G.,** Energy dosage for human transchest electrical ventricular defibrillation, *N. Engl. J. Med.,* 214, 1974.

36. **Gilman, S., Daully, G., Tennyson, V., Kramzer, L. T., Defenlini, R., and Carroll, J.,** Clinical morphological, biochemical and physiological effects cerebellar stimulation, in *Biomedical Engineering and Instrumentation,* Vol. 3, Hambrech, F. T. and Rewick, J. R., Eds., Marcel Dekker, New York, 1977.

37. **Maletzky, B. M.,** *Multiple-Monitored Electroconvulsive Therapy,* CRC Press, Boca Raton, Fla., 1981.

38. **Breggin, P. R.,** *Electroshock — Its Brain-Disabling Effects,* Spirnger-Verlag, Basel, 1979.

39. **Benz, R. and Zimmermann, U.,** Pulse length dependence of the electrical breakdown in lipid bilayer membranes, *Biochim. Biophys. Acta,* 597, 637, 1980.

40. **Vienken, J., Ganzer, R., Hampp, R., and Zimmermann, U.,** Electric field induced fusion of isolated vacuoles and protoplasts and metabolic provenience, *Physiol. Plant,* 52, 153, 1981.

41. **Vienken, J. and Zimmermann, U.,** Electric field-induced fusion: electro-hydraulic procedure for production of hetrokaryon cells in high yield, *FEBS Lett.,* 137, 11, 1982.

42. **Zimmermann, U.,** Electric field-mediated fusion and related electrical phenomena, *Biochim. Biophys. Acta ,* 694, 227, 1982.

43. **Zimmermann, U. and Scheurich, P.,** High frequency fusion of plant protoplasts by electric fields, *Planta,* 151, 26, 1981.
44. **Schwan, H. P. and Sher, L. D.,** Alternating current field-induced forces and their biological implications, *J. Electrochem. Soc.,* 116, 170, 1969.
45. **Hu, C. J. and Barnes, F. S.,** A simplified theory of pearl chain effects, *Radiat. Environ. Biophys.,* 12, 1975.
46. **Teissie, J., Knutson, V. P., Tsong, T. Y., and Lane, M. D.,** Electric pulse-induced fusion of 3T3 cells in monolayer culture, *Science,* 216, 537, 1982.
47. **Jaffe, L. F.,** Control of development of ionic currents, in *Biological Structures and Coupled Flows,* Optlaka, A. and Balaban, M., Eds., Academic Press, New York, 1983, 445.
48. **Rowley, B., McKenna, J. M., Chase, G. R., and Wolcott, L. E.,** The influence of electrical current on an infecting microorganism in wounds, in *Electrically Mediated Growth Mechanisms in Living Systems,* Vol. 238, Liboff, A. R. and Rinaldi, R. A., Eds., New York Academy of Sciences, 1974.
49. **Jaffe, L. F. and Nuccitelli, R.,** Electrical controls of development, in *Annual Review of Biophysics and Bioengineering,* Vol. 6, Academic Press, New York, 1977, 446.
50. **Weisenseel, M. H., Dorn, A., and Jaffee, L. F.,** *Plant Physiology,* in press.
51. **Jaffe, L. F.,** Control of plant development by steady ionic currents, in *Plant Membrane Transport: Current Conceptual Issues,* Spanswick, R. M., Lucas, W. J., and Dainty, J., Eds., Elsevier/North-Holland, Amsterdam, 1980, 381.
52. **Jaffe, L. F.,** Strong electrical currents leave the primitive streak of chick embryos, *Science,* 206, 569, 1979.
53. **Patel, N. and Poo, M.,** Orientation of neurite growth by extracellular electric fields, *J. Neurosci.,* 2, 483, 1982.
54. **Borgens, R. B., Roederer, E., and Cohen, M. J.,** Enhanced spinal cord regeneration in lamprey by applied electric fields, *Science,* 213, 611, 1981.
55. **Yamamoto, T.,** *Cytologia,* 14, 219, 1947.
56. **Jaffe, L. F.,** The role of ionic currents in establishing developmental pattern, *Philos. Trans. R. Soc. London, Ser.* B, 295, 553, 1981.
57. **Jaffe, L. F.,** Developmental currents, voltages, and gradients, in *Development Order: Its Origin and Regulation,* Alan R. Liss, New York, 1982, 183.
58. **Liboff, A. R. and Rinaldi, R. A., Eds.,** *Electrically Mediated Growth Mechanisms in Living Systems,* Vol. 238, New York Academy of Sciences, 1974.
59. **Bassett, C. A. L., Pawluk, R. J., and Pilla, A. A.,** Acceleration of fracture repair by electromagnetic fields and a surgically non invasive method, in *Electrically Mediated Growth Mechanisms in Living Systems,* Vol. 238, Liboff, A. R. and Rinaldi, R. A., Eds., New York Academy of Sciences, 1974, 242.
60. **Watson, J.,** The electrical stimulation of bone healing, *Proc. IEEE,* 67, 1339, 1979.

Chapter 2

EXTREMELY LOW FREQUENCY (ELF) AND VERY LOW FREQUENCY ELECTRIC FIELDS: RECTIFICATION, FREQUENCY SENSITIVITY, NOISE, AND RELATED PHENOMENA

Frank S. Barnes

TABLE OF CONTENTS

I. INTRODUCTION

Many of the physical phenomena at ELF frequencies are the same as those described in the previous chapter for direct current (DC). For example, the forces acting on neutral particles in a nonuniform ELF electric field are proportional to E^2 (see Part II, Chapter 1.) The effects of heating due to current flow are proportional to the current density squared and thus also are the same for both alternating current (AC) and DC. However, an important difference is that AC fields do not tend to lead to ion accumulations as long as the current densities are low enough to allow the medium to be treated as a linear system. This requirement for linearity is satisfied for biological fluids at low field levels and if average currents over large volumes of tissue are considered. However, it is rarely satisfied for tissue if we examine the local current crossing individual cell membranes. Thus, while it is both necessary and convenient to characterize biological materials by their linear conductivity σ and dielectric permittivity ϵ at low current densities, these coefficients are not adequate to predict possible important biological effects that involve the nonlinear properties of biological materials and fluids.

The application of an AC electric field to nonlinear systems which can be described by either a nonlinear resistance or capacitance leads to at least partial rectification of the input signal and the generation of harmonics. If two or more signal frequencies are applied, it also leads to frequency mixing of the form

$$f_0 = \pm mf_1 \pm nf_2 \tag{1}$$

where f_0 is the ouptut frequency, f_1 and f_2 are input frequencies, and m and n are integers. The rectified component of the AC current can in turn lead to ion accumulation at interfaces, which results in changes of ion concentration. These changes in ion concentration in turn can affect biological function. The partial rectification at the cell membrane interfaces is probably the most significant addition to the physics of interaction between ELF fields and biological material, beyond the effects discussed in the section on DC fields.

Another important additional effect is the dependence of the dielectric constant on frequency. Thus the electrophoretic or dielectrophoretic forces become both size- and frequency-dependent.

A third possibly important additional effect is excitation of frequency-sensitive biological systems in a resonant manner. By driving such systems near their resonant frequency, we may change the effective amplitude of the stimulating signal and change the frequency of the nerve cells firing.

In this chapter we will review the rectification process at the cell membrane in some detail. Additionally we will show that cell nonlinearities lead to frequency-dependent effects such as injection phase locking of pacemaker cells. We will also briefly examine some problems associated with safety, exposure of cells to very low ELF fields, and application of large ELF fields to biological systems.

II. RECTIFICATION BY CELL MEMBRANES

For many passive cell membranes an approximate relation for the transmembrane current can be derived from the Nernst equation[1]

$$I = I_0 \left(e^{\frac{V_M}{\eta V_T}} - 1 \right) \tag{2}$$

where I_0 and η are constants, and V_M is the voltage across the membrane. V_T is the voltage equivalent temperature and is given by kT/q where k is the Boltzmann constant, T is the absolute temperature, and q is the charge on the electron. If we apply an AC signal across the membrane of the form $V_M = V_0 + V_1\cos\omega t$, the resulting current can be approximated for small values of V_M (i.e., $qV_M \ll \eta kT$) by a Taylor series yielding

$$I = \frac{I_0}{\eta V_T} \left(V_0 + \frac{V_0^2}{2\eta V_T} + \frac{1}{4}\frac{V_1^2}{\eta V_T} + V_1\cos\omega t + V_0 V_1\cos\omega t + \frac{V_1^2}{4\eta V_T}\cos2\omega t \ldots \right) \quad (3)$$

It is to be noted that the third term in the expression is the first approximation to the fraction of the applied AC voltage V_1, which yields a DC current component ΔI

$$\Delta I = \frac{I_0}{4}\left(\frac{V_1}{\eta V_T}\right)^2 \quad (4)$$

or an offset voltage V_{DC} given by

$$V_{DC} \approx \frac{I_0}{4}\left(\frac{V_1}{\eta V_T}\right)^2 R_m \quad (5)$$

where R_m is the membrane impedance. This predicted voltage offset for an applied AC current has been measured by Montaigne and Pickard.[2] In their experiments an AC signal was applied to a large plant cell by a strip line, and the measured voltage shift was obtained through microelectrodes located outside the applied AC fields. For an applied AC field of about 0.2 V they measured a DC offset of 1 to 2 × 10^{-4} V. For frequencies above 2.5 kHz the effects of the membrane capacitance must be taken into account and the effective driving voltage is reduced to

$$(V_1)_{eff} = \frac{\sqrt{2}a\sigma_e E_{1\,rms}}{[(\sigma_e + aG)^2 + (a\omega C)^2]^{1/2}} \quad (6)$$

where a is the cell radius, σ_e is the conductivity of the medium, $E_{1\,rms}$ is the electric field strength in the medium surrounding the cell, G is the membrane conductance per unit area, ω is the frequency, and C is the membrane capacitance per unit area.[3] This leads to the usual RC roll-off in the measured DC offset with increasing frequency. It is to be noted that the DC effect gets still smaller at higher frequencies (above 1 MHz) because of transit time limitations for ion flow across the membrane.[4]

Rectification has also been demonstrated in thin lipid membranes.[5] In these systems, both the conductivity of the membrane and the ion concentration differences across it can be controlled. The Nernst equation was shown to apply to the I vs. V curve over a range of voltages from −60 to +40 mV. Depending on the ion concentration and membrane doping, the values of η ranged from 1 to 0.25.

A different treatment of the nonlinear response of passive cell membranes to an applied AC field has been carried out by Francechetti and Pinto[6] and by Casaleggio et al.[7] Both these groups have expanded the Nernst equation in a Volterra series which takes into account memory of the preceding state of the cell. They have also treated the cell in spherical rather than plane geometry. The inclusion of a spherical cell requires that the total current into and out of the cell be equal to zero and thus loops are formed circulating through the cell membrane (see Figure 1.) All of the theoretical treatments predict a DC component which varies as the square of the input signal V_i.

Cain[8] has considered the effects of an AC field on nonlinearities of the nerve cell by

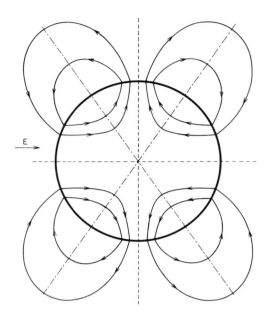

FIGURE 1.   Induced DC current distribution in a spherical cell.[7]

numerical analysis of the Hodgkin Huxley equation.* He applied a voltage $V_m = V_0 + V_1$ cosωt [u(t) − u(t − τ)] across the membrane where u(t) and u(t − τ) are unit step functions which define an AC pulse of length τ. For the case when AC frequency is large compared to the pulse length, if a 7-mV depolarizing pulse is also applied to the membrane, the action potential is obtained as shown in Figure 2. He has assumed coefficients appropriate to the giant squid axon. Increasing V_1 first delays and then suppresses the action potential. If no depolarizing pulse is applied, the predicted changes in g_{NA} and g_K and the deviation V from the resting potential are shown in Figure 3 for a 10-msec AC pulse with $V_i = 25$ mV. Note applied AC frequency is assumed high enough not to be resolved in these figures. The predictions for the changes in the Na^+ and K^+ conductances g_{NA} and g_K are shown in Figure 3 for a 10-msec pulse with $V_1 = 25$ mV. From these results it is clear that low frequency AC signals can induce substantial changes in the operating characteristics of nerve cells at moderate to high levels of applied voltage. Although the appropriate coefficients were not measured in order to make a direct comparison between theory and experiments, Wachtel's[10] results on *Aplysia* at frequencies above the lock-in range would appear to support Cain's theoretical predictions.

Wachtel[10] has made a series of measurements which demonstrate the nonlinear characteristics of pacemaker cells from *Aplysia*. First he measured the current input through a microelectrode which changed the firing rate of the cell. The current threshold for a minimum detectable change was approximately 6×10^{-10} A at frequencies between 0.8 and 1 Hz (see Figure 4). The natural firing rate for this cell is about 0.8 Hz, and an increasing current is required to synchronize the cell to the injected signal as the frequency deviates from the natural firing rate. A theory for injection-locking of electronic oscillators[11] predicts that the signal required for locking an oscillator to an external signal increases linearly as the difference between the two frequencies $\Delta\omega$ increases. The signal required for lock-in according to this theory is given by

---

* Bisceglia and Pinto[9] have applied a Volterra series expansion to the Hodgkin Huxley equations. This approach gives an alternate method to Cain's, of computing the current shifts resulting from applied AC signals.

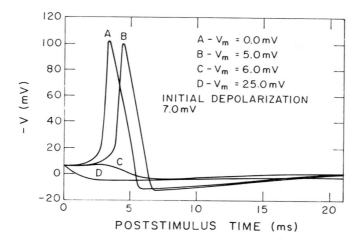

FIGURE 2. Computed membrane action potentials in response to an initial membrane depolarization of 7 mV for different values of V_m. Curves are solutions to the Hodgkin-Huxley equations. (From Cain, C. A., *Bioelectromagnetics*, 2, 23, 1981. With permission.)

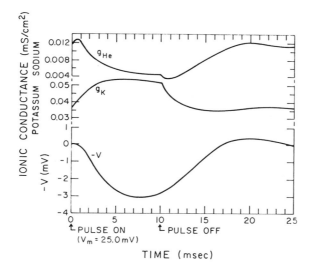

FIGURE 3. Response of model axon to a pulsed oscillating component of membrane electric field (10-msec pulse, $V_m = 25$ mV). The membrane potential and the sodium and potassium conductances are shown. These curves are solutions of the Hodgkin-Huxley equations. (Note: 1 mS/cm^2 = 10 S/m^2.) (From Cain, C. A., *Bioelectromagnetics*, 2, 23, 1981. With permission.)

$$I_1 \simeq |A\Delta\omega_0| \, I \qquad (7)$$

where I_1 is the injected signal current and I is the peak unperturbed oscillator current. A = $\dfrac{d\phi}{d\omega}$ is the rate of change of phase with respect to frequency in the unperturbed oscillator. $\Delta\omega_0$ is equal to the difference between the frequency of the free running oscillator and the injected signal. This expression is applicable as long as $\Delta\omega_0 \leqslant \dfrac{2\pi}{\tau}$ where τ is the time

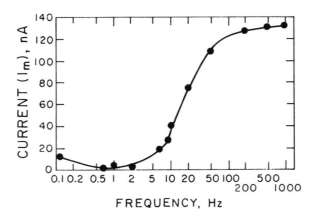

FIGURE 4. Intracellular (transmembrane) currents I_m (in nA) needed at different frequencies to produce firing-pattern changes (in pacemaker neuron). Note the detectable changes take on different forms at different frequencies. (From Wachtel, H., in *Proc. 18th Annu. Hanford Life Sci. Symp.*, Technical Information Center, Department of Energy, Richland, Wash., 1978, 132. With permission.)

constant for adjusting the gain of the circuit. The time constant τ for the *Aplysia* cells varied between $\tau = 0.1$ to 0.5 sec and this corresponds to a maximum measured lock in frequency of about 10 Hz. The results shown in Figure 4 show the threshold for 1 to 1 locking up to about 2 Hz. In the range from 2 to 10 Hz Wachtel observed a lower threshold for subharmonic locking than 1 to 1 locking. At frequencies above 80 Hz, he observed a constant shift in the firing rate of the neuron in response to the injected transmembrane AC. The natural firing rate would be restored by also injecting a transmembrane DC equal in amplitude to about 1% of the peak-to-peak value of the AC current. This DC current was in the depolarizing direction, making the exterior of the cell more negative with respect to the cell cytoplasm to increase the firing rate (i.e., to restore it to its natural value). Apparently the applied transmembrane AC current was partially rectified so as to hyperpolarize the membrane (making the interior of the cell more negative with respect to the external fluid). The details of how the applied field modifies the ion flow are only partially understood, but one characteristic is an increase in the conductivity of K^+ which increases its flow out of the cell. Wachtel[10] also injected low frequency currents into the seawater surrounding the cell preparation through external electrodes. In this case, the minimum current densities flowing in the vicinity of the cell preparation for injection-locking were estimated to be about 10^{-2} A/m² and there was about a 10 to 1 variation between the maximum and minimum sensitivity for changes in angle between the applied field and the cells. At frequencies above 100 Hz a minimum of about 0.35 A/m² was necessary to get a detectable change in firing rate.

Typical mammalian neurons are a factor of 10 smaller than the *Aplysia* cells and the membrane time constants (i.e., membrane resting resistance times membrane capacitance) are in the order of 10 msec or a factor of 100 faster. Thus, we would expect similar lock-in phenomena and rectification effects to occur at proportionately higher frequencies in man.

An increase in the sensitivity to electromagnetic fields has also been shown in isolated frog hearts for signals which approach the natural resonant frequency or firing rate.[12] In these experiments the heart firing rate was shown to increase as much as 30% when a signal in the vicinity of 10 to 20 V/m was applied through the Ringer's solution to the isolated frog hearts at frequency between 0.5 and 1 Hz. The natural firing rate of these excised

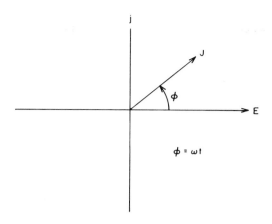

FIGURE 5. Steady state vector characterization of electric fields E vs. current density J in polar form. ϕ is the phase angle between the sinusoidal electric field E and the resulting current density J.

hearts started out at approximately 1 Hz and dropped to about 0.5 Hz over a period of 2 hr where they remained stable for at least 5 hrs. To get a 30% increase in firing rate at 60 Hz, it was necessary to apply field strengths of 60 to 80 V/m. Thus, we have additional evidence that electric fields with repetition rates near the natural biological signaling frequencies are more likely to induce changes than those of higher frequencies, and that signal strengths required for a given shift increase approximately linearly up to some cut-off as shown in Figure 4.

In addition to the nonlinear conductances associated with the Na^+ and K^+ currents, membranes also exhibit nonlinear (i.e., potential-dependent) and frequency-dependent capacitances and inductances. It is sometimes useful to think of these effects in terms of a phasor diagram as shown in Figure 5, where the electric field vector E is rotating at velocity ω and ϕ is the phase angle between E and the current density J. If there is, for example, a fixed time delay between the field activation of a current gate and the current flow, then depending on the frequency, J may be in any of the four quadrants and appear capacitive or inductive or even present a negative resistance to an external driving source.

Nonlinear inductive effects seem to be associated with the time delay for the onset of the K^+ currents under excitation in a typical excitable membrane, and have been studied in the giant squid axon.[13] The nonlinear capacitive effects are difficult to measure at frequencies below a few kilohertz. Extra care needs to be exercised to minimize the series resistance and the end effects of the wire being used to measure the capacitance or inductance. Additionally, corrections must be made in the calculations of the membrane capacitance to take into account the appropriate variations in the frequency response which these terms introduce. However, when this is done, it can be shown that the membrane capacitance has both frequency- and voltage-dependent terms. The capacitance of giant squid axons is shown in Figures 6 and 7 as a function of frequency and membrane voltage.

Variation of the capacitance of these membranes with frequency and amplitude differs from that of a simple bilipid membrane which has nearly constant capacitance. The variation appears to be associated with changes in the conformation of the proteins associated with the Na^+ conductance channels. Nonlinearity in conductance and capacitance can be induced into a bilipid membrane by the addition of Alamethicin. The nonlinear inductance or capacitance may also generate both sum and different frequencies if two signals are applied. For the case of the single signal, a DC term is added to the current density which is proportional to membrane potential and the square of the applied AC signal.[14] Because of

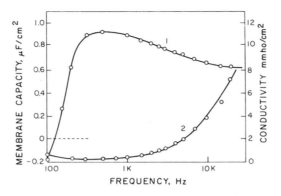

FIGURE 6. Membrane capacitance (Curve 1) and conductivity (Curve 2) of squid giant axon at various frequencies. Note anomalous behavior at low frequencies. (Note: 1 μF/cm² = 0.01 F/m².) (From Takashima, S., *Biophys. Chem.*, 11, 447, 1980. With permission.)

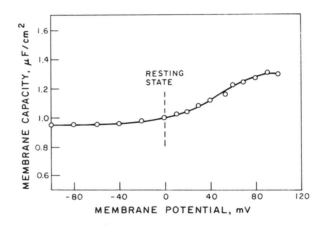

FIGURE 7. Membrane capacitance of squid giant axon at various membrane potentials. Membrane potential was shifted by injecting currents. The abscissa shows actual potential across the membrane in mV. (From Takashima, S., *Biophys. Chem.*, 11, 447, 1980. With permission.)

their small size, the effects thus far observed, due to the nonlinear membrane capacitance, appear likely to be more important in providing an understanding of the possible gating mechanism in membranes than as a mechanism for introducing rectification.

III. Ca^{++} AND OTHER "WINDOWING" EFFECTS

The lowest level changes in membrane properties reported so far are changes in the efflux of Ca^{++} from the surface of chick brains upon the application of electric fields in the range of 5 to 7.5 V/m across an air gap at 16 Hz.[15-1] Blackman et al.[16,17] report increases in the Ca^{++} efflux with peaks at applied fields of 6 and 40 V/m and no enhancement in the ranges 1 to 2, 10 to 30, and 60 to 70 V/m. It is to be noted that these are external fields in the exposure system and the fields at the cell surface are estimated to be much smaller ($\simeq 10^{-5}$

V/m). Bawin and Adey[15] report an approximate 15% decrease in Ca^{++} efflux for applied fields of 10 V/m at 16 Hz. These effects are also frequency-sensitive with peak increases occurring at multiples of about 15 Hz. The details of how these weak fields perturb the brain surface are not yet understood. However, several features of the results are important. First, the estimated fields at the cell membranes are very low (10^{-5} V/m). Second, the changes in the Ca^{++} efflux occur only over narrow ranges of frequency and field strength.

Field strength windowing effects have also been measured for the increase in steroidogenic response of rat adrenal cortical tissue to adrenocorticotropic hormone (ACTH) in rats at 60 Hz.[18] The peak response occurred at an exposure system field of 10 kV/m and an estimated internal tissue field 1.7×10^{-4} V/m and not at 5, 100, and 1000 kV/m. The exposure times were from 7 to 9 hr. Low-intensity electric fields of $\simeq 3 \times 10^{-2}$ V/m generated by a magnetic field have also been reported to increase the release of H-noradrenaline, 27% from a colonal nerve cell line, at 500 Hz.[19]

The existence of windowing effects in frequency and amplitude, and of sensitivity to low-intensity fields greatly expands the range of field strengths and frequencies which must be explored when looking for the effects of electrical fields on biological processes. Simple upper or lower bounds of frequency and field strength may not be adequate to describe an effect. Also, the disappearance of an effect at a particular level does not necessarily mean that it will not reappear at either lower or higher values of field strength or frequency. "Window effects" are discussed in greater detail in Part III, Chapter 5.

IV. NATURAL FIELDS AND NOISE OF ENVIRONMENTAL AND INTERNAL ORIGIN

In approaching the problems of setting safety standards and using electric fields for therapy, it is useful to know what are the natural levels of exposure. There are two types of natural fields to which we are all exposed on a regular basis. The first of these is external atmospheric electrical noise and the second is fields generated inside the body. The natural external electric fields at low frequencies are largely generated by lightning strikes and fluctuations in the atmospheric charge.[20] The level of natural AC fields in the atmosphere falls very rapidly from a DC value of about 130 V/m to about 10^{-4} V/m in a 1-Hz bandwidth at 60 Hz.[21] These very low levels of the natural fields are one of the reasons why electronic communications in the ELF band are useful for ships at sea and submarines. However, because of the very low level of the natural atmospheric fields at frequencies above a few Hertz, there is very little reason for biological organisms to develop natural protection against perturbations at these frequencies. It also means that biological systems can communicate internally while using very low signal power levels and still maintain a good signal-to-noise ratio.

The signals generated within the body are the result of nerve firings and other cell activity. A typical nerve cell fires with an action potential of 50 to 100 mV and transmits a current pulse about 0.4 msec long.[21a] The rise time for this current spike is approximately 0.1 msec and the fall time is about 0.5 msec. Each pulse is followed by a refractory period which is typically on the order of 1 to 3 msec. The longitudinal fields along the exterior of a nerve cell membrane are estimated to have a maximum value of about 5×10^{-2} V/m during an action potential[21a] when the cell is surrounded by a relatively high conductivity fluid of 5 S/m.

In looking at the natural fields in the body we have two concerns. The first is how large an external signal it takes to perturb the on-going natural signal which is being used to communicate or control some biological process. The second is how much of the signal field typically leaks away from active nerve fibers or bundles to form a background noise environment for surrounding tissue and processes. With respect to the first of these questions, it is interesting to look on the microscopic level at the electrical noise, i.e., the fluctuations that occur fundamentally as a result of the electrical process itself.

The first of several sources of noise which is always present is blackbody radiation, or Johnson noise which is given by[22]

$$P_N = kTB \tag{8}$$

P_N is the noise power, k is the Boltzman constant, T is the absolute temperature, and B is the bandwidth. The voltage equivalent of this noise power which can be delivered to a matched load, or the mean squared voltage fluctuation $\overline{V}_n^2$ across a resistance R is given by [23,24]

$$\overline{V}_n^2 = 4kTBR \tag{9}$$

or

$$\overline{i}_n^2 = \frac{4kTB}{R} \tag{10}$$

The second source of noise which is also present is the shot noise which is given by

$$\overline{i}_n^2 = 2qI_{DC}B \tag{11}$$

where $\overline{i}_n^2$ is the mean squared current fluctuation. This noise comes about because of the discreteness of the electronic charge q. It gives an AC fluctuation $\overline{i}_n^2$ which is proportional to the average value of the current I_{DC}. A third source of noise is $\frac{1}{f}$ noise. This noise may be generated by many processes. However, a critical characteristic of models which lead to noise of this kind is that they require memory of about one state variable per decade of time over which the model is used to generate noise with a power density spectrum $S(f) = \frac{C}{f}$, where C is a constant.[25] This means that we can expect to find this kind of noise for processes which evolve with time. $\frac{1}{f}$ noise is the dominant source of fluctuations at low frequencies in such diverse phenomena as transistors, quartz crystal oscillators, and the weather. It is also generated by the flow of ion currents through an orifice[26] and thus is a fundamental part of the transport of current through channels in membranes. Measurements of the noise voltage across a 10-μm hole in a 6-μm mylar film showed that for a wide range of ionic concentrations the voltage noise spectral density $S(f)$ is given by

$$\frac{S_\phi(f)}{\phi^2} = \frac{a}{bnr^3f} \tag{12}$$

where b is a numerical geometric factor, n is the density of ions in the solution, r is the radius of the hole, a is a constant, and ϕ is the applied voltage. The data showed $2.5 < a < 40$ with a mean value of 10 for a wide range of solutions including HCl, KCl, $AgNO_3$, etc. with concentrations from 0.05 to 5 mol.

For natural membranes this noise has been shown to take on the form of

$$S_E(f) = \frac{C_E}{f^\alpha} \tag{13}$$

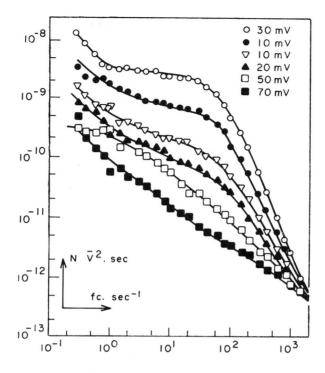

FIGURE 8.   Voltage noise spectra of a frog node of Ranvier at different levels of membrane potential. (From Verveen, A. A. and DeFelice, L. J., *Prog. Biophys. Mol. Biol.*, 28, 189, 1974. With permission from Pergamon Press.)

where $0.7 < \alpha < 1.2$ with a mean close to $\alpha = 1$. For the frog node of *Ranvier* the noise is a function of the membrane voltage as shown in Figure 8.[27] The dominant source of this noise appears to be the K^+ current and it has a minimum when the membrane is biased so that this K^+ current is biased to zero.

To get an estimate of the size of these noise sources, let us consider a pacemaker cell from the abdominal ganglion of *Aplysia*. This cell fires 20-msec pulses at about 1 Hz/sec. It has a resting voltage of about -50 mV, and a resistance R measured with a microelectrode between the inside of the cell and the surrounding solution of approximately 10^6 Ω. If we assume a system bandwidth of 100 Hz and $T = 300$ K, the Johnson noise voltage would be $\overline{V}_n \approx 3 \times 10^{-6}$ V. This gives a resting potential to noise $\overline{V}_n$ ratio of about 4×10^4. The peak current flow in these cells is estimated to be about 10^{-7} A, and thus the estimated shot noise current $\overline{i}_n \approx 2 \times 10^{-12}$ A, and the ratio of the peak current to the noise current is about 2×10^4. We do not have available the value S(f) for the *Aplysia*. However, $\overline{V}_\phi = \sqrt{S(f)B}$ where B is the bandwidth. If it is assumed that the maximum value of the noise is the same as that of the frog node of *Ranvier*, then for a bandwidth of 1 Hz we get $\overline{V}_\phi = 1.4 \times 10^{-5}$ V, at a center frequency of 1 Hz from the curve for -50 mV of Figure 8. This is about a factor of 10^3 greater than the Johnson noise. It is likely that $\frac{1}{f}$ is the largest source of noise at the cell membranes for frequencies below 160 Hz.[28]

In addition to the electrical noise generated by currents and voltages which are part of the single cell operation, electrical signals propagate through the body as the result of the incomplete confinement of electrical signals propagating in nerve cells. In some sense these signals may be thought of as noise if they are not pertinent to the activity in that portion of the body through which they are propagating. If on the other hand, they are utilized by tissue within the organism at some distance from the source, they must be thought of as signals. In the brain the fraction of these signals which reach the scalp is called the EEG (electroencephalogram). The EEG is obtained by placing two or more contacting electrodes on the scalp and measuring the voltage between them. For electrodes placed 5 cm apart, the peak-to-peak voltages range up to 30 μV.[29] The author views this voltage as the integral of the vector sum of the leakage fields from the firing of the nerve cells in the brain between the two electrodes. Since there is a very large number of cells firing, most of the 50-mV signals from an isolated nerve are canceled by summing over many like cells firing at different times and by the attenuation caused by propagation through the lossy tissue. Estimates of surface potential along a nerve fiber range from 3×10^{-4} to 5×10^{-2} V/m, and the corresponding current densities external to the nerve cells[21a,30] range from 5×10^{-2} to 4 A/m². The EEG voltage has a strong periodic component (particularly during sleep) near 10 Hz, which is known as the alpha (α) wave. A peak amplitude of this component may be as large as 50 μV when measured at the skin on the scalp.

At the surface of the chest, a signal may be recorded betweeen two electrodes known as the ECG or EKG (electrocardiogram). This signal results from the highly coordinated firing of the cells in the heart and has a definite wave shape which is closely related to the operation of the heart. The peaks of the so-called R wave in this signal may range up to 2.5 mV and are typically 0.5 to 1.5 mV, depending on the placement of the electrodes, the amount of body fat, etc. The pulse repetition rate is usually in the range of from 1 to 2 Hz and the ''QRS spike'' of the typical cardiogram is 40 msec long. Again, the signal measured at the surface of the skin is the result of leakage from electrically active cells located at a distance. The estimated current density[31] near the firing heart cell ranges up to 1 A/m². In this case the shape of the signal reaching the skin is so closely related to the activity of the heart that it provides detailed information on heart function.

V. MAN-MADE FIELDS IN TYPICAL LIVING ENVIRONMENT

Most of us are exposed to low-level man-made ELF fields on a continuous basis. We live a large part of our lives surrounded by a grid of wires that delivers the energy we need to power lights, electric motors, and many of the conveniences that make modern living possible. The majority of our exposure comes from fringing fields generated by the power distribution systems. In a typical suburban house these background 60-Hz fields range from 1 to 20 V/m and may be considerable larger near appliances where they may range up to 250 V/m. Table 1[32] shows some typical values for appliances. The corresponding current densities induced in the body for some of these fields are given in Table 2. It is also interesting to note that fields on the order of 150 to 500 V/m have been measured in a 1/4-in. spacing between an electric blanket and the human body.*

The scaling from external fields to fields inside the body depends on the geometry and the electrical properties of the body, as discussed in the Introduction and Part I of this volume. Figure 9 shows several possible exposure fields and approximate current distributions.[32] The details of scaling are so case-dependent as to require computer calculations when accurate values are needed; however, for many cases we can simply use a factor of 10^{-6} as a very rough approximation for scaling the electric fields in free space to those in the tissue at 60 Hz. Additionally, typical induced body currents in the legs (see Figure 9c) are 15 μA/kV/m of external field.

VI. THE EFFECTS OF ELECTRIC FIELDS ON BIOLOGICAL SYSTEMS

Many of the effects of exposure to DC and low frequency fields on whole plants and

---

* Measurements made by the author.

Table 1
SUMMARY OF REPRESENTATIVE ELECTRIC FIELDS AND CURRENTS FROM HOUSEHOLD APPLIANCES

Fields

Electric blanket	250 V/m
Hair dryer	40 V/m, 10—25 G
Electric train	60 V/m, 0.01—0.1 G
Food mixer	50 V/m, 1—5 G

Leakage currents

ANSI std fixed appliance	750 μA
ANSI std cord connected	500 μA
Coffee mill	380 μA
Refrigerator	40 μA
Sewing machine	34 μA
Coffee pot	6 μA

Induced currents

Heating pad[a]	18 μA
Electric blanket[a]	7—27 μA

[a]   Electric fields were measured 30 cm from appliance, magnetic a few centimeters; induced current was measured through grounded arm and leakage current through body to ground.

From Bridges, J. E. and Preache, M., *Proc. IEEE,* 69, 1092, 1981. With permission. ©1981 IEEE.

Table 2
COMPARISON OF BODY CURRENT AND RELATED CURRENT DENSITIES IN TERMS OF EQUIVALENT ELECTRIC FIELD

	Current		
	(μA)	(μA/cm²)	(kV/m)
Sinusoidal waveforms			
Cord-connected household appliance	20—500	0.5—12[a]	1.5—38
Man in 8 kV/m field	120	3[a]	8
Electric blanket	7—25	2—40[b]	0.50—1.7
Man in 0.16 kV/m field	2.2	0.05[a]	0.150
Nonsinusoidal waveforms-medical devices			
Electric Anesthesia (100 Hz square-wave)	10,000-	71,000[c]	670
Pacemaker electrode in myocardium[d,e]	6,000	20,000	400
Pacemaker electrode implanted in abdomen[d,f]	6,000	300	400

[a]   Passing through a 40-cm² ankle.
[b]   1/4 in. from electric wire in blanket.
[c]   Next to electrode.
[d]   Peak pulse current ~10^{-3} sec duration repeated every 0.8 sec.
[e]   0.3 cm² electrode area.
[f]   20 cm² electrode area.

From Verveen, A. A. and Derksen, H. E., *Proc. IEEE,* 56, 906, 1968. With permission. © 1968 IEEE.

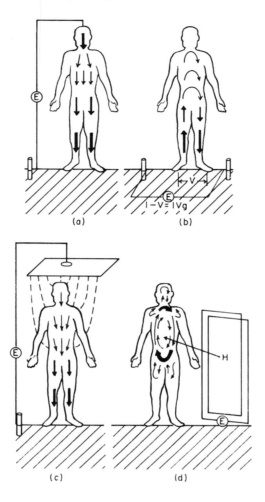

FIGURE 9.   Internal current density distribution for an erect primate. (a) to (d) indicate four possible ways wherein current can be caused to flow in an erect primate. The direction of arrows indicates the approximate direction of the internal body current flow and the width of the arrow suggests the current densities. The paths for displacement current flow are suggested by the dashed lines. (a) Conduction case; (b) step potential case; (c) electric field induced flow; (d) magnetic field induced flow. (From Bridges, J. E. and Preache, M., *Proc. IEEE,* 69, 1092, 1981. With permission. ©1981 IEEE.)

animals are difficult to assess, particularly at low levels. This is fundamentally true for at least two reasons: biological systems are very complex, and they contain a great deal of negative feedback that tends to null out perturbations which would damage them or change their performance. For example, increased energy dissipation, which would raise man's internal temperature, is compensated for by increased blood flow and sweating, which reduce the measurable temperature changes to a small fraction of what one would calculate assuming a constant energy dissipation system. Thus, the fact that many experiments on the effects of electric fields have not shown positive changes in growth, reproduction, blood chemistry, etc.[33] does not mean that we are not disturbing the system. It is possible that we have perturbed the system only a little less than the ability of the body to compensate. However, we can get some guidance concerning the effects of electric fields on complex living systems by working up from the cell level to more complex structures. For example, the slime mold *Physarium polycephalum* exposed to 75-Hz fields of 0.2 mT and 0.7 V/m shows a lengthening of the mitotic cycle by as much as 20%.[34] This effect was very slow to develop (80 to 100 days at 63 Hz and 100 to 120 days at 75 Hz). Recovery to normal mitotic cycles occurs after 30 to 60 days. These are very slowly developing changes in growth characteristics.

At higher levels of exposure a variety of effects have been shown. An examination of nerves taken from rats which had been exposed to fields of 100 kV/m for 30 days at 60 Hz has shown increases in the C-fiber conduction velocity, a decreased rate of fatigue, and shifts in the strength duration curve which suggest that the exposure had increased neural excitability.[35]

In another study,[36] experiments on dogs showed a significant change in blood chemistry for exposures at 25 kV/m and 50 Hz for 8 hr/day over a period of 6 weeks. These changes included an increase in reticulocytes, a decrease in hemoglobin concentration, an increase in serum protein, a decrease in neutrophils, and increase in lymphocytes. In separate experiments on rats, the growth rate was slowed upon exposure to fields of both 25 and 100 kV/m. These rats also showed an increase in neutrophils and a decrease in lymphocytes.

In experiments carried out on honeybees whose hives were located under a 765-kV transmission line and exposed to fields of 0.6 to 4.5 kV/m, important deleterious effects were shown, including a 64% increase in the over-winter mortality rate by comparison to shielded hives in the same area.[37] In the same study, those hives with exposures ranging from 1.1 to 8.3 kV/m showed almost no honey production over the summer and showed a loss of bee weight. Furthermore, all hives exposed at this level failed to survive the following winter. More recent experiments show that electrical shocks on entering the hive were the source of these effects.

This whole group of experiments showed the ability of large kilovolt per meter ELF electric fields to disrupt biological systems when applied for extended periods of time.

At levels as low as 2.5 V/m, 10-Hz signals have been reported to reduce the Circadian rhythm in man by more than 0.6 hr in a shielded environment. These signals were also shown to prevent the internal desynchronization and to resynchronize the length of sleep and activity periods.[38] Experiments on monkeys trained to estimate the passage of a time interval of 5 sec showed that exposure to 7-Hz fields at a level above 10 V/m caused the monkeys to underestimate the delay.[39] These results coupled with the Ca^{++} efflux experiments suggest the ability of very weak fields at frequencies close to natural biological system frequencies to influence behavior.[40]

In all of the above experiments, exposure to ELF electric fields (and in one case simultaneous exposure to magnetic fields) produced currents inside various organisms which did not make electrical contact with any external source of power. The fields inside the organisms were therefore in general, much smaller than the external fields, as explained in the Introduction to this book. When direct electrical contact with a conductor is established the situation becomes very different, because the large series resistance of air no longer limits current flow through the biological system. In view of the obvious safety problems associated with shocks and burns from contact with 60-Hz lines, a number of studies have been done to set standards for insulation. For direct contact with a metal wire connected to a 60-Hz generator, the minimum detectable current by humans is about 0.1 mA.[41] The average value at 60 Hz for just perceptible current is 1.1 mA for men and 0.7 mA for women. Dry skin has a resistance per unit area of about 10 to 30 Ω/m^2. About 100 V is needed for a detectable sensation with a contact resistance of 105 Ω. However, if the skin is moist, the resistance drops to about 0.1 Ω/m^2, and for a break in the skin, a typical body impedance for current from one major extremity to another is between 100 and 500 Ω. At voltages higher than 200 V, the skin resistance breaks down and the value of 500 Ω is often assumed for calculating the current flow through the body. Currents strong enough to freeze muscle or ''let go'' currents range upward from 5 mA, and average 16 mA and 10.5 mA for women. The maximum safe current for men is taken to be 9 mA and for women 6 mA. At still higher currents of 3 to 6 A, violent muscle contractions occur which may include arrest of both heart function and breathing.

For very short, high voltage pulses, the threshold for burns is estimated at 50 J. Electrical burns are estimated to constitute about 4% of the burns in the U.S. and about 2400 cases were seen in hospital emergency rooms in 1977.[42] The burns occurred in the areas where the current density is highest. Current densities tend to be the highest in regions of small cross-section which are often adjacent to one of the contacts. The relatively high resistance of bone tends to exclude the current and direct it into lower resistance arteries, nerves, and tissues. Tissue resistivities are functions of both temperature and current density. In general, the resistivity increases with temperature for muscle, nerves, and arteries, and decreases for fat. Burn damage tends to occur first along blood vessels and nerves.

As a result of this experimental background, the Americal National Standards Institute (ANSI) in November 1970 issued a standard of 0.5 mA allowable leakage current for two-wire portable appliances and 0.75 mA for heavy cord connected major appliances such as stoves, refrigerators, etc. This factor of 10 to 20 below the "let go current" seems adequate for normal use. However, it is clearly above what may be a safe operating level in a hospital situation where the leakage current might couple into a catheter supplying fluids to a patient. For patients with pacemakers or catheters in the vicinity of the heart, currents which are picked up by pacemaker leads or induced into a catheter fluid in the range of 20 to 800 μA may cause ventricular fibrillation. Thus, the standards for hospital equipment should be at least 100 times lower than for the general population.[41]

VII. SUMMARY AND CONCLUSIONS

We have reviewed a number of mechanisms by which ELF fields can affect biological systems. In addition to heating, which is discussed in Part II, Chapter 1, we have indicated that the nonlinearities of membrane resistance and capacitance can generate DC currents which could in turn upset ion balances. These nonlinearities also provide a vehicle for phase locking of pacemaker cells and a partial explanation for the increased sensitivity of biological systems to electric fields at frequencies near their natural firing rates. However, there is as yet no clear link between these nonlinear physical mechanisms and the Ca^{++} efflux effects. By way of comparison, it is interesting to note that the applied fields of about 10 V/m in the Ca^{++} efflux experiments are of about the same order of magnitude as those observed as fringing fields at 60 Hz in a number of suburban residences, and that this is about 10^5 times the noise fields of about 10^{-4} V/m generated in the atmosphere. At 10 Hz the noise from the atmosphere reaches approximately 3×10^{-3} V/m and is well below the levels where the Ca^{++} efflux changes are seen and the values which have been reported to control the Circadian rhythm.

Transforming the Ca^{++} efflux signals to estimated levels inside the tissue puts them at about 10^{-5} V/m, or about a factor of 60 below the 600 μV/m observed for a maximum EEG signal, but this still may be large enough to generate lock-in effects if the applied signals are sufficiently close to natural resonance frequencies.

There are still large gaps in our knowledge of the effects of ELF fields on living systems. This is particularly true for long-term, low-level effects. At levels above 25 kV/m there are some clear warnings[35-37] that long-term exposures may be dangerous. Additionally, there are a few indications that fields as low as a few volts per meter may cause changes in biological systems.[15-18,39] We are, however, still a long way from being able to define how the fields induce changes and which changes are beyond the capacity of the natural feedback systems to correct.

ACKNOWLEDGMENTS

The author wishes to express appreciation to Maria Stuchly, Ross Adey, Mike Marron, and Charles Polk for their many helpful comments and suggestions. He also wishes to express his appreciation to the Office of Naval Research under Contract N00014-81-K-0387 for financial support of his work in this area.

REFERENCES

1. **Barnes, F. S. and Hu, C. L.,** Model for some non thermal effects of radio and microwave fields on biological membranes, *IEEE Trans. Microwave Theory Technol.,* 25, 742, 1977.
2. **Montaigne, K. and Pickard, W. F.,** Offset of the vacuolar potential of characean cells in response to electromagnetic radiation over the range of 250 Hz to 250 kHz, *Bioelectromagnetics,* 5, 31, 1984.
3. **Pickard, W. F. and Rosenbaum, F. J.,** Biological effects of microwaves at the membrane level: two possible athermal electrophysiological mechanisms and a proposed experimental test, *Math. Biosci.,* 39, 235, 1978.
4. **Pickard, W. F. and Barsoum, Y. H.,** Radio-frequency bioeffects at the membrane level: separation of thermal and athermal contributions in the Characeae, *J. Membrane Biol.,* 61, 39, 1981.
5. **Kalkwarf, D. R., Frasco, D. L., and Brattain, W. H.,** Current rectification and action potentials across thin lipid membranes, in *Physical Principles of Biological Membranes,* Snell, F., Wolken, J., Iverson, G., and Lam, J., Eds., Gordon & Breach, New York, 1970.
6. **Fzanceschetti, G. and Pinto, I.,** Cell membrane nonlinear response to applied electromagnetic field, IEEE-MTT, submitted.
7. **Casaleggio, A., Marconi, L., Morgavi, G., Ridella, S., and Rolando, C.,** Current flow in a cell, with a non-linear membrane, stimulated by an electric field, *IEEE-MTT,* submitted.
8. **Cain, C. A.,** Biological effects of oscillating electric fields: role of voltage sensitive ion channels, *Bioelectromagnetics,* 2, 23, 1981.
9. **Bisceglia, and Pinto, I.,** Volterra series solution of Hodgkin-Huxley equation, *Bioelectromagnetics,* submitted.
10. **Wachtel, H.,** Firing-pattern changes and transmembrane currents produced by extremely low frequency fields in pacemaker neurons, in *Proc. 18th Annu. Hanford Life Sci. Symp.,* Technical Information Center, Department of Energy, Richland, Wash., 1978, 132.
11. **Adler, R.,** A study of locking phenomena in oscillators, *Proc. IRE,* 34, 351, 1946.
12. **Kloss, D. A. and Carstensen, E. L.,** Effects of ELF electric fields on the isolated frog heart, *IEEE Trans. Biomed. Eng.,* 30, 347, 1983.
13. **Takashima, S.,** Non-linear dielectric properties of nerve membranes, *Biophys. Chem.,* 11, 447, 1980.
14. **Berkowitz, G. C. and Barnes, F. S.,** The effects of nonlinear membrane capacity on the interaction of microwave and radio frequencies with biological materials, *IEEE Trans. Microwave Theory Technol.,* 27, 204, 1979.
15. **Bawin, S. M. and Adey, W. R.,** Sensitivity of calcium binding in cerebral tissue to weak environmental electric fields oscillating at low frequencies, *Proc. Natl. Acad. Sci. U.S.A.,* 73, 1999, 1976.
16. **Blackman, C. F., Benane, S. G., Kinney, L. S., Joines, W. T., and House, D. E.,** Effects of ELF fields on calcium-ion efflux from brain tissue in vitro, *Radiat. Res.,* 92, 510, 1982.
17. **Blackman, C. F., Benane, S. G., Joines, W. T., and House, D. E.,** Effects of ELF fields between 1 and 120 Hz on the efflux of calcium ions from brain tissue in vitro, presented at BEMS 5th Annu. Sci. Session Program, Boulder, Colo, 1983.
18. **Lymangrover, J. R., Keku, E., and Seto, Y. J.,** 60 Hz electric field alters the steroidogenic response of rat adrenal tissue in vitro, *Life Sci.,* 32, 691, 1983.
19. **Dixey, R. and Reins, G.,** H-noradrenaline release potentiated in a colonal nerve cell line by low-intensity pulsed magnetic fields, *Nature,* 296, 253, 1982.
20. **Polk, C.,** Sources, propagation amplitude and temporal variations of extremely low frequency (0—100 Hz) electromagnetic fields, in *Biological and Clinical Effects of Low-Frequency Magnetic and Electric Fields,* Llaurado, J. G., Sances, A., Jr., and Battocletti, J. H., Eds., Charles C Thomas, Springfield, Ill, 1974, 21.
21. **Smith, E. K.,** The natural radio noise source environment, invited paper, IEEE EMC Symp., Santa Clara, 1982, 266.

21a. **Plonsey, R.,** *Bioelectric Phenomena,* McGraw-Hill, New York, 1969.

22. **Yariv, A.,** *Introduction to Optical Electronics,* Holt, Reinhart & Winston, New York, 1976.

23. **MacDonald, D. K. C.,** *Noise and Fluctuations: An Introduction,* John Wiley & Sons, New York, 1962.

24. **Beck, A. H. W.,** *Statistical Mechanics, Fluctuations, and Noise,* Edward Arnold, London, 1967.

25. **Pehtig, R.,** *Dielectric and Electronic Properties of Biological Materials,* John Wiley & Sons, New York, 1979.

26. **Verveen, A. A. and DeFelice, L. J.,** Membrane noise, *Prog. Biophys. Mol. Biol.,* 28, 189, 1974.

27. **Sichenga, E. and Verveen, A. A.,** The dependence of the 1/f noise intensity of the diode of Ranvier on membrane potential, *Proc. 1st. Eur. Biophys. Congr.,* Vol. 5, Verlag Wiener Medizinischen Akademie, Vienna, 1971, 219.

28. **Verveen, A. A. and Derksen, H. E.,** Fluctuation phenomena in nerve membrane, *Proc. IEEE,* 56, 906, 1968.

29. **Nuney, P. L. and Katznelson, R. D.,** *Electric Fields of the Brain — the Neurophysics of EEG,* Oxford University Press, New York, 1981.

30. **Bernhardt, J.,** The direct influence of electromagnetic fields on nerve and msucle cells of man within the frequency range of 1 Hz to 30 MHz, *Radiat. Environ. Biophys.,* 16, 309, 1979.

31. **Beckwith, J. R. and McGuire, L. B.,** *Basic Electrocardiography and Vector Cardiography,* Raven Press, New York, 1982.

32. **Bridges, J. E. and Preache, M.,** Biological influences of power frequency electric fields: a tutorial review from a physical and experimental viewpoint, *Proc. IEEE,* 69, 1092, 1981.

33. **Sheppard, A. R. and Eisenbud, M.,** *Biological Effects of Electric and Magnetic Fields of Extremely Low Frequency,* New York University Press, 1977.

34. **Greenebaum, E., Goodman, E. M., and Marron, M. T.,** Long-term effects of weak 45—75 Hz electromagnetic fields on the slime mold physarum polycephalum, Proc. Biol. Eff. USNC/URSI Annu. Symp., Boulder, Colo., 1975.

35. **Jaffe, R. A., Phillips, R. D., and Kaune, W. T.,** Effects of chronic exposure to a 60 Hz electrical fields on synaptic transmission and peripheral nerve function, biological effects of extremely low frequency electromagnetic fields, *Proc. 18th Annu. Hanford Life Sci. Symp.,* Technical Information Center, Department of Energy, Richland, Wash. 1978, 277.

36. **Ceretelli, P., Veicsteinas, A., Margonato, V., Cantone, A., Viola, D., Malagati, C., and Previ, A.,** 1000 kV project research on the biological effects of 50 Hz electric fields in Italy, biological effects of extremely low frequency electromagnetic fields, *Proc. 18th Annu. Hanford Life Sci. Symp.,* Technical Information Center, Department of Energy, Richland, Wash. 1978, 241.

37. **Greenberg, B., Kunich, J. C., and Bindokas, V. P.,** Effects of high voltage transmission lines on honeybees, biological effects of extremely low frequency electromagnetic fields, *Proc. 18th Annu. Hanford Life Sci. Symp.,* Technical Information Center, Department of Energy, Richland, Wash. 1978, 74.

38. **Wever, R. A.,** *The Circadian System of Man — Results of Experiments under Temporal Isolation,* Springer-Verlag, Basel, 1979.

39. **Gavalas, R. J., Walter, D. O., Hamer, J., and Adey, W. R.,** Effect of low-level, low-frequency electric fields on EEG behavior in *Macaca nemestrina, Brain Res.,* 18, 491, 1970.

40. **Adey, W. R.,** Evidence for cooperative mechanisms in the susceptibility of cerebral tissue to environmental and intrinsic electric fields, in *Functional Linkage in Biomolecular Systems,* Schmitt, F. O. et al, Eds., Raven Press, New York, 1975.

41. **Dalziel, C. F.,** Electric shock, in *Advances in Biomedical Engineering,* Vol 3, Kenedi, R. M., Ed., Academic Press, New York, 1973, 223.

42. **Mehn, W. H.,** The human considerations in bioeffects of electric fields, biological effects of extremely low frequency electromagnetic fields, *Proc. 18th Annu. Hanford Life Sci. Symp.,* Technical Information Center, Department of Energy, Richland, Wash, 1978, 21.

Chapter 3

EXTREMELY LOW FREQUENCY (ELF) ELECTRIC FIELDS: EXPERIMENTAL WORK ON BIOLOGICAL EFFECTS*

Morton W. Miller

TABLE OF CONTENTS

*   This paper is based on work performed under Contract No. DE-ACO2-76EVO3490 with the U.S. Department of Energy at The University of Rochester Department of Radiation Biology and Biophysics and has been assigned Report No. DOE/EV/03490-2366.

I. INTRODUCTION

The exposure of living organisms to extremely low frequency (ELF) electric and magnetic fields is primarily the result of industrialization, such ELF fields being arbitrarily defined as having a sinusoidal frequency < 100 Hz. Polk[1] has indicated, for example, that the ambient 60-Hz electric field is about 0.001 V/m. Conversely, electrical devices produce in their vicinity far greater fields; e.g., the maximum electric field of a large 60-Hz transmission line can be as high as 10 kV/m at ground level; and 60-Hz electric fields in homes are reported to be in the range of 1 to 10 V/m.[2] Thus electrification of the country, indeed most of the world, has resulted in exposure of people, animals, and plants to ELF fields considerably above ambient levels.

That ELF fields can induce biological effects is probably known personally by most of us. Who among us has not at some time been shocked by electric currents in the home — e.g., a child sticking a metal object into an electric socket, an adult using a faulty electric device, etc.? These exposures are unintentional, but prove quite convincingly that ELF fields can cause biological effects. That the effect is related to the field and not the current is discussed later in this chapter. ELF fields are also used on patients in a number of routine medical procedures, e.g., electroshock convulsive therapy and electroanesthesia.

The purpose of this chapter is to review the ELF-bioeffects literature with a view toward understanding mechanisms of action of ELF fields on biological systems. Much of the research in this area has been accomplished at ELF frequencies of 45 to 75 Hz, but some important and exciting findings were observed at lower frequencies.

II. SOME PHYSICAL CONSIDERATIONS CONCERNING ELF ELECTRIC FIELDS

A. Field Perturbation due to Conductor Shape

An electric field is defined in terms of a force on an electric charge (newtons per coulomb), or in terms of the work per unit charge, per unit distance (volts per meter). Perhaps the simplest way to generate a reasonably uniform electric field is to apply a potential (voltage) difference between two parallel metal sheets. It will be stipulated for purposes of discussion that the electrodes are separated in space by 1 m and the potential difference between them is 100 V. The electric field between the electrodes is then 100 V/m and would be, in essence, the same intensity anywhere between the electrodes.

A flat object such as a sheet of metal when parallel to the electrodes has almost no effect on the external field, but a conducting sphere has a concentrating effect around the poles of the sphere. A prolate spheroid, placed with its ends facing the electrodes, experiences an even greater field concentrating effect at its poles. Thus, for a conducting sphere the electric field is enhanced three times greater than that of the unperturbed field which existed before introduction of the sphere. For a prolate spheroid, having a 10:1 axial ratio with its long axis parallel to the field, the field just outside the pole is 50 times the field in the absence of the conducting object.[3]

This field concentrating effect, especially for long, thin pointed objects can be an important consideration in evaluating ELF electric field-bioeffects. Kaune[4] has illustrated the effect of shape and size of a man, a pig, and a rat exposed to an initially unperturbed 60-Hz electric field of 10 kV/m (Figure 1). For the man whose shape resembles that of a prolate spheroid, the 60-Hz electric field at his head is 180 kV/m; for the pig the topmost part of the body experiences a field of 67 kV/m, and for the supine rat, 37 kV/m. Of course, when the rat rears — as they are prone to do — it assumes a more elongated ''prolate spheroid'' shape and with it a concomitant greater electric field enhancement. Kaune[4] has shown that for a rat in a supine or reared position (Figure 2) the induced short circuit current is 1.6 or 3.2 μA, respectively.

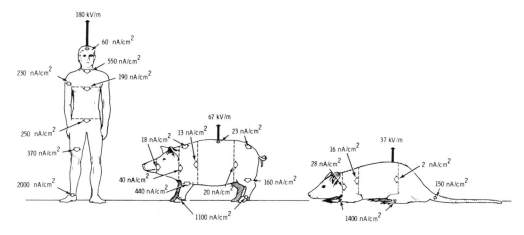

FIGURE 1. Grounded man, pig, and rat exposed to vertical 60-Hz, 10-kV/m electric field. Relative body sizes are not to scale. Surface electric field measurements for man and pig and surface electric field estimates for rats are shown. Estimated axial current densities averaged over selected sections through bodies are shown. Calculated current densities perpendicular to surface of body are shown for man and pig. (From Kaune, W. T., *Bioelectromagnetics, 2*, 403, 1981b. With permission.)

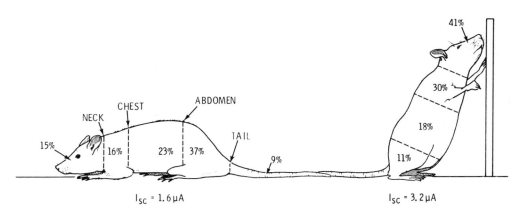

FIGURE 2. Grounded rat model exposed to vertical 60-Hz, 10-kV/m electric field. Conducting body of model is divided into five segments by planes through neck, chest, abdomen, and tail. Fractions of total short circuit current originating within each of five sections are shown as percentages. (From Kaune, W. T., *Bioelectromagnetics, 2*, 403, 1981b. With permission.)

Thus, in attempting to "scale" exposures from a pig or a rat to man it is obvious that different initially unperturbed exposures may have to be used to achieve comparable current densities in different mammalian bodies. For instance, for an initially unperturbed electric field of 10 kV/m the current densities in the thorax area for a man, pig, and rat are 1.9, 0.13, and 0.16 mA/m^2, respectively. Thus, to induce current densities in the rat thorax, which would be comparable to those induced in a man standing in a 10 kV/m, 60-Hz electric field, one would have to use much higher initially unperturbed electric fields for the rat.

Unfortunately, the scaling is not always comparable between a man and a rat or pig. Some areas of the man have relatively higher induced current densities than those for the rat if the exposures are adjusted so as to be comparable, say, for the thorax. For example, at an applied field strength of 100 kV/m for the rat, the current density in the paws will be about 37 mA/m^2, or about twice that for the ankles of a man standing in an initially unperturbed electric field of 10 kV/m. The scaling becomes progressively more complicated when different positions of a rat are considered, since the relative distribution of the current densities likewise changes (Figure 2).

Knowledge of these relative current densities and their corresponding fields is important if one is ever to scale from animal data to projected effects in people. Of course, before one can scale one has to have a biological effect and an understanding of a physical mechanism of action in the organism under study.

Plant material likewise obeys this fundamental principle of field enhancement in relation to conductor shape. Johnson et al.[5] compared the 60-Hz electric field sensitivity of various shaped leaves to the onset of leaf tip corona. Corona occurs when the electric field strength at the surface of the object exceeds the dielectric strength of the surrounding medium. Leaf tip corona is generally limited to the very tip of the leaf, and can be easily recognized by photographic or visual techniques.

One would naturally expect that pointed leaves would have a greater field concentrating effect than nonpointed, rounded leaves. From McKee's laboratory it was reported that "plant tissue-damage severity was dependent upon leaf-tip geometry and morphology. Fleshy, rounded, or blunt plant parts showed no damage in 50 kV/m fields, whereas plant parts with a pointed shape exhibited minor damage at field intensities of 20 to 22 kV/m... Tissue damage was observed to have the same electric field inception level as visible corona discharges about the damaged parts."[6]

B. Estimation of Internal Fields

For material placed in an electric field, consideration must be given as to whether they are in parallel or in series with each order for estimating internal electric fields. This topic is discussed by Polk (Introduction) and by Foster and Schwan (Part I) and also elsewhere by Schwan,[7,8] and the interested reader should consult these references for detailed, relevant theoretical and biophysical aspects of this problem. Consider, for example, the problem of determining the electric field in various parts of the erect human body when it is exposed to a vertical, 60-Hz electric field of 10 kV/m. The field will be considerably "enhanced" at the top of the person's head and shoulder, and one would predict therefore that the field in the tissues would be greatly enhanced above that of a flat slice of tissue exposed to the same field. Deno[9,10] has shown the following current densities (ampere per square meter) in various parts of the body for people exposed to a 60-Hz electric field of 1 kV/m in air: neck, 0.591×10^{-3}; waist, 0.427×10^{-3}; and ankle, 3.35×10^{-3}.

If we assume that the conductivity of soft tissue is about 0.2 S/m[7,8] we can estimate the 60-Hz electric fields induced in the corresponding portions of the body when exposed to a 60-Hz electric field of 10 kV/m. The fields would be about 0.03, 0.02, and 0.1 V/m for the neck, waist, and ankles, respectively. These values, however, will vary somewhat from person to person, depending upon their dimensions. For example Kaune and Phillips[11] have reported a slightly lower current density in the waist of a human standing in a 10-kV/m electric field; this would translate to an electric field of about 0.013 V/m, which is not quite half that from Deno's measurements. Kaune and Phillips however, used a waist area considerably larger than that used by Deno; hence the smaller current density.

From the measurements of Deno[9,10] and Kaune and Phillips,[11] it is seen that about one third of the total body current enters through the head, with the remainder entering via the shoulders and other extremities. All of the current flows through the ankles to ground, thus producing in them the highest current densities and largest fields — the ankles are very narrow and have nearly all the current flowing through them.

III. ELECTRICAL PROPERTIES OF BIOLOGICAL MATERIALS

A. Dielectric Properties of Cells

To understand how an electric field affects a biological cell or organism it is important to know first the electrical properties of biological specimens at the microscopic level. This type of information is extensively discussed by Foster and Schwan in Part I of this volume. Briefly, all tissues are composed of cells and extracellular fluid. The cells themselves have two distinct parts from a dielectric point of view: the outer membrane and the inner cytoplasm and nucleus (the latter includes the genetic information). The extracellular fluid is a solution of salts and organic molecules. The outer membrane has high resistance, and as a consequence, almost all of the current flows around the cell in the highly conducting extracellular fluid. Thus, in effect the outer cell membrane "protects" the interior of the cell from electric fields.

This leaves the outer membrane itself to consider. In addition to acting as the gateway to the cell, the outer membrane is also the site of a number of physiological processes that are important to cell function. The outer membrane is controlled in part by electric fields. It is most common to describe the fields in the outer membrane in terms of the potential drop across the outer membrane, i.e., the membrane potential. While induced fields inside the cell are orders of magnitude smaller than extracellular fields, the largest part of the potential drop across the cell occurs at the outer membrane.

Thus, three conclusions can be drawn. First, direct effects on the contents of the cells are very unlikely because intracellular fields are always extremely small. Second, for the extracellular fluids, which contain only molecules in solution, there are no known significant biological effects at fields lower than those necessary to produce heating. Finally, if there are any effects, the most probable site of action would be the outer membrane of the cell.

An examination of the field strength (10 to 100 V/m) in the extracellular fluid necessary to cause neural stimulation gives some gauge of the level that might be expected to produce effects on membranes of somatic, nonneural cells. For example, electric field effects at the cellular level are known for two plant species: reductions in root growth rates occurred with 60-Hz electric fields of the order of 200 to 300 V/m (in the root growth medium).[12-15] The effects correlate with the magnitude of the electric field, not the current density.[13]

B. Transmembrane Potentials and Tangential Fields

When a cell is exposed to an ELF electric field there will be simultaneously (a) an induced ELF transmembrane potential which will be maximal at the poles of the cell (relative to the direction of the field) and (b) an external tangential ELF field which will be maximal at the equator of the cell. It has also been suggested that relatively weak tangential fields may excite cooperative interactions among structures along the cell surface.[16]

Inoue et al.[17] devised an experimental protocol to compare the effectiveness of the transmembrane and the tangential electric fields for inducing a biological effect. Briefly, the level of applied electric field was kept constant (360 V/m) but the sizes of the exposed cells were varied. Under these conditions, the tangential electric field, being independent of cell size, is constant for all cells, but the magnitude of the induced transmembrane field, being proportional to cell diameter, will vary. Cells of large dimension in the direction of the applied field were more affected than cells of smaller size.

The excitation of nerves and muscle tissue provides us with excellent examples of the effects of electric fields acting via induced transmembrane potentials. We know that potentials great enough to depolarize the membrane act by producing transient changes in the permeability of the membrane for certain ions. We know that ELF fields are even more effective than direct current (DC) fields in the stimulation of peripheral nerves. In contrast, it is difficult to find any clear examples of effects of ELF tangential fields on cells. In most

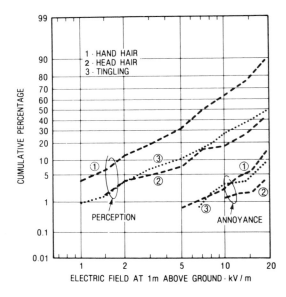

FIGURE 3.   Combination of all results obtained on perception and annoyance of hair erection and tingling (136 persons). (Copyright ©1982, Electric Power Research Institute, EPRI EL-2500, *Transmission Line Reference Book -345 kV and Above,* 2nd ed. Reprinted with permission.)

exposures it is difficult to modify one of these fields without a corresponding proportional change in the other. Of the information to date, however, all data are consistent with the transmembrane field hypothesis; the present data which discriminates between these alternate hypotheses are the vertical vs horizontal field exposures of *Pisum* roots,[13,14,18] and comparisons between *Pisum* and *Vicia*,[12,17] and comparisons among 12 species with widely differing cellular diameters.[17]

IV. EFFECTS AND MECHANISMS OF ACTION

There is no doubt that alternating electric fields can induce biological effects. There is reasonably rich and diverse literature to support this contention. There are, however, a sizable number of phenomenologically oriented reports which do not provide insights into the mechanism by which the electric field produces an effect. These reports will not be treated extensively or exhaustively. The analysis given below focuses on known mechanisms of action of electric fields and biological effects; these include hair stimulation, induction of a significant transmembrane potential, and for want of a better descriptor, induction of effects by ELF fields below perception levels.

A. Hair Movement
The mechanism of ELF electric field-induced hair movement (piloerection) is understood reasonably well. With DC electric fields hair can be made to stand erect. With alternating electric fields of sufficient magnitude, hairs can be made to vibrate. The mechanism for this vibration appears to involve a force on the electric charge on the hair surface. Hair will vibrate at 120 or 60 Hz depending upon whether fixed charges as well as induced charges are present.

The amount of hair vibration is related directly to electric field intensity. Figure 3 shows the relationship between field strength in air and the response of human volunteers (cumulative percentage reporting hair vibration) from a study conducted under the test transmission line at Project UHV, Pittsfield, Mass.[10] Curve 1 shows the greatest sensitivity to the electric field; the data were derived from subjects holding their hands over their heads. With this configuration the electric field enhancement would be considerably greater than in the normal standing position. Curves 2 and 3 represent normal erect position, with Curve 2 representing head hair stimulation and Curve 3 "tingling sensation in part of the body covered with clothes." The medium values of electric fields for the threshold of perception range from 7 kV/m for hand hair (remember this field strength is the initially unperturbed value at ground level) to 23 kV/m for head hair (see Table 8.10.4, p. 376, Deno and Zafanella[10]). Women appear to be less sensitive than men to hair vibration; the 50% value for perception for hand hair was 17.5 kV/m compared to 6.7 kV/m for men. The difference insensitivity is thought to be due to the relatively greater abundance of hair on a man's hand.

Results similar to those obtained at Project UHV were also reported by Cabanes and Gary[19] for volunteers exposed to a graded series of 50-Hz electric fields under laboratory conditions at the Renardieres High Voltage Laboratory of Electricé de France. Participating in the study were 75 volunteers who were asked to report whether or not they perceived the electric field through hair vibration. Each volunteer assumed three positions for each electric field intensity: both arms kept close to the body, one arm stretched horizontally, and one arm raised above the head. Most of the participants were males. In the last category 50% of the volunteers reported perception at an electric field intensity of about 12 kV/m; slightly higher intensities were required, as would be expected, for the hand held horizontally (50% being about 18 kV/m) and much higher intensities for arms held at the side (50% being about 25 kV/m).

The study by Cabanes and Gary[19] had two additional parts. The first dealt with the mechanism of perception. A two-plate electrode exposure device was used (30-cm separation distance) and volunteers placed their arms between the plates. At a field of 50 kV/m hairs were observed to vibrate at 100 Hz; when the forearm was then rubbed with a synthetic material (to increase the surface charge), the hair was then observed to vibrate at 50 Hz. When the arm was shaved, no sensation was perceived.

Similar results were obtained with animals exposed to 50-Hz, 50-kV/m electric fields. The body hair of a rat or mouse did not vibrate noticeably unless it was first brushed. The vibrissae (i.e., the long hairs near the nose), however, noticeably vibrated with a 10-mm, peak-to-peak amplitude. Stern and Laties[20,21] have shown that rats are able to detect 60-Hz electric fields between 4 and 10 kV/m,[20] and that hair clipping sometimes reduces the likelihood of field detection.[21] Hair clipping did not, however, result in complete hair removal. Hair vibration has also been observed in swine. Kaune and colleagues[22] exposed miniature swine to 60-Hz electric fields; above 50 kV/m there was ear flicking and head-shaking; for an anesthetized animal the vibration of the ear hairs continued until the field strength was decreased to 40 kV/m.

B. Induced Transmembrane Potential

There are several avenues of evidence suggesting that a significant induced transmembrane potential is a mechanism operating at the cellular level for electric field-induced biological perturbations. First, there are the "classical" studies dealing with neural stimulation whereby acute exposures to electric fields of sufficient magnitude cause neural excitation. The results are characterized by a definable electric field strength threshold and are discussed by Schwan.[7]

Second, for long term exposures, there is a series of papers by Miller and colleagues[12-15,17-18] using plant roots exposed to 60-Hz electric fields. Roots of *P. sativum* were exposed to 60-Hz electric fields ranging from 70 to 490 V/m; the conductivity of the inorganic root growth medium was 0.07 S/m. The threshold for root growth rate reduction was 290

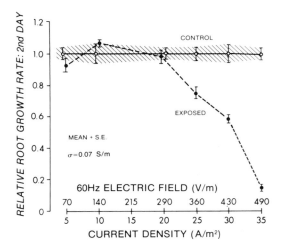

FIGURE 4.   Second-day root growth determinations from roots exposed to 60-Hz aqueous environments ranging from 70 to 490 V/m. (From Robertson, D., Miller, M. W., Cox, C., and Davis, A. T., *Bioelectromagnetics*, 2, 329, 1981b. With permission.)

V/m; above that intensity root growth rate decreased with increasing electric field intensity until at 490 V/m growth was essentially arrested.[14] Below 290 V/m there was no effect on root growth rate (Figure 4). The diameters of root cells from the area of the root meristem were measured and found to vary between 11 and 25 μm. Thus, the threshold field to produce growth effects corresponded to membrane potentials of 3 to 8 mV and the debilitating effects observed at about 500 V/m occurred at membrane potentials in the order of 6 to 12 mV. These values are within an order of magnitude of the typical resting potentials of a wide range of cell types.

That the cells within the root were exposed to the electric field was established through the measurement of the electrical conductivity of the intact roots.[23] It was determined that the low frequency conductivity of the roots is comparable to the suspending medium and changed as the conductivity of the growth medium changed — i.e., the root was a weak ionic sponge — and was not surrounded by an insulating macromembrane. Thus, the cells in the root are exposed to the applied electric field.

That the growth rate reduction in *P. sativum* roots is related to the applied field and not the current density was determined in a series of experiments which varied the ratio of field to current density (Figure 5).

Growth rate data are shown for three different media with conductivities of 0.07, 0.035, and 0.14 S/m. A field strength of 430 mV (and an associated current density of 30 A/m²; σ = 0.07 S/m) noticeably affected root growth rate; under this condition one cannot distinguish between a current- or field-related mechanism. However, when the conductivity of the exposure medium was changed so that a field strength of 430 V/m yielded a current density of only 15 A/m², the growth rate reduction persisted (under this condition the field intensity was high and the current density low). When conductivity of the medium was altered so that a field intensity of 215 V/m produced a current density of 27 A/m² (a relatively low field but a high current density), there was no effect on root growth rate.

Inoue et al.[17] determined growth rates and cell diameters from roots of 12 species of plant roots exposed to a 60-Hz electric field of 360 V/m in an aqueous inorganic nutrient medium (conductivity 0.07 to 0.09 S/m). Under these conditions the induced transmembrane potential

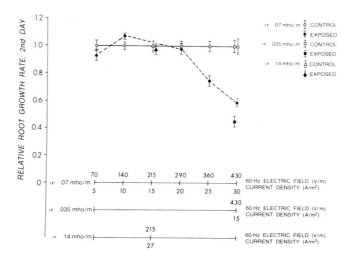

FIGURE 5. Summary of root growth rate experiments with mean of second-day growth rate determinations from roots exposed to 60-Hz aqueous environments ranging from 70 to 430-V/m electric fields and 5- to 30-A/m² current densities. (From Robertson, D., Miller, M. W., Cox, C., and Davis, A. T., *Bioelectromagnetics,* 2, 329, 1981b. With permission.)

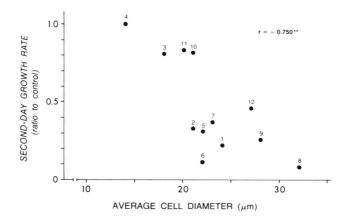

FIGURE 6. Relationship between second-day growth rate and average root cell diameters for 12 species exposed continuously to a 60-Hz, 360 V/m (2.5-A/m²) current density.[17]

would change in relation to cell size but the tangential field should be independent of cell size. The degree of growth depression ranged from near 100 to near 0% of control (Figure 6). Cell diameters ranged from 13.5 to 31.8 μm as an average value for procambial, cortical, and meristem cells. Sensitivity as determined by root growth rate reduction, to electric field, increased with relative increase in cell size. Sensitivity also increased with increase in 60-Hz induced transmembrane potentials; the transmembrane potential threshold for growth reduction was about 5 to 6 mV and the potential for near-complete cessation of growth was about 10 to 11mV.

It was concluded that the effects observed are consistent with a membrane related mechanism, not a tangential field mechanism, and that there was a narrow range of induced membrane potentials (a few millivolt) between threshold and debilitation.

The electric field strength (>300 V/m) needed to cause an effect on *Pisum* root growth rates is comparable to that reported by Marsh[24] to affect bipolar regeneration in flatworms or to that reported by Friend et al.[25] to affect amoeba shape. These two reports — Marsh for flatworms, and Miller and colleagues[12,14,18] for *Pisum* roots — are the only two studies that conclusively show through dose response relations an effect of electric fields on growth and development. Jaffe and Nuccitelli[26] have reviewed the literature dealing with DC field effects on plant and animal cells, and find, as will be discussed below, that small changes in the transmembrane potential can produce significant effects on growth and development. A review of the DC electric and magnetic field bioeffects literature is also included in the present volume (Part II, Chapters 1 and 4).

Two additional pieces of information support the postulate that the effects observed with *Pisum* roots are related to the induced transmembrane potential. The first is that when the roots were exposed to a vertical field, the growth rate was affected more than with exposure to a horizontal field of the same strength.[18] Cells in the root are much longer than broad (i.e., their length is generally much greater than their diameter).[27] Thus, if the exposure is parallel to the axis of the root, a correspondingly greater induced membrane potential would be expected.

It appears that the observed reductions in root growth rate are not due to electrolysis products. First, the production of electrolytic products is related to electrode potential which in turn is related to current density. For a given electric field, the higher the conductivity of the medium the greater the current flow and the greater will be the potential drop at the electrode. From Figure 5 it is clear that since a high current density (27 A/m^2, E_0 = 215 V/m) would produce more electrolytic products but has less of an effect on growth reduction than does a lower current density (15 A/m^2, E_0 = 430 V/m) where the field is quite high, it is unlikely that electrolytic products are an important factor in these experiments. Second, when roots were exposed to a vertical electric field, they had sensitivity which was different from that when exposed to a horizontal field.[18] If electrolytic products were being produced and were responsible for the growth rate reduction, then it is unlikely that there would be a difference between a vertical and a horizontal exposure in terms of growth rate effects. Third, a flow-through system was designed[28] such that growth medium flowed at a relatively high rate through three consecutive areas — Control 1 (upstream control), Exposed, and Control 2 (downstream control). Each area had a set of roots and the growth rates for each were compared to one another. It was reasoned that if toxic electrolytic products were being produced and were affecting root growth rates, then roots of Control 2 should have a growth rate different from that of Control 1. The electrodes were energized to produce a 60-Hz electric field strength at 430 V/m (a value known to reduce growth rates). In five separate experiments the exposed roots always displayed a growth rate reduction, but the two sets of control roots, plus those of an isolated control, had comparable growth rates.

An additional clue that effects at the cellular level are related to the electric field (and not its associated current density) is based on the observation that electric signals one or two orders of magnitude smaller than those needed to stimulate neuron firing may also exercise some influence on nerve cells. Wachtel[29] exposed a pacer neuron from a mollusk *(Aplysia)* to electric fields whose frequencies ranged from 3 to 1000 Hz; the field intensity was adjusted so as to determine the threshold for perturbing the natural firing rate of the neuron (see also Part II, Chapter 2). The experimental set-up involved the placement of the neuron in salt water (conductivity about 3 S/m) in contact with Ag:AgCl electrodes. The experiments were initially conduced with the long axis of the neuron parallel to the field force lines; the neuron was about 2 mm long. Changes in the firing pattern response to the applied fields were observed to be dependent upon intensity and frequency (Figure 7). The neuronal firing rate was most sensitive to a 0.5-Hz electric field; the threshold for the effect was at a current density of about 10 mA/m^2 (or a field in the medium of about 3 mV/m).

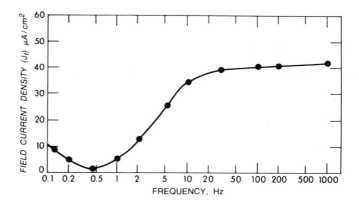

FIGURE 7. Field current densities (in microamperes per square centimeter) needed to produce firing-pattern changes in the *Aplysia* pacer neuron as a function of electric field frequency. (From Wachtel, H., in Biological Effects of Extremely Low Frequency Electromagnetic Fields, Phillips et al., Eds., NTIS, U.S. Department of Commerce, Springfield, Va., 1979, 132. With permission.)

The effect was, however, dependent upon the orientation of the neuron in the field (e.g., when oriented at right angles with respect to the field force lines, a tenfold increase in field strength was needed to produce the effect). This latter observation suggests that it is the field and not the current which causes the effect. It would appear, as suggested by Schmitt et al.[30] that neurons in the process of generating an action potential can have it modified by relatively weak fields (i.e., insufficiently strong to generate an action potential but sufficient to modify one in the formative process).

Weak fields are apparently also detected by certain fish which use them as a means for locating prey and for determining orientation.[54] First, some fish possess organs that are known to be used for electric field detection, and second, laboratory experiments have shown that both concealed prey and energized electrodes are detected by electrosensitive predators.

The electrosensitive organs of fish are usually located along the snout and have been termed the ''ampullae of Lorenzini''.[55] Their structures are somewhat varied but are characterized electrically as being a low pass filter that is not affected by the electric organ discharges of the fish. Sharks and rays are known to have such ampullae and they also occur (but are smaller in size) in some fresh water fish. In sharks the ampullae are quite long and may reach a length of 10% of the body length. There can also be lateral, ampullary-like organs. Microscopically, an ampulla consists of a canal which at the exterior protrudes to the skin and ends below the skin in some form of bulbous sac which contains sensory cells. The canal is filled with a jelly-like substance. There may be clusters of ampullae.

There have been two traditional ways of testing for sensitivity to electric fields. One is to use electrically derived signals of another fish or to fabricate such signals by electrodes. Exactly how an electrosensitive fish uses these signals is not known, but knowledge of some of the steps involved is available. It appears that the sensory cells operate at a level close to excitability. The path to the sensory cells is via the sac jelly, which is of a low resistance. The basic mechanism of excitation of the sensory cells (which are centrally innervated) is like that of the electrical step in other excitable tissues in that a threshold field strength exists for excitation. It should be pointed out that no such cells (ampullae of Lorenzini) have been found in mammals.

There is an apparent frequency-dependency to electric field detection. Electrosensitive fish will readily respond to DC and AC fields, particularly to 1-, 2-, and 4-Hz fields, but as the frequency increases the intensity of the field must be increased to be effective. In essence, the organs are responsive to fields ranging from DC to about 8 Hz. Thresholds for perception vary among species and range from 125 V/m to 10 μV/m[58] In a now classic experiment, Kalmijn[56] showed that a shark can detect a buried sole, will not be aware of the sole if it is electrically shielded, and will strike at an energized electrode simulating the electric field of the flatfish.

V. SUGGESTED BIOEFFECTS AT FIELDS BELOW PERCEPTION LEVELS

In some instances of exposure there is a strong possibility that the organism being exposed cannot perceive that the field is on. For example with classical work on neural stimulation it was observed that a tissue field strength of about 10 V/m was needed to affect excitation. With such an exposure the organism, through neural stimulation, would perceive the field. At exposures considerably < 10 V/m in the extracellular fluid, neural stimulation would not be expected to occur, and the organism would not be expected to perceive that the field is on.

There are several examples given below of exposure where it is highly unlikely that the organism would perceive the applied field; these are reported behavioral modifications in mammals and calcium efflux changes in in vitro brain slice preparations.

A. Behavioral Modification in Mammals

Gavalas-Medici and Magdaleno[31,32] have reported that weak electric fields in air (7 to 75 Hz, 1 to 1000 V/m) influence the interresponse timing of monkeys. Such exposure conditions would produce electric fields (estimated) in the brain of about 0.002 to 0.2 μV/m. Briefly, monkeys were trained to time the pressing of a lever so that there would be a timed limit range (5 to 7.5 sec) between the first and second lever pressings for a correct response. The 2.5- sec interval was termed the interresponse time (IRT). During each test session each monkey sat on a nonconducting stool between electrodes and performed at will a number of IRT trials. For each correct response the monkey received a small amount of juice as a reward. The authors indicate that a field-frequency effect (a "window") is observed in the results from the 10-V/m, 7-Hz exposure.[31] However, the IRTs from this particular regime are included within the larger range for those from the rest of the exposures at 1 V/m where there was "no obvious ordering of responses relative to either frequency or voltage."[31] Thus, the absence of a particular regime producing a consistent, shortest IRT argues strongly against the effectiveness of such fields to modify behavior.

Is there evidence elsewhere that weak fields elicit behavioral responses in monkeys? Not really. The most extensive investigations in this area have been those by deLorge (Table 1). Rhesus monkeys have been exposed to a variety of low frequency, weak field conditions and assessed for behavioral modifications (Table 1). None of these regimens produced modified behavior. Of the 10-Hz, 7-V/m field regimen deLorge[33] writes "When the animals were exposed to 10-Hz fields, statistically significant effects were observed, but they were not clinically significant because the effects did not occur in both subjects nor in either subject when the experiment was repeated." The results possibly depict in some sense the Cheshire cat phenomenon proposed by Graves et al.[34] — i.e., now you see it, now you don't.

B. Calcium Efflux

The subject of calcium efflux in relation to exposure to weak fields contains data that appear contradictory, confirmatory, or provocative — depending upon one's perspective. This area has received considerably more attention in exposures dealing with very high frequencies, has been reviewed by Adey,[38,39] Myers and Ross,[40] and Greengard et al.,[41] and

Table 1
SYNOPSIS OF BEHAVIORAL STUDIES BY
deLORGE[33,35-37]

Author	NAMRL Report No. (see Ref.)	Electric field (V/m)	Frequency (Hz)	Behavioral modification
deLorge	1155	4	75	No
	1179	3—7	45	No
		3—7	10	No
	1196	3—25	60	No
		5—10	10	No
	1203	9.2—8.4	45	No
		9.2	15	No

is reviewed in Part III, Chapter 4. In the ELF frequency range the data are somewhat limited. Initially Bawin and Adey[42] reported that weak sinusoidal electric fields (10 and 56 V/m, at 6 and 16 Hz) decreased the calcium efflux from in vitro, freshly isolated chick and cat cerebral tissues. To do these experiments the animals were first decapitated and then the appropriate brain portions quickly removed, sliced, immersed, and incubated in a radioactive calcium solution to "load" the cells, and then further incubated in a nonradioactive solution; the amount of calcium effluxed from the sytem was then determined. During the efflux part of the experiment some slices were exposed to electric fields by putting the culture vessels between energized electrodes. There were four exposure frequencies (1, 6, 16, and 32 Hz) and four electric field strengths (5, 10, 56, and 100 V/m in air). Statistically significant differences were noted for the tissues exposed to the 6- and 16-Hz and 10- and 56-V/m fields; all others were not significantly different from controls. Thus, the authors proposed "frequency and amplitude windows for the selective inhibition of calcium release from cerebral tissues."

There are several interesting facets of this research. First, since the field strengths reported in the research concern fields in air, it is obvious that at the tissue level the fields are extremely small. The authors note for their exposed tissues that the fields at the cellular level would be 10^{-7} V/m — a very low field when one compares this to the lowest level needed to affect the firing rate of Aplysia neurons (10^{-3} V/m), the growth rates of roots (Robertson et al.[14] [300 V/m]), or the regeneration of flatworms (Marsh[24] [300 V/m]). Second, it has been explained by Myers and Ross[40] that there is no indication of how many brain cells are viable in this type of experimentation with excised tissue.[42] Third, the authors state that "Extreme counts in any set more than 1.5 standard deviations away from the mean... were eliminated." It is obvious that as the numbers of samples and variability of the data are decreased the sensitivity of the statistical test diminishes and the probability of a false positive increases.

In an attempt to independently verify the results of Bawin and Adey,[42] Blackman and colleagues[43] decapitated chicks, removed and sliced the forebrains, loaded them with $^{45}Ca^{2+}$ for 30 min, and then exposed them for 20 min to 16-Hz electric fields ranging from 1 to 70 V/m. The amount of $^{45}Ca^{2+}$ effluxed to the medium during the exposure was determined. Half of each brain served as a control and the other half as an exposed specimen. The results indicated that tissue exposed to 16-Hz fields of 5, 6, 7.5, 35, 40, 45, and 50 V/m had $^{45}Ca^{2+}$ efflux yields significantly greater than those of the control. There was no effect at 1, 2, 3.5, 10, 20, and 30 V/m (16 Hz) or at 40 V/m at 1 and 30 Hz.

The experiments by Blackman et al.[43] may be taken to suggest that "windows" involving specific combinations of frequency and amplitude exist. However, the results of the two experiments described above are diametrically opposite, Bawin and Adey[42] indicating that calcium efflux is decreased, Blackman et al.[43] suggesting it is increased. Additionally, Bawin and Adey indicate that a "window" exists at 16 Hz 10 V/m, but for Blackman et al.[43] no effects were seen at that frequency/amplitude combination. Rather, Blackman's "window" is from 5 to 7.5 and 35 to 50 V/m, with increased calcium effluxes observed at both "windows." Bawin and Adey[42] also reported an effect at 56 V/m — a decreased calcium efflux. Recently, Blackman et al.[43a] reported experiments which seem to indicate that the different results obtained in their and the Bawin-Adey studies may be due to differences of magnetic field levels in the two exposure systems.

C. Human Circadian Rhythms and Other Parameters

Wever[44] has studied circadian rhythms of humans placed in underground, steel-reinforced concrete bunkers in Bavaria. There were two bunkers, each closely resembling the other in appearance, but somewhat different in construction. The bunkers were initially constructed of steel-reinforced concrete and placed in the side of a hill such that they were underground. Each bunker contained a living room, a bathroom, and a kitchen. In the living room were a bed, reading lamp, and chair. One of the bunkers (Room II) was built with additional (i.e., extra) steel reinforcement; additionally, it also contained electrodes in the inner walls so that artificial electric fields could be introduced. There was also an attempt made to eliminate any external cues of time through sound-deadening materials and absence of external light. The general purpose of the chambers was to isolate individuals from their normal temporal cues (including cues potentially obtained from electromagnetic sources), and then to determine if circadian rhythms were affected.

As a first approximation, it appears reasonably certain that the magnetic field of earth inside the extra-shielded bunker was reduced. Wever indicated that the magnetic field outside the bunker area was 4.1×10^{-5} T, inside the bunker without additional shielding 3.7×10^{-5} T (Room I), and inside the extra-shielded bunker 4×10^{-6} T (Room II). As Wever noted, this result suggests that the natural magnetic field penetrated the nonshielded room nearly undiminished but was almost completely absent in the specially shielded bunker. There were no measurements made of the electric or magnetic fields associated with the electric fixtures inside the bunkers.

In general, two types of circadian rhythms were studied: rectal temperature and activity (sleeping, awake). Rectal temperatures normally have a high and a low value at specific times of the day; so do activity and rest periods. There are two claims made by Wever about his data. First, the human activity cycle in the shielded bunker is initially stable but then by about day 14 becomes desynchronized or "free running". For the stable period the average activity cycle is 25.7 hr and for the "free running" period, 33.4 hr. During the latter period the rectal temperatures maintain a periodicity comparable to that of the earlier period. This type of result led Wever to postulate the presence of at least two independent "oscillators" in the body.

To explain the apparent differences in responses derived from the use of regular and specially shielded bunkers, Wever postulated "that normal electromagnetic fields which penetrate into Room I but not (or much more weakly) into Room II are responsible for these differences." To test the postulate, Wever imposed a vertical electric field (10 Hz, square wave, 2.5 V/m) for a portion of the confinement period and reported that in the absence of the artificial field the activity patterns were somewhat longer than with the field. The results are shown in Figure 8.

However, there appear to be two apparent activity cycles, the first from day 1 through day 5, the second from day 6 through day 18. The change in activity cycles appears to begin before field imposition. Thus, it is not clear that the applied field caused a change in circadian periodicity.

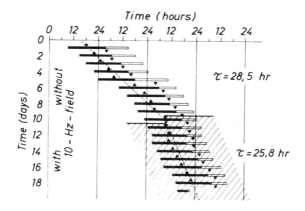

FIGURE 8. Free-running circadian rhythm of a subject living under strict isolation from environmental time cues; during the first section protected from natural and artificial electromagnetic fields; during the second section under the influence of a continuously operating electric 10-Hz field. Each bar is divided into rest (black) and activity (clear) periods, and the relative timing of the elevated (arrow head up) and depressed (arrow head down) rectal temperatures is noted. (From Wever, R., *Pflugers Arch.*, 302, 97, 1968. With permission.)

Additionally, for an exposure regime involving "without field", "with field", and "without field", in that order for 11, 11, and 7 days, respectively, none of the periodicities varied. All are 24.7 hr. In another comparable experiment, the periodicities were 26.6, 25.8, and 36.7 hr for 6-, 9-, and 5-day regimens, respectively.

There are some obvious difficulties in interpreting these data. First, it appears that human periodicities may vary somewhat with both time and the individual. A clear causal relationship between any of the experimental electromagnetic field conditions and human periodicity is absent. Second, the methodology is unclear and there is no quantitative evaluation of when periodicity changes occurred. Each day is divided into rest and activity periods; however, activity can include naps, and subjects were not constrained to rest during the rest periods. Whether the subjects were active or resting during the "active" and "rest" periods is not accurately recorded. Third, there are internal inconsistencies (as noted above) in the data and it is not readily apparent that a change in periodicity occurred with a change in the electromagnetic environment.

Consider also a report by Hamer[45] that human reaction time can be affected by low level electric (2 to 12 Hz) fields (in air, about 1 V/m). Participating as subjects in the project were 29 undergraduate students. The purpose of the project was to determine if subjects responded differently at relatively low (e.g., 6 Hz) and relatively high (e.g., 12 Hz) frequency fields. "The normal scatter in RTs (reaction times) of different subjects was controlled by testing all subjects at two frequencies, but not necessarily at the same pair of frequencies." Use of the same frequency pair for all subjects was considered unreasonable since there was not an *a priori* reason for the existence of unique frequencies best suited to all subjects. A crude criterion for frequency selection was based on each subject's RT: if his RT was fast, higher frequencies were used (e.g., 12 and 6); if his RT was slow, lower frequencies were used (e.g., 6 and 2). Thus, a person could be exposed to a 6-Hz field and it could be "high" or "low", depending on the subject. There was an attempt made to "mask any unconscious bias of the research assistant" by having test intervals varied with an "adequate rest period between tones." "To reduce variability, it was made standard procedure to discard both sets of a subject's data if the difference in mean RT was greater than 50 ms." The results of these procedures were that the higher frequency resulted in a somewhat longer RT than the lower frequency; the difference between the two was about 2 msec, and the standard deviation was about 30 msec. The difference in RT was reported to be significantly different ($p < 0.05$).

There are, however, two additional studies on human RT and exposure to very low frequency, low level electric fields. The exposure regimen of Konig[46] involved 3- and 10-Hz electric fields of 1 to 5 V/m. The 3-Hz regimen produced "an increase in reaction time — i.e., a decrease in performance." The 10-Hz regimen produced "a decrease in reaction time — i.e., an increase in performance." The number of trials was small, and as the authors point out, "the small number of cases did not allow for statistically significant results." However, the results were opposite to those reported by Hamer.[45] For Hamer, the lower frequency resulted in the smaller RT; for Konig, the lower frequency resulted in the longer RT. Persinger et al.[47] exposed humans to 0.3- and 30-V/m electric fields of 3 and 10 Hz and noted no differences in reaction times between exposed and control groups.

In an additional report dealing with circadian rhythms, Halberg et al.[48] investigated the effects of ELF fields (45 to 75 Hz, 0.4 to 2 G, and 1 to 180 V/m) on silk tree leaflets, flour beetles, and mice; none showed a conclusive effect.

VI. BIOEFFECTS OF FIELDS INVOLVING PERCEPTION LEVELS

For many reports of bioeffects there is a strong possibility that the organism being exposed can perceive that the electric field is "on". For instance, much research has been done on rodents exposed to 60-Hz electric fields at 100 kV/m. At this field strength, it appears very likely that the animals are aware of the presence of the field; Cabanes and Gary,[19] as mentioned above, showed that for rats and mice exposed to a 50-Hz electric field at 50 kV/m the vibrissae vibrated with an excursion of about 10 mm. Hair vibrations of this amplitude are certainly perceived by the animals. More recently, Stern and colleagues[20] demonstrated that for rats the threshold of detection of a 60-Hz electric field was generally between 4 and 10 kV/m,[20] and that detection was affected by animal shaving.[49] Kaune et al.[22] exposed pigs to 60-Hz electric fields and noticed ear flicking and head shaking at unperturbed fields above about 50 kV/m, a level at which hair vibration was observed to occur. Recently, Rosenberg et al.[50,51] indicated that for mice exposed to 60-Hz electric fields during the inactive phase (lights-on) of the circadian cycle, arousal responses occurred with field strengths of 50 kV/m or higher, but "rarely occurred with exposures at lower field strengths." The authors indicated that "the 'response threshold' of 50 kV/m...correlates closely with level expected to produce hair movements in rodents." Thus, it is obvious that if one is testing for effects of electric fields with animals, then surface stimulation, as through hair vibration, can be one mechanism causing effects.

Several studies have been conducted with people under laboratory conditions; they are presented in Table 3. However, before discussing the results in this general category it is important to recognize the care and rigor that are absolutely necessary for unequivocal substantiation of an effect caused by an applied electric (or magnetic) field. The report by Tucker and Schmit[52,53] is both illustrative and instructive.

The objective of Tucker and Schmit's[52,53] research was to determine whether or not humans could detect a 60-Hz magnetic fields of 7.5×10^{-4} T. Two Helmholtz coil pairs were built. The subjects, one at a time, sat in a chair between the coils and were asked to make a response after presentation of a cue; perception was the dependent variable.

The investigators soon learned that the mere presence of an operator visible to the test subject provided cues as to when the field was on or off. "No poker face is good enough to hide, statistically, knowledge of a true answer, and even such feeble clues as changes in building light, hums, vibrations and relay chatter are converted into low but significant statistical biases.'' The operator was soon moved to a remote room. For the initial large coil experiments (7.5×10^{-4} T) no special precautions were taken to isolate the subject from coil hum or associated vibrations; 12 of 35 individuals reported that they experienced a vague but definite ''hearing'' of something or a feeling of ''vibration''.

A modification was then made to the exposure chamber; it was vacuum-sealed and the coils were further isolated by additional wooden support and a plastic, compliant air cell sheeting installed. A few of the subjects still obtained very high perception levels; these few subjects again indicated that ''something'' could be heard. A re-examination of the exposure booth revealed that one plastic bubble ''had deflated, allowing the coil to rest more directly on its support.'' When the problem was corrected the high perceptions scores disappeared, but there appeared to be some residual but significant amount of detection. Ear plugs and a 120-Hz masking noise were then used and all scores dropped to a level explainable by chance.

The rigor of the experiments by Tucker and Schmit[53,53] is obvious. At any one point before the last analysis, had the investigation been halted, it would have been possible to claim that the subjects had detected the field. Dogged persistence and the finding and eliminating of confounding cues led to a proper conclusion about the sensitivity of the subjects to the potential stimulus.

A. Animal Studies

There is a sizeable number of studies in which rodents and other animals were exposed to ''high-field-strength'' electric fields. A sample of these studies is given in Table 2. All of the studies are phenomenologically oriented and unfortunately, appear not to have been based on hypothesis testing. The research appears to have been guided by the question as to whether or not the particular exposure conditions of a particular organism would produce an effect. Kaune et al.,[22] as mentioned above, attempted to define the threshold of electric field perception for pigs. Four pigs were used; it was observed that an exposure of 30 kV/m was an approximate threshold for detection. However, the authors were not sure that it was the electric field that was the cue for detection of the animals since audible signals were also present during exposure. In the Marino et al.[57] experiments with rats exposed to 60-Hz electric fields at 15 kV/m for 28 days, the one reasonably consistent effect is that the exposed rats drank less water; this effect was statistically significant in five of six experiments. Phillips et al.,[58] however, exposed rats to 60-Hz electric fields at 100 kV/m and did not observe any differences in water consumption between exposed and control rats. Why this disparity in results? To gain some insight into possible reasons one must scrutinize the exposure arrangements. Marino et al.[57] protocol involved the placement of rats between the electrodes. The animal were ''electrically floating'', the metal cage top served as the electric ground, and the water bottle was electrically bonded to the cage top. Thus, whenever the animals drank they grounded themselves (through the snout). There is a strong possibility of microschocks to the animals in this exposure system. In the Hilton and Phillips[59] exposure chamber the animals were on the ground plane and the water nozzles were electrically bonded to the plane so that the animals and spigot were at the same potential. Under these conditions the likelihood of transient microshocks is reduced.

Cerretelli et al.[60] reported no effects on animals exposed to a 50-Hz, 10-kV/m electric field. Serum chemistry changes (increased neutrophils and decreased lymphocytes) were reported for rats exposed briefly or chronically to a 50-Hz, 100-kV/m electric field (the experiment was done once); such an effect was not noted by Ragan et al.[61] for rats chronically exposed to a 60-Hz, 100-kV/m electric field or in the Cerretelli et al.[60] experiment with dogs exposed to 50-Hz, 25-kV/m electric fields. Graves et al.[34] reported a possible initial ''orientation'' of mice to 60-Hz, 50-kV/m electric fields. An elevated corticosterone level observed after a 5-min initial exposure (the new environment — done only once) was not

Table 2
ILLUSTRATIVE ANIMAL LABORATORY EXPERIMENTS

Author(s)	Setting	Results
Phillips et al.[58]	Rats and/or mice, 60 Hz, 100 kV/m Hematology, *Immunology,+ growth,* bone growth,* bone fracture repair,+ endocrinology,* cardiovascular function,* neuro-physiology,+ reproduction,* behavior,+ perception+	Negative* +Some positive results
Stern et al.[20]	Rats, 60 Hz, 100 kV/m	Detection threshold ~4 kV/m
Kaune et al.[22]	Field perception (pigs)	Threshold 30 kV/m
Grissett[70]	Growth, physiology (monkeys, 147 weeks, 75 Hz, 2 G, 20 kV/m)	Exposed males slightly heavier
Smith et al.[80]	Growth (rats, 60 Hz, 20 kV/m, 30 days)	Negative
Seto[81]	Growth (rats, 60 Hz, 20 kV/m)	Negative
Knickerbocker et al.[63]	Growth, reproduction (mice, 60 Hz, 160 kV/m, P_1, F_1)	Negative
Mathewson et al.[82]	Growth (rats, 45 Hz, 100 kV/m, 28 days)	Negative
Kreuger and Reed[83]	Growth (mice, 45—75 Hz, 100 kV/m, 28 days)	Negative
deLorge et al.[33,35-37,84-86]	Behavior, physiology (monkeys, 7—75 kV/m, 3—10 V/m)	Negative
Graves et al.[34]	Physiology, behavior (mice, 60 Hz 50 kV/m, 6 weeks; pigeons, 21 kV/m)	Corticosterone level elevated first 5 min at 50 kV/m pigeon perception
Marino et al.[57]	Growth, physiology (rats, 60 Hz, 15 kV/m, 28 days)	Decreased water consumption
Cerretelli et al.[60]	Cardiac, physiology, growth immunology (mice, rabbits, rats, dogs, 50 Hz, 100 kV/m, 2 months)	No effects at 10 kV/m
Marino et al[87,88]	Growth, reproduction, 3 generations (mice, 60 Hz, 3.5—15 kV/m)	Increased and decreased weights, increased and decreased mortality
Marino et al.[89]	Bone fracture healing (rats, 60Hz, 1—5, kV/m, 14 days)	Healing depressed 5 kV/m
McClanahan and Phillips[73]	Bone growth, fracture repair (rats, 60 Hz, 100 kV/m, 30 days)	Negative for bone growth; fracture repair slight retard
Rosenberg et al.[50,51]	Arousal, 60 Hz, 0—100 kV/m (rats)	50-kV/m threshold

observed with analyses from chronic (2 to 6 weeks) exposure. Fam[62] has reported that exposure of mice to very intense 60-Hz electric fields (240 kV/m) did not affect growth and development of males but had some retarding effect on females; however, no effect on breeding capacity or number of progeny born or progeny body weight was observed. Field exposure apparently did produce differences in a few blood chemistry analyses, but had no apparent effect on the histological appearance of organs. The animals were ''electrically floating'' and it is not clear how the animals received their water.

In a much earlier study, Knickerbocker et al.[63] examined the effects of intense 60-Hz electric fields on mice. At 1000 kV/m the animals were paralyzed; at 700 kV/m the hair of the animals was in "constant erection" and the animals were in severe discomfort. At 200 kV/m the animals were highly aroused but at 160 kV/m they were not "markedly disturbed". However, at 160 kV/m the field was "sufficiently strong that if the animal stood up on his hind legs *corona could be heard,* but this seldom occurred" (emphasis added). Thus, it is obvious that at very high electric field strengths, there can be very substantial perturbations in the field, particularly as the animal rears (as mice and rats are prone to do). Kaune and Phillips[11] experimentally verified, as discussed earlier, that for rats in the supine and erect positions the fields over various parts of the body and the induced electric current and its distribution are dramatically different; in essence, rats in the supine position have maximal fields across their back, erect rats have maximal fields at the tips of their noses, and the total induced current in the erect rat is twice that of the supine rat.

A long-term multigeneration teratology study on Hanford miniature pigs exposed continuously to 60-Hz, 30 kV/m electric fields was completed in 1983 at the Pacific Northwest Laboratories. The results of the experiment, which has been done only once and apparently will not be repeated, are presently available only in abstracts. The pigs were also assessed for changes in growth, hematology and serum chemistry profiles, humoral and cell-mediated immunity, blood hormone levels, peripheral nerve function, behavior, reproductive function, and prenatal developments. Phillips[64,65] writes "Few remarkable differences were observed between electric-field exposed and control populations except in behavior and in reproduction and development. Statistically significant differences between exposed and control populations were: 1) Fewer prenatal deaths among fetuses from exposed F_0 bred to unexposed boars at 4 months of exposure, and as a consequence, more live fetuses/litter; 2) an increased incidence of morphological malformations and a smaller body mass among fetuses from exposed F_0 bred to unexposed boars at 18 months of exposure; 3) an impaired mating performance of F_1 bred at 18 months of exposure; 4) an increased incidence of birth defects and a lower body mass among (F_2) from F_1 bred at 18 months of exposure." There were no differences in fetal anomalies between exposed, rebred F_1 sows mated to unexposed boars.

It is not possible to state categorically that the electric field exposure caused the developmental anomalies. The number of animals tested was small (e.g., 16 pregnant sows in the F_1 exposed group and 8 pregnant sows in the control), there appeared to be a high incidence of birth anomalies in litters from controls, and a severe dysentery swept the herd a few months before the second breeding of the F_0.

A similar breeding schedule was then begun with rats[64,65] and guinea pigs[66] exposed to 60-Hz, 100 kV/m electric fields. The rat experiment was done twice and there were no consistent differences between control and exposed groups. The guinea pig study produced results only for the F_0 (first and second breeding) before a fire terminated the experiment; there were no statistically significant effects for either breeding cycle. This information, as with the pig teratogenesis study, is presently available only in abstract format.

Greenberg et al.[67,68] observed that bee hive colony behavior and honey production were affected by 60-Hz electric fields of 7, 5.5, and 4.1 kV/m, but not by fields of 1.8 kV/m or lower. Procedures which reduced or eliminated fields inside the hive also mitigated against behavioral and productivity changes. It was not known if the causative factor(s) was the applied field or transient discharges from within the hive. However, in preliminary investigations which eliminated shocks at the hive porches, no effects on the propolization were evident with fields up to 100 kV/m.[69]

Grissett and colleagues[70] have undertaken a large screening study of monkeys exposed to 76 Hz, 2 G, 20 V/m electromagnetic fields; these fields were chosen to represent a worst case simulation of the fields associated with the proposed submarine communications antenna of the Navy, known originally as Project SANGUINE (more recently SEAFARER and now called simply ELF communication system). This study is included for discussion in the category of "bioeffects involving perception levels" because the exposure system has a strong likelihood for injecting sizeable amounts of current (e.g., 5 mA) into the gonads of the exposed male monkeys. Pubescent monkeys were matched (control, exposed) and subsequently examined weekly for a variety of growth and physiological parameters. The most significant result after 147 weeks of continuous exposure was that the exposed male monkeys had a slightly but significantly faster growth rate than their control counterparts. The difference in weight occurred during the first year.[71]

Phillips' project[58] at the Pacific Northwest Laboratories using rats and mice involved a massive number of animals and assays. Additionally, the exposure facilities were engineered to eliminate or reduce many of the potentially confounding variables that had not been carefully controlled in other investigations. For example, control and exposed animals were exposed to the same sounds, potential vibrations of electrodes, and other environmental variables; the experiments were also done "blind" and were usually replicated. In short, the possibility of secondary confounding variables is greatly reduced in these experiments. In general, the results are overwhelmingly negative. Some of the information has at this writing already been published in the peer-reviewed literature.[59,61,72-79] It should be borne in mind that this large study was established as a screening procedure to obtain information useful for estimating whether or not high voltage transmission line electric fields posed a hazard to human health and welfare.

It is reasonable to conclude that there are some biological effects in the Phillips et al.[58] data. For behavioral studies it was noted that rats appeared to avoid field strengths > 75 kV/m but preferred to be in field strengths of 50 kV/m as opposed to the no-field condition.[76] Fracture repair in experimentally broken leg bones was retarded in rats exposed to 100 kV/ m, but there was no effect on normal developmental bone processes.[73] Both effects could well be mediated by a surface stimulation. For example, rats exposed to fields at 100 kV/ m may be slightly more active due to continuous surface stimulation caused by the high field. Since the fractured leg is uncasted, a slightly greater movement could be responsible for the slight delay in fracture healing. Neuroendocrine responses in rats exposed to 60-Hz electric fields appear to include reduction in the normal night-time increase of melatonin concentration in the pineal gland[78] and effects on serotonin metabolism.[90,91]

The interpretation of any positive results in this category of studies is presently problematic. First, the majority of the studies in negative. Second, the absence of hypothesis testing in any of them precludes direct evaluation of any possible physical mechanism of action causing an effect. Third, it is readily apparent (whether or not the investigation considered if the animals perceived the fields) that for the studies in this category in all likelihood, the animals were generally aware of the presence of the field by some surface or peripheral stimulation. The question arises, then, as to whether or not the positive effects — e.g., retarded bone healing for rats in fields of 5[89] and 100 kV/m[73] are due to surface phenomena (e.g., hair vibration which stimulated the animal to increased movement) or to some effect caused by the electric fields within the tissues (i.e., cellular stimulation). Kaune and Phillips[11] have indicated that for a rat in a 10-kV/m, 60-Hz electric field, the current density in the paw is about 14 A/m^2. Thus, for a 100-kV/m electric field the current density would be about 140 A/m^2. Away from the paw, and toward the body, the current density would be much less. From Ohm's law one can estimate the approximate electric field in the area of the fracture. Thus, for a current density of < 14 mA/m^2 and a soft tissue conductivity of 0.2 S/m, the approximate field at the tissue level is < 0.07 V/m — a level without an effect on bone growth[73] and at least an order of magnitude of less than that demonstrated to affect neural stimulation[7] or root growth rates.[24] Tissue inhomogeneities would not appear to be a major factor in estimating the induced electric fields in the fracture area of the leg. The tissues

Table 3
ILLUSTRATIVE HUMAN LABORATORY EXPOSURES INVOLVING ELECTRIC FIELDS, MAGNETIC FIELDS, ELECTRIC CURRENTS, AND BIOLOGICAL ASSAYS

Author(s)	Setting	Results
Hauf[93]	1—20 kV/m[a], 50 Hz, 3 hr Behavior, blood	Negative (some slight differences but within normal range)
Manntell[94]	30 kV/m[a], 50 Hz, 75 min Behavior	Negative
Johannson et al.[95]	30 kV/m[a], 50 Hz, 24 hr Behavior, blood (including triglycerides)	Negative
Eisemann[96]	200 μA, 50 Hz, 3 hr Behavior, blood	Negative
Amon[97]	20 kV/m[a], 50 Hz, 5 hr Physiology	Negative
Rupilius[98]	3 G, 20 kV/m[a], 50 Hz Behavior, blood (including triglycerides)	Negative
Carstensen et al.[92]	60 Hz, 1 V/m (in eye) Electrophosphene stimulation	Positive

[a] Electric field in air.

(bone, muscle) are arranged somewhat parallel to each other and thus there would be comparable fields in the bone and soft tissue.

The noted effects on neuroendocrine responses in rats[91] are not understood at the present time, and no effort to confirm them in another laboratory has thus far been made.

B. Human Studies

In general, there have been effects reported in people from exposure to ELF electric or magnetic fields. First, it is obvious as mentioned earlier, that there can be "surface effects" leading to perception of electric fields. Cabanes[19] had reported that for erect people with both arms kept close to the body, 5% of the individuals perceived a 50-Hz field of 5 kV/m; 60% perceived the field at 27 kV/m. With one arm stretched horizontally, 5% of the people perceived a field of 2 kV/m and 90% perceived the 27-kV/m field. With one arm raised above the head, 5% perceived a field of 2 kV/m and all perceived the field at 27 kV/m. In studying the mechanism of perception an exposure chamber was fabricated; essentially, there were two metal plate electrodes separated by 30 cm — "just long enough so the forearm of a volunteer" could be exposed. When the arm was shaved there was no perception of the field, "even under the effects of a local field of 100 kV/m." In addition to perception from surface stimulation, neural perception of electric fields has occurred with volunteers grasping energized electrodes; perception resulted when the current density in the forearm was of the order of 1 A/m^2 (and corresponding extracellular applied electric fields of about 10 V/m). These results are discussed in more detail by Schwan.[7] There has been, however, little success in finding effects other than perception in volunteers exposed to air electric and magnetic fields (Table 3). However, Carstensen et al.[92] recently reported that the threshold for perception of ocular electrophosphenes is approximately 1 V/m at 60 Hz, or about an order of magnitude lower than the sensitivity threshold of peripheral nerves. The current (administered by an eye-cup electrode) necessary to achieve this effect was within the range of other reports[99-102] regarding power frequency thresholds for phosphene stimulation. The human eye thus appears to be relatively sensitive to electric field exposure.

VII. EPIDEMIOLOGICAL STUDIES

An epidemiological investigation ideally involves individuals who have not been exposed to a particular agent and others who have. In all other variables there should be reasonable conformity between the two groups. For example, the two sets would be matched for age, sex, background, etc. Epidemiological investigations can be either retrospective — i.e., comparing sets of people which have been exposed — or prospective — i.e., comparing sets of people which are being exposed. In most instances there are compromises drawn as to sample composition — one can rarely obtain exactly matched groups — and relatively large sample sizes are required in order to obtain meaningful results. For instance, thousands of individuals have been involved in studies which dealt with the issue of smoking and cancer, and in this instance it was relatively easy to obtain sets of individuals who did not smoke and to compare them to individuals who did. Additionally, for those who did smoke there was information as to the relative amount of smoking — the number of smokes per day, the length of time one has smoked, etc.

For electric field exposure it appears virtually impossible to produce a meaningful epidemiological study. First, because of the electrification of much of the world, it is virtually impossible to find individuals who have not been exposed to 50- or 60-Hz electric fields. Consider the difficulty in attempting to deal with the question of which person, a high voltage transmission lineman or a housewife, is exposed to strong electric fields. The transmission lineman works in environments where there obviously are strong electric fields. He may wear protective shielding and work in a shielded box. However, there are monitors available for providing information on time-averaged, induced current density levels in the body. As a "ball park" estimate of total body currents one might consider the amount of current induced in the body of a grounded person standing in the maximum electric field of a large transmission line; e.g., assuming an electric field of 10 kV/m, the induced total body current is about 160 µA.[9,10] Consider next the housewife. The appliances in the home generally have small electric fields associated with them but the currents that leak from them ("leakage current") can be very sizeable.[103] The housewife may or may not be grounded. The American National Safety Institute (ANSI), which operates the Underwriter's Laboratory, permits as "safe," leakage current levels of up to 500 and 750 µA for portable and fixed appliances, respectively,[104] although in reality these levels are seldom reached under normal conditions. Thus, it is obvious that individuals engaged in activities at home, e.g., cooking and cleaning, could have very substantial conduction body currents. The body currents of a person at home can be considerably larger than those induced in the person standing in the maximum field of an extra-high voltage transmission line. The time duration of such currents in the person at home may be, and probably is, much less than in the transmission line worker since we normally do not have extended contact with operating appliances. Of course, a sewing machine operator probably has considerable contact with the device for extended periods and may have a larger time exposure to body currents than the lineman.

Second, with regard to the type of occupation involved, what type of individual should serve as a control for, for example, a lineman? Ideally, one would like to have a person engaged in the same type of work but not exposed to electric fields — i.e., a lineman working on unenergized lines. Such an individual would be engaged in all the types of activities of a lineman working on "live" wires but would not have the field exposure. In this way the investigator would have a matched sample (provided other variables were not confounding). A less satisfactory procedure would be to have comparisons between different categories of lineman, e.g., those working on high and low voltage transmission lines, or those working on lines vs. individuals working in switchyards. To some extent, preliminary results have been reported on this topic. Sazanova[105] reported the results of some behavioral and physiological tests of 54 people working in open switchyards. Two groups were compared: operating personnel ("working under electric field exposure of not more than two hours daily" [29 persons]) and maintenance personnel ("exposed not less than five hours daily" [25 persons]). A series of tests was conducted on data from both groups, before and after work daily for 6 days. The measured parameters included temperature, pulse, blood pressure, RT and quantitative error, critical flicker frequency, and reaction of the adductor muscle of the thumb to electrical stimulation.

Sazanova[105] presented 1-day measurements. No significant changes were seen in temperatures or critical flicker frequency. A small difference was seen in the pulse rates at the end of the day, due to a slight increase in pulse rate from the beginning of the day in the operating personnel. The average blood pressure in the maintenance personnel was significantly lower than that of the operating personnel both at the beginning and the end of the day. An increase in RT and quantitative error was seen in both groups from the beginning to the end of the day, with a greater change in the maintenance personnel. An increase in the sensitivity of electric stimulation of the adductor muscle of the thumb was observed in the maintenance workers from the beginning to the end of the day; no such change was observed in the operating personnel.

To attribute the above differences to the effects of electric fields, two conditions must be met. First, the two groups must be similar in age, sex, physical condition, etc. Second, the work and work environment for the two groups must be similar in all respects except for the presence or absence of the electric field. The information provided by Sazanova[105] does not indicate that either of these conditions was met. While data were presented indicating age distribution and work experience of the personnel taken as a whole (maintenance and operating personnel), no information is given as to sex or as to age distribution or work experience of the personnel in each group. Thus, the difference in blood pressure between the two groups could be solely due to differences in age and/or sex and have no relationship to the work environment.

It is also doubtful that the type of work and the work environments (other than the electric field) are similar for the two groups. It is stated that "Workers for both groups carry out light work with elements of intellectual and physical stress." However, it is also indicated that the operating personnel work indoors while the maintenance personnel work anywhere in the switchyard and that the maintenance work includes repair of air breakers, disconnects, current transformers, etc. The differences seen in RT and quantitative error and in sensitivity of the thumb to electrical stimulation could be indicative of greater fatigue in the maintenance workers at the end of the day.

In summary, a number of physiological differences was seen between operating personnel with relatively low electric field exposure and maintenance personnel with relatively high electric field exposure. However, the results did not indicate abnormalities in either group, and it is impossible to exclude a large number and variety of associated factors other than electric field exposure, such as spark discharge, as the cause of the observed differences. A similar criticism applies to the report by Asanova and Rakov,[106] who reported on subjective evaluations of 400 to 500-kV substation employees (electricians, signalmen, and secondary personnel) and to a similar type of analysis of the subjective parameters of personnel from 500- and 110/120-kV substations.

Studies in the West have not revealed medical problems associated with working in the environment of high voltage transmission lines; for some workers the tasks have involved barehand work on live wires. The results are summarized in Table 4.

Table 4
ILLUSTRATIVE WORKING SITUATIONS: AC
TRANSMISSION LINE EXPOSURE INVESTIGATION
FOR BIOLOGICAL EFFECTS

Author(s)	Setting	Results
Asanova and Rakov[106]	400/500 kV switchyards	Positive
Kouwenhoven et al.,[113] Singewald et al.[114]	9-year survey, 345 and 765 kV transmission lineman	Negative
Roberge[115]	735-kV Switchyard personnel	Negative
Strumza[116]	Residents near, far from, 200—400 kV	Negative
Krumpe and Tockman,[117] Houck[118]	5-year survey for SEAFARER, controls	Negative
Hodges and Mitchell[119]	765 kV, crops	Negative
Knave[120]	5-year survey, 400 kV substantion workers, controls	Negative
Sazanova[105]	400/500 kV switchyards	Positive
Stopps and Janischewskyi[121]	High-voltage equipment and TL workers	Negative
Amstutz and Miller[122]	765 kV, farm animals	Negative

A recent occupational epidemiological investigation of workers at high voltage substations in Sweden[107] indicated a statistically significant reduction in percent normal births of offspring from mothers married to 400-kV switchyard workers relative to that of normal births associated with other 400-kV workers plus controls. When the same statistical test (Chi-square) is applied to the 400-kV switchyard workers vs. controls, there is no difference.

A number of surveys has suggested an association between working in high strength electric and/or magnetic fields and the incidence of cancer.[108-111] None of these studies has measured an electric or a magnetic field; each is, in essence, a letter to the editor, and each has some sort of statement to the effect that electric or magnetic fields may cause cancer. They are also not internally consistent, for example, for the Wright et al.[109] and Coleman et al.,[111] reports the proportionate mortality ratios (PMRs) for telegraph operators are 0 and 246, respectively. Consider, for example, the leukemia mortality study by Milham[108] of white males in the state of Washington from 1950 to 1979. There is a total of 196 deaths over a 30-year period for 11 different job categories, or 0.59 deaths per year per category. This amounts to being an extraordinarily small study. Aluminum workers are reported to have a very high PMR, but aluminum smelting also may involve the production of some fairly toxic fumes. Welders, in this study, do not have a high incidence of leukemia; their PMR is 56 to 67 and is the lowest in the 11 job categories. Yet, arc welders hold in their hand a device which undoubtedly has a very high magnetic field associated with its operation; arc welding generally involves electric currents up to 50 A. Thus, if one tries to "guestimate" exposure to say, a magnetic field, welders should be "high" for exposure, yet they have a low PMR. In addition to the Washington State study, Petersen and Milham[112] also did a comparable study for California white males from 1959 to 1961: there were no significant differences for lineman (PMR 45), electricians (PMR 100), electrical engineers (PMR 92), and welders and flame cutters (PMR 118). On the other hand, foresters and conservationists had a PMR of 209 for the incidence of leukemia.

VIII. CONCLUSION

That living organisms can be affected by applied ELF electric fields is well known; the induced effects include, for example, reductions in root growth rates, perception of buried prey by predatory fish, and modulation of a pacer neuron. Effects at the cellular level, in general, appear to be related to the applied field — not its associated current, and to be mediated through a change in transmembrane potential. Thresholds for inducing a biological effect with respect to the extracellular field strength vary considerably. Certain fish have specific electrosensitive organs and can detect fields as low as 10^{-6} V/m. The *Aplysia* pacer neuron has a threshold of about 3×10^{-3} V/m for pacer firing rate modulation. For electrophosphene stimulation in the human eye, fields of about 1 V/m are needed and pea root meristem cells have a threshold level of about 300 V/m for affecting root growth rates. At the whole animal level, perception can be achieved by surface stimulation such as hair vibration, which typically requires fields of a few thousand volts per meter in air for animals and man. Neural perception in humans can also be achieved through direct contact with energized electrodes; a tissue electric field of the order of 10 V/m is necessary to achieve this sensory effect. The coupling of animal tissues to an electric field applied through air is however, very poor, and the resulting tissue fields are many orders of magnitude smaller than the applied field in air.

ACKNOWLEDGMENT

I thank my colleague, Prof. Edwin L. Carstensen, for his many helpful comments and discussion during the preparation of this review.

REFERENCES

1. **Polk, C.,** Sources, propagation, amplitude and temporal variations of extremely low frequency (0—100 Hz) electromagnetic fields, in *Biologic and Clinical Effects of Low-Frequency Magnetic and Electric Fields,* Llaurado, J. G. et al., Eds., Charles C. Thomas, Springfield, Ill. 1974, 21.
2. **Miller, D. A.,** Electric and magnetic fields produced by commercial power systems, in *Biologic and Clinical Effects of Low-Frequency Magnetic and Electric Fields,* Llaurado, J. G., et. al., Eds., Charles C. Thomas, Springfield, Ill., 1974, 62.
3. **Barnes, H. C., McElroy, A. J., and Charkow, J. H.,** *IEEE Trans. Power Appar. Syst., 80,* 482, 1967.
4. **Kaune, W. T.,** Power-frequency electric fields averaged over the body surfaces of grounded humans and animals, *Bioelectromagnetics,* 2, 403, 1981b.
5. **Smith, M., D'Andrea, J. A., Gandhi, O. P., and Johnson, C. C.,** Behavioral and physiological effects on rats of chronic exposure to strong 60-Hz electric fields, 1977 Int. Symp. Biological Eff. Electromagnetic Waves, Airlie, Va., 132.
6. **Johnson, J. G., Poznaniak, D. T., and Mckee, G. W.,** Prediction of damage severity on plants due to 60-Hz high-intensity electric fields, in *Biological Effects of ELF Electromagnetic Fields,* Phillips, R. D., Gillis, M. F., Kaune, W. T., and Mahlum, D. D., Eds., Conf. 781016, Proc. 18th Annu. Hanford Life Sci. Symp., Richland, Wash. 1979, 172.
7. **Schwan, H. P.,** Biological hazards from exposure to ELF electric fields and potentials, U.S. Naval Weapons Laboratory, Tech. Rep. 2713, Dahlgren, Ill., 1972.
8. **Schwan, H. P.,** Biophysics of the interaction of electromagnetic energy with cells and membranes, in *Biological Effects and Dosimetry of Nonionizing Radiation,* Grandolf, M., et al., Eds., Plenum Press, New York, 1983. 213.

9. **Deno, D. W.,** Monitoring of personnel exposed to a 60-Hz electric field, in *Biological Effects of ELF Electromagnetic Fields,* Phillips, R. D., Gillis, M. F., Kaune, W. T., and Mahlum, D. D., Eds., Conf. 181016, Proc. 18th Annu. Hanford Life Sci. Symp., Richland, Wash., 1979, 93.

10. **Deno, D. W. and Zaffanella, L. E.,** Field effects of overhead transmission lines and stations, in *Transmission Line Reference Book. 345KV and above,* 2nd ed., LaForest, J. J., Ed., Project UHV, Technical Resource Operations, Large Transformer Division, General Electric Co., Pittsfield, Mass, 1982, 329/625.

11. **Kaune, W. T. and Phillips, R. D.,** Comparison of the coupling of grounded humans, swine and rats to verticel, 60-Hz electric fields, *Bioelectromagnetics,* 1, 117, 1980.

12. **Miller, M. W., Carstensen, E. L., Kaufman, G. E., and Robertson, D.,** 60-Hz electric field parameters associated with the perturbation of a eukaryotic system, in *Biological Effects of ELF Electromagnetic Fields,* Phillips. R. D., Gillis, M. F., Kaune, W. T., and Mahlum, D. D., Eds., Conf. 781016, Proc. 18th Annu. Hanford Life Sci. Symp., Richland, Wash. 1979, 109.

13. **Roberston, D., Miller, M. W., and Carstensen, E. L.,** Relationship of 60-Hz electric field parameters to the inhibition of growth of *Pisum sativum* roots, *Radiat. Environ. Biophys.,* 19, 227, 1981a.

14. **Roberston, D., Miller, M. W., Cox, C., and Davis, A. T.,** Inhibition and recovery of growth processes in roots of *Pisum sativum* L. exposed to 60-Hz electric fields, *Bioelectromagnetics,* 2, 329, 1981b.

15. **Inoue, M., Miller, M. W., Cox, C., and Cartsensen, E. L.,** Growth rate and mitotic index effects in *Vicia faba* L. roots exposed to 60-Hz electric fields, *Bioelectromagnetics,* 6, 293, 1985.

16. **Frohlich, H.,** Possibilities of long- and short-range electric interactions of biological systems, *Neurosci. Res. Program Bull.,* 15, 67, 1977.

17. **Inoue, M., Miller, M. W., and Carstensen, E. L.,** On the relationship between sensitivity to 60-Hz electric fields and induced transmembrane potentials in plant root cells, *Radiat. Environ. Biophys.,* in press.

18. **Miller, M. W., Dooley, D. A., Cox, C., and Carstensen, E. L.,** On the mechanism of 60-Hz electric field induced effects in *Pisum sativum* L. roots: vertical field exposures, *Radiat. Environ. Biophys.,* 22, 203, 1983.

19. **Cabanes, J. and Gary, C.,** Direct perception of electric field, CIGRE Report, Stockholm, 1981, 233.

20. **Stern. S., Laties, V. G., Stancampiano, C. V., Cox, C., and deLorge, J. O.,** Behavioral detection of 60-Hz electric fields by rats, *Bioelectromagnetics,* 4, 215, 1983a.

21. **Stern, S. and Laties, V. G.,** Behavioral effects of 60-Hz electric fields. I. Detection of 60-Hz electric fields by rats, U.S. Department of Energy Contractors' Review (Abstr.), November 1982.

22. **Kaune, W. T., Phillips, R. D., Hjeresen, D. L., Richardson, R. L., and Beamer, J. L.,** A method for the exposure of miniature swine to vertical 60Hz electric fields, *IEEE Trans. Biomed. Eng.,* 25, 1978.

23. **Miller, M. W., Carstensen, E. L., Roberstson, D., and Child, S. Z.,** 60 Hz electric field parameters associated with the perturbation of a eukaryotic cell system, *Radiat. Environ. Biophys.,* 18, 289, 1980.

24. **Marsh, G.,** The effect of sixty-cycle AC current on the regeneration axis of *Dugesia, J. Exp. Zool.,* 169, 65, 1968.

25. **Friend, A. W., Finch, E. D., and Schwan, H. P.,** Low frequency electric field induced changes in the shape and mobility of amoebas, *Science,* 187, 1975.

26. **Jaffe, L. F. and Nuccitelli, R.,** Electrical controls of development, *Annu. Rev. Biophys. Bioeng.,* 6, 445, 1977.

27. **Easu, K.,** *Plant Anatomy,* 2nd ed., John Wiley & Sons, New York, 1965, 116.

28. **Inoue, M., Miller, M. W., Cox, C., and Brayman, A. A.,** Absence of an electrolytic contaminant effect from a 60-Hz electric field exposure sufficient to reduce root growth rate of *Pisum sativum* L., *Environ. Exp. Bot.,* 25, 89, 1985.

29. **Wachtel, H.,** Firing-pattern changes and transmembrane currents produced by extremely low frequency fields in pacemaker neurons, *Biological Effects of Extremely Low Frequency Electromagnetic Fields,* Phillips, R. D., Gillis, M. F., Kaune, W. T., and Mahlum, D. D., Eds., Conf. 781016, Technical Information Center, U.S. Department of Energy, National Technical Information Service, U. S. Department of Commerce, Springfield, Va., 1979, 132.

30. **Schmitt, F. O., Der, P., and Smith, B. H.,** Electronic processing of information by brain cells, *Science,* 193, 114, 1976.

31. **Gavalas-Medici, R. and Magdaleno, S. R.,** An evaluation of possible effects of 45 Hz and 75 Hz electric fields on neurophysiology and behavior of monkeys, Technical Report ONR Contract N00014, 1975, 296.

32. **Gavalas-Medici, R. and Day-Magdaleno, S. R.,** Extremely low frequency, weak electric fields affect schedule-controlled behavior of monkeys, *Nature,* 261, 256, 1976.

33. **deLorge, J.,** Experiment 3, Naval Aerospace Medical Research Laboratory, NAMRL-1196, AD774106, 1973.

34. **Graves, H. B., Long, P. D., and Pozniak, D.,** Biological effects of 60-Hz alternating-current fields: a Cheshire cat phenomenon? in *Biological Effects of ELF Electromagnetic Fields,* Phillips, R. D., Gillis, M. F., Kaune, W. T., and Mahlum, D. D., Eds., Conf. 781016, Proc. 18th Annu. Hanford Life Sci. Symp. Richland, Wash., 1979, 184.

35. **deLorge, J.,** Operant behavior of Rhesus monkeys in the presence of extremely low frequency-low intensity magnetic and electric fields: Experiment 1, Naval Aerospace Medical Research Laboratory, NAMRL-1155, AD754058, 1972.

36. **deLorge, J.,** Operant behavior of Rhesus monkeys in the presence of extremely low frequency-low intensity magnetic and electric fields: Experiment 2, Naval Aerospace Medical Research Laboratory, NAMRL-1179, AD764532, 1973.

37. **deLorge, J.,** A psychological study of Rhesus monkeys exposed to extremely low frequency-low intensity magnetic fields, Naval Aeropsace Medical Research Laboratory, NAMRL-1203, AD000078, 1974.

38. **Adey, W. R.,** The influence of impressed electrical fields at eeg frequencies on brain and behavior, in *Behavior and Brain Electrical Activity,* Burch, N. and Altshuler, H. L., Eds., Plenum Press, New York, 1974, 363.

39. **Adey, W. R.,** Tissue interactions with nonionizing electromagnetic fields, *Physiol. Rev.,* 61, 435, 1981.

40. **Myers, R. D. and Ross, D. H.,** Radiation and brain calcium. A review and critique, *Neurosci. Biobehav. Rev.,* 5, 503, 1981.

41. **Greengard, P., Douglas, W. W., Nairn, A. C., Nestler, E. J., and Ritchie, J. M.,** Effects of electromagnetic radiation on calcium in the brain, Aeromedical Review 2-82, FY7624-81-00065, F33615-78-0-0617; Report # SAM-TR-82-15, U.S. Air Force School of Aeropsace Medicine (RZP), Aerospace Medical Division (AFSC), Brooks Air Force Base, Tex., 1982.

42. **Bawin, S. M. and Adey, W. R.,** Sensitivity of calcium binding in cerebral tissue to weak environmental electric fields oscillating at low frequency, *Natl. Acad. Sci. U.S.A.,* 78, 1999, 1976.

43. **Blackman, C. F., Benane, S. G., Kinney, L. S., Jounes, W. T., and House, D. E.,** Effects of ELF fields on calcium-ion efflux from brain tissue *in vitro, Radiat. Res.,* 92, 510, 1982.

43a. **Blackman, C. F., Benane, S. G., House, D. E., Rabinowitz, J. R., and Joines, W. T.,** A role for the magnetic component in the field-induced efflux of calcium ions from brain tissue, Open Symp. Interaction Electromagnetic Fields Biological Syst., XXI Gen. Assembly Int. Union Radio Sci., Florence, August 27 to 30, 1984, 22.

44. **Wever, R.,** The circadian multi-oscillator system of man, *Int. J. Chronbiol.,* 3, 19, 1975.

45. **Hamer, J. R.,** Effects of low level, low frequency electric fields on human reaction time, *Commun. Behav. Biol.,* 2, 217, 1968.

46. **Konig, H. L.,** Behavioral changes in human subjects associated with ELF electric fields, in *ELF and VLF Electromagnetic Field Effects,* Persinger, M. A., Ed., Plenum Press, New York, 1974, 81.

47. **Persinger, M. A., Lafreniere, G. F., and Mainprize, D. N.,** Human reaction time variability changes from low intensity 3-Hz and 10-Hz electric fields: interactions with stimulus pattern, sex and field intensity, *Int. J. Biometeorol.,* 19, 56, 1975.

48. **Halberg, F., Cutkomp, L., Nelson, W., and Sothern, R.,** Circadian rhythm in plants, insects and mammals exposed to ELF magnetic and/or electric fields and currents, University of Minnesota, Minneapolis, ADO19958, 1975.

49. **Stern, S. and Laties, V.,** Behavioral effects of 60 Hz electric fields. I. Detection of 60 Hz electric fields by rats, in Project Resumes. Biological Effects from Electric Fields Associated with High Voltage Transmission Lines, U.S. Department of Energy, Contractors' Review, 1981, 11.

50. **Rosenberg, R. S., Duffy, P. H., and Sacher, G. A.,** Effects of intermittent 60-Hz high voltage electric fields on metabolism, activity and temperature in mice, *Bioelectromagnetics,* 2, 291, 1981.

51. **Rosenberg, R. S., Duffy, P. H., Sacher, G. A., and Ehret, C. F.,** Relationship between field strength and arousal response in mice exposed to 60-Hz electric fields, *Bioelectromagnetics,* 4, 181, 1983.

52. **Tucker, R. D. and Schmit, O. H.,** Tests for human perception of 60-Hz moderate strength magnetic fields, in *Biological Effects of Electric and Magnetic Fields Associated with Proposed Project Seafarer,* NAS/NRC Comm. Rep., 1977, 413.

53. **Tucker, R. D. and Schmit, O. H.,** Tests for human perception of 60 Hz moderate strength magnetic fields, *IEEE Trans. Biomed. Eng.,* 25, 509, 1978.

54. **Bullock, T. H.,** General introduction. An essay on the discovery of sensory receptors and the assignment of their functions together with an introduction to electroreceptors, in *Handbook of Sensory Physiology, Electroreceptors and Other Specialized Receptors in Lower Vertebrates,* Vol 3, Fessard, A., Ed., Springer-Verlag, Basel, 1975, 1.

55. **Fessard, A., Ed.,** *Handbook of Sensory Physiology: Electroreceptors and other Specialized Receptors in Lower Vertebrates,* Vol. 3, Fessard, A., Ed., Springer-Verlag, Basel, 1975, 1.

56. **Kalmijn, A. J.,** The detection of electric fields from inanimate and animate sources other than electric organs, in *Handbook of Sensory Physiology: Electroreceptors and Other Specialized Receptors in Lower Vertebrates,* Vol. 3, Fessard, A., Ed., Springer-Verlag, Basel, 1975, 147.

57. **Marino, A. A., Berger, T. J., Austin, B. P., Becker, R. O., and Hart, F. X.,** *In vivo* bioelectrochemical changes associated with exposure to extremely low frequency electric fields, *Physiol. Chem. Phys.,* 9, 433, 1977.

58. **Phillips, R. D., Anderson, L. E., and Kaune, W. T.,** Biological effects of high-strength electric fields on small laboratory animals, Interim Report, Contract DE-ACO2-76RLO-1830, U.S. Department of Energy, 1981, 534.

59. **Hilton, D. I. and Phillips, R. D.,** Growth and metabolism of rodents exposed to 60-Hz electric fields, *Bioelectromagnetics,* 2, 381, 1981.

60. **Cerretelli, P., Veicsteinas, A., Margonato, V., Cantone, A., Viola, D., Malaguti, C., and Previ, A.,** 1000-KV project: research on the biological effects of 50-Hz electric fields in Italy, in *Biological Effects of Extremely Low Frequency Electromagnetic Fields,* Phillips, R. D., Gillis, M. F., Kaune, W. T., and Mahlum, D. D., Eds., Conf. 781016, Proc. 18th Hanford Life Sci. Symp., Richland, Wash., 1979, 241.

61. **Ragan, H. A., Buschbom, R. L., Pipes, M. J., Phillips, R. D., and Kaune, W. T.,** Hematologic and serum chemistry studies in rats exposed to 60-Hz electric fields, *Bioelectromagnetics,* 4, 79, 1983.

62. **Fam, W. Z.,** Long-term biological effects of very intense 60Hz electric field on mice, *IEEE Trans. Biomed. Eng.,* 27, 376, 1980.

63. **Knickerbocker, G., Kouwenhover, W., and Barnes, H.,** Exposure of mice to a strong AC electric field — an experimental study, *IEEE Trans. Power Appar. Syst.,* 86, 498, 1967.

64. **Phillips, R. D.,** Reproduction and development of miniature pigs exposed to 60-Hz electric fields, in Project Resumes, DOE/EPRI Contractors' Review (Abstr.), Kansas City, Mo., November 1983.

65. **Phillips, R. D.,** Reproduction and development of rats exposed to a 60-Hz electric field, in Project Resumes, DOE/EPRI Contractors' Review (Abstr.), Kansas City, Mo., November 1983.

66. **Sasser, L. B.,** Morphologic development and reproductive function in guinea pigs exposed to a 60-Hz electric field, in Project Resumes, DOE/EPRI Contractors' Review (Abstr.), Kansas City, Mo., November 1983.

67. **Greenberg, B., Kunich, J. C., and Bindokas, V. P.,** Effects of high-voltage transmission lines on honeybees, in Biological Effects of Extremely Low Frequency Electromagnetic Fields, Phillips, R. D., Gillis, M. F., Kaune, W. T., and Mahlum, D. D., Eds., Conf. 781016, Proc. 18th Annu. Hanford Life Sci. Symp., Richland, Wash. 1979, 74.

68. **Greenberg, B., Bindokas, V. P., and Gauger, J. R.,** Biological effects of a 765-KV transmission line: exposure and thresholds in honeybee colonies, *Bioelectromagnetics,* 2, 315, 1981.

69. **Greenberg, B., Bindokas, V. P., and Gauger, J. R.,** Effects of a 765-kV, 60 Hz transmission line on honey bees, in Project Resumes, DOE/EPRI Contractors' Review, Kansas City, Mo., November, 1983.

70. **Grissett, J. D.,** Enhanced growth in pubescent male primates chronically exposed to extremely low frequency fields, in, Biological Effects of Extremely Low Frequency Electromagnetic Fields, Phillips, R. D., Gillis, M. F., Kaune, W. T., and Mahlum, D. D., Eds., Conf. 781016, Proc. 18th Annu. Hanford Life Sci. Symp. Richland, Wash. 1979, 348.

71. **Lotz, G. and Saxton, J. L.,** Growth and sexual maturation of *Rhesus* monkeys chronically exposed to ELF electric and magnetic fields, 23rd Hanford Life Sci. Symp., Richland, Wash., (Abstr.), 42, 1984.

72. **Free, M. J., Kaune, W. T., Phillips, R. D., and Cheng, H. C.,** Endocrinological effects of strong 60-Hz electric fields on rats, *Bioelectromagnetics,* 2, 105, 1981.

73. **McClanahan, B. J. and Phillips, R. D.,** The influence of electric field exposure on bone growth and fracture repair in rats, *Bioelectromagnetics,* 4, 11, 1983.

74. **Jaffe, R. A., Laszewski, B. L., Carr, D. B., and Phillips, R. D.,** Chronic exposure to a 60-Hz electric field: effects on synaptic transmission and peripheral nerve function in the rat, *Bioelectromagnetics,* 1, 131, 1980.

75. **Jaffe, R. A., Laszewski, B. L., and Carr, D. B.,** Chronic exposure to a 60-Hz electric field: effects on neuromuscular function in the rat, *Bioelectromagnetics,* 2, 227, 1981.

76. **Hjeresen, D. L., Kaune, W. T., Decker, J. R., and Phillips, R. D.,** Effects of 60-Hz electric fields on avoidance behavior and activity of rats, *Bioelectromagnetics,* 1, 299, 1980.

77. **Hilton, D. I. and Philllips, R. D.,** Cardiovascular response of rats exposed to 60-Hz electric fields, *Bioelectromagnetics,* 1, 55, 1980.

78. **Wilson, B. W., Anderson, L. E., Hilton, D. I., and Phillips, R. D.,** Chronic exposure to 60-Hz electric fields: effects on pineal function in the rat, *Bioelectromagnetics,* 2, 371, 1981.

79. **Morris, J. E. and Phillips, R. D.,** Effects of 60-Hz electric fields on specific humoral and cellular components of the immune system, *Bioelectromagnetics,* 3, 341, 1982.

80. **Smith, M., D'Andrea, J. A., Gandhi, O. P., and Johnson, C. C.,** Behavioral and physiological effects in rats of chronic exposure to strong 60-Hz electric fields, Int. Symp. Biological Eff. Electromagnetic Waves, URSI, Airlie, Va., (Abstr.), 1977, 132.

81. **Seto, J. Y.,** CNS and endocrine studies of power frequency electric field effects, School of Engineering, Tulane University, ERBL Report TR-27, 1979.

82. **Mathewson, N. S., Oosta, G. M., Oliva, S. A., Levin, S. G., and Diamond, S. S.,** Influence of 45-Hz vertical electric fields on growth, food and water consumption, and blood constituents of rats, *Radiat. Res.,* 79, 468, 1979.

83. **Krueger, A. P. and Reed, E. J.,** A study of the biological effects of certain ELF electromagnetic fields, *Int. J. Biometeor.,* 19, 194, 1975.

84. **deLorge, J.,** Operant behavior of Rhesus monkeys in the presence of extremely low frequency-low intensity magnetic and electric fields, Naval Aerospace Medical Research Laboratory, NAMRL-1196, AD774106, 1973.

85. **deLorge, J.,** Operant behavior of Rhesus monkeys in the presence of extremely low frequency-low intensity magnetic fields: Experiment 1, Naval Aerospace Medical Research Laboratory, NAMRL-1155. AD754058, 1972.

86. **deLorge, J.,** Experiment 2, Naval Aerospace Medical Research Laboratory, NAMRL-1179, AD764532, 1973.

87. **Marino, A. A., Becker, R. O., and Ullrich, B.,** The effect of continuous exposure to low frequency electric fields on three generations of mice: a pilot study, *Experientia,* 32, 565, 1976.

88. **Marino, A. A., Reichmanis, M., Becker, R. O., Ullrich, B., and Cullen, J. M.,** Power frequency electric field induces biological changes in successive generations of mice, *Experientia,* 36, 309, 1980.

89. **Marino, A. A. Cullen, J. M., Reichmanis, M., and Becker, R. O.,** Power frequency electric fields and biological stress: a cause-and-effect relationship, in *Biological Effects of Extremely Low Frequency Electromagnetic Fields,* Phillips, R. D., Kaune, W. T., Gillis, M. F., and Mahlum, D. D., Eds. Conf. 781016, Proc. 18th Annu. Hanford Life Sci. Symp., Richland, Wash. 1978, 258.

90. **Wilson, B. W., Anderson, L. E., Hilton, D. I., and Phillips, R. D.,** Chronic exposure to 60-Hz electric fields: effects on pineal function in the rat. Erratum, *Bioelectromagnetics,* 4, 293, 1983.

91. **Anderson, L. E., Hilton, D. I., Chess, E. U., and Wilson, R. W.,** Neuroendocrine responses in rats exposed to 60-Hz electric fields. Abstr. Bull. 23rd Handford Live Sci. Symp., Richland, Wash., Battelle Pacific Northwest Laboratories, October 2 to 4, 1984, 15.

92. **Carstensen, E. L., Buettner, A., Genberg, V. L., and Miller, M. W.,** Sensitivity of the human eye to power frequency electric fields, *IEEE Trans. Biomed. Eng.,* 32, 561, 1985.

93. **Hauf, R.,** Electric and magnetic fields at power frequencies, with particular reference to 50 and 60 Hz., in *Nonionizing Radiation Protection,* Seuss, M. J., Ed., World Health Organization, Copenhagen, 1982, 175.

94. **Mantell, B.,** Untersuchungen über die Wirkung eines magnetischen Wechselfeldes 50 Hz auf den. Menschen, Inaugural dissertation, Albert-Ludwigs-University, Freiberg, West Germany, 1975, 50.

95. **Johansson, R., Lundquist, A. G., Lundquist, S., and Scuka, V.,** Is there a connection between the electricity in the atmosphere and the function of man? III. 50Hz field variations in FAO Report C2621-45, C2627-H5, Geneva, 1973.

96. **Eisemann, B.,** Tests on the long-term effects of low alternating currents (50 Hz) on man, Inaugural dissertation, Albert-Ludwigs-University, Freiburg, West Germany, 1975, 64.

97. **Amon, G.,** Intermediärer Stoffwechsel, Elektrolythaukshalt und Vernalten der Homone bei der einwirkung elektrischer Felder auf den Menschen, Inaugural M.D. dissertation, Albert-Ludwigs-University, Freiburg, West Germany, 1977, 77.

98. **Rupilius, J. P.,** Investigations on the effect of electric and magnetic 50-Hz AC-field on humans, Inaugural M.D. dissertation, Albert-Ludwigs-University, Freiburg i. Br., 1976, 97.

99. **Adrian, D. J.,** Auditory and visual sensations stimulated by low-frequency electric currents, *Radio Sci.,* 12, 243, 1977.

100. **Brindley, G. S.,** The site of electrical excitation of the human eye, *J. Physiol.,* 127, 189, 1955.

101. **Brindley, G. S.,** A new interaction of light and electricity in stimulating the human retina, *J. Physiol.,* 171, 514, 1967.

102. **Crapper, D. R. and Noel, W. K.,** Retinal excitation and inhibition from direct electrical stimulation, *J. Neurosphysiol.,* 26, 924, 1963.

103. **Bridges, J. E. and Preache, M.,** Biological influences of power frequency electric fields — a tutorial review from a physical and experimental viewpoint, *Proc. IEEE,* 69, 1092, 1981.

104. American National Standard for Leakage Current for Appliances, A.N.S.I., C101.1, New York, 1973.

105. **Sazanova, T. E.,** A physioloigcal assessment of work conditions in 400—500 kV open switching yards, Scientific Publications of the Institute of Labor Protection of the All-Union Central Council of Trade Unions, Issue 46, Profizdat, 1967.

106. **Asanova, T. P. and Rakov, A. I.,** The state of health of persons working in the electric field of outdoor 400 KV and 500 KV switchyards (transl.), *Gig. Tr. Prof. Zabol.,* 10, 50, 1966.

107. **Norstrom, S., Birkee, E., and Gustavsson, L.,** Reproductive hazards among workers at high voltage stations, *Bioelectromagnetics,* 4, 91, 1983.

108. **Milham, S.,** Mortality from leukemia in workers exposed to electrical and magnetic fields, *N. Eng. J. Med.,* 307, 249, 1982.

109. **Wright, W. E., Peters, J. M., and Mack, T. M.,** Leukemia in workers exposed to electrical and magnetic fields, *Lancet,* 20, 1160, 1982.

110. **McDowall, M. E.,** Leukemia mortality in electrical workers in England and Wales, *Lancet,* 19, 246, 1983.

111. **Coleman, M., Bell, J., and Skeet, R.,** Leukemia incidence in electrical workers, *Lancet,* 30, 1106, 1983.

112. **Peterson, G. R. and Milham, S.,** Occupational mortality in the state of California, NIOSH-80-104, 80, 1980.

113. **Kouwenhoven, W. B., Langworthy, O. R., Singewald, M. L., and Knickerbocker, G. G.,** Medical evaluation of man working in AC electric fields, *IEEE Trans. Power Appar. Syst.,* 86, 506, 1967.

114. **Singewald, M. L., Langworthy, O. R., and Kouwenhoven, W. B.,** Medical follow-up study of high voltage lineman working in AC electric fields, *IEEE Trans. Power Appar. Syst.,* 92, 1307, 1973.

115. **Roberge, P. F.,** Study on the state of health of electrical maintenance workers on Hydro-Quebec's 735-KV power transmission system, Report, Health Department, Hydro-Quebec, Montreal, 1976, 29.

116. **Strumza, M. V.,** Influence sur la santé humain de la proximité des conducteurs d'électricité a haute tension. Resultats d'une enquête sur la "consommation medicale". *Arch. Mal. Prof. Med. Trav. Secur. Soc.,* 31, 269, 1970.

117. **Krumpe, P. E. and Tockman, M. S.,** Evaluation of the health of personnel working near project Sanguine beta test facility from 1971 to 1972, in *Biological and Clinical Effects of Low Frequency Magnetic and Electric Fields,* Llaurado, J. G., Sances, A., and Battocletti, J. H., Eds., Charles C. Thomas, Springfield, Ill., 1974, 98.

118. **Houck, W.,** The continuing medical surveillance of personnel exposed to extremely low frequency (ELF) electromagnetic fields, Naval Aerospace Medical Research Laboratory, NAMRL-1225, 1976.

119. **Hodges, T. K. and Mitchell, C. A.,** Growth and Yield of Field Crops in the Proximity of an Ultra-High Voltage Electric Transmission Test Line, Study conducted for the Indiana and Michigan Electric Co. and American Electric Power Service Corp., 1979.

120. **Knave, B., Gamberale, F., Bergstrom, S., Birke, E., Iregren, A., Kolmoclin, Hedman, B., and Wennberg, A.,** Long-term exposure to electric fields. A cross-sectional epidemiologic investigation on occupationally exposed high voltage substation workers, *Electra,* 65, 41, 1979.

121. **Stopps, G. J. and Janischewskyi, W.,** Epidemiological Study of Workers Maintaining HV Equipment and Transmission Lines in Ontario, Report to Canadian Electrical Assoc., Montreal, 1979, 123.

122. **Amstutz, H. E. and Miller, D. B.,** A study of farm animals near 765 KV transmission lines, *Bovine Practitioner,* 15, 51, 1980.

Chapter 4

BIOLOGICAL EFFECTS OF STATIC MAGNETIC FIELDS

Richard B. Frankel

TABLE OF CONTENTS

I. INTRODUCTION

Claims for the biological effects of magnetism date from the discovery of magnetism in rocks by the ancient Greeks, Chinese, and Central American Olmecs.[1-4] The pre-Homeric Greeks mined Fe_3O_4 in the province of Magnesia in Asia Minor, whence comes the names magnet and magnetite. Explanations for the powers of attraction exerted by pieces of magnetite on each other and on iron metal included the notion, thought to originate with Thales of Miletus, that magnets were alive and attracted iron metal by animating it. Both magnetite and the more weakly magnetic hematite α-Fe_2O_3 were thought to have medicinal qualities, and were even prescribed for Queen Elizabeth I of England by her physician William Gilbert of Colchester.

In 1600 this same William Gilbert published his monumental treatise, *De Magnete Magneticisque Corporibus et de Magno Magnete Tellure* (On the Magnet and Magnetic Bodies and on the Great Magnet the Earth). In addition to comparing magnetic forces with life forces, he proclaimed that the earth itself is a giant magnet. The publication summarized his 16 years of study of the interactions of small iron needles with spheres of magnetite called terrellas, or "little earths". In 1570, Robert Norman, a London compass maker, had discovered that magnetized needles made neutrally buoyant with pieces of cork oriented downward at the angle of 70° as well as northward when suspended in water. Gilbert reached his conclusion about the magnetism of earth by comparing the magnetic inclination of the needle with his observations that the iron needles on the surface of the terrella are inclined from the horizontal at angles that increase from 0 to 90° as one passes from the equator to the poles.

Gilbert's work was a triumph of experimental science and provided a mechanism for the operation of the magnetic compass. The fact that magnetized pieces of magnetite would orient in astronomically significant directions had long been used, at least by the Chinese and possibly by the Olmecs, for geomancy and divination, but was apparently not used for navigational purposes until the 11th century A.D. The first western accounts of this use are by Alexander Neckam and Petrus Peregrinus de Maricourt in the 13th century. From this use of magnetite came the name "lodestone" because it indicated the direction of the pole or lodestar. Iron needles magnetized by stroking with magnetite were subsequently employed. The invention and development of the magnetic compass was one of the technological developments that allowed navigation of the oceans and ushered in the great Age of Exploration. Christopher Columbus, in fact, discovered magnetic declination, namely, that magnetic north and geographic north do not coincide, and the difference varies with longitude.

An inversion of the Greek notion that magnets are animate was popularized by Franz Anton Mesmer, an Austrian physician, who arrived in Paris in the 1770s with a radical theory of human health.[5] This theory, referred to as "animal magnetism", was based on the same medieval Hermetic philosophy that underlies astrology, alchemy, and magic. The theory envisions a human being as a microcosm corresponding to the world macrocosm. Mesmer believed that the human body had magnetic poles that correspond to those of the earth or cosmic, magnetic poles proclaimed by Gilbert. He postulated a subtle or imperceptible fluid that flowed from the cosmic magnetic poles through the body. Smooth, uninterrupted flow constitutes health. Blockage of the flow results in disease. Health can be restored by removing blockages and restoring smooth flow. This was accomplished by rubbing various parts of the body with magnets. This treatment was apparently efficacious because Mesmer became famous and was even immortalized in Mozart's opera "Cosi Fan Tutti". Along the way, he and his disciples invented group therapy, and by discovering that painted pieces of wood worked as well as magnets, the placebo. As Mesmer's fame and fortune increased, the animosity of the French medical establishment increased until the French Royal Academy of Sciences appointed a commission, including Benjamin Franklin (then the American ambassador to France) and Lavoisier, to investigate him. The commission declared him to be a quack and a fraud. Thus a cloud was cast over animal magnetism that persists to this day. This is in contrast with the brilliant success of "animal electricity", namely the experiments of Galvani and Volta of the same era that underlay the development of the understanding of electrical phenomena.

However, significant discoveries in the last few years require a reevaluation of animal magnetism, although not Mesmer's version. It is now known that many organisms precipitate inclusions of magnetic material. This phenomenon will be a chief focus of this review.

Biological effects of magnetic fields is a subject that includes many different topics. In order to make the treatment tractable, we have arbitrarily separated time-varying from static field effects. Static field effects will be covered here; time-varying field phenomena including magneto-phosphenes and effects of induced currents will be covered in a companion chapter.

Even the subject of static field effects covers many topics. These include the use of magnetic fields in spectroscopies of biological material, including nuclear magnetic resonance (NMR), electron paramagnetic resonance (EPR), and recoilless nuclear gamma resonance (Mössbauer effect), and magnetic susceptibility and magnetization measurement. Magnetic fields have also been used to orient cells or cell fragments in suspension. Applications of magnetism in physiology and clinical medicine include NMR imaging, magnetic targeting and modulation of drug delivery, magnetic separation of biological materials, use of magnetism in surgical procedures, and noninvasive measurement of blood flow. A rapidly growing area spanning AC and DC regimes is the measurement of magnetic fields generated by the human body, and the use of those measurements in medicine and physiology. A large area involves mutagenic, mitogenic, metabolic, morphological, and developmental effects of exposure of organisms or biological materials to intense DC magnetic fields or to null field conditions. Another important area includes behavioral effects of magnetic fields, including effects on orientation, migration, and homing and the involvement of the geomagnetic field in the activities of organisms. Biomineralization of magnetic materials is especially significant for this latter topic, but could also play a role in other interactions of organisms with electromagnetic fields.

II. PHYSIOLOGICAL AND MEDICAL APPLICATIONS OF MAGNETISM

Magnetic properties and spectroscopy of biological materials have been extensively studied. Especially significant advances in NMR have been made in the last few years.[6] These include increased resolution and sensitivity, pulsed programming, Fourier decomposition of complex spectra allowing study of whole organisms or perfused organs, and observation of the time development of chemical intermediates, e.g., metabolites containing phosphorus, in metabolic pathways.[7-9]

Superconducting magnets with bores large enough to accommodate human bodies have allowed development of NMR imaging systems with resolution comparable and even superior to X-ray and positron computerized tomography and ultrasound techniques.[10,11] The basis of the method is that when a magnetic field gradient is superposed on a static, homogeneous magnetic field, nuclei such as protons will resonate at different frequencies at each point along the gradient. The magnitude of the signals at each frequency is proportional to the number of protons at that point. By switching the gradient to different directions and recording the frequency spectra, the three- or two-dimensional proton density map in any plane can be reconstructed by computer. Measurement of relaxation times also allows discrimination of chemical differences, e.g., different types of tissue. Fields of the order of 5×10^{-2} to 1.5 T with gradients of the order of 10^{-2} T/m are used for periods of the order of $1/2$ hr. Switching of the gradients will involve changing magnetic fields as high as 2 T/sec. Considering the rapid development of this diagnostic method and its enthusiastic reception by the medical community, it will shortly be the major cause of human exposure to intense magnetic fields (above 1 T).

The development of superconducting quantum interference devices (SQUIDS) with very high sensitivity has spurred the study of the very weak magnetic fields generated by electrical processes in the human body.[12-14] Applications include magnetocardiography, magneto-encephalography, measurement of pulmonary activity, and detection of body iron stores due to asbestos inhalation or diseases such as Thalassemia, which result in hemochromatosis.[15]

Magnetic devices have been used in several surgical procedures including repair of giant retinal tears,[16] bougienage of esophogeal atresia in infants,[17] and in the treatment of aneurysms.[18] Magnetic agitation has been used to modulate the release of macromolecules such as insulin from polymers with magnetic inclusions which are implanted in the body.[19]

The voltage induced when an electrolyte flows perpendicular to a magnetic field, known as the magnetohydrodynamic effect, has been used for a number of years to noninvasively meter blood flow.[20,21] The voltage induced across a blood vessel of diameter d is

$$\epsilon(\text{volts}) = Bvd \tag{1}$$

where B is the magnetic flux density in tesla and v is the perpendicular flow velocity. For v = 0.6 m/sec and d = 0.025 m (human aorta), 15 mV would be generated in a 1-T field. These potential also show up in electrocardiograms of animals in magnetic fields (see Section IV).

Magnetic microcarriers have been used to target drugs to specific locations in the body and in cell separation and immunological assays.[23,24] These applications are based on the translation of magnetic particles relative to the diamagnetic fluid background in magnetic field gradients. The potential energy of a magnetic dipole of moment m in a magnetic field with flux density B

$$E_m = -\overrightarrow{m} \cdot \overrightarrow{B} \tag{2}$$

If the field is homogeneous, the dipole tends to align in the field but no translation occurs. If there is a magnetic field gradient in the x direction, the dipole will experience a force in the x direction

$$F = m \, dB/dx \tag{3}$$

If the magnetic material is ferromagnetic or ferrimagnetic, m is generally independent of B for fields greater than a few tenths of a tesla. For paramagnetic materials at sufficiently high temperatures this is not the case and

$$m = \chi VH \tag{4}$$

where χ is the magnetic susceptibility, V is the volume, and H is the magnetic field intensity. Then Equation 3 becomes

$$F_x = \chi VH \, dB/dx \tag{5}$$

ignoring the smaller susceptibility of opposite sign of the diamagnetic background fluid. For small particles in a viscous medium (including water at ambient temperatures), viscous forces are more important than inertial forces and the particle quickly reaches its terminal velocity

$$v = F_x/6\pi\eta r \tag{6}$$

where r is the radius of the particle and η is the viscosity (10^{-3} nsec/m² = 0.01 poise for water at room temperature). Thus the velocity due to magnetic forces depends on the nature of the material, the magnitude of the field gradient, and the size of the particle. Since F_x and m are proportional to volume of the magnetic material and therefore increase as r^3, v increases as r^2. However, the size of magnetic microcarriers is limited by the need for a large surface-to-volume ratio, and by the tendency of large particles to coagulate due to interparticle forces.

To employ magnetic drug microcarriers, large field gradients must be generated outside the body by suitably shaped, magnetized pole pieces. In the immunological procedures, the microcarriers, to which specific antibodies or antigens are chemically attached, can be separated by the use of high gradient magnetic separation (HGMS) filters consisting of fine stainless steel wires in a magnetic field strong enough to magnetize the wires. Gradients as high as 10^4 to 10^5 T/m can be generated within a few diameters of the wires. As the fluid flows through the filter the magnetic particles and anything attached to them are trapped on the wires. This method has been applied to the separation of cells and proteins[25 28] and to removal of microorganisms from water.[29,30]

III. ORIENTATION OF CELLS AND BIOMOLECULES IN INTENSE MAGNETIC FIELDS AND MAGNETIC FIELD EFFECTS ON CHEMICAL REACTIONS

Alignment of molecular aggregates with sufficient diamagnetic anisotropy will occur in intense magnetic fields (see Appendix). The energy of cylindrically symmetric molecular aggregates in an external magnetic field B is given by

$$E_m = -\frac{1}{2} [\chi_\| + (\chi_\| - \chi_\perp) \cos^2\theta] \, V \, H \cdot B \tag{7}$$

where V is the volume and $\chi_\|$ is the least negative susceptibility ($\chi_\| - \chi_\perp$) > 0. θ is the angle between the $\chi_\|$ direction and the field direction. $\chi_\|$ is usually parallel or perpendicular to the axis of cylindrical symmetry. If $\chi_\|$ is parallel to the axis, the aggregate will align with the axis perpendicular to the field. If $\chi_\|$ is perpendicular to the axis the aggregate will align with the axis parallel to the field. Since the alignment in the field is opposed by the randomizing forces due to thermal energy, the average alignment $<\cos^2\theta>$ of an ensemble of rods in the field at ambient temperatures will be an exponential function of the ratio of magnetic to thermal energies

$$<\cos^2\theta> = \frac{\int \cos^2\theta \, e^{-E_m/kT} \, dV}{\int e^{-E_m/kT} dV} \tag{8}$$

Experimental results have been reviewed comprehensively by Maret and Dransfeld.[31,32] Muscle fibers,[33] chloroplasts,[34] retinal elements,[35,36] sickled erythrocytes,[37] bacteriophage fibers,[38], membranes,[39,40] and macromolecules including nucleic acids[31] have diamagnetic anisotropy and have been aligned in intense, homogeneous magnetic fields. Highly oriented structures can result from polymerization in an intense field. Torbet et al.[41] produced oriented fibrin gels from fibrin monomers in a field of 11 T. The fibrin monomers were produced by enzymatic cleavage of fibrinogen. Murthy and Yannas[42] prepared oriented collagen films by heat precipitation in a magnetic field. Collagen in solution dissociates into monomers at low temperatures and precipitates at higher temperature ($\sim$ 37 K). The monomers are apparently not aligned. Alignment of dimers, trimers, etc. occurs as they form from monomers with increasing temperature in the field.

In addition to polymer alignment, intense magnetic fields may interact with biomaterials in other ways. Sperber et al.[43] observed oriented growth of pollen tubes in intense fields and suggests redistribution of membrane proteins that regulate intercellular concentrations of Ca^{2+}. Audus[44,45] had previously observed oriented growth, or magnetotropism, in oat shoots and cress roots in inhomogeneous magnetic fields, with growth occurring in the direction of decreasing field intensity. Labes[46] and Aceto et al.[47] proposed interaction of magnetic fields with cell membranes as a plausible mechanism for physiological effects. They noted that membranes have liquid, crystal-like properties and are close to phase transitions at physiological temperatures. Magnetic field could affect membrane fluidity or other properties. Magnetic orientation of diamagnetically anisotropic domains in artificial phospholipid bilayers has been reported.[48] Magnetic fields also affect the fluid-gel transition in agarose gels,[49] possibly by alignment of the monomers in the fluid phase.

Magnetic fields of the order of 10^{-3} to 10^{-2} T can affect chemical reactions by influencing the electronic spin states of reaction intermediates.[50-53] These effects can have biological consequences.[54,55] A relatively simple chemical illustration of the effect involves homolytic cleavage of a chemical bond to produce two radicals.* Since the electrons in the chemical bond are spin-paired in an $S = 0$ or singlet state,[55a] these electrons on the nascent radicals will also have overall singlet character as the radicals separate. Separation is a diffusion-controlled process and there is a high probability that the two radicals will reencounter each other. If the electrons retain their overall singlet character, a reencounter is likely to produce recombination. If the electrons have overall triplet ($S = 1$) character, the bond will not reform and the radicals will eventually separate and perhaps participate in other chemical reaction. The transition from singlet to triplet can result from the interaction of the odd electrons of the radicals with the nuclear magnetic moment(s) of the atom(s) on which they have high probability density. This interaction, the magnetic hyperfine interaction, is equivalent to a local magnetic field at the electrons produced by the nuclei. Different local magnetic fields cause the electrons on the radicals to precess at different rates, which destroys singlet phasing and results in triplet formation. However, an applied magnetic field will decouple the electrons and the nuclei, suppressing formation of the triplet state. This enhances the recombination rate and suppresses the other chemical reactions. The decoupling of the electrons and the nuclei will occur when the intensity of the applied field exceeds the effective magnetic field produced by the hyperfine interaction. Then the electrons will precess in phase about the applied field rather than at different rates about the local field. This condition is typically satisfied for fields of the order of 10^{-3} to 10^{-2} T.

A variation of this scheme is proposed to account for the effects of magnetic fields on electron transport in photosynthetic, purple bacterial membranes.[51,55,56] The effects are observed when elements of the transport chain are electrochemically reduced, forcing back transfer of the photoexcited electron. The electron transport sequence can be summarized as follows:

1. $^S(A) + photon \rightarrow {}^S(A)*$
2. $^S(A) * + (B) \rightarrow {}^S(A^+ + B^-)$
3. $^S(A^+ + B^-) \rightleftarrows {}^T(A^+ + B^-)$
4. $^S(A^+ + B^-) \rightarrow {}^S(A) + (B)$
5. $^T(A + B) \rightarrow {}^T(A) + (B)$

(A) corresponds to bacteriochlorophyll dimer and (B) corresponds to bacteriopheophytin. S and T stand for singlet and triplet, respectively, and * stands for the photoexcited state. (1) Bacteriochlorophyll absorbs a photon resulting in electron excitation. (2) Electron transfer

---

* Homolytic cleavage is the breaking of a covalent, single bond so that one electron from the bond is left on each fragment, resulting in two free radicals with single, unpaired electrons.

to bacteriopheophytin occurs, resulting in positive and negative charges on donor and acceptor, respectively. (3) The positive ion-negative ion pair are initially in a singlet state which can evolve into a triplet state via the hyperfine interaction mechanism. In the blocked transport chain, the ion pair decays by back transfer of the electron to a less energetic state of bacteriochlorophyll. (4) If the ion pair is in the singlet state, back transfer populates the singlet ground state of bacteriochlorophyll. (5) If the ion pair is in the triplet state, back transfer populates an intermediate energy triplet state of bacteriochlorophyll which is detected by a delayed fluorescence method. It is found that the amount of bacteriochlorophyll triplet produced following flash excitation is magnetic field-dependent. Because the photosynthetic apparatus is highly structured and membrane-bound, exchange interactions between the ions also play a role in the formation of the triplet state in (3). Although the experimental conditions cited above are nonphysiological, mechanisms of this kind could conceivably play a role in electron transport in viable biological systems.

IV. MUTAGENIC, MITOGENIC, MORPHOLOGICAL, AND DEVELOPMENTAL EFFECTS OF MAGNETIC FIELDS

A large number of papers have been published on this topic and a number of bibliographies, reviews, and symposia have appeared.[57-75] Moreover, a number of interaction mechanisms have been proposed.[51,76-82] However, at this time this area remains an empirical science with little elucidation of effects in terms of mechanism. Only some of the more recent reports will be cited here.

Mutagenic effects of chronic exposure to DC magnetic fields have been investigated. Mahlum et al.[83] exposed male mice to a magnetic field of up to 1T for 28 days. The mice were subsequently mated to two females per week for up to 8 weeks and the resulting embryos were assayed for viability 10 days later. No significant differences were reported between exposed and sham-exposed (control) groups. Kale and Baum[84] exposed fruitfly (Drosophila melanogaster) male eggs, larvae, pupae, and adults to fields up to 3.7 T for up to 7 days. After mating with females, broods were tested for induction of genetic damage by the sex-linked, recessive lethal test. No evidence for induction of mutations under the conditions of exposure were reported. Skopek et al.[84a] exposed Salmonella and cultured human lymphoblasts to 10-T magnetic fields for 4 hr. Cells were surveyed for toxic and mutagenic effects with a forward mutation assay. No effects of magnetic field exposure were found for either cell type when compared with sham-exposed controls.

Morphological and developmental effects have been investigated. Sikov et al.[85] reported no effects on the development of mice after intrauterine exposure to 1 T during gestation. An earlier study by Nahas et al.[86] indicated that exposure of rodents to 0.02- to 0.12-T fields for 1 month resulted in no toxic or histopathological effects. Brewer[87] studied guppies (Lebistes reticulatus) chronically exposed to a 0.05-T magnetic field and reported reduction in spawn rate and gestation period in successive generations exposed to the field. However, field effects were not permanent; reproduction eventually returned to normal when fish were removed from the magnetic field. Mild et al.[88] studied development of frog (Xenopus laevis) embryos exposed at 0.25 T for periods up to 1 week, at temperatures just above the threshold for development in the embryos. If the effect of the magnetic field is equivalent to a reduction in temperature,[47] exposed embryos should not have developed. However, no differences between development of exposed embryos and unexposed controls were reported. Previous studies had indicated effects of magnetic field exposure on development of frog embryos.[89] Strand et al.[89a] report enhancement of fertilization following exposure of trout sperm, ova, or both to 1-T magnetic fields.

Frazier et al.[99] investigated mammalian cell cultures continuously exposed to magnetic fields of 0.1 or 0.3 T through 80 cell doublings. No mitogenic effects of the field were reported when doubling times of exposed cells were compared to controls. Differences in plating efficiencies between exposed cells and controls were cited and ascribed to an as yet unexplained increased clumping of exposed cells. Exposure of frozen cells at 1 T did not result in changes in cell morphology[90] as reported earlier.[91] Cultured cells from human bronchogenic carcinoma and from Burkitt lymphoma were exposed to DC magnetic fields up to 1.15 T by Chandra and Stefani.[92] They report that growth characteristics were unaffected by exposure. In vivo exposure of mouse tumors did not cause retardation of growth of the tumor. Leitmannova et al.[93] reported changes in morphology of aged red blood cells in magnetic fields.

Moore[94] studied growth of five species of bacteria and a yeast in DC and slowly varying magnetic fields up to 0.09 T. He reports stimulation or retardation of growth depending on the field strength, frequency, and organism. A number of previous studies had indicated that growth of bacteria and yeasts could be altered by static magnetic fields.[95 97]

Electrophysiological effects of static magnetic fields have been investigated. Blatt and Kuo[98] reported no change in the action potential in the interpodal cells of the fresh water alga *Nitella* exposed to fields up to 2 T. These cells have bioelectrical activity and previous studies[98a] had indicated a reduction of the action potential in magnetic fields. Extended exposure at 1.6 T was not toxic for cells. Edelman et al.[99] reported an increase with time in the amplitude of the compound action potential of stimulated frog sciatic nerve when fields up to 0.71 T were applied perpendicular to the nerve. Fields applied parallel to the nerve produced no changes. However, Gaffey and Tenforde[100] report no changes in electrical conduction in frog sciatic nerve in fields up to 2 T and suggest that results of Edelman et al. are due to changes in temperature. Semm et al.[101] reported electrical changes in cells in the pineal glands of guinea pigs when exposed to magnetic fields of the order of 10^{-4} T. Raybourn[101a] reports that 10^{-3} to 10^{-2}-T fields acutely reduce electroretinographic responses in turtle retina, but do not reduce retinal sensitivity. The effect might involve magnetic field effects on chemical reactions (see Section III).

Static, magnetic fields affect electrocardiograms in mammals.[22,102,103] Alterations in the T wave of rats, rabbits, and baboons are reported at exposures above 0.3 T, and are due to the potentials associated with the flow of blood in the magnetic field. There are apparently no chronic physiological effects associated with this phenomenon.

There are reports of alteration of enzyme activity in vitro by magnetic fields.[104,105] For example, Haberditzl[104] claims that fields up to 7.8 T diminished the activity of glutamic dehydrogenase, while a 6-T field enhanced the activity of catalase. Nonuniform fields produced larger effects than uniform fields. Weissbluth and co-workers[106] reported no effects of intense magnetic fields up to 22 T on the activities of several enzymes.

Ripamonti et al.[107] studied the effect of magnetic field exposures up to 12.5 T on the responses of the contractile protozoan *Spirostomum ambiguum* to the toxic substance 2,2′-dipyridyldisulfide. Magnetic field exposure reportedly diminished the ability of the organism to survive the drug and lengthened the extension phase of the contraction cycle. It was hypothesized that interactions of the magnetic field with cellular membranes alters the regulation of Ca^{2+} transients. There were no toxic effects of exposure to magnetic fields without the drug. Bücking et al.[108] had previously reported that magnetic field exposure affected the force of contraction of isolated muscle, which also involves regulation of intracellular Ca^{2+}.

deLorge[109] reports that low-intensity magnetic fields have no measurable effects on operant behavior in monkeys. However, experiments at high magnetic fields showed a suppression of a learned response above a threshold between 4.6 and 7.0 T. Davis et al.[109a] report no behavioral alterations in mice exposed to 1.5-T magnetic fields. Further data on the effects of very intense magnetic fields come from NMR studies on perfused, whole organisms. Fossell et al.[9] note that exposure of perfused rat hearts at 6.4 T does not alter either pressure development or heart rate.

A survey of workers exposed to intense magnetic fields was conducted by Beischer[110] who found no adverse effects of short exposures to fields up to 0.5 T. Epidemiological studies of magnetic field effects in humans are presently being conducted by Budinger et al.[111] and by the National Radiological Protection Board in the U.K. Reviews concerned especially with potential hazards of magnetic field exposure associated with NMR imaging have been published.[74] Interim magnetic field exposure guidelines have been discussed.[112]

Finally, nature has conducted an experiment over geologic times on life in a substantial magnetic field. Magnetotactic bacteria[113] are sediment-dwelling bacteria that contain particles of Fe_3O_4 that orient them in the geomagnetic field (see Section V). These particles produce strong intracytoplasmic magnetic fields and field gradients in the bacterium. The fields can be as large as several tenths of a tesla near the surface of the particles. Thus these bacteria carry out all of their cellular and metabolic functions in intracellularly generated magnetic fields.

V. MAGNETIC FIELD EFFECTS ON ORIENTATION AND HOMING

At the end of the last century, Kreidl[114] published an experiment on magnetic field effects on orientation in crabs. The experimental design was contrived to produce an effect and so does not correspond to the natural circumstances of crabs, but provides a paradigm for effects in other organisms. Crabs periodically molt and subsequently form a new exoskeleton. In the process of molting they also lose their statoliths, dense particles that form part of the vestibular system. They subsequently pick up particles of sand to serve as new statoliths. Kreidl took newly molted crabs and placed them in an aquarium in which the only available particles were magnetic. The crabs indeed placed magnetic particles in their ear labyrinths. When Kreidl approached a crab with magnetic statolith with a bar magnet the crab adopted an orientation that could be correlated with the resultant of the magnetic and gravitational forces on the particles. Electrophysiological responses from crayfish with ferrite statoliths stimulated by magnetic fields have subsequently been reported.[114a]

Since Kreidl's experiment, magnetic field effects have been observed in the orienting behavior of a very diverse group of normal organisms, including bacteria,[113] algae,[115] snails,[116] planaria,[117] honeybees,[118] salmon,[119] salamanders,[120] homing pigeons,[121] robins,[122] mice,[123] and humans.[124-126] In addition, training experiments on pigeons,[127] skates,[128] and tuna[129] have demonstrated the ability of these organisms to sense magnetic fields. Two interaction mechanisms have been elucidated: (1) detection by the organism of the electric field induced by Faraday effect as the organism moves through the magnetic field; (2) interaction of the magnetic field with magnetic material in the organism.

The first mechanism is apparently operative in marine sharks, skates, and rays which are sensitive to electric fields as low as 5×10^{-7} V/m in sea water.[130] The animals detect the electric fields through the ampullae of Lorenzini, which are long, conductive channels that connect electrically sensitive cells in the snout with pores on the skin. Flowing ocean currents or motion of the animal through the geomagnetic field induce voltage gradients with sign and magnitude depending on orientation, which are in general above the sensitivity threshold of the animal. Kalmijn[128] demonstrated that skates could be trained to use magnetic fields of the order of the geomagnetic field as an orientational cue. Brown et al.[7131] used electro-physiological measurements to show that the ampullae of Lorenzini can detect variations in the geomagnetic field. Jungerman and Rosenblum[132] have considered the possibility of the magnetic induction mechanism for an animal in air. They concluded that a circular, electrically conducting loop millimeters in diameter, would be required to overcome thermal noise, with voltages induced by changes in magnetic flux in the loop as the animal changes its heading.

Evidence for orientation by the second mechanism was obtained for homing pigeons in the classical experiment of Keeton.[133] Keeton glued small bar magnets to the backs of the heads of a group of homing pigeons and compared their homing ability with that of a group of control birds carrying brass weights. Under sunny skies both groups oriented and homed equally well when released from unfamiliar sites many miles from the home loft, but under overcast skies when the birds could not see the sun, the orientation of the birds carrying magnets was disrupted, whereas control birds oriented normally. Subsequently, Walcott and Green[134] used Helmholtz coils attached to the heads of pigeons to change the orientation of the birds under overcast conditions. The orientation depended on the direction of the magnetic field, as determined by the direction of current in the coils. Pigeon orientation is also affected by magnetic anomalies and magnetic storms.[135,136] The experimental results suggest that in addition to a magnetic compass, a homing pigeon may have a magnetic "map".[137] The results have been reviewed by Walcott,[121] Gould,[118,137] Able,[138] and Griffin.[138a] Magnetic orientation in migratory birds has been reviewed by Able[138] and by Wiltschko.[139] Although attempts to observe magnetic sensitivity in pigeons by cardiac response have not been successful, Bookman[127] was able to train pigeons to detect the presence of magnetic fields in a flight cage.

Walcott et al.[140] dissected pigeons with nonmagnetic tools and found magnetic material in head and neck sections. Most of the magnetic material was localized in a piece of tissue between the dura and the skull. Each pigeon had an inducible, remanent moment* of 10^{-8} to 10^{-9} A·m^2, which disappeared at 575°C, indicating Fe_3O_4. Presti and Pettigrew[141] found magnetic material in the neck musculature of pigeons and migratory, white-crowned sparrows but did not find localized magnetic materials in the heads. Although the connection between the magnetic material and magnetic sensitivity has not been established definitely, it is suggested. Elucidation of anatomical structure is clearly required. Yorke,[142] Kirschvink and Gould,[143] and Presti and Pettigrew,[141] have speculated on the role of Fe_3O_4 in a magnetic sensor. Yorke points out that if a pigeon can somehow measure the total magnetization of its ensemble of magnetic particles, there is enough magnetic material present to indicate the field direction with high accuracy.

A possible connection between Fe_3O_4 and magnetic field effects on behavior is also found in honeybees. The behavioral effects have been reviewed by Martin and Lindauer[144] and Gould.[118] Honeybee workers communicate the location of a food source to other workers in a hive by means of a "waggle dance" on a vertical honeycomb. The angle between the direction of the dance and the vertical direction indicates the angle between the food source and the sun. Consistent errors in the dance angle occur which vanish when the magnetic field in the hive is nulled by means of external coils. In anomalous situations where bees are made to dance on horizontal surfaces, after an initial period of disorientation they dance along the eight magnetic compass directions (N, NE, E, SE, etc.).[118,144,145] If the geomagnetic field is nulled, the dances become disoriented again. There is also evidence that bees can use the daily variations in the geomagnetic field to set their circadian rhythms.[144]

Gould et al.[146] have found that honeybees also contain Fe_3O_4. They measured an average, induced remanent moment of about 2×10^{-9} A·m^2 per bee, distributed between single-domain and superparamagnetic-sized particles, mostly localized to the abdomen. Recently, Kuterbach et al.[147] found bands of cells around the abdominal segments that contain numerous iron-rich granules. The granules are primarily a hydrous iron oxide, which can be a precursor in the precipitation of Fe_3O_4 (see below).

In addition to pigeons and honeybees, Fe_3O_4 appears to be widely distributed in the biological world.[148] Magnetic inclusions have been found in organisms as diverse as dolphins,[149] butterflies,[150] tuna,[151] green turtles,[152] marine crustacea,[153] bacteria,[154] and humans.[155] The first identification of Fe_3O_4 in an organism was by Lowenstam,[156] who found

---

* Magnetic dipole moment that persists following exposure to an intense magnetic field (see Reference 163).

it in the tooth denticles of a group of mollusks, called "chitons". Fe_3O_4 is very hard as well as magnetic and this is advantageous to chitons as they scrape algae off rocks. This is an illustration of the fact that the presence of Fe_3O_4 in an organism does not necessarily mean that the organism has a magnetic detector. In addition to hardness, Fe_3O_4 is also apparently the densest material that can be mineralized by organisms, and this might play a role in certain cases.

The best documentation to date for the connection between magnetically sensitive behavior and the presence of Fe_3O_4 is for aquatic bacteria that orient and swim along magnetic field lines.[113,157] This behavior is termed "magnetotaxis." Magnetotactic bacteria were discovered serendipitously in the early 1970s by Richard Blakemore.[113] He initially found bacteria from both fresh water and marine muds that accumulated at the north side of drops of water placed on a microscope slide. These bacteria were attracted and repelled by the north and south poles of a bar magnet, respectively. Subsequently, Blakemore and Kalmijn[158] used homogeneous magnetic fields produced by Helmholtz coils to show that New England bacteria swim along magnetic field lines in the field direction, that is, in the direction indicated by the north-seeking end of a magnetic compass needle. When the field produced by the coils is reversed by reversing the direction of current flow, the bacteria respond immediately by executing U-turns and continuing to swim in the field direction. Killed cells orient along the field lines and rotate when the field direction is reversed, but do not move along the field lines. Thus magnetotactic bacteria from New England behave as self-propelled magnetic dipoles and are predominantly north-seeking.[158,159]

Magnetotactic bacteria are easy to find in the sediments of almost any aquatic environment.[157,159] In addition to world-wide dispersion, the diversity of morphological types suggests that the phenomenon is a feature of a number of bacterial species. Two characteristics unify these species. They are apparently all anaerobic or microaerophillic[157] and they all contain magnetosomes,[160] which are unique intracytoplasmic structures consisting of enveloped Fe_3O_4[154,161] (Figure 1). One species, *Aquaspirillum magnetotacticum,* has been isolated and grown in pure culture in a chemically defined medium,[162] In this species there are typically 20 to 25 cuboidal Fe_3O_4 particles about 500 Å on a side per cell. The particles are arranged in a chain which is fixed along the axis of motility of the bacterium. Magnetosomes are arranged in one or two chains in most other bacterial species as well. Since only soluble ferric iron is available in the growth medium,[162] the presence of intracytoplasmic Fe_3O_4 in *A. magnetotacticum* implies a bacterial biomineralization process. In fact, since total cellular iron is about 2% of the cellular dry weight, these bacteria are prodigious manufacturers of Fe_3O_4. If iron is withheld from the growth medium, these bacteria grow without magnetosomes; these cells are nonmagnetotactic.* Thus the magnetotactic response is definitely correlated with the presence of the magnetosomes.

Fe_3O_4 has an inverse spinel structure and is ferrimagnetic with a Curie temperature of 580 °C.[163] Fe_3O_4 particles of 500-Å dimensions are single, magnetic domains with a permanent magnetic moment approaching the saturation magnetization of bulk Fe_3O_4, 480×10^3 A/m. Larger ferrimagnetic particles form magnetic domains, reducing the magnetostatic energy and the remanent magnetic moment. The upper size limit for single magnetic domains is approximately the width of a domain wall d_w, which is a function of the exchange and anisotropy energy of the material

$$d_w = \left(\frac{kT_c}{Ka^3}\right)^{1/2} a \qquad (9)$$

---

* Nonmagnetotactic cells grow as well as magnetotactic cells in the homogeneous culture medium which provides all essential nutrients.

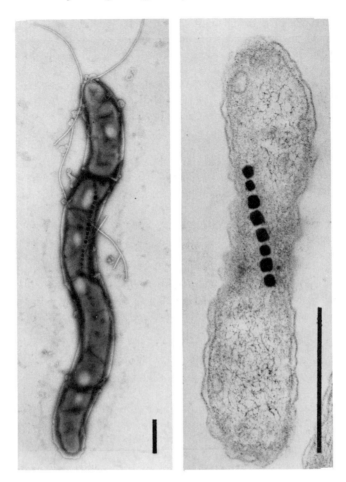

FIGURE 1.   Magnetotactic spirillum showing chain of enveloped Fe_3O_4
particles. (Left) whole spirillum; (right) thin section. Bar in each photo is
1-μm.[167]

where k is Boltzman's constant, T_c is the Curie temperature, K is the anisotropy energy per
unit volume, and a is the atomic spacing. Substituting values for Fe_3O_4[163] yields $d_w \simeq 500$
Å. More precise calculations by Butler and Banerjee[164] for cubic particles yield $d_w = 760$
Å. On the other hand, if the particle dimension is less than a certain value d_s, it will be
superparamagnetic at room temperature; i.e., thermal energy will cause transitions of the
magnetic moment between equivalent, easy magnetic axes of the particle with a consequent
loss of the time-averaged remanent moment.[165] The transition probability is a function of
the anisotropy energy and the thermal energy and the most probable transition time between
orientations is

$$\tau \sim \tau_0 \exp(KV/2kT) \tag{10}$$

where τ_0 is a constant of the order of 10^{-9} sec and V ($= d_s^3$) is the particle volume. Particles
of dimensions > 350 Å are stable for times > 10^6 years; hence d_s < 350 Å. Thus particles
of Fe_3O_4 with dimensions 350 Å < d < 760 Å are permanent, single magnetic domains
with remanent magnetization of 4.8×10^5 A/m. We can assume that each 500-Å particle
produced by a bacterium has a moment of 6.0×10^{-17} A·m^2.

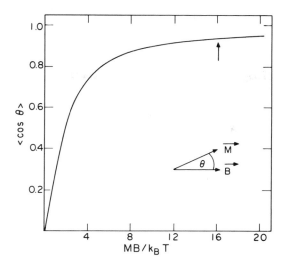

FIGURE 2. Langevin function (Equation 14) plotted as a function of mB/kT. Arrow indicates the value of the function for typical magnetotactic bacteria.

When the single-domain particles are organized in a chain as they are in *A. magnetotacticum,* the interactions between the particle moments will cause them to be oriented parallel to each other along the chain direction.[166] Thus the moment of the entire chain will be equal to the sum of the individual particle moments. For chains of 22 particles, this gives a total remanent moment $m = 1.3 \times 10^{-15}$ A·m^2. Since the particles are fixed in the bacterium by the magnetosome envelope, the bacterium is, in effect, a swimming magnetic dipole.

The simplest hypothesis for magnetotaxis is passive orientation of the swimming bacterium along the magnetic field lines by the torque exerted by the field on the magnetic moment.[167,168] Thermal energy, on the other hand, will tend to disorient the bacterium during swimming. The energy of the bacterial moment in a magnetic field B is

$$E_m = -\vec{m} \cdot \vec{B}$$
$$= -mB \cos\theta \tag{11}$$

where θ is the angle between $\vec{m}$ and $\vec{B}$. The thermally averaged orientation of an ensemble of moments, or equivalently, the time-averaged orientation of a single moment

$$<\cos\theta> = \frac{\int \cos\theta \; e^{-E_m/kT} \; dV}{\int e^{-E_m/kT} \; dV} = L(\alpha) \tag{12}$$

$L(\alpha)$ is the Langevin function

$$L(\alpha) = \coth(\alpha) - \frac{1}{\alpha} \; ; \; \alpha = mB/kT \tag{13}$$

and is plotted in Figure 2. If we consider *A. magnetotacticum* in the earth's magnetic field of 0.5×10^{-4} T at room temperature, then $\alpha \sim 16$ and $<\cos\theta> > 0.9$. Because the

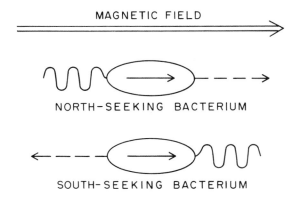

FIGURE 3.   Schematic drawings of north- and south-seeking magnetotactic bacteria. The arrow in the cell represents the magnetic dipole moment of the magnetosome chain, with the arrowhead indicating the north-seeking pole.

Langevin function asymptotically approaches 1 as α increases, the orientation would not improve significantly if there were more particles and the moment per bacterium were larger. Thus each bacterium is in effect a biomagnetic compass optimized to the geomagnetic field at room temperature.[6167]

For passively oriented bacteria, the migration velocity along the magnetic field lines is

$$v_H = v_0 <\cos\theta> \tag{14}$$

where v_0 is the forward velocity of the swimming bacterium and θ is the angle between the axis of motility and the magnetic field. If v_0 is independent of B and the magnetic moment is parallel to the axis of motility

$$v_H = v_0 L(\alpha) \tag{15}$$

providing that the velocity is averaged over a time which is long compared to the rotational diffusion time (typically, 1 sec). This is the basis of a method for measuring the magnetic moment of individual bacteria.[169] Determination of the width of the U-turns executed by swimming bacteria following reversal of the magnetic field direction[170] also gives a measure of the magnetic moment.[171] Static light scattering[172] and magnetically induced birefringence techniques[173] have been used to determine the average moment per cell in suspensions of live or dead bacteria. For diverse types of cells the moments range from 10^{-15} to 5×10^{-13} A·m^2 per cell.

A. magnetotacticum is bipolarly flagellated, i.e., it has a flagellum at each end of the cell and can swim in either direction along the magnetic field lines. However, many other magnetotactic bacterial species one observes in sediments are asymmetrically flagellated and have unidirectional motility. As noted above, these bacteria from New England swim along magnetic field lines in the field direction. Based on the passive orientation hypothesis, this occurs if the bacterial moment is oriented in the cell forward with respect to the flagellum (Figure 3). Then the bacterium will propel itself in the field direction when the moment is oriented in the field, and will be north-seeking in the geomagnetic field. If the bacterial moment were oriented in the cell rearward with respect to the flagellum, the cell would propel itself opposite to the field direction when the moment was oriented in the field, and hence would be south-seeking in the geomagnetic field.

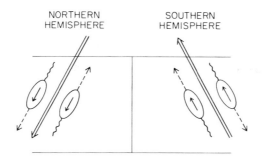

FIGURE 4. Swimming directions of north- and south-seeking bacteria along the inclined geomagnetic field lines in each hemisphere. The arrow in each cell represents the magnetic dipole moment of the magnetosome chain, with the arrowhead indicating the north-seeking pole.

South-seeking bacteria have been produced in the laboratory by subjecting them to magnetic pulses of AC magnetic fields which are strong enough to overcome the magnetic interaction forces between the particles in the chain and cause their moments to rotate and reorient along the chain in the opposite direction.[158] Field strengths of several times 10^{-2} T are required, consistent with magnetic measurements on freeze-dried cells[174] and in agreement with estimates based on the "chain of spheres" model of Jacobs and Bean,[166] who considered the magnetic properties of a chain of single-domain particles in a different context before the discovery of magnetotactic bacteria.

The predominance of north-seeking bacteria in the Northern Hemisphere is due to the inclination of the geomagnetic field.[175] Since many sediment-dwelling bacteria are anaerobic or microaerobic,* it is advantageous for them to have mechanisms that prevent them from swimming up toward the toxic, higher oxygen concentration at the water surface, and keep them in the sediments. Since the geomagnetic field is approximately dipolar, the magnetic field lines at the surface of the earth are inclined at an angle that increases with latitude. The total flux density at geomagnetic latitude θ is approximately

$$B_G = 0.3 \ (\sin^2\theta + 1)^{1/2} \times 10^{-4} \ \text{T} \tag{16}$$

and the inclination I from the horizontal is given by

$$\tan I = 2 \tan\theta \tag{17}$$

In the Northern Hemisphere the field is inclined downwards, pointing straight down at the north magnetic pole. In the Southern Hemisphere the field is inclined upwards, at an angle increasing with latitude, pointing straight up at the south magnetic pole. At the geomagnetic equator the field is horizontal.

Because of the inclination of the field lines, north-seeking bacteria migrate downward in the North Hemisphere and upward in the Southern Hemisphere (Figure 4). South-seeking bacteria migrate upward in the Northern Hemisphere and downward in the Southern Hemisphere. At the equator, both polarity types migrate horizontally. Because downward directed motion is advantageous, north-seeking bacteria should be favored in the Northern Hemisphere

---

* Anaerobic bacteria live in the absence of free oxygen. Microaerobic bacteria tolerate or require low concentrations of oxygen.

and south-seeking bacteria should be favored in the Southern Hemisphere. At the equator neither polarity would be favored.

Examination of bacteria in sediments from various places in the world confirms this hypothesis. In contrast to New England (inclination, 70° N) and other Northern Hemisphere locales, magnetotactic bacteria in fresh water and marine sediments in Australia and New Zealand (inclination, 70° S) are almost exclusively south-seeking.[176,177] These bacteria have chains of particles and can be remagnetized to north-seeking polarity.[176] At the geomagnetic equator in Brazil (inclination, 0°) both north- and south-seeking bacteria are present in roughly equal numbers.[178] Thus the vertical component of the geomagnetic field selects the predominant cell polarity in natural environments, with downward directed motion advantageous for, and upward directed motion detrimental to, survival of the organisms. At the geomagnetic equator where motion is directed horizontally, both polarities benefit because horizontally directed motion reduces harmful upward migration.

The role of the vertical magnetic field component has also been confirmed in laboratory experiments.[176,178] When a sediment sample from New England, initially containing north-seeking bacteria, is placed in a coil that produced a field of twice the magnitude and opposite sign to the ambient vertical field, the polarity of the bacteria in the sample inverted over several weeks, i.e., over many bacterial generations. If a sample is placed in a coil that cancels the vertical component of the ambient magnetic field, the population in the sample tends toward equal numbers of both polarities, again over many generations. Equal numbers of both polarities also result when samples initially containing all north- or all south-seeking bacteria are placed in an enclosure that cancels the ambient magnetic field. Further experiments in null field by Blakemore[157] confirm the role of oxygen. When samples with tight stoppers are placed in the zero field enclosure, bacteria of both polarities are ultimately found in the sediment and in the water column up to the surface. When the sample bottles are loosely stoppered, allowing diffusion of air, bacteria are found in the sediments but not in the water column.

While the ability to synthesize Fe_3O_4 and construct magnetosomes is certainly genetically encoded, the polarity of the magnetosome chain cannot be encoded. If a bacterium that lacks magnetosomes starts to synthesize them de novo, there is equal probability that when the particles grow to permanent, single-domain size, the chain will magnetize with north-seeking pole forward as with south-seeking pole forward; a population of these bacteria will consist of 1:1 north- and south-seekers. If, however, the daughter cells inherit some of the parental magnetosomes during cell division, they will inherit the parental polarity. As they synthesize new magnetosomes at the ends of their inherited chains, the magnetic field produced by the existing particles will magnetize the new particles in the same direction. Thus, north-seeking bacteria can produce north-seeking progeny and south-seeking bacteria can produce south-seeking progeny. However, there are mechanisms by which some progeny with the opposite polarity can be produced in each generation. For example, if in the cell division process, some of the daughter cells inherit no parental magnetosomes, these cells will synthesize them *de novo* and about one half those cells will end up with the polarity opposite to that of the parental generation. So in New England, where north-seeking bacteria are found and predominate, some south-seekers are produced in each population division. Under normal circumstances, these south-seekers are unfavored by being directed upwards towards the surface when they are separated from the sediments, and their total population remains low compared to the north-seeking population. However, when the vertical magnetic field is inverted, as in the experiment described above, these south-seekers are suddenly favored and their progeny eventually predominate as the previously favored north-seeking population declines in their newly unfavorable circumstances. When the vertical component is set equal to zero, neither polarity is favored and the north- and south-seeking populations eventually equalize.

We can envision a similar process occurring in natural environments during reversals or excursions of the geomagnetic field. During these processes the vertical component changes sign over thousands of years. This would be accompanied by a change in the predominant polarity of the magnetotactic bacteria population in that locale.

Other possible advantages of magnetotaxis to bacteria involve rapid migration along magnetic field lines. This could be useful for population dispersal, as an escape response, or in outrunning chemical diffusion. There are also consequences of magnetotaxis and Fe_3O_4 synthesis that may or may not be advantageous. Magnetic bacteria that are within 4 μm of each other will experience magnetic forces greater than the forces of Brownian motion. Fe_3O_4 has a density of 5.1, hence precipitation increases the average density of the bacteria, helping them to stay down in the sediments even when they are not swimming, and may serve some metabolic functions as well.[157]

Finally, as mentioned in Section IV, magnetotactic bacteria live their lives and carry out all their cellular functions in a magnetic field of their own making which varies from over 0.1 T at the surface of the particles to 0.01 T or less at the periphery of the cell, depending on shape of the bacterium, and the location of the particles. The field due to a dipole moment m at points with coordinates r, θ, varies as r^{-3} and $(\mu_0/4\pi)(1 + 3 \cos^2\theta)^{1/2}$ when r$\gg\ell$, the length of the dipole. $\theta = 0$ corresponds to the dipole moment direction. If we consider the moment m $= 10^{-15}$ A·m^2 in a magnetotactic bacterium as a point dipole, then the field at the surface of a sphere of radius 1 μm around the moment would vary from 2 $\times$ 10^{-4} to 1 $\times$ 10^{-4} T as θ varies from 0 to 90°.

When r $< \ell$, the calculation is more difficult. One approach is to replace the individual Fe_3O_4 particles of dimension d by equivalent current cylinders. This calculation[179] gives fields up to about 0.5 T between the particles and up to 0.3 T at the ends of the particle chain. At points on a line perpendicular to one end of the particle chain the field falls from 0.22 T at the edge of the last particle to 0.03 T at d, 0.008 T at 2d, 0.003 T at 3d, and 0.002 T at 4d.

Progress has been made in elucidating the Fe_3O_4 biomineralization process.[180,181] Mössbauer spectroscopic studies of magnetic and nonmagnetic strains and cell fractions of *A. magnetotacticum* have revealed several iron-containing materials in the cells in addition to Fe_3O_4. One of these is a high-density hydrous iron oxide that is spectroscopically similar to the mineral ferrihydrite. It appears that Fe_3O_4 precipitation occurs following partial reduction of a ferrihydrite precursor.

Reduction of a ferrihydrite precursor to Fe_3O_4 also occurs in the radular teeth of the marine chiton.[148,182,183] Iron is transported to the superior epithelial cells of the radulae as ferritin. Then iron is transferred to a preformed organic matrix on the tooth surface as ferrihydrite. Finally, the ferrihydrite is reduced to Fe_3O_4. The resulting Fe_3O_4 particles have dimensions of the order of 0.1 μm. Thus the Fe_3O_4 precipitation process appears to be similar in magnetotactic bacteria and in chitons, and may be similar in other organisms as well. The facility with which bacteria can be manipulated will allow further elucidation of the precipitation process and eventual understanding of how the bacteria control the size, shape, and number of particles.

VI. CONCLUSION

Life evolved in the geomagnetic field and it should not be surprising that some and perhaps many species are magnetically sensitive. The discovery of Fe_3O_4 in diverse organisms provides a new basis for understanding the interaction of organisms with the geomagnetic field. Inclusions of magnetic materials in organisms could also play a role in the other magnetic field and nonionizing radiation effects.

Effects of high-static fields remains an important area for research, because of increasing exposure of humans to higher fields in connection with NMR imaging and other technologies involving high magnetic field devices. Epidemiological studies are important for establishing guidelines for human, but more work needs to be done at both the cellular and subcellular levels to elucidate interactive mechanisms. High field alignment of macromolecules and magnetic field interactions with membranes are promising areas but their connection with physiological effects in intact cells and organisms needs to be elucidated.

Finally, there is the question of synergistic effects. Healthy cells or organisms might be able to adapt successfully to the physiological stress imposed by static magnetic fields, but this might not be the case for organisms that are coping with additional stresses, such as disease, environmental factors, etc. This possibility should be considered especially in setting guidelines for high field exposure.

ACKNOWLEDGMENTS

This work was partially supported by the Office of Naval Research. The Francis Bitter National Magnet Laboratory is sponsored by the National Science Foundation. I gratefully acknowledge discussions with R. Blakemore, S. Foner, and C. Rosenblatt, and T. Rossell, T. Tenforde, A. Sheppard, and A. Dawson for comments on the manuscript. I also thank G. Lynch for editorial assistance.

APPENDIX: ELECTROMAGNETIC UNITS AND DEFINITIONS

Purcell[184] gives an excellent discussion of magnetic field concepts; for a review of magnetic measurements see Foner[185] and Morrish.[163] In discussing the interactions of static magnetic fields with materials most workers use the centimeter-gram-second (cgs)-electromagnetic units (emu). In the SI system, the magnetic flux density

$$\vec{B} = \mu_0 \vec{H}$$

and

$$\vec{B} = \mu_0(\vec{H} + \vec{M}) \tag{A1}$$

in vacuum and in a material medium, respectively. $\vec{H}$ is the magnetic field intensity and the magnetization per unit volume. The permeability of free space

$$\mu_0 = 4\pi \times 10^{-7} \text{ H/m} \tag{A2}$$

where H/m (henry per meter) is equivalent to weber per meter per ampere. The volume magnetization of diamagnetic and paramagnetic materials is related to the magnetic field intensity by the magnetic susceptibility

$$\vec{M} = \chi\vec{H} \tag{A3}$$

In the SI system, χ is a dimensionless quantity. In magnetically ordered materials, $\vec{M}$ is a complex function of H and can have finite values even at $\vec{H} = 0$.

In the cgs-emu $\mu_0 = 1$ and

$$\vec{B} = \vec{H} + 4\pi\vec{M} \tag{A4}$$

Some of the relations between emu and SI units are as follows

Magnetic flux den- $1\ G = 10^{-4}\ T$
sity B

Magnetic field den- $1\ Oe = (10^3/4\pi)\ A/m$
sity H

Magnetic moment m $1\ emu = 10^{-3}\ A\cdot m^2$

Magnetization, $1\ emu/cm^3 = 10^3\ A/m$
magnetic
moment per unit vol-
ume M

Magnetic suscepti- $\chi(emu) = (1/4\pi)\chi(MKS)$
bility χ

In the emu system, Faraday's law of magnetic induction is

$$\epsilon = -10^{-8}\ d\Phi/dt \qquad (A5)$$

where the emf (electromotive force) ϵ (volts) is induced in a conducting loop of area A normal to B, and the magnetic flux

$$\Phi = BA \qquad (A6)$$

The emf can be produced by a time-varying field in a stationary loop or by a loop whose normal component is changing with respect to a static magnetic field. The factor of 10^{-8} also appears in the equation for the flow potential when Equation 1 is written with v and d in centimeter per second and centimeter respectively.

Magnetic moments, magnetization, and magnetic susceptibility of materials are expressed on a unit weight, unit volume, per mole, per atom, or molecule basis. Magnetic moments and magnetization have the units emu, gauss, cgs, or ergs per gauss in the emu system, with $1\ emu = 1\ G = 1\ cgs = 1\ erg/G$. For example, the saturation magnetic moment per unit volume, or magnetization, of Fe_3O_4 is 480 emu/cm³, or 92 emu/g. The conversion factor is the density. Magnetic moments of atoms and molecules are often expressed as Bohr Magnetons (μB) with $1\ \mu B = 0.927 \times 10^{-20}$ ergs/G. The electron has a magnetic moment of 1 μB. Fe_3O_4 has a magnetic moment of 4 μB per formula unit.

The free energy of magnetic dipoles or of materials with permanent, macroscopic magnetization $\vec{m}$ oriented at angle θ in a magnetic field of flux density B is

$$E_m = -\vec{m} \cdot \vec{B} = -mB\cos\theta \qquad (A7)$$

In a homogeneous magnetic field the free energy is a minimum when $\cos\theta = 1$, i.e., $\vec{m}$ is parallel to $\vec{B}$. In an inhomogeneous magnetic field additional lowering of the free energy comes from translational motion of the material toward increasing field strength. The translational force along the gradient is

$$F_x = m\frac{dB}{dx} \qquad (A8)$$

where x is the magnetic field gradient direction.

For materials without a permanent dipole moment, a magnetic field will induce a moment per unit volume $\vec{M}$, where

$$\vec{M} = \bar{\bar{\chi}} \vec{H} \tag{A9}$$

$\bar{\bar{\chi}}$ is the susceptibility tensor with units emu/G or ergs/G^2. For diamagnetic materials $\bar{\bar{\chi}}$ is small and negative and is generally independent of temperature. For example, H_2O has an isotropic volume magnetic susceptibility $\chi_v = -0.4 \times 10^{-7}$ emu/(G-cm^3). In paramagnetic materials, i.e., materials with unpaired electrons, χ is larger in magnitude, positive in sign, and is generally temperature-dependent. At ambient temperatures typical paramagnets have susceptibilities that follow the Curie Law

$$\chi = N\mu_{eff}^2/3k_BT \tag{A10}$$

where k_B is Boltzman's constant, N is the number density of paramagnetic atoms, and μ_{eff} is the effective magnetic moment per atom. From quantum mechanics, the saturation magnetic moment of an atom or molecule is proportional to the spin and orbital angular momentum

$$\mu = g \, S \, \mu_B \tag{A11}$$

where μB is the Bohr magneton and g is the proportionality or g-factor. If we consider spin angular momentum only, $g \simeq 2$. The effective magnetic moment

$$\mu_{eff} = [g^2 \, S \, (S \, + \, 1) \, \mu_B^2]^{1/2} \tag{A12}$$

In a hypothetical example, if every water molecule had an unpaired electron spring (S = 1/2), the magnetic moment per molecule would be 1 μB and the paramagnetic susceptibility per cubic centimeter at 300 K would be 2.2×10^{-6} emu/G. The total susceptibility would be the sum of paramagnetic and diamagnetic contributions of 2.16×10^{-6} emu/G. In some cases the diamagnetic susceptibility contribution can be larger than the paramagnetic contributions. This is often the case in large biological molecules that have a single or a few paramagnetic atoms.

In a magnetic field the free energy is given by

$$E_m = -\frac{1}{2} \vec{H} \cdot \chi \cdot \vec{H} \tag{A13}$$

where the magnitude and direction of the induced moment depends on the orientation of the molecule in the field. If the susceptibility is isotropic, the induced moment is always parallel to H and

$$E_m = -\frac{1}{2} \chi H^2 \tag{A14}$$

There are no rotational or translational forces in a homogeneous magnetic field. In an inhomogeneous magnetic field, the material will experience a translational force in the direction of increasing or decreasing field strength depending on the sign of χ

$$F_x = \chi \, VH \, dH/dx \tag{A15}$$

where χ is the susceptibility per unit volume and V is the volume. Diamagnetic materials move in the direction of decreasing field strength while paramagnetic materials move toward increasing field strength. As discussed in Section II, this is the basis of a method for separating diamagnetic from paramagnetic materials. In suspensions or solutions, the force depends on the difference in susceptibility between the material and the medium.

Anisotropic materials require two or at most three independent parameters to specify the magnetic susceptibility. In the most general case,

$$E_m = -\frac{1}{2}(\chi_x H_x^2 + \chi_y H_y^2 + \chi_z H_z^2) \tag{A16}$$

where x, y, and z denote the eigenvectors of the diagonalized susceptibility tensor. These three vectors often correspond to molecular symmetry axes. In addition to translational forces in inhomogeneous fields, there are rotational forces in homogeneous fields. For example, benzene has an inplane susceptibility $\chi_\perp = -4.5 \times 10^{-7}$ emu/G-cm³ $(= -5.7 \times 10^{-6}$ in SI units) and a susceptibility normal to the plane $\chi_\perp = -12 \times 10^{-7}$ emu/G-cm³ $(= -1.5 \times 10^{-5}$ in SI units).[186] The molecule will experience a torque tending to align its plane parallel to the field direction. In general, any molecule or molecular assembly with anisotropic diamagnetism will tend to align so that the least negative susceptibility direction is parallel to the applied field.

In anisotropic paramagnets the highest (positive) susceptibility direction will tend to align parallel to the field. If $x_\parallel$ is the susceptibility along the minimum energy direction and θ is the angle between that direction and the applied field

$$E_m = -\frac{1}{2}\chi_\parallel H^2 - \frac{1}{2}\Delta\chi H^2 \cos^2\theta$$

$$\Delta\chi = \chi_\parallel - \chi_\perp \tag{A17}$$

where by definition $\Delta\chi > 0$ and E_m is minimized when $\theta = 0$. The degree of orientation of an ensemble of molecules at a given field strength and temperature can be calculated from statistical mechanics. The angular distribution function $F(\theta)$, which specifies the probability that a molecule has an equilibrium orientation at angle θ with respect to H, can in general be expanded in terms of $\cos^n\theta$. For molecules with cylindrical symmetry, odd terms in n vanish. Then a convenient measure of the degree of orientation is the average value of the second Legendre polynomial

$$\overline{P_2(\cos\theta)} \equiv \overline{(3/2 \cos^2\theta - 1/2)} \tag{A18}$$

From statistical mechanics,

$$P_2(\cos\theta) = \frac{\int \left(\frac{3}{2}\cos^2\theta - \frac{1}{2}\right)\exp\left(-\frac{1}{2}\Delta\chi H^2 \cos^2\theta/kT\right)d\cos\theta}{Z} \tag{A19}$$

where Z is the partition function. Complete alignment means $\overline{\cos^2\theta} = 1$ and $\overline{P_2(\cos\theta)} = 1$. For random orientation in three dimensions $\overline{\cos^2\theta} = 1/3$, hence $\overline{P_2(\cos\theta)} = 0$. Even for molecules such as benzene with large diamagnetic anisotropies, thermal excitation will overcome magnetic alignment and the equilibrium alignment will be small. However, if N molecules are contained in an ordered array or aggregate, Δx for the single molecule could be replaced by $N\Delta\chi$ in the exponential in Equation A17, resulting in substantial alignment.

REFERENCES

1. **Stoner, E. C.,** Magnetism, in *Encyclopedia Britannica,* Vol. 14, Benton, Chicago, Ill., 1972.
2. **Needham, J.,** *Science and Civilization in China,* Vol. 4(Part 1), Cambridge University Press, London, 1962.
3. **Carlson, J. B.,** Lodestone compass: Chinese or Olmec primacy?, *Science,* 189, 753, 1975.
4. **Malmstrom, V. H.,** Knowledge of magnetism in pre-Columbian Mesoamerica, *Nature,* 259, 390, 1976.
5. **Darnton, R.,** *Mesmerism,* Schocken, New York, 1970.
6. **Shulman, R. G.,** *Biological Applications of Magnetic Resonance,* Academic Press, New York, 1979.
7. **Shulman, R. G.,** NMR spectroscopy of living cells, *Sci. Am.,* 248(1), 86, 1983.
8. **Ackerman, J. H., Grove, T. H., Wong, G. G., Gadion, D. G., and Radda, G. K.,** Mapping of metabolites in whole animals by ^{31}P NMR using surface coils, *Nature,* 283, 167, 1980.
9. **Fossell, E. T., Morgan, H. E., and Ingwall, J. S.,** Measurement of changes in high-energy phosphates in the cardiac cycle using gated ^{31}P nuclear magnetic resonance, *Proc. Natl. Acad. Sci. U.S.A.,* 77, 3654, 1980.
10. **Mansfield, P. and Morris, P. G.,** *NMR Imaging in Biomedicine,* Academic Press, New York, 1982.
11. **Kaufman, L., Crooks, L. E., and Margulis, A. R., Eds.,** *Nuclear Magnetic Resonance Imaging in Medicine,* Igaku-Shoin, New York, 1982.
12. **Cohen, D.,** Magnetic fields of the human body, *Phys. Today,* 28, 34, 1975.
13. **Williamson, S. J. and Kaufman, L.,** Magnetic fields of the human brain, *J. Magn. Magn. Mater.,* 15—18, 1548, 1980.
14. **Erne, S. N., Hahlebohm, H. D., and Lübbig, H., Eds.,** *Biomagnetism,* de Gruter, Berlin, 1982.
15. **Farrell, D. E., Tripp, J. H., Zanzucchi, P. E., Harns, J. W., Brittenham, G. M., and Muir, W. A.,** Magnetic measurement of human iron stores, *IEEE Trans. Magn.,* 16, 1818, 1980.
16. **Lobel, D., Hale, J. R., and Montgomery, D. B.,** A new magnetic technique for the treatment of giant retinal tears, *Am. J. Ophthalmol.,* 85, 699, 1978.
17. **Hendren, W. H. and Hale, J. R.,** Electromagnetic bougienage to lengthen esophageal segments in congenital esophageal atresia, *N. Engl. J. Med.,* 293, 428, 1975.
18. **Hale, J. R.,** Medical applications of magnetic devices, *IEEE Trans. Magn.,* 11, 1405, 1975.
19. **Hsieh, D. S. T., Langer, R., and Folkman, J.,** Magnetic modulation of release of macromolecules from polymers, *Proc. Natl. Acad. Sci. U.S.A.,* 78, 1863, 1981.
20. **Kolin, A.,** Magnetic fields in biology, *Phys. Today,* 21, 39, 1968.
21. **Salles-Chuna, S. X., Battocletti, J. H., and Sances, A.,** Steady magnetic fields in non-invasive electromagnetic flowmetry, *Proc. IEEE,* 68, 149, 1980.
22. **Gaffey, C. T. and Tenforde, T. S.,** Alterations in the rat electrocardiogram induced by stationary magnetic fields, *Bioelectromagnetics,* 2, 357, 1981.
23. **Widder, K. J., Senyei, A. E., and Sears, B.,** Experimental methods in cancer therapeutics, *J. Pharm. Sci.,* 71, 379, 1982.
24. **Hirschbein, B. L., Brown, D. W., and Whitesides, G. M.,** Magnetic separations in chemistry and biochemistry, *Chem. Tech.,* March, 172, 1982.
25. **Owen, C. S.,** High gradient magnetic capture of cells and ferritin-bound particles, *IEEE Trans. Magn.,* 18, 1514, 1982.
26. **Owen, C. S., Babu, V. M., Cohen, S. W., and Maurer, P. H.,** Magnetic enrichment of antibody-secreting cells, *J. Immunol. Methods,* 51, 171, 1982.
27. **Melville, D., Paul, F., and Roath, S.,** Direct separation of red cells from whole blood, Nature, 255, 706, 1975; Fractionation of blood components using high gradient magnetic separation, *IEEE Trans. Magn.* 18, 1680, 1982.
28. **Kronick, P. L., Campbell, G. L., and Joseph, K.,** Magnetic microspheres prepared by redox polymerization used in a cell separation based on gangliosides, *Science,* 200, 1074, 1978.
29. **Kurinobu, S. and Uchiyama, S.,** Recovery of plankton from red tide by HGMS, *IEEE Trans. Magn.,* 18, 1526, 1982.
30. **deLatour, C.,** Magnetic separation in water pollution control, *IEEE Trans. Magn.,* 9, 314, 1973.
31. **Maret, G. and Dransfield, K.,** Macromolecules and membranes in high magnetic fields, *Physica,* 86—88B, 1077, 1977.
32. **Maret, G. and Dransfeld, K.,** Biomolecules and polymers in high steady magnetic fields, in *Applications of Strong and Ultrastrong Magnetic Fields,* Herlach, F., Ed., Springer-Verlag, Basel, 1981.
33. **Arnold, W., Steele, R., and Mueller, H.,** On the magnetic asymmetry of muscle fibers, *Proc. Natl. Acad. Sci. U.S.A.,* 44, 1, 1958.
34. **Geacintov, N. E., van Nostrand, F., Pope, M., and Tinkel, J. B.,** Magnetic field effects on the chlorophyll fluorescence in *Chlorella, Biochim. Biophys. Acta,* 226, 486, 1971.
35. **Chagneux, R., Chagneux, H., and Chalazonitis, N.,** Decrease in magnetic anisotropy of external segments of the retinal rods after a total photolysis, *Biophys. J.,* 18, 125, 1977.

36. **Chambre, M.,** Diamagnetic anisotropy and orientation of α-helix in frog rhodopsin and meta II intermediate, *Proc. Natl. Acad. Sci. U.S.A.,* 75, 5471, 1978.
37. **Murayama, M.,** Orientation of sickled erythrocytes in a magnetic field, *Nature,* 206, 420, 1965.
38. **Torbert, J. and Maret, G.,** Fibers of highly oriented Pfl bacteriophage produced in a strong magnetic field, *J. Mol. Biol.,* 134, 843, 1979.
39. **Hong, F. T.,** Photoelectric and magneto-orientation effects in pigmented biologic membranes, *J. Coll. Interface Sci.,* 58, 471, 1977.
40. **Neugebauer, D. C., Blaurock, A. E., and Worcester, D. L.,** Magnetic orientation of purple membranes demonstrated by optical measurements and neutron scattering, *FEBS Lett.,* 78, 31, 1977.
41. **Torbet, J., Freyssinet, J. M., and Hudry-Clergon, G.,** Oriented fibrin gels formed by polymerization in strong magnetic fields, *Nature,* 289, 91, 1981.
42. **Murthy, N. S. and Yannas, I. V.,** Liquid crystallinity in collagen solutions and fromation of collagen fibrils, private communication, 1982.
43. **Sperber, D., Dransfeld, K., Maret, G., and Weisenseel, M. H.,** Oriented growth of pollen tubes in strong magnetic fields, *Naturwissenschaften,* 68, 40, 1981.
44. **Audus, L. J.,** Magnetotropism: a new plant-growth response, *Nature,* 185, 132, 1960.
45. **Audus, L. J. and Whish, J. C.,** Magnetrotropism, in *Biological Effects of Magnetic Fields,* Barnothy, M. F., Ed., Plenum Press, New York, 1964, 170.
46. **Labes, M. M.,** A possible explanation for the effect of magnetic fields on biological systems, *Nature,* 211, 968, 1966.
47. **Aceto, H., Jr., Tobias, C. A., and Silver, I. L.,** Some studies on the biological effects of magnetic fields, *IEEE Trans. Magn.,* 6, 368, 1970.
48. **Gaffney, B. J. and McConnell, H. M.,** Effect of magnetic field on phospholipid membranes, *Chem. Phys. Lett.,* 24, 310, 1974.
49. **Kalkwarf, D. R. and Langford, J. C.,** Response of agarose solutions to magentic fields, in Biological Effects of Extreme Low Frequency Electromagnetic Fields, Phillips, R. D. and Gillis, M. F., Eds., Conf. 781016, U.S. Department of Energy, 1979, 408.
50. **Atkins, P. W.,** Magnetic field effects, *Chem. Br.,* 12, 214, 1976.
51. **Schulten, K.,** Magnetic field effects in chemistry and biology, *Adv. Solid State Phys.,* 22, 61, 1982.
52. **Brocklehurst, B.,** Spin correlations in the geminate recombination of radical ions in hydrocarbons, *J. Chem. Soc.,* Faraday Trans. 2, 72, 1869, 1976.
53. **Bube, W., Haberkorn, R., and Michel-Beyerle, M. E.,** Magnetic field and isotope effects induced by hyperfine interactions in a steady-state photochemical experiment, *J. Am. Chem. Soc.,* 100, 5953, 1978.
54. **Swenberg, C. E.,** Theoretical remarks on low magnetic field interaction with biological systems, in *Magnetic Field Effects in Bioloigcal Systems,* Tenforde, T., Ed., Plenum Press, New York, 1979, 88.
55a. **Pohl, H. A.,** *Quantum Mechanics for Science and Engineering,* Prentice-Hall, Englewood Cliffs, N.J., 1967, 81.
55. **Hoff, A. J.,** Magnetic field effects on photosynthetic reactions, *Q. Rev. Biophys.,* 14, 599, 1981.
56. **Blankenship, R. E., Schaafsma, T. J., and Parson, W. W.,** Magnetic field effects on radical pair intermediates in bacterial photosynthesis, *Biochim. Biophys. Acta,* 461, 297, 1977.
57. **Barnothy, M. F., Ed.,** *Biological Effects of Magnetic Fields,* Vol. 1, Plenum Press, New York, 1964; *Biological Effects of Magnetic Fields,* Vol. 2, Plenum Press, New York, 1969.
58. **Kholadov, Y. A.,** *Magnetism in Biology,* Nauka, Moscow, 1970.
59. **Presman, A. S.,** *Electromagnetic Fields and Life,* Plenum Press, New York, 1970.
60. **Kaufman, G. E. and Michaelson, S. M.,** Critical review of the biological effects of electric and magnetic fields, in *Biological and Clinical Effects of Low-Frequency Magnetic and Electric Fields,* Llaurado, J. G., Sances, A., Jr., and Battocletti, J. H., Eds., Charles C. Thomas, Springfield, Ill. 1974, 49.
61. **Persinger, M. A., Ed.,** *ELF and VLF Electromagnetic Field Effects,* Plenum Press, New York, 1974.
62. **Silver, I. L. and Tobias, C. A.,** Magnetic fields and their biological effects, in *Space Radiation Biology and Related Topics,* Tobias, C. A. and Todd, P. W., Eds., Academic Press, New York, 1979, 258.
63. **Sheppard, A. R. and Eisenbud, M.,** *Biological Effect of Electric and Magnetic Fields of Extremely Low Frequency,* New York University Press, 1977.
64. Committee on Biosphere Effects of Extremely-Low-Frequency-Radiation, Biological Effects of Electric and Magnetic Fields Associated with Proposed Project Seafarer, U.S. National Academy of Sciences, Washington, D. C., 1977.
65. **Dubrov, A. P.,** *The Geomagnetic Field and Life (Geomagnetobiology),* Plenum Press, New York, 1978.
66. **Ketchen, E. E., Porter, W. E., and Bolton, N. E.,** The biological effects of magnetic fields on man, *Am. Ind. Hyg. Assoc. J.,* 39, 1, 1978.
67. **St. Lorant, S. J.,** Biomagnetism: a review, in *Proc. 6th Int. Conf. Magn. Technol.,* Vol 6, Alpha Press, Bratislava, 1978, 337.
68. **Taylor, L. S. and Cheung, A. Y.,** Eds., *The Physical Basis of Electromagnetic Interactions with Biological Systems,* Food and Drug Administration, U.S. Department of Health, Education and Welfare, Washington, D. C., 78-8055, 1978.

69. **Lie, R.,** *The Biological Effects of Magnetic Fields, a Bibliography,* Los Alamos Scientific Laboratory, Los Alamos, N.M. 7723-Ms, 1979.

70. **Phillips, R. D., Gillis, M. F., Kaune, W. T., and Mahlum, D. D.,** Eds., Biological Effects of Extremely Low Frequency Electromagnetic Fields, Conf. 781016, U.S. Department of Energy Symposium Series 50, 1979.

71. **Tenford, T. S. Ed.,** *Magnetic Field Effects on Biological Systems,* Plenum Press, New York, 1979.

72. **Kinn, J. B. and Postow, E.,** Index of Publications on Biological Effects of Electromagnetic Radiation (0 — 100 GHz), U.S. Environmental Protection Agency, 600/9-81-011, 1981.

73. **Kleinstein, B. H.,** Biological Effects of Non-Ionizing Electromagnetic Radiation: a Digest of Current Literature, Vols. 1—5, U.S. Department of Commerce, 1974 to 1982.

74. **Budinger, T. F.,** Potential medical effects and hazards of human NMR studies, in *Nuclear Magnetic Resonance Imaging in Medicine,* Kaufman, L., Crooks, L. E., and Margulis, A. R., Eds., Igaku-Shoin, New York, 1982, 207.

75. **Easterly, C. E.,** Biological Effects of Static Magnetic Fields: A Selective Review with Emphasis on Risk Assessment, Oak Ridge National Laboratory Report, Oak Ridge, Tenn., TM-7860, 1982.

76. **Liboff, R. L.,** A biomagnetic hypothesis, *Biophys. J.,* 5, 845, 1965; Neuromagentic thresholds, *J. Theor. Biol.,* 83, 427, 1980.

77. **Wikswo, J. P., Jr. and Barach, J. P.,** An estimate of the steady magnetic field strength required to influence nerve conduction, *IEEE Trans. Biomed. Eng.,* 27, 722, 1980.

78. **Kim, Y. S.,** Some possible effects of static magnetic fields on cancer (review), *T. I. T. J. Life Sci.,* 6, 11, 1976.

79. **Cope, F. W.,** Biological sensitivity to weak magnetic fields due to biological superconductive Josephson junctions?, *Physiol. Chem. Phys.,* 5, 173, 1973.

80. **Neurath, P. W.,** Simple theroetical models for magnetic interactions with biological units, in *Biological Effects of Magnetic Fields,* Vol. 1, Barnothy, M. F., Ed., Plenum Press, New York, 1964, 25.

81. **Roberts, A. M.,** Motion of *Paramecium* in static electric and magnetic fields, *J. Theor. Biol.,* 27, 97, 1970.

82. **Leask, M. J. M.,** A physiochemical mechanism for magnetic field detection by migratory birds and homing pigeons, *Nature,* 267, 144, 1977.

83. **Mahlum, D. D., Sikov, M. R., and Decker, J. R.,** Dominant lethal studies in mice exposed to direct current magnetic fields, in *Biological Effects of Extremely Low Frequency Electromagnetic Fields,* Phillips, R. D. and Gillis, M. F., Eds., Conf. 781016, U.S. Department of Energy, 1979, 474.

84. **Kale, P. G. and Baum, J. W.,** Genetic effects of strong magnetic fields in *Drosophila melanogaster:* homogenous fields ranging from 13,000 to 37,000 gauss, *Environ. Mutagen.,* 1, 371, 1979.

84a. **Skopek, T., Fairfax, S., and Thilly, W. G.,** unpublished results.

85. **Sikov, M. R., Malhum, D. D., Montgomery, L. D., and Decker, J. R.,** Development of mice after intrauterine exposure to direct current magnetic fields, in *Biological Effects of Extremely Low frequency Electromagnetic Fields,* Phillips, R. D. and Gillis, M. F., Eds., U.S. Department of Energy, 1979, 402.

86. **Nahas, G. G., Boccalon, H., Berryer, P., and Wagner, B.,** Effects on rodents of a one month exposure to magnetic fields (200—1200 gauss), *Adv. Space Environ. Med.,* 46, 1161, 1975.

87. **Brewer, H. B.,** Some preliminary studies of the effects of a static magnetic field on the life cycle of *Lebistes reticulatus* (guppy), *Biophys. J.,* 28, 305, 1979.

88. **Mild, K. H., Sandström, M., and Lovtrup, S.,** Development of *Xenopius laevis* embryos in a static magnetic field, *Bioelectromagnetics,* 2, 199, 1981.

89. **Neurath, P. W.,** High gradient magentic field inhibits embryonic development of frogs, *Nature,* 219, 1358, 1968; *Biological Effects of Magnetic Fields,* Barnothy, M. F., Ed., Plenum Press, New York, 1969, 177.

89a. **Strand, J. A., Abernethy, C. S., Skalski, J. R., and Genoway, R. G.,** Effects of magnetic field exposure on fertilization success in rainbow trout, *Salmo gairdneri, Bioelectromagnetics,* 4, 295, 1983.

90. **Frazier, M. E., Andrews, T. K., and Thompson, B. B.,** In vitro evaluation of biomagnetic effects, in *Biological Effects of Extremely Low Frequency Electromagnetic Fields,* Phillips R. D. and Gillis, M. F., Eds., Conf. 781016, U.S. Department of Energy, 1979, 417.

91. **Malinin, G. I., Gregory, W. D., and Morelli, L.,** Evidence of morphological and physiological transformation of mammalian cells by strong magnetic fields, *Science,* 194, 844, 1976.

92. **Chandra, S. and Stefani, S.,** Effect of constant and alternating magnetic fields on tumor cells in vitro and in vivo, in *Biological Effects of Extremely Low Frequency Electromagnetic Fields,* Phillips, R. D. and Gillis, M. F., Eds., Conf. 781016, U.S. Department of Energy, 1979, 436.

93. **Leitmannova, A., Stösser, R., and Glaser, R.,** Changes in the shape of human erythrocytes under the influence of a static homogeneous magnetic field, *Acta Biol. Med. Ger.,* 36, 931, 1977.

94. **Moore, R. L.,** Biological effects of magnetic fields: studies with microorganisms, *Can. J. Microbiol.,* 25, 1145, 1979.

95. **Gerencer, V. F., Barnothy, M. F., and Barnothy, J. M.,** Inhibition of bacterial growth by magnetic fields, *Nature,* 196, 539, 1962.

96. **Varga, A.,** Proteinbiosynthese bei Mikroorganismen unter Einwirkung von ausseren electromagnetischen Felden, *Fortschr. Exp. Theor. Biophys.,* 20, 1, 1976.

97. **Schaarschmidt, B., Lamprecht, I., and Müller, K.,** Influence of a magnetic field on the UV sensitivity in yeast, *Z. Naturforsch.,* 290, 447, 1974.

98. **Blatt, F. J. and Kuo, Y.,** Absence of biomagnetic effects in *Nitella. Biophys. J.,* 16, 441, 1976.

98a. **Arajs, S., Yehlin, L. C. L., and Farrington, T. E.,** Behavior of the action potential of *Nitella clavata* cells in the presence of uniform magnetic fields, *Proc. Int. Conf. Mang. Mang. Mater.,* American Institute of Physics, New York, 1975, 759.

99. **Edelman, A., Teulon, J., and Puchalska, I. B.,** Influence of the magnetic field on frog sciatic nerve, *Biochem. Biophys. Res. Commun.,* 91, 118, 1979.

100. **Gaffey, C. T and Tenforde, T. S.,** Electrical properties of conducting frog sciatic nerve exposed to high dc magnetic fields, *Bioelectromagnetics,* 1, 208, 1980; Bioelectric properties of frog sciatic nerves during exposure to stationary magnetic fields, *Radiat. Environ. Biophys.,* in press, 1983.

101. **Semm, P., Schneider, T., and Vollrath, L.,** Effects of an earth-strength magentic field on electrical activity of pineal cells, *Nature,* 288, 607, 1980.

101a. **Raybourn, M. S.,** The effects of dc magnetic fields on vertebrate photoreception, *Science,* 220, 715, 1983.

102. **Beischer, D. E. and Knepton, J. C.,** Influence of strong magnetic fields on the electro-cardiogram of squirrel monkeys, *Aerosp. Med.,* 35, 939, 1964.

103. **Gaffey, C. T., Tenforde, T. S., and Dean, E. E.,** Alterations in the electrocardiograms of baboons exposed to dc magnetic fields, *Bioelectromagnetics,* 1, 209, 1980; **Tenforde, T. S., Gaffey, C. T., Moyer, B. R., and Budinger, T. F.,** Cardiovascular alterations in *Macaca* monkeys exposed to stationary magnetic fields: experimental observations and theoretical analysis, *Bioelectromagnetics,* 4, 1, 1983.

104. **Haberditzl, W.,** Enzyme activity in high magnetic fields, *Nature,* 213, 72, 1967.

105. **Akoyunoglou, G.,** Effect of a magnetic field on carboxydismutase, *Nature,* 202, 452, 1964.

106. **Weissbluth, M.,** Enzyme-substrate reactions in high magnetic fields, in *Magnetic Field Effects in Biological Systems,* Tenforde, T., Ed., Plenum Press, New York, 1979, 44.

107. **Ripamonti, A., Ettienne, E. M., and Frankel, R. B.,** Effect of homogeneous magnetic fields on responses to toxic stimulation in *Spirostomum ambiguum, Bioelectromagnetics,* 2, 187, 1981; *Bioelectromagnetics,* 3, 391, 1982.

108. **Bücking, J., Herbst, M., and Pointer, P.,** The influence of a strong magnetic field on muscular contraction, *Radiat. Environ. Biophys.,* 11, 79, 1974.

109. **deLorge, J.,** Effects of magnetic fields on behavior in nonhuman primates, in *Magnetic Field Effects in Biological Systems,* Tenforde, T., Ed., Plenum Press, New York, 1979, 32.

109a. **Davis, H. P., Mizumori, S. J. Y., Allen, H., Rosenzweig, M. R., Bennett, E. L., and Tenforde, T. S.,** Behavioral studies with mice exposed to dc and 60-Hz magnetic fields, *Bioelectromagnetics,* in press, 1984.

110. **Beischer, D. E.,** Human tolerance to magnetic fields, *Astronautics,* 7, 24, 1962.

111. **Budinger, T. F., Wong, P. D. C., and Yen, C. K.,** Magnetic field effects on humans: epidemiological design study, in *Biological Effects of Extremely Low Frequency Electromagnetic Fields,* Phillips, R. D. and Gillis, M. F., Eds., Conf. 781016, U.S. Department of Energy, 1979, 379.

112. **Alpen, E. L.,** Magnetic field exposure guidelines, in *Magnetic Field Effects in Biological Systems,* Tenforde, T., Ed., Plenum Press, New York, 1979, 19.

113. **Blakemore, R. P.,** Magnetotactic bacteria, *Science,* 190, 377, 1975.

114. **Kreidl, A.,** Wertere Beitrage zur Physiologie des Ohrlabyrinthes: Versuche an Krebsen, *Sitzungsber. Akad. Wiss. Wien, Math. Naturwiss. K.,* 102, 149, 1893.

114a. **Ozeki, M., Takahata, M., and Hisada, M.,** Afferent response pattern of the crayfish statocyst with ferrite grain statolith to magnetic field stimulation, *J. Comp. Physiol.,* 123, 1, 1978.

115. **Lins de Barros, H. G. P., Esquivel, D. M. S., Danon, J., and de Oliveira, L. P. H.,** Magnetotactic algae, *Anal. Acad. Bras. Cienc.,* 54, 258, 1982.

116. **Brown, F. A., Jr., Barnwell, F. H., and Webb, H. M.,** Adaptation of the magneto-receptive mechanism of mud snails to geomagnetic field strength, *Biol. Bull.,* 127, 221, 1964.

117. **Brown, F. A., Jr. and Park, Y. H.,** Duration of an after effect in planarians following a reversed horizontal magnetic vector, *Biol. Bull.,* 128, 347, 1965.

118. **Gould, J. L.,** The case for magnetic sensitivity in birds and bees (such as it is), *Am. Sci.,* 68, 256, 1980.

119. **Quinn, T. P.,** Evidence for celestial and magnetic compass orientation in lake migrating sockeye salmon fry, *J. Comp. Physiol.,* 137, 243, 1980.

120. **Phillips, J. B.,** Use of the earth's magnetic field by orienting cave salamander *(Eurycea lucifuga), J. Comp Physiol.,* 121A, 273, 1977.

121. **Walcott, C.,** Magnetic oreintation in homing pigeons, *IEEE Trans. Magn.,* 16, 1008, 1980.

122. **Wiltschko, W. and Wilstschko, R.,** Magnetic compass of European robins, *Science,* 176, 62, 1972.

123. **Mather, J. and Baker, R.,** Magnetic sense of direction in woodmice for route-based navigation, *Nature,* 291, 152, 1981.

124. **Baker, R. R.,** Goal orientation by blindfolded humans after long-distance displacement: possible involvement of a magnetic sense, *Science,* 210, 555, 1980.

125. **Baker, R. R.,** *Human Navigation and the Sixth Sense,* Hodder and Stoughton, London, 1981.

126. **Gould, J. L. and Able, K. P.,** Human homing: an elusive phenonenon, *Science,* 212, 1061, 1981.

127. **Bookman, M. A.,** Sensitivity of the homing pigeon to an earth-strength magnetic field, in *Animal Migration, Navigation and Homing,* Schmidt-Koenig, K. and Keeton, W. T., Eds., Springer-Verlag, Basel, 1978, 127.

128. **Kalmijn, A. J.,** Electric and magnetic field detection in elasmobranch fishes, *Science,* 218, 916, 1982.

129. **Walker, M. and Kirschvink, J. L.,** private communication, 1982.

130. **Kalmijn, A. J.,** The detection of electric fields from inanimate and animate sources other than electric organs, in *Handbook of Sensory Physiology,* Vol. 3, Fessard, A., Ed., Springer-Verlag, Basel, 1974, 147.

131. **Brown, H. R. and Ilyinsky, O. B.,** The ampullae of Lorenzini in the magnetic field, *J. Comp. Physiol.,* 126, 333, 1978; **Brown, H. R., Ilyinsky, O. B., Muravejko, V. M., Corshkov, E. S., and Fonarev, G. A.,** Evidence that geomagnetic variations can be detected by Lorenzini ampullae, *Nature,* 277, 648, 1979.

132. **Jungerman, R. L. and Rosenblum, B.,** Magnetic induction for the sensing of magnetic fields by animals — an analysis, *J. Theor. Biol.,* 87, 25, 1980.

133. **Keeton, W. T.,** Magnets interfere with pigeon homing, *Proc. Natl. Acad. Sci. U.S.A.,* 68, 102, 1971.

134. **Walcott, C. and Green, R.,** Orientation of homing pigeons altered by a change in the direction of an applied magnetic field, *Science,* 184, 180, 1974.

135. **Keeton, W. T., Larkin, T. S., and Windsor, D. M.,** Normal fluctuations in the earth's magnetic field influence pigeon orientation, *J. Comp. Physiol.,* 95, 95, 1974.

136. **Walcott, C.,** Anomalies in the earth's magnetic field increase the scatter of pigeons vanishing bearings, in *Animal Migration, Navigation and Homing,* Schmidt-Koenig, K. and Keeton, W. T., Eds., Springer-Verlag, Basel, 1978.

137. **Gould, J. L.,** The map sense of pigeons, *Nature,* 296, 205, 1981.

138. **Able, K. P.,** Mechanisms of orientation, navigation and homing, in *Animal Migration, Orientation and Navigation,* Gauthreaux, S. A., Jr., Ed., Academic Press, New York, 1980, 283.

138a. **Griffin, D. R.,** Ecology of migration: is magnetic orientation a reality?, *Q. Rev. Biol.,* 57, 293, 1982.

139. **Wiltschko, W.,** Further analysis of the magnetic compass of migratory birds, in *Animal Migration, Navigation and Homing,* Schmidt-Koenig, K. and Keeton, W. T., Eds., Springer-Verlag, Basel, 1978, 302.

140. **Walcott, C., Gould, J. L., and Kirschvink, J. L.,** Pigeons have magnets, *Science,* 205, 1027, 1979.

141. **Presti, D. and Pettigrew, J. D.,** Ferromagnetic coupling to muscle receptors as a basis for geomagnetic field sensitivity in animals, *Nature,* 285, 99, 1980.

142. **Yorke, E. D.,** Two consequences of magnetic material found in pigeons, *J. Theor. Biol.,* 89, 533, 1981.

143. **Kirschvink, J. L. and Gould, J. L.,** Biogenic magnetite as a basis for magnetic field detection in animals, *Biosystems,* 13, 181, 1981.

144. **Martin, H. and Lindauer, M.,** Der Einfluss der Erdmagnetfeld und die Schwereorientierung der Honigbiene, *J. Comp. Physiol.,* 122, 145, 1977.

145. **Kirschvink, J. L.,** The horizontal magnetic dance of the honeybee is compatible with a single-domain ferromagnetic magnetoreceptor, *Biosystems,* 14, 193, 1981.

146. **Gould, J. L., Kirschvink, J. L., and Deffeyes, K. S.,** Bees have magnetic remanence, *Science,* 201, 1026, 1978.

147. **Kuterbach, D. A., Walcott, B., Reeder, R. J., and Frankel, R. B.,** Iron-containing cells in the honeybee *(Apis mellifere), Science,* 218, 695, 1982.

148. **Lowenstam, J. A.,** Minerals formed by organisms, *Science,* 211, 1126, 1981.

149. **Zoeger, J., Dunn, J. R., and Fuller, M.,** Magnetic material in the head of the common Pacific dolphin, *Science,* 213, 892, 1981.

150. **Jones, D. S. and MacFadden, B. J.,** Induced magnetization in the monarch butterfly *(Danus plexippus), J. Exp. Biol.,* 96, 1, 1981.

151. **Walker, M. M. and Dizon, A.,** Identification of magnetite in tuna, *Trans. Am. Geophys. Un.,* 62, 850, 1981.

152. **Perry, A., Bauer, G. B., and Dizon, A. S.,** Magnetite in the green turtle, *Trans. Am. Geophys. Un.,* 62, 850, 1981.

153. **Buskirk, R. E.,** Magnetic material in marine crustacea, *Trans. Am. Geophys. Un.,* 62, 850, 1981.

154. **Frankel, R. B., Blakemore, R. P., and Wolfe, R. S.,** Magnetite in freshwater magnetotactic bacteria, *Science,* 203, 1355, 1979.

155. **Kirschvink, J. L.,** Ferromagnetic crystals (magnetite?) in human tissue, *J. Exp. Biol.,* 92, 333, 1981.

156. **Lowenstam, H. A.,** Magnetite in the denticle capping in recent chitons *(Polyplacophora), Geol. Soc. Am. Bull.,* 73, 435, 1962.

157. **Blakemore, R. P.,** Magnetotactic bacteria, *Annu. Rev. Microbiol.,* 36, 217, 1982.
158. **Kalmijn, A. J. and Blakemore, R. P.,** The magnetic behavior of mud bacteria, in *Animal Migration, Navigation and Homing,* Schmidt-Koenig, K. and Keeton, W. T., Eds., Springer-Verlag, Basel, 1978, 344.
159. **Moench, T. T. and Konetzka, W. A.,** A novel method for the isolation and study of magnetotactic bacterium, *Arch. Microbiol.,* 119, 203, 1978.
160. **Balkwill, D. L., Maratea, D., and Blakemore, R. P.,** Ultrastructure of a magnetotactic spirillum, *J. Bacteriol.,* 141, 1399, 1980.
161. **Towe, K. M. and Moench, T. T.,** Electron-optical characterization of bacterial magnetite, *Earth Planet Sci. Lett.,* 52, 213, 1981.
162. **Blakemore, R. P., Maratea, D., and Wolfe, R. S.,** Isolation and pure culture of a freshwater magnetic spirillum in chemically defined medium, *J. Bacteriol.,* 140, 720, 1979.
163. **Morrish, A. H.,** *The Physical Principles of Magnetism,* John Wiley & Sons, New York, 1968.
164. **Butler, R. F. and Banerjee, S. K.,** Theoretical single-domain grain size range in magnetite and titanomagnetite, *J. Geophys. Res.,* 80, 4049, 1975.
165. **Bean, C. P. and Livingston, J. D.,** Superparamagnetism, *J. Appl. Phys.,* 30, 1205, 1959.
166. **Jacobs, I. S. and Bean, C. P.,** An approach to elongated fine-particle magnets, *Phys. Rev.,* 100, 1060, 1955.
167. **Frankel, R. B. and Blakemore, R. P.,** Navigational compass in magnetic bacteria, *J. Magn. Magn. Mater.,* 15—18, 1562, 1980.
168. **Frankel, R. B.,** Magnetotactic bacteria, *Comments Mol. Cell. Biophys.,* 1, 293, 1982; Magnetic guidance of organisms, *Annu. Rev. Biophys. Bioeng.,* 13, 85, 1984.
169. **Kalmijn, A. J.,** Biophysics of geomagnetic field detection, *IEEE Trans. Magn.,* 17, 1113, 1981.
170. **Bean, C. P.,** Dynamics of a magnetotactic bacterium in a magnetic field, private communication, 1979; **Purcell, E. M.,** Calculation of bacterial U-turn, private communication, 1979.
171. **Esquivel, D. M. S., Lins de Barros, H. G. P., Farina, M., Arago, P. H. A., and Danon, J.,** Microorganismes magnetotactiques de la region de Rio de Janeiro, *Biol. Cell.,* 47, 227, 1983.
172. **Rosenblatt, C., Torres de Araujo, F. F., and Frankel, R. B.,** Light scattering determination of magnetic moments of magnetotactic bacteria, *J. Appl. Phys.,* 53, 2727, 1982.
173. **Rosenblatt, C., Torres de Araujo, F. F., and Frankel, R. B.,** Birefrigence determination of magnetic moments of magnetotactic bacteria, *Biophys. J.,* 40, 83, 1982.
174. **Denham, C. R., Blakemore, R. P., and Frankel, R. B.,** Bulk magnetic properties of magnetotactic bacteria, *IEEE Trans. Magn.,* 16, 1006, 1980.
175. **Blakemore, R. P. and Frankel, R. B.,** Magnetic navigation in bacteria, *Sci. Am.,* 245(6), 58, 1981.
176. **Blakemore, R. P., Frankel, R. B., and Kalmijn, A. J.,** South-seeking magnetotactic bacteria in the Southern Hemisphere, *Nature,* 286, 384, 1981.
177. **Kirschvink, J. L.,** South-seeking magnetic bacteria, *J. Exp. Biol.,* 86, 345, 1980.
178. **Frankel, R. B., Blakemore, R. P., Torres de Araujo, F. F., Esquivel, D. M. S., and Danon, J.,** Magnetotactic bacteria at the geomagnetic equator, *Science,* 212, 1269, 1981.
179. **Weggel, R. J. and Frankel, R. B.,** unpublished results.
180. **Frankel, R. B., Papaefthymiou, G. C., Blakemore, R. P., and O'Brien, W. D.,** Fe_3O_4 precipitation in magnetotactic bacteria, *Biochim. Biophys. Acta,* 763, 147, 1983.
181. **Frankel, R. B., Papaefthymiou, G. C., and Blakemore, R. P.,** Mössbauer spectroscopy of iron biomineralization products in magnetotactic bacteria, in *Magnetite Biomineralization and Magnetoreception: A New Biomagnetism,* Kirschvink, J. L., Jones, D. S., and MacFadden, B. J., Eds., Plenum Press, New York, 1985.
182. **Towe, K. M. and Lowenstam, H. A.,** Ultrastructure and development of iron mineralization in the radular teeth of *C. stelleri* (Mollusca), *J. Ultrastructure Res.,* 17, 1, 1967.
183. **Kirschvink, J. L. and Lowenstam, H. A.,** Mineralization and magnetization of chiton teeth: paleomagnetic, sedimentologic and biologic implications of organic magnetite, *Earth Planet, Sci. Lett.,* 44, 193, 1979.
184. **Purcell, E. M.,** *Electricity and Magnetism* (Berkeley Physics Course), Vol. 2, McGraw-Hill, New York, 1965.
185. **Foner, S.,** Review of magentometry, *IEEE Trans. Magn.,* 17, 3358, 1981.
186. **Schmatz, T. G., Norris, C. L., and Flygare, W. H.,** Localized magnetic susceptibility anisotropies, *J. Am. Chem. Soc.,* 95, 7961, 1973.

ADDITIONAL REFERENCES

1. **Beason, R. C. and Nichols, J. E.,** Magnetic orientation and magnetically sensitive material in a transequitorial migratory bird, *Nature,* 309, 151, 1984.

2. **Blakemore, R. P., Short, K. A., Bazylinski, D. A., Rosenblatt, C., and Frankel, R. B.,** Microaerobic conditions are required for magnetite formation within *Aquaspirillum magnetotacticum, Geomicrobiol. J.,* 4, 53, 1985.

3. **Jafary-Asi, A. H., Solanki, S. N., Aarholt, E., and Smith, C. W.,** Dielectric measurements on live biological materials under magnetic resonance conditions, *J. Biol. Phys.,* 11, 15, 1982.

4. **Mann, S., Frankel, R. B., and Blakemore, R. P.,** Structure, morphology and crystal growth of bacterial magentite, *Nature,* 310, 407, 1984.

5. **Mann, S., Moench, T. T., and Williams, R. J. P.,** A high resolution electron microscopic investigation of bacterial magnetite: implications for crystal growth, *Proc. R. Soc. London, Ser. B.,* 221, 385, 1984.

6. **Ofer, S., Nowik, I., Bauminger, E. R., Papaefthymiou, G. C., Frankel, R. B., and Blakemore, R. P.,** Magnetosome dynamics in magnetotactic bacteria, *Biophys. J.,* 46, 57, 1984.

7. **Sperber, D., Oldenbourg, R., and Dransfeld, K.,** Magnetic field induced temperature change in mice, *Naturwissenschaften,* 71, 100, 1984.

8. **Spormann, A. M. and Wolfe, R. S.,** Chemotactic, magnetotactic and tactile behavior in a magnetic spirillum, *FEMS Microbiol. Lett.,* 22, 171, 1984.

9. **Walker, M. M., Kirschvink, J. L., Chang, S. B. R., and Dizon, A. E.,** A candidate magnetic sense organ in the yellowfin tuna *Thunnus albacares, Science,* 224, 751, 1984.

10. **Williamson, S. J., Romani, G. L., Kaufman, L., and Modena, I., Eds.,** *Biomagnetism,* Plenum Press, New York, 1983.

11. **Semm, P.,** Neurobiological investigations on the magnetic sensitivity of the pineal gland in rodents and pigeons, *J. Comp. Biochem. Physiol.,* 76, 683, 1984.

12. **Welker, H. A., Semm, P., Willig, R. P., Kommentz, J. C., Wiltschko, W., and Vollrath, L.,** Effects on an artificial magnetic field on serotonin N-acetyltransferase activity and melatonin content of the rat pineal gland, *Exp. Brain Res.,* 50, 426, 1983.

13. **Cremer-Bartels, G., Krause, K., and Kuchle, H. J.,** Influence of low magnetic field strength variations on retinal and pineal glands of quail and humans, *Graefe's Arch. Clin. Exp. Ophthalmol.,* 220, 248, 1983.

14. **Kirschvink, J. L., Jones, D. S., and MacFadden, B. J., Eds.,** *Magnetic Biomineralization and Magnetoreception in Organisms,* Plenum Press, New York, 1985.

15. **Semm, P., Nohr, D., Demaine, C., and Wiltschko, W.,** Neural basis of the magnetic compass: interactions of the visual, magnetic and vestibular inputs in the pigeon's brain, *J. Comp. Physiol.,* 155A, 283, 1984.

16. **Davis, H. P., Mizumori, S. J. Y., Allen, H., Rosenzweig, M. R., Bennett, E. L., and Tenforde, T. S.,** Behavioral studies with mice exposed to DC and 60 Hz magnetic fields, *Bioelectromagnetics,* 5, 147, 1984.

17. **Blackman, C. F., Benane, S. G., Rabinowitz, J. R., House, D. E., and Jones, W. T.,** A role for the magnetic field in the radiation-induced efflux of calcium ions from brain tissue in vitro, *Bioelectromagnetics,* 6, 327, 1985.

18. **Torres de Araujo, F. F., Pires, M. A., Frankel, R. B., and Bicudo, C. E. M.,** Magnetite and magnetotaxis in algae, *Biophys. J.,* in press, 1986.

19. **Matsuda, T., Endo, J., Osakube, N., and Tonomura, A.,** Morphology and structure of biogenic magnetite particles, *Nature,* 302, 411, 1983.

20. **Klinowska, M.,** Cetacean live stranding sites relate to geomagnetic topography, *Aquatic Mammals,* 11, 27, 1985.

Chapter 5

INTERACTION OF ELF MAGNETIC FIELDS WITH LIVING MATTER

T. S. Tenforde

TABLE OF CONTENTS

I. INTRODUCTION

The principal topic of this chapter is the interaction with living systems of time-varying magnetic fields in the extremely low frequency (ELF) range below 300 Hz. Magnetic fields in the ELF range are present throughout the environment and originate from both natural and man-made sources.[1] The naturally occurring, time-varying fields in the atmosphere have several origins, including diurnally varying fields on the order of 30 nT associated with solar and lunar influences on ionospheric currents. The largest time-varying, atmospheric magnetic fields arise intermittently from intense solar activity and thunderstorms, and reach intensities on the order of 0.5 μT during a large magnetic storm. Superimposed on the magnetic fields associated with irregular atmospheric events is a weak ELF field resulting from the Schumann resonance phenomenon. These fields are generated by lightning discharges and propagate in the resonant atmospheric cavity formed by the surface of the earth and the lower boundary of the ionosphere. The five lowest resonant frequencies are below 40 Hz, and the magnetic flux densities per unit bandwidth associated with the Schumann resonance phenomenon vary from 0.25 to 3.6 pT/Hz$^{1/2}$.

The time-varying, ELF magnetic fields originating from man-made sources generally have much higher intensities than the naturally occurring atmospheric fields, and in some occupational settings reach levels that approach 0.1 T. Two sources of ELF fields that have been topics of considerable public interest are high-voltage transmission lines and land-based naval communication systems. The field at ground level beneath a 765-kV, 60-Hz power line carrying 1 kA per phase is 15 μT.[2] The surface field associated with the ELF antennae that are proposed for use in submarine communications is 20 μT.[3] Household appliances operated from a 60-Hz line voltage produce local fields in their immediate vicinity with intensities as high as 2.5 mT.[1] However, the magnetic field strength decreases rapidly as a function of distance from the surfaces of household devices. The video display terminals present in most modern offices generate local ELF magnetic fields with intensities up to 2 μT.[4] A number of industrial processes that involve induction heating produce ELF magnetic fields of high intensity within the occupational environment. For example, based on a survey of electrosteel and welding industries in Sweden, it was reported that the local fields near 50-Hz ladle furnaces reached intensities of 8 mT, and intensities as high as 0.07 T were measured near induction heating devices operating in the 50-Hz to 10-kHz range.[5]

In this chapter, a summary and critical evaluation will be given of the published literature describing effects of time-varying, ELF magnetic fields on living systems. Selected aspects of this subject have also been summarized in several review articles and monographs published in recent years.[6-9] In an effort to provide a framework for the description of ELF magnetic field bioeffects, the published literature has been divided into four major categories. These categories include ELF magnetic field interactions with the visual system, nervous system, various cellular and tissue systems, and carcinogenic effects. Brief summaries are also given of ELF magnetic field interaction mechanisms and various medical applications of these fields.

II. THEORETICAL PRINCIPLES

The interaction of time-varying, ELF magnetic fields with living tissues occurs principally through the induction of electric currents in accord with Faraday's law. To illustrate the relevant physical principles, consider a purely sinusoidal, time-varying magnetic field $\vec{B}$ with frequency f and amplitude $\vec{B}_0$

$$\vec{B} = \vec{B}_0 \sin(2\pi ft) \tag{1}$$

Because of the weakly diamagnetic nature of animal tissues, their magnetic permeability is nearly identical to that of air (within approximately one part in 10^6), and an applied ELF magnetic field is transmitted into and through the body tissues with negligible attenuation. In the ELF frequency range there is also no attenuation of the magnetic field at the air/tissue interface due to wave reflection or skin depth, as discussed by Polk in the Introduction to this monograph.

If the sinusoidal field B is incident upon a circular loop of radius r and area $S = \pi r^2$, with a parallel orientation of the field vector and the normal to the surface, then by Faraday's law the magnitude of the peak induced potential V_{peak} around the loop is given by

$$V_{peak} = \left|\frac{d(\vec{B} \cdot \vec{S})}{dt}\right|_{peak} = 2\pi^2 r^2 f |\vec{B}_0| \tag{2}$$

The magnitude of the peak induced electric field $\vec{E}_{peak}$ is equal to the peak potential divided by the loop circumference

$$|\vec{E}|_{peak} = \frac{V_{peak}}{2\pi r} = \pi r f |\vec{B}_0| \tag{3}$$

A third quantity of interest in the analysis of ELF field interactions is the magnitude of the peak induced current density $\vec{J}_{peak}$ which can be calculated from Ohm's law

$$|\vec{J}|_{peak} = \sigma |\vec{E}|_{peak} = \pi r f \sigma |\vec{B}_0| \tag{4}$$

where σ = electrical conductivity.

It is of interest to use the above equations to calculate the magnitude of a time-varying, ELF magnetic field that would be expected to perturb the function of critical biological tissues such as the heart and the central nervous system (CNS). Using data from several sources, Bernhardt[10] has estimated that the endogenous current densities associated with electrical activity of the brain and heart have lower limits of 1 and 10 mA/m^2, respectively, and perturbations of normal biological functions might be expected to occur in the presence of ELF magnetic fields that induce tissue currents above these levels. Consider for illustration a 60-Hz, sinusoidal magnetic field that is normally incident upon a circular loop of tissue with a radius r = 0.06 m comparable to the human heart, and a conductivity σ = 0.2 S/m.[11] From Equation 4 the amplitude of the magnetic flux density that would induce a peak current density of 10 mA/m^2 is 4.4 mT. A similar calculation[10] for brain tissue with an average conductivity of 0.1 S/m and a loop radius r = 0.1 m comparable to the human cranium, leads to the prediction that a peak current density of 1 mA/m^2 is induced by a 60-Hz magnetic field with an amplitude $|\vec{B}_0|$ = 0.53 mT. Because ELF magnetic fields with intensities higher than 5 mT are present in the vicinity of certain types of instruments and industrial processes, the induction of tissue fields at levels that could potentially perturb biological functions is therefore possible.

As discussed in the Introduction to this monograph and in Part II, Chapter 4, ELF magnetic fields can also interact with biological tissues through magneto-orientation effects. This phenomenon arises from the interaction of a magnetic field with paramagnetic substances possessing permanent magnetic moments (e.g., the magnetite inclusions in magnetotactic bacteria), or with macromolecular assemblies in which the summed diamagnetic anisotropy is large (e.g., the photopigment molecules of retinal photoreceptors). The interactions of such systems with static magnetic fields have been well studied, and many of the interesting biomagnetic phenomena that have been characterized in such biological systems are summarized in Part II, Chapter 4. In considering the possible role of magneto-orientation phenomena in the biological interactions of time-varying magnetic fields, it is important to recognize that the frictional resistance to motion in biological tissues is high, and thus serves to damp out even low-frequency oscillations associated with time-varying, magnetic orientational forces. This is well illustrated by the fact that the orientation of diamagnetically anisotropic retinal photoreceptor outer segments in a 1-T static field occurs with a characteristic time of 4 sec in water.[12] A time-varying field with a frequency exceeding approximately 1 Hz would therefore be unable to induce a "flickering" orientational phenomenon in this system, because the frictional drag force would not allow the motion of the retinal rods to keep pace with the oscillating field. A similar conclusion can be drawn for the interaction of paramagnetic entities such as magnetotactic bacteria with an ELF, time-varying magnetic field. With the possible exception of quasi-static fields with frequencies in the range 0 to 1 Hz, it is therefore probable that magneto-orientation phenomena play little if any role in the interaction of ELF magnetic fields with living systems.

Returning to the electrical phenomena associated with time-varying magnetic field interactions, a factor that is often overlooked in biological investigations is the importance of the waveform in determining the response of living tissues to the field. Numerous types of magnetic field waveforms have been used in biological studies, including both sinusoidal and square-wave fields, and pulsed fields with burst repetition rates that lie in the ELF frequency range. For both square-wave and pulsed fields, two parameters of key importance are the rise and decay times of the signal, which determine the maximum time rates of change of the field, and hence the maximum instantaneous current densities that are induced in living tissue. For example, a sharply rising, square-wave magnetic field pulse will induce a peak current density in tissue that exceeds the value achieved with a sinusoidal field having the same rms intensity and fundamental frequency. Another factor that must be considered for waveforms with a rapid rise time is the skin depth. As discussed in the Introduction to this monograph, magnetic fields with a rise time < 10 nsec will be attenuated at the air/tissue interface due to the finite skin depth and reflection losses. Pulses with such short rise times, however, are seldom used in biological studies.

Another factor that is of key importance in determining the response of living systems to ELF magnetic fields with any type of waveform is the fundamental field frequency. The phenomenon of magnetophosphenes, which will be discussed in the next section of this chapter, is limited to time-varying magnetic fields with frequencies < 100 Hz. The mechanism underlying the loss of sensitivity at higher frequencies has not been elucidated, but it is conceivable that the visual system cannot process and respond to induced electrical currents with frequencies above 100 Hz. This hypothesis is supported by the fact that flicker fusion occurs in response to repetitive photic stimuli with frequencies above approximately 30 Hz. Although the frequency dependence of the biological response to time-varying magnetic fields has not been well characterized for systems other than the visual apparatus, it is conceivable that a similar dependence may exist in tissues such as the CNS and heart in which the endogenous electrical activity has dominant frequencies < 50 Hz.

III. MAGNETOPHOSPHENES

Among the various interactions of time-varying ELF fields with living tissues, perhaps the most widely known and well-documented effect is the production of visual sensations known as "phosphenes". In the broadest sense, the term "phosphene", which is derived from the Greek words "phos" (light) and "phainein" (to show), means the production of luminous sensations in the eye by agents other than light. These physical agents include mechanical pressure applied directly to the eye and electrical stimulation at ELF frequencies applied to the body surface in the region of the head through contact electrodes. In 1896 d'Arsonval first reported that phosphenes could also be produced by placing the head in an external magnetic field oscillating in the ELF frequency range.[13]

d'Arsonval's initial qualitative observations, which were made with a 42-Hz magnetic field, have been followed by numerous studies in which efforts have been made to define the visual characteristics of magnetophosphenes, their dependence on the frequency and intensity of the applied field, and the exact locus of the field effect within the complex visual pathway. In 1910 Thompson,[14] who was apparently unaware of d'Arsonval's earlier work, reported that a 50-Hz, 0.1-T (peak) field produced a colorless, flickering illumination over the entire visual field which was most intense in the peripheral region. Dunlap[15] confirmed Thompson's observations during the following year, and he further demonstrated that more intense magnetophosphenes were generated by a 25-Hz field relative to a 60-Hz field of comparable intensity. Shortly after Dunlap's experiments were published, Magnusson and Stevens[16] reported the results of phosphene studies in which they used pulsed DC and time-varying fields with frequencies ranging from 7 to 66 Hz. With pulsed fields they were able to elicit a flash of light that moved rapidly in a narrow band across the visual field. Using AC magnetic fields of various frequencies and approximately equal intensities, Magnusson and Steven found that the strongest magnetophosphenes were produced in the 20- to 30-Hz frequency range. They also found that field frequencies < 15 Hz led to a pulsating succession of flashes covering the entire visual field. In the 10- to 35-Hz frequency range, the magnetophosphenes were reported to form a continuous network of standing waves that flickered at a rate synchronized with the frequency of the applied magnetic field. Frequencies above 40 Hz generated magnetophosphenes with a more uniform luminosity, but which continued to exhibit flickering.

The first systematic study of the dependence of magnetophosphene properties on the intensity of the applied magnetic field was made by Barlow et al.[17] These investigators demonstrated that the duration of an induced phosphene was dependent upon the magnetic field intensity above a threshold level of 20 mT (rms). These workers also characterized a "fatigue" phenomenon in which the application of an ELF magnetic field stimulus for periods up to 1 min led to a subsequent refractory period during which a second magnetophosphene could not be elicited. With a 60-Hz, 79-mT (rms) field applied for 1 min, the subsequent refractory period lasted as long as 40 sec. Barlow et al.[17] also demonstrated that magnetophosphenes are produced by ELF magnetic fields applied in the region of the eye and not by fields directed towards the visual cortex in the occipital region of the brain. Together with the observation that pressure on the eyeball abolished visual sensitivity to ELF magnetic fields, this fact led Barlow and co-workers to suggest that the retina is the locus of excitation leading to the phosphene phenomenon. Another interesting finding by Barlow et al.[17,18] was that magneto- and electrophosphenes shared many properties in common, except that the threshold intensity for inducing phosphenes by ELF magnetic fields was relatively insensitive to background lighting conditions, whereas the threshold intensity for electrophosphene induction changed rapidly as a function of light or dark adaptation. In a later study, Seidel et al.[19] also reported that the types of light patterns induced in the visual field by ELF magnetic and electric fields were similar, but that the probability of occurrence of various classes of patterns was different for the two types of stimuli. A comprehensive description of the various types of phosphene patterns has been given by Oster.[20]

During the past several years, an extensive series of investigations of magnetophosphenes has been carried out by the Swedish scientists P. Lövsund, P. Å. Öberg, and S. E. G. Nilsson.[21-25] In this research, the relationship between the frequency of the applied field and the threshold intensity required to elicit phosphenes has been defined with precision for both ELF electric and magnetic field stimuli. A determination of the threshold magnetic field intensity required to elicit phosphenes was measured over the frequency range of 10 to 45 Hz.[21,23,24] The threshold vs. frequency curve was found to be a clear function of the state of dark adaptation, the intensity of superimposed background lighting, and the wavelength (and hence color) of the transmitted background illumination within the visual field. Following a 30-min dark adaptation period, the maximum sensitivity for magnetically induced phosphenes occurred at a frequency of 30 Hz. The minimum field intensity required for the generation of phosphenes was 10 mT at this frequency. In broad-spectrum light with intensities ranging from 0.1 to 130 cd/m^2, the sensitivity maximum occurred at 20 Hz. For low levels of background illumination (0.1 to 1.2 cd/m^2), the magnetic field threshold intensity was 10 mT. However, with background illumination at the 130-cd/m^2 level, the threshold field intensity increased to 14 mT. Immediately following removal of the 130-cd/m^2 light, the average threshold field intensity measured at frequencies of 20, 30, and 35 Hz dropped to 10 mT and then slowly rose to a plateau level of 16 mT during the ensuing 30-min dark adaptation period. During the same dark adaptation interval, the threshold value for light detection dropped by a factor of 10^4. Lövsund et al.[24] concluded from this observation that the visual elements responsible for slow dark adaptation processes lie before the "magnetoreceptors" in the optic pathway.

A very interesting series of experiments that related normal color vision to magnetophosphene sensitivity was also performed by Lövsund et al.[24] For subjects with normal color vision and for deutans who possess defects in green-sensitive photopigments, measurements were made of the magnetic field threshold intensity as a function of frequency when the background illumination was filtered to produce narrow wavelength bands centered at 443, 531, and 572 nm, which correspond to light absorption maxima of different cone photopigments. For all three wavelengths, normal subjects exhibited a local maximum in magnetic field sensitivity at 20 Hz, a local minimum at 30 Hz, and an increase in sensitivity at higher frequencies, that reached a greater level at 40 to 45 Hz than the sensitivity measured at 20 Hz (Figure 1). For broad-spectrum light, the local maximum and minimum at 20 and 30 Hz, respectively, were also observed, but the increase in sensitivity at higher frequencies was less pronounced and did not reach the same level as the sensitivity maximum at 20 Hz. For deutans the maximum sensitivity was observed at 20 Hz with the three wavelengths of background light, but the local minimum in sensitivity at 30 Hz was found only for 443-nm background illumination. With all three background colors, the sensitivity of deutans was less at 40 to 45 Hz than at 20 Hz, in distinct contrast to the observations made with subjects having normal color vision. From these observations, Lövsund et al.[24] concluded that visual processes mediated by green- and red-sensitive cones, which are responsive to flickering light frequencies as high as 60 Hz, may provide the pathways for enhanced reception by normal subjects of magnetophosphenes induced by fields with frequencies in the 40- to 45-Hz range. The cone pigment defects present in deutans would inhibit sensitivity to magnetophosphenes induced by fields in this range of frequencies.

Two other studies with visually defective subjects that implicate the retina as a site of generation of phosphenes by ELF magnetic fields were performed by Lövsund et al.[24] In a patient with Retinitis pigmentosa, in whom the photoreceptors and pigment epithelium were defective but the bipolar and ganglion cell layers of the retina were conserved, magnetophosphenes could be generated that had prolonged after-images. In a second patient in which both eyes had been removed as the result of severe glaucoma, phosphenes could not be induced by ELF magnetic fields, thereby precluding the possibility that magnetophosphenes can be initiated directly in the visual pathways of the brain.

The Swedish investigators have also extended the comparative studies of magneto- and electrophosphene phenomena that were initiated by Barlow et al.[17] The threshold field intensities required to elicit phosphenes were measured both for electric and magnetic fields over the frequency range 10 to 45 Hz.[23] Because the eddy currents induced by magnetic

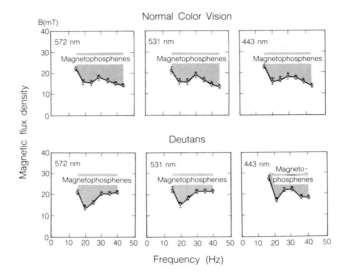

FIGURE 1.   The mean threshold magnetic flux densities required to elicit magnetophosphenes are plotted as a function of frequency for six subjects with normal color vision and for nine deutans with defects in color vision. Threshold values were determined at wavelengths of 443, 531, and 572 nm with a luminance of 3 cd/m². Error bars represent ± 1 S.E.M. From Lövsund, P., Öberg, P. Å., Nilsson, S. E. G., and Reuter, T., *Med. Biol. Eng. Comput.*, 18, 326, 1980. With permission.)

fields vary as the product of field frequency and peak intensity, the threshold electric field current required to generate phosphenes was divided by the frequency in order to permit a direct comparison with the threshold magnetic field intensity. When this "normalization" procedure was used, the curves of threshold stimuli vs. frequency for the generation of phosphenes by ELF electric and magnetic fields were qualitatively similar (Figure 2). With both types of fields, the maximum sensitivity occurred at a frequency of 20 Hz. However, the threshold intensity vs. frequency curve for electrophosphenes did not exhibit the local minimum in sensitivity at 30 Hz observed for magnetophosphenes. This observation suggests that some differences may exist in the retinal current paths involved in the generation of phosphenes by ELF electric and magnetic fields. The differences in sensitivity to magneto- and electrophosphenes that are associated with the state of dark adaptation[17] and the wavelength of light used for background illumination[23] also support this tentative conclusion.

In a series of studies on in vitro frog retinal preparations, Lövsund et al.[25] have made extracellular, electrical recordings from the ganglion cell layer of the retina immediately following termination of exposure to a 20-Hz, 60-mT field in the presence or absence of broad-spectrum background light. It was found that the average latency time for response of the ganglion cells to a photic stimulus was increased from 87 to 92 msec ($p < 0.05$) in the presence of the magnetic field. In addition, the ganglion cells that exhibited electrical activity during photic stimulation ("on" cells) ceased their activity during magnetic field stimulation (i.e., they became "off" cells). The converse behavior of ganglion cells was also observed. These observations indicate that stimulation of the retina by light and by an ELF magnetic field elicits responses in similar postsynaptic neural pathways.

An important electrophysiological finding by Lövsund et al.[25] was the observation that the electrical response of frog retinal ganglion cells to both photic and ELF magnetic field stimuli was blocked when either sodium aspartate or cobalt chloride was added to the Ringer's solution in the eyecup preparation. These compounds inhibit the transfer of information from

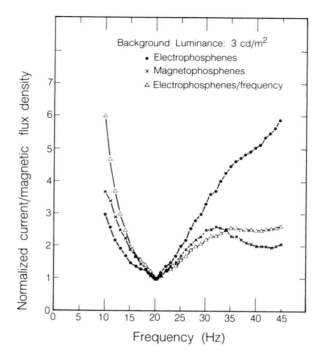

FIGURE 2. The mean threshold electric currents and magnetic flux densities for eliciting electro- and magnetophosphenes are plotted as a function of frequency for ten subjects with normal vision. The data have been normalized relative to the threshold values determined at 20 Hz, the frequency at which maximum sensitivity was observed for the induction of both electro- and magnetophosphenes. The threshold electric current divided by the field frequency has also been plotted (see text for explanation). All studies were conducted with a broad-spectrum white light background at a luminance of 3 cd/m². (From Lövsund, P., Öberg, P. Å., and Nilsson, S. E. G., *Med. Biol. Eng. Comput.*, 18, 758, 1980. With permission.)

the photoreceptors to the neuronal elements of the retina. The electrophysiological observations on chemically blocked retinal preparations appear to implicate the photoreceptors per se as the locus of ELF magnetic field stimulation. The origin of magnetic field responses within the receptors is consistent with the hypothesis of Knighton[26] that a transretinal electric current may act to polarize the photoreceptor synaptic membrane, and thereby alter the postsynaptic transmission of electrical information. One experimental observation made by Lövsund et al.[24] which appears to be inconsistent with this hypothesis is the ability of an applied ELF magnetic field to induce phosphenes in a patient with Retinitis pigmentosa, as described previously. The disparity in these observations, however, may be attributable to a small number of functional photoreceptors within the otherwise degenerated retina of the Retinitis patient. In this context, it is of interest to note that Kato et al.[27] found that electrophosphenes could be generated in patients with pigmentary retinal dystrophy, but a substantially larger stimulus intensity was required over the entire frequency range of 7 to 80 Hz than with subjects that had normal vision. Lövsund et al.[25] have also speculated that sensitivity to ELF magnetic fields may exist within both the photoreceptor and the neuronal elements of the retina, but that the former are stimulated with greater ease. In order to identify more clearly the retinal elements stimulated by ELF magnetic fields, it would be advantageous to conduct further studies on the induction of magnetophosphenes in subjects with well-defined pathological alterations of the retina.

IV. NERVOUS TISSUE INTERACTIONS AND ANIMAL BEHAVIOR

Several studies have been made of the electrical response of neurons to stimulation with time-varying magnetic fields. As discussed by Bernhardt,[10] the current densities induced by the field must exceed 1 to 10 mA/m^2 in order to have an appreciable effect on nerve bioelectric activity, and a threshold extracellular current density of about 20 mA/m^2 has been found experimentally with *Aplysia* pacemaker neurons stimulated by an ELF electric field.[28] In a subsequent study with *Aplysia*,[29] an induced current density of approximately 5 mA/m^2 produced by a 10-mT, 60-Hz sinusoidal field was ineffective in altering the spontaneous neuronal electrical activity. Ueno et al.[30] were also unable to alter the amplitude, conduction velocity, or refractory period of evoked action potentials in lobster giant axons by applying ELF magnetic fields with intensities of 1.2 T at 5 to 20 Hz, 0.8 T at 50 Hz, and 0.5 T at 100 Hz. However, using magnetic flux densities in the range of 0.2 to 0.8 T, Kolin et al.[31] were able to stimulate frog nerve-muscle preparations at field frequencies of 60 and 1000 Hz. Öberg[32] and Ueno et al.[33] were also able to stimulate contractions in frog nerve-muscle preparations by using pulsed magnetic fields with pulse durations < 1 msec. In addition, the excitation of frog sartorius and cardiac muscles[34] and the sciatic nerves of dogs and rabbits[35] has been reported to occur in response to pulsed magnetic fields. From these studies, it appears that sinusoidal ELF magnetic fields with intensities in the range generally used in the laboratory or encountered by humans in occupational settings are insufficient to alter the bioelectric properties of isolated neurons. However, direct magnetic stimulation of nerve and muscle tissues can be achieved by using pulsed fields with a large time rate of change of the magnetic flux density. It should also be borne in mind that the effects of ELF sinusoidal fields on complex, integrated neuronal networks such as those within the CNS may be considerably greater than the effects that occur in single neurons or nerve bundles. This amplification of a field effect could occur through a summation of the small responses evoked in individual neuronal elements.[36] An additive response mechanism may also underlie the production of magnetophosphenes through the stimulation of multiple neuronal elements of the retina by ELF magnetic fields.[37]

Turning next to the subject of potential effects of magnetic fields on behavior, reports from a number of laboratories suggest that the activity patterns and reaction rates of several species of animals are altered by low-intensity fields in the ELF frequency range.[38-49] These reports are summarized in Table 1, which does not include studies in which the behavior of birds and honeybees was found to be altered in the presence of combined ELF electric and magnetic fields.[50-53] Based on the simple theoretical principles discussed earlier in this chapter, the ELF magnetic field intensities used in many of the experiments described in Table 1 would not be expected to induce sufficiently large internal currents to directly influence the nervous systems of the various animal species in which positive effects were found. However, in the case of bees and avian species, the observed sensitivity to ELF time-varying fields may originate through a field interaction mechanism similar to that which is believed to underlie the response of these animals to low-intensity, static magnetic fields. As discussed in Part II, Chapter 4, deposits of magnetite crystals have been identified in bees and avians,[54,55] and the magnetic force interaction with these ferrimagnetic inclusions may produce autonomic responses. From a theoretical perspective, it is unlikely that a time-varying ELF field could orient or produce significant motion of the magnetite inclusions, as discussed earlier in this chapter. The time-varying force produced by the field may, however, trigger somatosensory responses. At present there is no convincing evidence to suggest that a similar interaction mechanism exists in mammalian species.

In addition to the relatively large number of positive findings of ELF magnetic field effects on the behavior of mammals including man, several extensive reports[56-64] have appeared during recent years in which no evidence could be found for a behavioral response of mice,

Table 1

BEHAVIORAL EFFECTS OF EXPOSURE TO TIME-VARYING, ELF MAGNETIC FIELDS

POSITIVE FINDINGS

Subject	Exposure conditions[a]	Results	Ref.
Human	0.1 and 0.2 Hz, 0.5—1.1 mT; acute exposures	Increased reaction time in 0.2-Hz field	38
Honeybee	60 Hz, 2.2—30 mT; 10-min exposures	Altered exploratory behavior	39
Rat	0.5 Hz, 0.3—3.0 mT, rotating field; exposure during entire gestational period	Decreased open-field activity and increased defecation when tested postnatally at 21—25 days	40
		Decreased avoidance of aversive electrical shock when tested postnatally at 30 days	41
		Suppressed rate of response to a conditioned stimulus preceding an aversive shock when tested postnatally at 70 days	42
	Same as Reference 40, but with adult animals exposed for 21—30 days	Increased ambulatory activity after removal from field	43
Duck eggs	0.5 Hz, 2—10 and 10—30 mT, rotating field; exposure for entire prenatal period	Increased ambulation and defecation rate when tested postnatally	44
Human	50 Hz, 10—13 μT; acute exposures	Increased latency of sensorimotor reactions	45
Mouse	60 Hz, 1.4—2.0 mT; 2-min aperiodic exposures over 2 days	Increased locomotor activity and aggression-related vocalization	46
Hamster	10^{-5} Hz, 0.8—26 μT; 26-hr schedule of high (14-hr) to low (12-hr) field switching over period of 4—5 months	Modified circadian rhythm in locomotor activity	47
Chicken	60 Hz, 2.4 mT; aperiodic exposures during 1-hr interval for 10 days	Increased variability of response to electric shock stimulus when 60-Hz magnetic field used as conditional stimulus	48

Species	Exposure	Finding	Ref.
Monkey	9—500 Hz, 0.1 mT (applied to cerebellum); 9-hr daily exposures for maximum of 19 days	Modification of threshold for excitation of motor neurons	49

NEGATIVE FINDINGS

Species	Exposure	Finding	Ref.
Human	45 Hz, 0.1 mT; 22.5-hr exposure	No effect on reaction time	56
Monkey	10, 15, 45, 60, and 75 Hz, 0.8—1.0 mT; fields applied in 4—13 daily sessions of 2—8 hr duration	No consistent influence on motor activity, reaction time, inter-response time, overall lever responding, or match-to-sample performance	57—62
Human	60 Hz, 1.06 mT over whole body, or 2.12 mT over head region; repetitive acute exposures	No perception of field	63
Mouse	60 Hz, 2.33 mT; 3-day continuous exposure	No change in memory retention, locomotor activity, or sensitivity to a neuropharmacologic agent	64

a   The magnetic fields were sinusoidal unless otherwise indicated.

monkeys, and humans to ELF magnetic fields with intensities in the range 1 to 2 mT. These behavioral studies are also summarized in Table 1. With the exception of the study by Beischer et al.[56] on human reaction time, the field intensities used in these experiments produced intracranial current densities that approached, or were slightly above, the value 1 mA/m², which has been predicted on theoretical grounds to be the lower threshold limit for producing a direct effect on the CNS.[10] It is interesting to note that among the various positive findings of ELF magnetic field effects on mammalian behavior, only one study with mice[46] used a field frequency and intensity that would be expected to induce internal body currents comparable to those present in the experiments where no field effects were observed. In examining the possible reasons underlying the disparity among experimental results obtained in different laboratories, it is important to assess the potential role of extraneous factors such as mechanical vibration and audible noise that may accompany the activation of magnet coils. The importance of these factors has been elegantly demonstrated by Tucker and Schmitt,[63] who found that perceptive individuals could sense the presence of a 60-Hz magnetic field through auxiliary clues. When these investigators developed an exposure chamber that provided extreme isolation from vibration and audible noise, none of the more than 200 individuals tested could detect 60-Hz fields with intensities of 1.1 mT over the whole body or 2.1 mT over the head region. The sensitivity of behavioral indices to adventitious factors such as changes in barometric pressure has also been discussed by deLorge,[59] who emphasized that the correlation of such variables to positive findings of apparent ELF field effects must be examined.

Another aspect of ELF magnetic field effects that should be considered in the context of behavioral alterations is the recent report of a correlation between the incidence of suicides and the intensity of residential, 50-Hz magnetic fields from power line sources.[65] Based on coroner and police records from various urban and rural regions within a 5000-km² area in the Midlands of England, a statistically significant increase in suicide rate was found among individuals that lived in residences where the 50-Hz field intensity exceeded 0.15 μT at the front entrance. A subsequent statistical analysis of the same data indicated that the cumulative probability ratio for the incidence of suicide increased above the null effect level of unity for residential, 50-Hz magnetic field intensities exceeding 15 nT.[66] However, oscillations occurred in the cumulative probability ratio as a function of increasing magnetic field intensity, and at 0.2 μT the ratio for the "urban" study group was consistent with the absence of any 50-Hz magnetic field effect. From an epidemiological perspective, the lack of a clear-cut dependence of the suicide incidence on magnetic field intensity suggests that the apparent correlation between these variables may be purely fortuitous. An extension of the studies initiated by Perry et al.[65] using a significantly larger population of individuals will be required before any firm judgment can be made regarding the proposed correlation between suicide incidence and ELF magnetic field exposure.

V. ELF MAGNETIC FIELDS AND CANCER

Eight reports have appeared in the literature during the past 5 years which suggest from epidemiological evidence that a link may exist between residential and occupational exposure to ELF magnetic fields and the incidence of cancer. The first of these reports was by Wertheimer and Leeper,[67] who found that cancer deaths (primarily leukemia) in children less than 19 years of age in the Denver, Colo. area was correlated with the presence of high-current, primary and secondary wiring configurations in the vicinity of their residences. This retrospective epidemiological study was based on 344 fatal childhood cancer cases during the period 1950 to 1973 and an equal number of age-matched controls chosen from birth records. The electrical power lines near the birth and death residences of the cancer cases and the residences of the controls were inspected and classified as being either high-current configurations (HCC) or low-current configurations (LCC), which were assumed to reflect the local intensity of the 60-Hz magnetic field within the homes of the subjects. The percentage of the cancer cases whose birth and death residences were near HCC was found to be significantly greater than for the residences of the control subjects, from which Wertheimer and Leeper concluded that an association may exist between the strength of magnetic fields from the residential power distribution lines and the frequency of childhood cancer. In a subsequent publication, these authors reported that a similar association exists for the incidence of adult cancer.[68] This later study was based on 1179 cancer cases (78% fatal cancers) in Denver, Boulder, and Longmont, Colo. during the period 1967 to 1977.

Following the initial report of Wertheimer and Leeper on childhood cancer, two other epidemiological studies have been made to determine whether a relationship exists between residential magnetic fields from power line sources and the incidence of leukemia in children. In the first of these studies, Fulton et al.[69] used methodology that was matched as closely as possible to that of Wertheimer and Leeper, including the designation of HCC and LCC power lines. This study involved 119 leukemia patients with ages of onset from 0 to 20 years, whose address histories were obtained from medical records at Rhode Island Hospital, and 240 control subjects chosen from Rhode Island birth certificates. In their study, Fulton et al.[69] concluded that no statistically significant correlation existed between the incidence of leukemia and the residential power line configurations. Wertheimer and Leeper[70] were critical of the study by Fulton et al.[69] on the basis that the control and case groups had not been matched for interstate migration, for years of occupancy of residences, or for the ages of the children at the time their residential addresses were determined from birth records and hospital medical records. In a subsequent analysis of the data obtained by Fulton et al.,[69] Wertheimer and Leeper[70] excluded cases and controls aged 8 and above in order to define a complete residential history for the remaining subjects (53 cases and 71 controls). In this subset of the total population studied by Fulton and associates, Wertheimer and Leeper found a weakly significant correlation (p $\sim$ 0.05) between the incidence of leukemia and residential HCC wiring configurations.

Another study of childhood leukemia incidence was conducted in the county of Stockholm by Tomenius et al.,[71] who analyzed the residential, 50-Hz magnetic fields for 716 cases that had a stable address from the time of birth to the time of leukemia diagnosis, and for 716 controls that were matched for age, sex, and birth location. These investigators evaluated the electrical wiring configurations near the residences of the study population, and also made measurements at the entrance door to each residence of the magnetic field intensity in the frequency range above 30 Hz. Among the residences where a magnetic field intensity exceeding 0.3 μT was recorded, the incidence of leukemia was greater by a statistically significant amount than the expected level. From the data presented by Tomenius et al.,[71] it is interesting to note that a statistically significant increase in the incidence of leukemia was not evident for the study population living in residences where the magnetic field intensity exceeded 0.4 μT. In addition, the arithmetic mean of the magnetic field intensity for the residences of all the leukemia cases did not exceed the mean value for the residences of the control group.

During 1982 and 1983, a total of four brief epidemiological reports were published in the format of letters to journal editors, all of which showed an apparent association between the incidence of adult leukemia and occupational exposure to ELF electric and magnetic fields. The following is a brief summary of these four reports: (1) Milham[72] analyzed 438,000 deaths among adult males in Washington State during the period 1950 to 1979, and encoded the data into 158 cause-of-death groups in each of 218 occupational classes. In 10 out of 11 occupations in which the exposure to electric and magnetic fields was assumed to be greater than average, the proportionate mortality ratio for leukemia was found to be elevated. The increase in leukemia mortality was significant at the level $p < 0.01$ for electricians, power-station operators, and aluminum workers. It is interesting to note that the highest magnetic field intensities to which the aluminum workers are exposed originate from DC electric current sources[73] rather than from power-frequency sources. (2) Wright et al.[74] used the Cancer Surveillance Program registry for Los Angeles County during the years 1972 to 1979 to study the proportional incidence ratios for all leukemias, acute leukemias, and acute myelogenous leukemia among white males in ten occupations that involved exposure to ELF electric and magnetic fields. At the significance level $p < 0.05$, a higher than average incidence of acute leukemia was found among power linemen, and of acute myelogenous leukemia among power linemen and telephone linemen. In the pooled data for all ten job classifications, the overall proportional incidence ratios for acute leukemia and acute myelogenous leukemia were significant at the level $p < 0.05$. (3) McDowall[75] analyzed the incidence of leukemia using occupational mortality data for males 15 and above that were collected in England and Wales during the period 1970 to 1973. For ten electrical occupations the overall incidence of leukemia was not greater than expected at a statistically significant level. However, a trend towards increased risk was noted in the data, and the relative risk of contracting acute myelogenous leukemia for the various electrical occupations that were studied was 1.6 to 4.0 in comparison with a randomly selected control population. (4) Coleman et al.[76] analyzed the incidence of leukemia in southeast England among men aged 15 to 74 working in the same ten electrical occupations that were studied by McDowall.[75] In the study by Coleman et al.,[76] the South Thames Cancer Registry, which encompasses a total population of 6.5 million, was used to determine the proportional registration ratio for leukemia during the period 1961 to 1979. For all ten electrical occupations taken together, there was a 17% excess of all types of leukemias above the expected level, which was significant at the $p < 0.05$ level. Two occupations for which the incidence of all leukemias was significantly above the expected level were electrical/electronic fitters and telegraph/radio operators, for which the excess incidence was 84 and 146%, respectively.

Two other studies have recently been published on the incidence of leukemia among electrical and telecommunications workers in both of which the Swedish Cancer-Environment Registry for the period 1961 to 1977 was used as an epidemiological database. In the first study by Wiklund et al.,[77] no increased risk was observed for telecommunications workers compared with the Swedish population as a whole. The second study by Vågerö and Olin[78] reported a slightly higher total incidence of cancer among male and female workers in electrical manufacturing industries as compared to the general population.

Overall, 8 of the 10 recent epidemiological studies that have sought to find an association between cancer incidence and residential or occupational exposure to ELF fields from electric power sources have obtained positive results. As pointed out by Liburdy,[79] a number of other studies on humans and laboratory animals have not found an association between exposure to electromagnetic fields and carcinogenesis. In assessing the literature in which positive correlations have been found, the following methodological deficiencies should be noted.

1. The sample populations in many of the epidemiological studies were small, and an increase in cancer incidence by a factor of 2 or less might be expected on the basis of chance alone. In these studies, it would have been informative if the authors had presented data on several nonexposed occupational groups in which the sample size was comparable to that of the exposed groups.
2. Control groups were frequently chosen in a nonblind manner involving subjective criteria, and the control population was often not matched with the exposed group on the basis of age, sex, socioeconomic class, or urban/rural residential status.
3. In all of the studies thus far reported, the magnetic field dosimetry was at best qualitative. In studies of residential, ELF magnetic fields, the neglect of local fields from

appliances may have led to incorrect conclusions concerning the peak and average exposure of individuals to power-frequency fields and the higher harmonics that emanate from electrical devices used within the home.[80]

4. Concomitant environmental factors of known carcinogenic potential (e.g., aryl hydrocarbons) were ignored in all of the epidemiological studies that have attempted to relate ELF fields and cancer incidence.

In view of the numerous deficiencies in the epidemiological studies conducted to date, it is not possible at this point in time to conclude that a definite association exists between the exposure of individuals to ELF magnetic (or electric) fields and their relative risk of contracting leukemia or other forms of cancer. In addition, the field levels to which humans are generally exposed are sufficiently low that it is difficult to conceive plausible mechanisms that might underlie a causal relationship between cancer incidence and ELF magnetic field exposure. To put this issue into clearer perspective, it is instructive to consider the internal potentials and currents induced in humans as the result of motion through the magnetic field of the earth. A straightforward calculation based on Faraday's law indicates that the motion of a human bending forward at the waist within the geomagnetic field will induce instantaneous internal currents comparable to those produced by exposure to an external 60-Hz sinusoidal field with an intensity of approximately 0.1 to 0.2 μT. This magnetic field intensity is comparable to the ambient power-frequency fields in many residences and occupational settings. Such considerations indicate the clear need for careful dosimetry in any attempt to detect a relationship between power-frequency magnetic fields and cancer. The conduct of prospective epidemiological studies with carefully matched control groups would also be of great value in assessing the validity of conclusions drawn from many of the retrospective studies that have been carried out during the past few years.

VI. SUMMARY OF CELLULAR AND TISSUE INTERACTIONS

During the past 2 decades, a large number of reports have appeared in the literature that describe the effects of time-varying, ELF magnetic fields on a variety of cellular and tissue systems. Many of these reports are summarized in Table 2, in which the literature citations have been ordered chronologically. Publications in which the magnetic field parameters and/or the exposure conditions were not described have been excluded from the table. In addition, reports of research that involved combined exposures to ELF electric and magnetic fields have not been included (e.g., Reference 3, 112, and 113) because of the obvious difficulty in delineating the relative effects of the two types of fields.

Despite the large number of test specimens that have been examined for sensitivity to ELF magnetic fields, it is difficult at present to draw firm conclusions concerning the bioeffects of these fields at the cellular and tissue levels as a result of several factors. (1) A wide range of intensities, frequencies, waveforms, and exposure durations have been used. Many of the earlier studies utilized sinusoidal fields oscillating at 15 to 75 Hz, but research during the last few years has focused increasingly on the bioeffects of square-wave or pulsed fields with complex waveforms. Among the studies conducted with purely sinusoidal fields, the field intensities have ranged from approximately 1 μT to 0.1 T, and the exposure durations have varied from 10 min to 1 to 4 weeks of either continuous or intermittent exposures. (2) Although the vast majority of the published literature describes positive bioeffects of time-varying, ELF magnetic fields, none of the findings listed in Table 2 have been verified by means of independent replication in other laboratories. (3) A number of apparent inconsistencies can be found in the comparison of data acquired on similar (but not identical) test specimens. For example, exposure to a low-intensity, ELF magnetic field was reported to produce an elevation in the serum triglyceride levels of human subjects,[56] but comparable effects were not observed in monkeys.[60]

Table 2
CELLULAR AND TISSUE INTERACTIONS OF TIME-VARYING, ELF MAGNETIC FIELDS

Test specimen	Exposure conditions[a]	Results	Ref.
Mouse	50 Hz, 20 mT; 6.5-hr single exposure or 6.5-hr daily for 15 days	Increased resistance to *Listeria* infection	81
Guinea pig brain mitochondria	60 Hz, 10 mT; 10—110-min exposures	No effect on respiration (oxidative phosphorylation)	82
Rat brain synaptosomes	60 Hz, 5—10 mT; 30-min exposure	Decreased uptake of norepinephrine at 0°C but not at 10, 25, or 37°C	82
Rat	0.5 Hz, 0.05—0.30 or 0.3—1.5 mT, rotating field; exposure during entire gestational period	Increased thyroid and testicle weights at 105—130 days of age; no change in thymus or adrenal weights relative to controls	83
Human	45 Hz, 0.1 mT; 22.5-hr exposure	Elevated serum triglycerides; no other effects on blood cell counts or serum chemistry	56
Monkey	15 and 45 Hz, 0.82—0.93 mT; fields applied in 5—8 daily sessions of 2-hr duration	No alteration in blood cell counts or serum chemistry (including triglycerides)	60
Guinea pig	50 Hz, 20 mT; 6.5-hr single exposure or 6.5-hr daily for 24 days	Pathomorphological changes in testes, kidneys, liver, lungs, nervous tissues, eyes, capillaries, and lymphatic system	84
Rat	50 Hz, 20 mT; 1—7 days exposure	Increase in adrenal 11-hydroxy corticosteroids	85
	50 Hz, 0.12 T; 3-hr exposure	Anti-inflammatory effects of field on carrageenan-induced edema and adjuvant-induced arthritis	86
Mouse	45 Hz, 0.1 mT; 24-hr exposure	No change in liver triglycerides	87
Rat	50 Hz, 20 mT; 24-hr exposure	Increased LDH activity and change in distribution in heart and skeletal muscles	88
Mouse neuroblastoma	60 Hz, 1.2 mT; 13-day exposure	Decreased tumor growth rate	89
Mouse mammary carcinoma	60 Hz, 0.16 T; 1-hr daily exposure for 1—4 days	No effect on tumor growth rate	90
Slime mold	75 Hz, 0.2 mT; 400-day exposure	Lengthened nuclear division cycle and respiration rate (decreased O_2 uptake)	91, 92, 93
Rat	50 Hz, 9.4 and 40 mT; 5 hr daily for 15 days	Altered brain metabolism at higher field intensity, including decreased rate of respiration, decreased levels of glycogen, creatine phosphate and glutamine, and increased DNA content	94
Mouse	60 Hz, 0.11 T; 23 hr daily for 7 days	Decreased body weight and increased water consumption; hematology, organ histology, and reproduction not affected	95
Bacteria	16.66 and 50 Hz, 0—2.0 mT; 10—12-hr exposure	Decreased growth rate	96
	60 and 600 Hz, 2 mT; 17—64-hr exposure	Decreased growth rate and cytolysis	97
Rat	50 Hz, 20 mT; 24-hr exposure	Mobilization of adrenal catecholamines	98
	50 Hz, 20 mT; 1—24-hr exposure	Pathomorphological changes in brain	99
Chicken embryo	10, 100, and 1000 Hz; 0.12, 1.2, and 12 μT; 0.5-msec rectangular pulses; 2-day exposure	Morphological abnormalities in nervous tissue, heart, blood vessels, and somites	100, 101

Table 2 (continued)
CELLULAR AND TISSUE INTERACTIONS OF TIME-VARYING, ELF MAGNETIC FIELDS

Test specimen	Exposure conditions[a]	Results	Ref.
Rat	50 Hz; 20, 40, and 70 mT; 6.5 hr; daily for 5 days, or 24-hr continuous exposure	Pathomorphological changes in brain tissue	102
Human	50 Hz, 5 mT; 4-hr exposure	No changes in ECG, EEG, hormones, blood cell counts, or blood chemistry	103
Mouse osteoblast cultures	Single bidirectional pulses at 72 Hz, or 4-kHz bursts of bidirectional pulses with 15-Hz repetition rate; 2-mT peak intensity; 3-day exposure	Reduced cAMP production in response to parathyroid hormone	104
Human lymphocytes	1, 3, 50, and 200 Hz; 2.3—6.5 mT; square-wave pulses; 3-day exposure	Inhibition of lectin-induced mitogenesis by 3- and 50-Hz fields	105
Dipteran salivary glands	15- and 72-Hz pulses as in Ref. 104 above; 5—90-min exposure	Increased RNA transcription	106
Rabbit pancreas	4-kHz bursts of bidirectional pulses with 15-Hz repetition rate; 2-mT peak intensity; 18-hr exposure	Reduced Ca^{++} content and efflux; reduced insulin release during glucose stimulation	107
Drosophila eggs	0.5-msec square-wave pulses at 100 Hz, or 50-Hz, 1.41-mT sinusoidal field; 2-day exposure	Decreased viability of eggs	108
Chicken embryos	0.5-msec bidirectional pulses at 100 Hz (4 different waveforms); 0.4—104-µT peak intensity; 2-day exposure	Teratogenic changes in nervous system, circulatory system, and foregut	109
Cultured chicken tibiae	1 Hz, 15—60-mT square-wave pulses; 7-day exposure	Decreased collagenous and noncollagenous protein synthesis; no alteration in glyco-soaminoglycan and DNA synthesis	110
Cultured human fibroblasts	15 Hz—4 kHz; 2.3—560 µT; 18—96-hr exposure	Increased DNA synthesis	111

[a] The magnetic fields were sinusoidal unless otherwise indicated.

Irrespective of the inadequacies in the existing database, a survey of the extant literature indicates that several aspects of the biochemistry and physiology of cells and organized tissues may be perturbed by exposure to time-varying, ELF magnetic fields. Briefly summarized, the reported bioeffects for which there is a growing body of evidence include

1. Decreased rate of cellular respiration[91,93,94]
2. Altered metabolism of carbohydrates, proteins, and nucleic acids[88,94,106,110,111]
3. Endocrine changes and altered hormonal responses of cells and tissues[82,83,85,98,104,107]
4. Decreased cellular growth rate[89,91-93,96,97]
5. Teratology and developmental effects[100,101,108,109]
6. Morphological tissue changes in adult animals, frequently reversible with time after exposure[84,99,102]
7. Altered immune response to various antigens and lectins[81,86,105]

It is also interesting to note that no consistent field-associated alterations have been observed in the hematological system,[56,60,95,103] which is known to be sensitive to ionizing radiation and high-intensity microwave fields.

With regard to the interaction mechanisms that may underlie the reported cellular and tissue effects of time-varying, ELF magnetic fields, an interpretive difficulty is once again posed by the large variety of waveforms, intensities, and frequencies that have been used by different investigators. In studies on the effects of pulsed magnetic fields with time rates of change from 5 to 100 T/sec,[104,106,107] the peak induced electric field and current density in the medium were > 100 mV/m and 10 mA/m^2, respectively. The induced currents exceed the endogenous electrical currents that are normally present in extracellular fluids in vivo. It has been suggested that the currents induced by pulsed magnetic fields may exert an electrochemical effect at the cell surface which in turn influences the membrane transport and intracellular concentration of calcium ions.[104,107] Because of the important role played by calcium ions in metabolism and growth regulation, this proposal deserves careful consideration in the context of ELF magnetic field effects at the cellular and tissue levels.

In several studies that used repetitive square waveforms with frequencies in the ELF range,[100,101,105,108-110] a high time rate of change of the magnetic flux density may have been present during the rising portion of the square wave. Unfortunately, the time constants associated with the square wave pulses were usually not stated by the authors of these reports, and it is therefore difficult to estimate the peak fields and currents that were induced in their experimental specimens. One exception is the report by Ubeda et al.[109] on embryological changes in chick eggs exposed to 0.5-msec bidirectional pulses repeated with a 100-Hz frequency. Four different waveforms with pulse rise times of 2, 42, and 100 μsec and pulse amplitudes ranging from 0.4 to 104 μT were used in this study. Although not all combinations of pulse rise times and amplitudes were used, the teratogenic effects observed in chick embryos appeared to be relatively insensitive to the peak value of dB/dt that occurred during the rising portion of the pulse. Unfortunately, the authors used only the lowest pulse amplitude in combination with the shortest pulse rise time, and the maximum value of dB/dt that was achieved with any combination of amplitude and rise time was 1 T/sec. This value is small in comparison with the maximum value of 52 T/sec that could have been achieved by using a 2-μsec rise time in combination with a 104-μT pulse amplitude.

Several investigations with ELF sinusoidal fields have involved the exposure of rodents to 50- and 60-Hz fields with intensities in the range 0.02 to 0.16 T.[81,84-86,88,90,94,95,98,99,102] The maximum current densities induced in the experimental subjects ranged from approximately 1 to 20 mA/m^2 in these studies, which may be sufficient to perturb certain biological functions as discussed above. However, many of the other studies on ELF sinusoidal fields that are summarized in Table 2 were conducted with low field intensities that induced current densities substantially < 1 mA/m^2 in the test specimens. In these studies, it is possible that the observed biological effects may have resulted from interaction mechanisms other than the induction of electrical currents in accord with Faraday's law. This possibility was recently suggested by Liboff et al.[111] on the basis of measurements of DNA synthesis in human fibroblasts exposed to sinusoidal magnetic fields with low intensities in the range 2.3 to 560 μT. An enhanced DNA synthesis rate was observed for field frequencies ranging from 15 Hz to 4kHz, and this effect was independent of the current density induced in the cell cultures. The lack of dependence on the time rate of change of the field (and hence the induced current density) was tested for values of dB/dt ($= 2\pi f\,|\vec{B}_0|$) that ranged from approximately 1.8×10^{-4} T/sec to 1.8 T/sec. As a possible explanation of the observations made by Liboff et al.,[111] Polk[114] has recently suggested an interaction mechanism in which the Lorentz force exerted on moving counterions produces motion and a resulting electric current that is independent of the field frequency. This interaction, which relies upon the preexistence of coherent motion within a surface-bound counterion atmosphere, could produce local currents that are several orders of magnitude larger than the Faraday currents induced on the cell surface by the time-varying magnetic fields used in the experiments of Liboff et al.[111]

VII. MEDICAL AND DIAGNOSTIC APPLICATIONS OF ELF MAGNETIC FIELDS

Magnetic fields have been used medically in several types of diagnostic and therapeutic procedures, and some relatively new techniques such as magnetoencephalography have considerable potential for future medical applications. In this section, a brief description is given of electromagnetic blood flow measurements, the use of pulsed magnetic fields for the facilitation of bone fracture reunion, and the measurement of endogenous magnetic fields as indicators of various biological functions. A large number of publications has appeared on each of these topics, and an effort will not be made here to provide an extensive review. Each of the techniques will be briefly described, citing only a limited number of relevant publications.

A. Electromagnetic Blood Flow Measurements

As a result of magnetic induction, a conductive fluid such as blood will develop a transverse electrical potential when flowing in the presence of a magnetic field. It can be shown from simple physical principles that the magnitude of the induced potential V_i for axisymmetric blood flow in a cylindrical vessel is given by[115,116]

$$V_i = |\vec{u}||\vec{B}| \, d \sin \theta \tag{5}$$

where $\vec{u}$ = mean axial flow velocity, d = vessel diameter, $\vec{B}$ = magnetic flux density (static or time-varying), and θ = angle between $\vec{u}$ and $\vec{B}$. The magnetically induced potential described by Equation 5 was first used by Kolin[115,117] as the basis of a method for measuring blood flow rate without opening a vessel or introducing foreign substances such as dyes into the circulation. The original electromagnetic flowmeter developed by Kolin employed a static magnetic field, which was later replaced by a 60-Hz field in order to avoid artifacts associated with polarization of the electrodes used to measure the induced potential. Present models of the blood flowmeter are constructed with small cuffs that fit closely around the blood vessel perimeter, and contain both the magnet coil and the pick-up electrodes. The use of flowmeters with either perivascular or intravascular transducers has become over the years a widely accepted clinical method for monitoring blood flow during surgical procedures.[118] This technique can be used for measuring the rate of blood flow in major vessels such as the portal vein[119] and the coronary artery[120] with a precision of 1 to 2% over a period of several hours. Other useful applications of the flowmeter principle include the measurement of microscopic variations in vascular diameter[121] and the determination of blood flow through large segments of an organ such as the pancreas.[122]

Several studies with experimental animals have also demonstrated the feasibility of exposing the whole body, or a large segment of the body such as the chest, to an external uniform magnetic field in order to measure blood flow rates in major vessels of the circulatory system.[123] This procedure is noninvasive insofar as the conventional surface leads used to measure the electrocardiogram can also detect the major blood flow potentials that are induced by an external magnetic field. The applied field can be either static or time-varying, although the former is technically simpler to use in this application. The measurement with surface electrodes of magnetically induced potentials associated with pulsatile blood flow into the aorta has been demonstrated for several species of animals, including monkeys,[124-126] baboons,[127] rabbits,[128] and rats.[129] A possible future medical application of this technique is the noninvasive measurement of blood flow within the atrial and ventricular chambers of the heart following the opening and closing of the mitral and tricuspid valves.

B. Facilitation of Bone Fracture Reunion by ELF Pulsed Magnetic Fields

The use of DC electrical stimulation to treat bone nonunions and pseudoarthroses dates back to the early part of the 19th century. As described by Bassett,[130] several investigators during the 19th century and the first half ot the 20th century reported success in the treatment of bone fractures by electrotherapy. Heightened interest in this area arose during the 1950s and 1960s as the result of several demonstrations of the piezoelectric properties of bone.[131,132]

Although some measure of success was achieved in treating bone nonunions by electrical stimulation, the use of DCs led to several undesirable side effects, including

1. Surgical trauma and a risk of infection through the implantation of electrodes in bone
2. The development of electrode polarization with time, which led to increased impedance and decreased current for a given applied voltage
3. Osteogenesis was found to be increased near the negative (cathodic) pole, but decreased near the positive (anodic) pole

These disadvantages of DC electrical stimulation were overcome by the recent introduction of pulsed magnetic field generators as a means of inducing ELF electrical currents into bone tissue.[133] By using magnetic coils placed about a limb containing a fractured bone, electric fields with a typical strength of 0.2 to 2.0 V/m can be induced within the bone tissue. In the usual configuration, two coils are placed about the limb and positioned such that the bone fracture lies along a line joining the centers of the coils, and hence along the magnetic field lines. Assuming the conductivity of bone to be 0.01 S/m at ELF frequencies,[134] the local current densities induced in bone by the pulsed magnetic field can be estimated to lie in the range of approximately 2 to 20 mA/m^2. Initial studies on bone fracture reunion in dogs demonstrated that a pulse repetition frequency of 65 Hz was more effective than 1 Hz,[133] and several subsequent studies have found that frequencies of 60 to 75 Hz are the most advantageous in facilitating fracture union and preventing pseudoarthroses.[130]

Following the initial demonstration of the efficacy of pulsed magnetic fields in achieving bone fracture reunion in experimental animals, several clinical trials have reported success in treating bone fractures and arthroses in humans by this method. In a 4-year clinical trial involving more than 100 patients, Bassett et al.[135] reported an 85% success rate in the treatment of long-established pseudoarthroses. The successful use of pulsed magnetic fields in the facilitation of bone healing in human subjects has subsequently been reported by several clinical groups.[136-139] The importance of continuing these clinical trials, including the use of dummy stimulators in control subjects, has been emphasized by Barker and Lunt.[140] In this context, the recent development of a lightweight, portable magnetic field stimulator by Watson and Downes[141] may prove very beneficial in facilitating the long-term treatment of patients with bone fractures.

The mechanism by which the weak ELF electric currents induced in bone tissue by pulsed magnetic fields exert an influence on fracture repair is under investigation in a number of laboratories. Evidence from in vitro studies on osteoblasts and chondrocytes indicate that the pulsed fields depress the intracellular concentrations of calcium ions and cyclic AMP.[104,130] These effects, in turn, can significantly influence cellular metabolism and stimulate growth. Studies by Hinsenkamp and Rooze[142] with in vitro cultures of limbs from mouse fetuses have demonstrated that electromagnetic stimulation leads to chondrocyte proliferation and an improved alignment of trabeculae and cartilage. Archer and Ratcliffe[110] recently demonstrated that cultured tibiae from chicken embryos exhibit a reduced collagen content following exposure to a pulsed magnetic field for 7 days. The interesting observation was also made in this study that the total synthesis of sulfated glycosoaminoglycans, which are major components of the extracellular matrix, was not affected by exposure to the pulsed magnetic field. The further elucidation of the macromolecular and developmental changes that accompany the stimulation of bone tissue by pulsed ELF magnetic fields remains a challenging area of research, which will ultimately lend useful insight into the mechanisms by which weak ELF fields interact with living cells.

Another aspect of pulsed magnetic field interactions that should be noted is their apparent stimulatory effect on certain nonosseous tissues. It has been reported, for example, that electromagnetic stimulation facilitates ligament healing in rabbits[143] and promotes the regeneration of peripheral nerves and the spinal cord following injury.[144-146] One final aspect of the clinical use of pulsed, ELF magnetic fields that deserves mention is the concern that has been expressed regarding the potential tumorigenic risk arising from the growth stimulatory effects of these fields.[134,147] This issue has not been fully resolved, but it has been reported that increased tumor incidence did not result from the continuous exposure of rats during 94 weeks of adult life to a 5-Hz pulsed field with a 5-nsec rise time produced by a 447-kV/m electric field generator.[148] This field exposure was also found not to produce untoward effects on the hematological system, fertility, or reproductive capacity.

C. Magnetic Fields of Biological Origin and Their Potential Use in Medical Diagnosis

Circulating electric currents from excitable tissues produce small magnetic fields in the ELF frequency range that can be measured externally. Two studies in the early 1960s using multiturn loops of copper wire as pick-up coils first demonstrated that the magnetic field produced by a nerve impulse could be detected.[149,150] In 1963 Baule and McFee[151] extended this work to show that the peak magnetic fields associated with cardiac bioelectric activity could be detected by coils at the body surface. Cohen[152] subsequently made similar measurements in a magnetically shielded room that substantially reduced the background fields and allowed a clear recording of the magnetocardiogram (MCG). By using signal-averaging techniques, Cohen[153] was also able to detect the alpha rhythm in the magnetoencephalogram (MEG) with a coil detector in the magnetically shielded room. However, the peak magnetic field of the brain measured at the body surface[154] ($\sim$ 3 pT) is substantially lower than the peak field of the heart[154] ($\sim$ 100 pT), and a direct measurement of the MEG was not possible with a coil detector. In the early 1970s this limitation was overcome by the development of the superconducting quantum interference device (SQUID), which is capable of measuring extraordinarily small magnetic fields.[155] By using this device in a magnetically shielded room, Cohen[156] was able to make the first direct measurement of the MEG without signal-averaging. During the past decade, the SQUID technique has been further developed and used in combination with first- and second-order gradiometer configurations of the detection coils. The use of a gradiometer coil configuration provides a substantial reduction in the level of interference from ambient magnetic noise. These external fields have a nearly uniform intensity over dimensions comparable to the human body, while the fields originating within the body are highly nonuniform over the same spatial dimensions. The SQUID gradiometer technique now provides a field detection sensitivity of approximately 5 to 10 fT/Hz$^{1/2}$ for ELF fields.[157]

The development of SQUID instrumentation for the measurement of biomagnetic fields has led to a rapid expansion of research in this area, as reviewed by Williamson and Kaufman.[158] A comprehensive monograph that discusses all of the major aspects of biomagnetism was published in 1983.[159] Four organs that have been a focus of biomagnetic research efforts are the heart, brain, lung, and liver. Various abnormal cardiac conditions such as arteriovenous block, myocardial infarction, and coronary artery disease have been analyzed from the MCG,[160-162] although it is not clear as yet that this technique provides unique information relative to the electrocardiogram. Studies of the MEG have received increased interest during the past decade, primarily because the MEG appears to give a more precise localization of current sources within the brain than is provided by the electroencephalogram (EEG).[163,164] The MEG and EEG also provide complementary information because of the orthogonal configuration of the electric and magnetic field vectors arising from the same current source.[165-167] Both the spatial and temporal features of visually and somatically evoked responses in the brain have been shown in several studies to be well characterized by the MEG.[168-172]

An application of biomagnetic measurements that appears to be particularly promising from a clinical viewpoint is the detection of ferromagnetic materials within the lung and other organs. Cohen[173] has demonstrated that the accumulation of small quantities of asbestos within the lung can be detected magnetically as a result of the adsorbed magnetite carried by asbestos particles. A SQUID magnetometer has also been used to detect an abnormally large quantity of iron within the liver in patients with Thalassemia major.[174] Although the total number of clinical applications up to the present time has been relatively small, the initial results have indicated the potential of biomagnetic measurements as a noninvasive tool for studying both normal and pathological conditions of various organs.

VIII. GENERAL SUMMARY AND CONCLUSIONS

It is evident from the literature survey presented in this chapter that a wide variety of effects of time-varying, ELF magnetic fields have been observed in cellular, tissue, and animal systems. Several of these effects have been well established through careful research in different laboratories. Notable examples are the experimental studies on magnetophosphenes, the induction of blood flow potentials, and the facilitation of bone fracture reunion. However, in many instances the reports of ELF magnetic field bioeffects must be viewed with caution, either because of a lack of independent verification of the experimental findings, or because the reported field effects may have resulted from the presence of confounding variables. Examples of ELF magnetic field effects where such considerations are clearly relevant are the findings of behavioral alterations in laboratory animals and the association between human cancer incidence and exposure to power-frequency fields. These reported effects, as well as many findings of ELF magnetic field effects in cellular and tissue systems, also present a difficult interpretive problem because the field intensities involved are so low that no plausible mechanism can be proposed at the present time to explain the observed effects. From this perspective there is a clear need for substantially more research of both a theoretical and an experimental nature to define in an unambiguous manner the interactions that occur between ELF magnetic fields and living organisms.

ACKNOWLEDGMENTS

The helpful comments of the editors, C. Polk and E. Postow, and the excellent secretarial assistance of K. Springsteen are gratefully acknowledged. Magnetic field research in the author's laboratory is supported by the Office of Energy Research, Health and Environmental Research Division of the U.S. Department of Energy under Contract No. DE-AC03-76SF00098 with the Lawrence Berkeley Laboratory.

REFERENCES

1. **Grandolfo, M. and Vecchia, P.,** Natural and man-made environmental exposures to static and ELF electromagnetic fields, in *Biological Effects and Dosimetry of Nonionizing Radiation: Static and ELF Electromagnetic Fields,* Grandolfo, M., Michaelson, S. M., and Rindi, A., Eds., Plenum Press, New York, 1985, 49.
2. **Scott-Walton, B., Clark, K. M., Holt, B. R., Jones, D. C., Kaplan, S. D., Krebs, J. S., Polson, P., Shepherd, R. A., and Young, J. R.,** Potential Environmental Effects of 765-kV Transmission Lines: Views Before the New York State Public Service Commission, Cases 26529 and 26559, 1976—1978, NTIS Rep. No. DOE/EV-0056, Springfield, Va., 1979, II-7.
3. **Grissett, J. D.,** Biological effects of electric and magnetic fields associated with ELF communications systems, *Proc. IEEE,* 68, 98, 1980.
4. **Stuchly, M. A., Lecuyer, D. W., and Mann, R. D.,** Extremely low frequency electromagnetic emissions from video display terminals and other devices, *Health Phys.,* 45, 713, 1983.
5. **Lövsund, P., Öberg, P. Å., and Nilsson, S. E. G.,** ELF magnetic fields in electrosteel and welding industries, *Radio Sci.,* 17(5S), 35S, 1982.
6. **Sheppard, A. R. and Eisenbud, M.,** *Biologic Effects of Electric and Magnetic Fields of Extremely Low Frequency,* New York University Press, 1977.
7. **Tenforde, T. S., Ed.,** *Magnetic Field Effects on Biological Systems,* Plenum Press, New York, 1979.
8. **Adey, W. R.,** Some fundamental aspects of biological effects of extremely low frequency (ELF), in *Biological Effects and Dosimetry of Nonionizing Radiation: Static and ELF Electromagnetic Fields,* Grandolfo, M., Michaelson, S. M., and Rindi, A., Eds., Plenum Press, New York, 1983, 561.
9. **Budinger, T. F.,** Nuclear magnetic resonance (NMR) *in vivo* studies: known thresholds for health effects, *J. Comp. Assist. Tomogr.,* 5, 800, 1981.
10. **Bernhardt, J.,** The direct influence of electromagnetic fields on nerve and muscle cells of man within the frequency range of 1 Hz to 30 MHz, *Radiat. Environ. Biophys.,* 16, 309, 1979.
11. **Schwan, H. P. and Kay, C. F.,** Specific resistance of body tissues, *Circ. Res.,* 4, 644, 1956.
12. **Hong, F. T., Mauzerall, D., and Mauro, A.,** Magnetic anisotropy and the orientation of retinal rods in a homogeneous magnetic field, *Proc. Natl. Acad. Sci. U.S.A.,* 68, 1283, 1971.
13. **d'Arsonval, M. A.,** Dispositifs pour la mesure des courants alternatifs à toutes frequences, *C. R. Soc. Biol. (Paris),* 3, (100), 451, 1896.
14. **Thompson, S. P.,** A physiological effect of an alternating magnetic field, *Proc. R. Soc. London, Ser. B,* 82, 396, 1909—1910.
15. **Dunlap, K.,** Visual sensations from the alternating magnetic field, *Science,* 33, 68, 1911.
16. **Magnusson, C. E. and Stevens, H. C.,** Visual sensations caused by changes in the strength of a magnetic field, *Am. J. Physiol.,* 29, 124, 1911—1912.
17. **Barlow, H. B., Kohn, H. I., and Walsh, E. G.,** Visual sensations aroused by magnetic fields, *Am. J. Physiol.,* 148, 376, 1947.
18. **Barlow, H. B., Kohn, H. I., and Walsh, E. G.,** The effect of dark adaptation and of light upon the electric threshold of the human eye, *Am. J. Physiol.,* 148, 376, 1947.
19. **Seidel, D., Knoll, M., and Eichmeier, J.,** Anregung von subjektiven Lichterscheinungen (Phosphenen) beim Menschen durch magnetische Sinusfelder, *Pflügers Arch.,* 299, 11, 1968.
20. **Oster, G.,** Phosphenes, *Sci. Am.,* Feb., 83, 1970.
21. **Lövsund, P., Öberg, P. Å., and Nilsson, S. E. G.,** Influence on vision of extremely low frequency electromagnetic fields, *Acta Ophthalmol.,* 57, 812, 1979.
22. **Lövsund, P., Öberg, P. Å., and Nilsson, S. E. G.,** A method for the study of retinal ganglion cell activity induced by ELF magnetic fields, *Radio Sci.,* 14(6S), 125, 1979.
23. **Lövsund, P., Öberg, P. Å., and Nilsson, S. E. G.,** Magneto- and electrophosphenes: a comparative study, *Med. Biol. Eng. Comput.,* 18, 758, 1980.
24. **Lövsund, P., Öberg, P. Å., Nilsson, S. E. G., and Reuter, T.,** Magnetophosphenes: a quantitative analysis of thresholds, *Med. Biol. Eng. Comput.,* 18, 326, 1980.
25. **Lövsund, P., Nilsson, S. E. G., and Öberg, P. Å.,** Influence on frog retina of alternating magnetic fields with special reference to ganglion cell activity, *Med. Biol. Eng. Comput.,* 19, 679, 1981.
26. **Knighton, R. W.,** An electrically evoked slow potential of the frog's retina. I. Properties of response, *J. Neurophysiol.,* 38, 185, 1975.
27. **Kato, S., Saito, M., and Tanino, T.,** Response of the visual system evoked by an alternating current, *Med. Biol. Eng. Comput.,* 21, 47, 1983.
28. **Wachtel, H.,** Firing pattern changes and transmembrane current produced by extremely low frequency fields in pacemaker neurons, in *Biological Effects of Extremely Low Frequency Electromagnetic Fields,* Phillips, R. D., Gillis, M. F., Kaune, W. T., and Mahlum, D. D., Eds., NTIS Rep. No. CONF-781016, Springfield, Va., 1979, 132.

29. **Sheppard, A. R.,** Results of exposure of *Aplysia* pacemaker neurons to ELF/60 Hz and DC magnetic fields, presented at 5th Annu. meet. Bioelectromagnetics Soc. Boulder, Colo., June 12 to 16, 1983, 25.

30. **Ueno, S., Lövsund, P., and Öberg, P. Å.,** On the effect of alternating magnetic fields on action potential in lobster giant axon, Proc. 5th Nordic meet. Med. Biol. Eng., Linköping, Sweden, 1981, 262.

31. **Kolin, A., Brill, N. Q., and Bromberg, P. J.,** Stimulation of irritable tissues by means of an alternating magnetic field, *Proc. Soc. Exp. Biol. Med.,* 102, 251, 1959.

32. **Öberg, P. Å.,** Magnetic stimulation of nerve tissue, *Med. Biol. Eng.,* 11, 55, 1973.

33. **Ueno, S., Matsumoto, S., Harada, K., and Oomura, Y.,** Capacitive stimulatory effect in magnetic stimulation of nerve tissue, *IEEE Trans. Mag.,* 14, 958, 1978.

34. **Irwin, D. D., Rush, S., Evering, R., Lepeschkin, E., Montgomery, D. B., and Weggel, R. J.,** Stimulation of cardiac muscle by a time-varying magnetic field, *IEEE Trans. Mag.,* 6, 321, 1970.

35. **Maass, J. A. and Asa, M. M.,** Contactless nerve stimulation and signal detection by inductive transducer, *IEEE Trans. Mag.,* 6, 322, 1970.

36. **Valentinuzzi, M.,** Notes of magnetic actions upon the nervous system, *Bull. Math. Biophys.,* 27, 203, 1965.

37. **Valentinuzzi, M.,** Theory of magnetophosphenes, *Am. J. Med. Electr.,* 1, 112, 1962.

38. **Friedman, H., Becker, R. O., and Bachman, C. H.,** Effect of magnetic fields on reaction time performance, *Nature,* 213, 949, 1967.

39. **Caldwell, W. E. and Russo, F.,** An exploratory study of the effects of an A.C. magnetic field upon the behavior of the Italian honeybee *(Apis mellifica), J. Genet. Psychol.,* 113, 233, 1968.

40. **Persinger, M. A.,** Open-field behavior in rats exposed prenatally to a low intensity-low frequency, rotating magnetic field, *Dev. Psychobiol.,* 2, 168, 1969.

41. **Persinger, M. A. and Foster, W. S.,** IV, ELF rotating magnetic fields: prenatal exposure and adult behavior, *Arch. Meteorol. Geophys. Bioklimatol. Ser. B,* 18, 363, 1970.

42. **Persinger, M. A. and Pear, J. J.,** Prenatal exposure to an ELF-rotating magnetic field and subsequent increase in conditioned suppression, *Dev. Psychobiol.,* 5, 269, 1972.

43. **Persinger, M. A., Ossenkopp, K.-P., and Glavin, G. B.,** Behavioral changes in adult rats exposed to ELF magnetic fields, *Int. J. Biometeorol.,* 16, 155, 1972.

44. **Ossenkopp, K.-P. and Shapiro, J.,** Effects of prenatal exposure to a 0.5-Hz low-intensity rotating magnetic field on White Peking ducklings, *Am. Zool.,* 12, 650, 1972.

45. **Medvedev, M. A., Urazaev, A. M., and Kulakov, I. U. A.,** Effect of a constant and low frequency magnetic field on the behavioral and autonomic responses of the human operator, *Zh. Vyssh. Nervn. Deyat. im I. P. Pavlova,* 26, 1131, 1976.

46. **Smith, R. F. and Justesen, D. R.,** Effects of a 60-Hz magnetic field on activity levels of mice, *Radio Sci.,* 12(6S), 279, 1977.

47. **Brown, F. A., Jr. and Scow, K. M.,** Magnetic induction of a circadian cycle in hamsters, *J. Interdiscipl. Cycle Res.,* 9, 137, 1978.

48. **Clarke, R. L. and Justesen, D. R.,** Behavioral sensitivity of a domestic bird to 60-Hz ac and to dc magnetic fields, *Radio Sci.,* 14(6S), 209, 1979.

49. **Delgado, J. M. R., Monteagudo, J. L., and Ramiriz, E.,** Non-invasive magnetic stimulation of the monkey cerebellum, presented at 5th Annu. meet. Bioelectromagnetics Soc. Boulder, Colo., June 12 to 16, 1983, 103.

50. **Southern, W. E.,** Orientation of gull chicks exposed to Project Sanguine's electromagnetic field, *Science,* 189, 143, 1975.

51. **Larkin, R. P. and Sutherland, P. J.,** Migrating birds respond to Project Seafarer's electromagnetic field, *Science,* 195, 777, 1977.

52. **Greenberg, B.,** Effects of 60-Hz electromagnetic fields on bees and soil arthropods, in *Magnetic Field Effects on Biological Systems,* Tenforde, T. S., Ed., Plenum Press, New York, 1979, 21.

53. **Greenberg, B., Bindokas, V. P., Frazier, M. J., and Gauger, J. R.,** Response of honey bees, *Apis mellifera L.,* to high-voltage transmission lines, *Environ. Entomol.,* 10, 600, 1981.

54. **Gould, J. L., Kirschvink, J. L., and Deffeyes, K. S.,** Bees have magnetic remanence, *Science,* 201, 1026, 1978.

55. **Walcott, C., Gould, J. L., and Kirschvink, J. L.,** Pigeons have magnets, *Science,* 205, 1027, 1979.

56. **Beischer, D. E., Grissett, J. D., and Mitchell, R. R.,** Exposure of man to magnetic fields alternating at at extremely low frequency, Rep. No. NAMRL-1180, Nav. Aerosp. Med. Res. Lab., Pensacola, Fla., 1973.

57. **deLorge, J.,** Operant behavior of rhesus monkeys in the presence of low-frequency low-intensity magnetic and electric fields: experiment 1, Rep. No. NAMRL-1155 (NTIS No. AD754058), Nav. Aerosp. Med. Res. Lab., Pensacola, Fla., 1972.

58. **deLorge, J.,** Operant behavior of rhesus monkeys in the presence of low-frequency low-intensity magnetic and electric fields: experiment 2, Rep. No. NAMRL 1179 (NTIS No. AD764532), Nav. Aerosp. Med. Res. Lab., Pensacola, Fla., 1973.

59. **deLorge, J.,** Operant behavior of rhesus monkeys in the presence of low-frequency low-intensity magnetic and electric fields: experiment 3, Rep. No. NAMRL 1196 (NTIS No. AD774106), Nav. Aerosp. Med. Res. Lab., Pensacola, Fla., 1973.

60. **deLorge, J.,** A psychobiological study of rhesus monkeys exposed to extremely low frequency-low intensity magnetic fields, Rep. No. NAMRL 1203 (NTIS No. AD/A000078), Nav. Aerosp. Med. Res. Lab., Pensacola, Fla., 1974.

61. **deLorge, J.,** Effects of magnetic fields on behavior in nonhuman primates, in *Magnetic Field Effects on Biological Systems,* Tenforde, T. S., Ed., Plenum Press, New York, 1979, 37.

62. **deLorge, J.,** Behavioral studies of monkeys in electric and magnetic fields at ELF frequencies, in *Biological Effects and Dosimetry of Nonionizing Radiation: Static and ELF Electromagnetic Fields,* Grandolfo, M., Michaelson, S. M., and Rindi, A., Eds., Plenum Press, New York, 1985, 129.

63. **Tucker, R. D. and Schmitt, O. H.,** Tests for human perception of 60 Hz moderate strength magnetic fields, *IEEE Trans. Biomed. Eng.,* 25, 509, 1978.

64. **Davis, H. P., Mizumori, S. J. Y., Allen, H., Rosenzweig, M. R., Bennett, E. L., and Tenforde, T. S.,** Behavioral studies with mice exposed to DC and 60-Hz magnetic fields, *Bioelectromagnetics,* 5, 147, 1984.

65. **Perry, F. S., Reichmanis, M., Marino, A. A., and Becker, R. O.,** Environmental power-frequency magnetic fields and suicide, *Health Phys.,* 41, 267, 1981.

66. **Smith, C. W.,** Comments on the paper "Environmental power-frequency magnetic fields and suicide", *Health Phys.,* 43, 439, 1982.

67. **Wertheimer, N. and Leeper, E.,** Electrical wiring configurations and childhood cancer, *Am. J. Epidemiol.,* 109, 273, 1979.

68. **Wertheimer, N. and Leeper, E.,** Adult cancer related to electrical wires near the home, *Int. J. Epidemiol.,* 11, 345, 1982.

69. **Fulton, J. P., Cobb, S., Preble, L., Leone, L., and Forman, E.,** Electrical wiring configurations and childhood leukemia in Rhode Island, *Am. J. Epidemiol.,* 111, 292, 1980.

70. **Wertheimer, N. and Leeper, E.,** Re: "Electrical wiring configurations and childhood leukemia in Rhode Island", *Am. J. Epidemiol.,* 111, 461, 1980.

71. **Tomenius, L., Hellström, L., and Enander, B.,** Electrical constructions and 50 Hz magnetic field at the dwellings of tumour cases (0—18 years of age) in the county of Stockholm, presented at the Inst. Symp. Occup. Health Saf. Min. Tunnelling, Prague, June 21 to 25, 1982.

72. **Milham, S., Jr.,** Mortality from leukemia in workers exposed to electrical and magnetic fields, *N. Engl. J. Med.,* 307, 249, 1982.

73. **Tenforde, T. S., Geyer, A. B., Fujita, T. Y., Bristol, K. S., Goulding, F. S., and Budinger, T. F.,** Magnetic field monitoring with a portable dosimeter, presented at 5th Annu. Meet. Bioelectromagnetics Soc., Boulder, Colo., June 12 to 16, 1983, 36.

74. **Wright, W. E., Peters, J. M., and Mack, T. M.,** Leukemia in workers exposed to electrical and magnetic fields, *Lancet,* 2(8308), 1160, 1982.

75. **McDowall, M. E.,** Leukemia mortality in electrical workers in England and Wales, *Lancet,* 1(8318), 246, 1983.

76. **Coleman, M., Bell, J., and Skeet, R.,** Leukemia incidence in electrical workers, *Lancet,* 1(8332), 982, 1983.

77. **Wiklund, K., Einhorn, J., and Eklund, G.,** An application of the Swedish cancer-environment registry. Leukaemia among telephone operators at the telecommunications administration in Sweden, *Int. J. Epidemiol.,* 10, 373, 1981.

78. **Vågerö, D. and Olin, R.,** Incidence of cancer in the electronics industry: using the new Swedish Cancer Environment Registry as a screening instrument. *Br. J. Ind. Med.,* 40, 188, 1983.

79. **Liburdy, R. P.,** Carcinogenesis and exposure to electrical and magnetic fields, *N. Engl. J. Med.,* 307, 1402, 1982.

80. **Miller, M. W.,** Re: "Electrical wiring configurations and childhood cancer", *Am. J. Epidemiol.,* 112, 165, 1980.

81. **Odintsov, Y. N.,** The effect of a magnetic field on the natural resistance of white mice to *Listeria* infection, *Tr. Tomsk. Nauchno-Issled. Inst. Vaktsiny Syvorotok,* 16, 234, 1965.

82. **Riesen, W. H., Aranyi, C., Kyle, J. L., Valentino, A. R., and Miller, D. A.,** A pilot study of the interaction of extremely low frequency electromagnetic fields with brain organelles, in *Compilation of Navy Sponsored ELF Biomedical and Ecological Research Reports,* Vol. 1, Tech. Memo. No. 3 (IITRI Proj. E6185), Nav. Med. Res. Dev. Command, Bethesda, Md., 1971.

83. **Ossenkopp, K.-P., Koltek, W. T., and Persinger, M. A.,** Prenatal exposure to an extremely low frequency-low intensity rotating magnetic field and increases in thyroid and testicle weight in rats, *Dev. Psychobiol.,* 5, 275, 1972.

84. **Toroptsev, I. V., Garganeyev, G. P., Gorshenina, T. I., and Teplyakova, N. L.**, Pathologoanatomic characteristics of changes in experimental animals under the influence of magnetic fields, in Influence of Magnetic Fields on Biological Objects, Kholodov, Yu. A., Ed., NTIS Rep. No. JPRS 63038, Springfield, Va., 1974, 95.

85. **Udintsev, N. A. and Moroz, V. V.**, Response of the pituitary-adrenal system to the action of a variable magnetic field, *Byull. Éksp. Biol. Med.*, 77, 51, 1974.

86. **Mizushima, Y., Akaoka, I., and Nishida, Y.**, Effects of magnetic field on inflammation, *Experientia*, 21, 1411, 1975.

87. **Beischer, D. E. and Brehl, R. J.**, Search for effects of 45 Hz magnetic fields on liver triglycerides in mice, Rep. No. NAMRL-1197, Nav. Aerosp. Med. Res. Lab., Pensacola, Fla., 1975.

88. **Udintsev, N. A., Kanskaia, N. V., Shchepetil'nifova, A. I., Ordina, O. M., and Pichurina, R. A.**, Dynamics of cardiac and skeletal muscle lactate dehydrogenase activity following a single exposure to an alternating magnetic field, *Byull. Éksp. Biol. Med.*, 81, 670, 1976.

89. **Batkin, S. and Tabrah, F. L.**, Effects of alternating magnetic field (12 Gauss) on transplanted neuroblastoma, *Res. Commun. Chem. Pathol. Pharmacol.*, 16, 351, 1977.

90. **Chandra, S. and Stefani, S.**, Effect of constant and alternating magnetic fields on tumor cells *in vitro* and *in vivo*, in Biological Effects of Extremely Low Frequency Electromagnetic Fields, Phillips, R. D., Gillis, M. F., Kaune, W. T., and Mahlum, D. D., Eds., NTIS Rep. No. CONF-781016, Springfield, Va., 1979, 436.

91. **Goodman, E. M., Greenebaum, B., and Marron, M. T.**, Bioeffects of extremely low-frequency electromagnetic fields: variation with intensity, waveform, and individual or combined electric and magnetic fields, *Radiat. Res.*, 78, 485, 1979.

92. **Greenebaum, B., Goodman, E. M., and Marron, M. T.**, Extremely-low-frequency fields and the slime mold *Physarum polycephalum:* evidence of depressed cellular function and of internuclear interaction, *Radio Sci.*, 14(6S), 103, 1979.

93. **Greenebaum, B., Goodman, E. M., and Marron, M. T.**, Magnetic field effects on mitotic cycle length in *Physarum, Eur. J. Cell Biol.*, 27, 156, 1982.

94. **Kolodub, F. A. and Chernysheva, O. N.**, Special features of carbohydrate-energy and nitrogen metabolism in the rat brain under the influence of magnetic fields of commercial freuquency, *Ukr. Biokhim. Zh. (Kiev)*, 3, 299, 1980.

95. **Fam, W. Z.**, Biological effects of 60-Hz magnetic field on mice, *IEEE Trans. Mag.*, 17, 1510, 1981.

96. **Aarholt, E., Flinn, E. A., and Smith, C. W.**, Effects of low-frequency magnetic fields on bacterial growth rate, *Phys. Med. Biol.*, 26, 613, 1981.

97. **Ramon, C., Ayaz, M., and Streeter, D. D., Jr.**, Inhibition of growth rate of *Escherichia coli* induced by extremely low-frequency weak magnetic fields, *Bioelectromagnetics*, 2, 285, 1981.

98. **Sakharova, S. A., Ryzhov, A. I., and Udintsev, N. A.**, Mechanism of the sympathoadrenal system's response to the one time action of a variable magnetic field, *Kosm. Biol. Aviakosmicheskaya Med.*, 15, 52, 1981.

99. **Toroptsev, I. V. and Soldatova, L. P.**, Pathomorphological reactions of cerebrocortical neural elements to alternating magnetic field, *Arkh. Patol. (Moscow)*, 43, 33, 1981.

100. **Delgado, J. M. R., Monteagudo, J. L., Garcia, M. G., and Leal, J.**, Teratogenic effects of weak magnetic fields, *IRCS Med. Sci.*, 9, 392, 1981.

101. **Delgado, J. M. R., Leal, J., Monteagudo, J. L., and Garcia, M. G.**, Embryological changes induced by weak, extremely low frequency electromagnetic fields, *J. Anat.*, 134, 533, 1982.

102. **Soldatova, L. P.**, Sequence of pathomorphological reactions to the effect of alternating magnetic fields, *Arkh. Anat. Gistol. Embriol.*, 83, 12, 1982.

103. **Sander, R., Brinkmann, J., and Kuhne, B.**, Laboratory studies on animals and human beings exposed to 50 Hz electric and magnetic fields, Paper 36-01, Int. Congr. Large High Voltage Elect. Syst., Paris, September 1 to 9, 1982.

104. **Lubin, R. A., Cain, C. D., Chen, M. C.-Y., Rosen, D. M., and Adey, W. R.**, Effects of electromagnetic stimuli on bone and bone cells *in vitro:* inhibition of responses to parathyroid hormone by low-energy low-frequency fields, *Proc. Natl. Acad. Sci. U.S.A.*, 79, 4180, 1982.

105. **Conti, P., Gigante, G. E., Cifone, M. G., Alesse, E., Ianni, G., Reale, M., and Angeletti, P. U.**, Reduced mitogenic stimulation of human lymphocytes by extremely low frequency electromagnetic fields, *FEBS Lett.*, 162, 156, 1983.

106. **Goodman, R., Bassett, C. A. L., and Henderson, A. S.**, Pulsing electromagnetic fields induce cellular transcription, *Science*, 220, 1283, 1983.

107. **Jolley, W. B., Hinshaw, D. B., Knierim, K., and Hinshaw, D. B.**, Magnetic field effects on calcium efflux and insulin secretion in isolated rabbit islets of Langerhans, *Bioelectromagnetics*, 4, 103, 1983.

108. **Ramiriz, E., Monteagudo, J. L., Garcia-Gracia, M., and Delgado, J. M. R.**, Oviposition and development of *Drosophila* modified by magnetic fields, *Bioelectromagnetics*, 4, 315, 1983.

109. **Ubeda, A., Leal, J., Trillo, M. A., Jimenez, M. A., and Delgado, J. M. R.,** Pulse shape of magnetic fields influences chick embryogenesis, *J. Anat.,* 137, 513, 1983.

110. **Archer, C. W. and Ratcliffe, N. A.,** The effects of pulsed magnetic fields on chick embryo cartilaginous skeletal rudiments *in vitro, J. Exp. Zool.,* 225, 243, 1983.

111. **Liboff, A. R., Williams, T., Jr., Strong, D. M., and Wistar, R., Jr.,** Time-varying magnetic fields: effect on DNA synthesis, *Science,* 223, 818, 1984.

112. **Coate, W. B.,** Project Biological Effects Test Program, Pilot Studies, Final Report, Hazelton Laboratories, NTIS Rep. No. AD75058, Springfield, Va., 1970.

113. **Marron, M. T., Goodman, E. M., and Greenebaum, B.,** Mitotic delay in the slime mold *Physarum polycephalum* induced by low intensity 60 and 75 Hz electromagnetic fields, *Nature,* 254, 66, 1975.

114. **Polk, C.,** Time varying magnetic fields and DNA synthesis: magnitude of forces due to magnetic fields on surface bound counterions, presented at 6th Annu. Meet. Bioelectromagnetics Soc., Atlanta, July 15 to 19, 1984.

115. **Kolin, A.,** An alternating field induction flow meter of high sensitivity, *Rev. Sci. Instrum.,* 16, 109, 1945.

116. **Tenforde, T. S.,** Mechanisms for biological effects of magnetic fields, in *Biological Effects and Dosimetry of Nonionizing Radiation: Static and ELF Electromagnetic Fields,* Grandolfo, M., Michaelson, S. M., and Rindi, A., Eds., Plenum Press, New York, 1985, 71.

117. **Kolin, A.,** Improved apparatus and technique for electromagnetic determination of blood flow, *Rev. Sci. Instrum.,* 23, 235, 1952.

118. **Mills, C. J.,** The electromagnetic flowmeter, *Med. Instrum.,* 11, 136, 1977.

119. **Conti, F., Stefanoni, G., and Donadelli, G.,** A new method for measuring venous blood flow with the aid of an electromagnetic flowmeter, *J. Cardiovasc. Surg.,* 17, 363, 1976.

120. **Portnoi, V. F. and Dvortsin, G. F.,** Experimental basis of a new method of determination of coronary blood flow with the aid of electromagnetic flowmeter, *Kardiologiya,* 16, 118, 1976.

121. **Kolin, A. and MacAlpin, R. N.,** Induction angiometer. Electromagnetic magnification of microscopic vascular diameter variations *in vivo, Blood Vessels,* 14, 141, 1977.

122. **Gooszen, H. G., Van Schilfgaarde, R., and Terpstra, J. L.,** Arterial blood supply of the left lobe of the canine pancreas. II. Electromagnetic flow measurements, *Surgery,* 93, 549, 1983.

123. **Okai, O. and Oshima, M.,** Magnetorheography: calculation of blood flow from surface induced potentials, *Jpn. Heart J.,* 16, 694, 1975.

124. **Beischer, D. E. and Knepton, J. C.,** Influence of strong magnetic fields on the electrocardiogram of squirrel monkeys *(Saimiri sciureus), Aerosp. Med.,* 35, 939, 1964.

125. **Beischer, D. E.,** Vectorcardiogram and aortic blood flow of squirrel monkeys *(Saimiri sciureus)* in a strong superconductive electromagnet, in *Biological Effects of Magnetic Fields,* Vol. 2, Barnothy, M., Ed., Plenum Press, New York, 1969, 241.

126. **Tenforde, T. S., Gaffey, C. T., Moyer, B. R., and Budinger, T. F.,** Cardiovascular alterations in *Macaca* monkeys exposed to stationary magnetic fields: experimental observations and theoretical analysis, *Bioelectromagnetics,* 4, 1, 1983.

127. **Gaffey, C. T., Tenforde, T. S., and Dean, E. E.,** Alterations in the electrocardiograms of baboons exposed to DC magnetic fields, *Bioelectromagnetics,* 1, 209, 1980.

128. **Togawa, T., Okai, O., and Oshima, M.,** Observation of blood flow E.M.F. in externally applied strong magnetic fields by surface electrodes, *Med. Biol. Eng.,* 5, 169, 1967.

129. **Gaffey, C. T. and Tenforde, T. S.,** Alterations in the rat electrocardiogram induced by stationary magnetic fields, *Bioelectromagnetics,* 2, 357, 1981.

130. **Bassett, C. A. L.,** Pulsing electromagnetic fields: a new approach to surgical problems, in *Metabolic Surgery,* Buchwald, H. and Varco, R. L., Eds., Grune & Stratton, New York, 1978, 255.

131. **Fukada, E. and Yusuda, I.,** On the piezoelectric effect in bone, *J. Phys. Soc. Jpn.,* 12, 1158, 1957.

132. **Bassett, C. A. L. and Becker, R. O.,** Generation of electric potentials by bone in response to mechanical stress, *Science,* 137, 1063, 1962.

133. **Bassett, C. A. L., Pawluk, R. J., and Pilla, A. A.,** Acceleration of fracture repair by electromagnetic fields. A surgically noninvasive method, *Ann. N. Y. Acad. Sci.,* 238, 242, 1974.

134. **Lunt, M. J.,** Magnetic and electric fields produced during pulsed-magnetic-field therapy for non-union of the tibia, *Med. Biol. Eng. Comput.,* 20, 501, 1982.

135. **Bassett, C. A. L., Pilla, A. A., and Pawluk, R. J.,** A non-operative salvage of surgically-resistant pseudoarthroses and non-unions by pulsing electromagnetic fields, *Clin. Orthop.,* 124, 128, 1977.

136. **Watson, J. and Downes, E. M.,** The application of pulsed magnetic fields to the stimulation of bone healing in humans, *Jpn. J. Appl. Phys.,* 17, 215, 1978.

137. **Hinsenkamp, M. G.,** Electromagnetic stimulation of nonunion, *Rev. Med. Bruxelles Nouv.,* 1(3), 19, 1982.

138. **Bassett, C. A. L., Mitchell, S. N., and Gaston, S. R.,** Pulsing electromagnetic field treatment in ununited fractures and failed arthrodeses. *JAMA,* 247, 623, 1982.

139. **Bigliani, L. U., Rosenwasser, M. P., Caulo, N., Schink, M. M., and Bassett, C. A. L.,** The use of pulsing electromagnetic fields to achieve arthrodesis of the knee following failed total knee arthroplasty. A preliminary report, *J. Bone Jt. Surg.,* 65, 480, 1983.

140. **Barker, A. T. and Lunt, M. J.,** The effects of pulsed magnetic fields of the type used in the stimulation of bone fracture healing, *Clin. Phys. Physiol. Meas.,* 4, 1, 1983.

141. **Watson, J. and Downes, E. M.,** Light-weight battery-operable orthopaedic stimulator for the treatment of long-bone nonunions using pulsed magnetic fields, *Med. Biol. Eng. Comput.,* 21, 509, 1983.

142. **Hinsenkamp, M. G. and Rooze, M. A.,** Morphological effect of electromagnetic stimulation on the skeleton of fetal or newborn mice, *Acta Orthop. Scand.,* Suppl. 196, 39, 1982.

143. **Frank, C., Schachar, N., Dittrich, D., Shrive, N., Dehaas, W., and Edwards, G.,** Electromagnetic stimulation of ligament healing in rabbits, *Clin. Orthop.,* May (175), 1983, 263.

144. **Wilson, D. H. and Jagadeesh, P.,** Experimental regeneration in peripheral nerves and the spinal cord in laboratory animals exposed to a pulsed electromagnetic field, *Paraplegia,* 14, 12, 1976.

145. **Raji, A. R. M. and Bowden, R. E. M.,** Effects of high peak pulsed electromagnetic field on degeneration and regeneration of the common peroneal nerve in rats, *Lancet,* 2(8295), 444, 1982.

146. **Raji, A. R. and Bowden, R. E.,** Effects of high-peak pulsed electromagnetic field on the degeneration and regeneration of the common peroneal nerve in rats, *J. Bone Jt. Surg.,* 65, 478, 1983.

147. **Bassett, C. A. L., Becker, R. O., Brighton, C. T., Lavine, L., and Rowley, B. A.,** Panel discussion: to what extent can electrical stimulation be used in the treatment of human disorders?, *Ann. N.Y. Acad. Sci.,* 238, 586, 1974.

148. **Baum, S. J., Ekstrom, M. E., Skidmore, W. D., Wyant, D. E. and Atkinson, J. L.,** Biological measurements in rodents exposed continuously throughout their adult life to pulsed electromagnetic radiation, *Health Phys.,* 30, 161, 1976.

149. **Seipel, J. H. and Morrow, R. D.,** The magnetic field accompanying neuronal activity: a new method for the study of the nervous system, *J. Wash. Acad. Sci.,* 50, 1, 1960.

150. **Gengerelli, J. A., Holter, N. J., and Glassock, W. R.,** Magnetic fields accompanying transmission of nerve impulses in the frog's sciatic, *J. Psychol.,* 57, 202, 1961.

151. **Baule, G. M. and McFee, R.,** Detection of the magnetic field of the heart, *Am. Heart J.,* 66, 95, 1963.

152. **Cohen, D.,** Magnetic fields around the torso: production by electrical activity of the human heart, *Science,* 156, 652, 1967.

153. **Cohen, D.,** Magnetoencephalography: evidence of magnetic fields produced by alpha rhythm currents, *Science,* 161, 784, 1968.

154. **Cohen, D.,** Magnetic fields of the human body, *Phys. Today,* August, 34, 1975.

155. **Zimmerman, J. E., Thiene, P., and Harding, J. T.,** Design and operation of stable rf-biased superconducting point-contact quantum devices, and a note on the properties of perfectly clean metal contacts, *J. Appl. Physiol.,* 41, 1572, 1970.

156. **Cohen, D.,** Magnetoencephalography: detection of the brain's electrical activity with a superconducting magnetometer, *Science,* 175, 664, 1972.

157. **Romani, G. L., Williamson, S. J., and Kaufman, L.,** Biomagnetic instrumentation, *Rev. Sci. Instrum.,* 53, 1815, 1982.

158. **Williamson, S. J. and Kaufman, L.,** Biomagnetism, *J. Mag. Mag. Mat.,* 22, 129, 1981.

159. **Williamson, S. J., Romani, G.-L., Kaufman, L., and Modena, I.,** Eds., *Biomagnetism,* Plenum Press, New York, 1983.

160. **Awano, I., Muramoto, A., and Awano, N.,** An approach to clinical magnetocardiology, *Tohoku J. Exp. Med.,* 138, 367, 1982.

161. **Savard, P., Cohen, D., Lepeschkin, E., Cuffin, B. N., and Madias, J. E.,** Magnetic measurement of S-T and T-Q segment shifts in humans. I. Early repolarization and left bundle branch block, *Circ. Res.,* 53, 264, 1983.

162. **Cohen, D., Savard, P., Rifkin, R. D., Lepeschkin, E., and Strauss, W. E.,** Magnetic measurement of S-T and T-Q segment shifts in humans. II. Exercise-induced S-T segment depression, *Circ. Res.,* 53, 274, 1983.

163. **Reite, M. and Zimmerman, J.,** Magnetic phenomena of the central nervous system, *Annu. Rev. Biophys. Bioeng.,* 7, 167, 1978.

164. **Cohen, D. and Cuffin, B. N.,** Demonstration of useful differences between magnetoencephalogram and electroencephalogram, *Electroencephalogr. Clin. Neurophysiol.,* 56, 38, 1983.

165. **Reite, M., Zimmerman, J. E., Edrich, J., and Zimmerman, J.,** The human magnetoencephalogram: some EEG and related correlations, *Electroencephalogr. Clin. Neurophysiol.,* 40, 59, 1976.

166. **Gutman, A. M. and Morgenshtern, V. Ya.,** Possible mechanism of the genesis of the magnetoencephalogram, *Biofizika,* 22, 529, 1977.

167. **Plonsey, R.,** The nature of sources of bioelectric and biomagnetic fields, *Biophys. J.,* 39, 309, 1982.

168. **Brenner, D., Williamson, S. J., and Kaufman, L.,** Visually evoked magnetic fields of the human brain, *Science,* 190, 480, 1975.

169. **Brenner, D., Kaufman, L., and Williamson, S. J.,** Application of a squid for monitoring magnetic response of the human brain, *IEEE Trans. Mag.,* 13, 365, 1977.

170. **Brenner, D., Lipton, J., Kaufman, L., and Williamson, S. J.,** Somatically evoked magnetic fields of the human brain, *Science,* 199, 81, 1978.

171. **Okada, Y. C., Williamson, S. J., and Kaufman, L.,** Magnetic field of the human sensorimotor cortex, *Int. J. Neurosci.,* 17, 33, 1982.

172. **Okada, Y. C., Kaufman, L., and Williamson, S. J.,** The hippocampal formation as a source of the slow endogenous potentials, *Electroencephalogr. Clin. Neurophysiol.,* 55, 417, 1983.

173. **Cohen, D.,** Ferromagnetic contamination in the lungs and other organs of the human body, *Science,* 180, 745, 1973.

174. **Farrell, D. E., Tripp, J. H., Zanaucchi, P. E., Harris, J. W., Brittenham, G. M., and Muir, W. A.,** Magnetic measurement of human iron stores, *IEEE Trans. Mag.,* 16, 818, 1980.

Part III — Effects of Radio Frequency (Including Microwave) Fields

Chapter 1

EXPERIMENTAL RADIO AND MICROWAVE DOSIMETRY

Maria A. Stuchly and Stanislaw S. Stuchly

TABLE OF CONTENTS

I. INTRODUCTION

Interactions of radio frequency (RF) fields with biological tissues and bodies are complex functions of numerous parameters. Dosimetric studies attempt to quantify these interactions. Radiowaves in free space are characterized by the frequency, intensity of the electric (E) and magnetic (H) fields, their direction, and polarization. However, only the fields inside the tissues and biological bodies can interact with them, and therefore it is necessary to determine these fields for any meaningful and general quantification of biological data obtained experimentally.

Fields inside biological bodies exposed to known external electromagnetic fields can be calculated by solving Maxwell's equations subject to given boundary conditions. A biological body is an inhomogeneous, lossy dielectric, whose macroscopic electrical properties are described by the complex permittivity. The inhomogeneity of the dielectric properties and the complexity of the shape make a solution, or sometimes even a full formulation of the problem, a forbidding task. Only much simplified models can be analyzed. The other approach is an experimental one, which is also subject to considerable limitations.

The intensity of the internal fields depends on the parameters of the external field, the frequency, intensity, and polarization, size, shape, dielectric properties of the exposed body, spatial configuration between the exposure source and the exposed body, and the presence of other objects in the vicinity. With a complex dependence on so many parameters, it is apparent that the internal fields in a mouse and a man exposed to the same external field can be dramatically different, and so will be their biological response, regardless of physiological differences. Conversely, different exposure conditions, e.g., different frequencies, may induce similar fields inside such diverse shapes as a mouse and a man.

A. Specific Absorption Rate (SAR)

The internal fields can be quantified in various ways. The magnetic permeability of most tissues is practically equal to that of free space, and all known and anticipated interactions at higher RFs occur through mechanisms involving the E field (including the currents induced by the H field). Therefore, the E-field vector, or its distribution throughout the exposed body, fully describes the exposure field-tissue interactions. Some additional information may be needed for full quantification, e.g., frequency characteristics of the exposure field, modulation characteristics, and modulation frequency. A dosimetric measure that has been widely adopted is the SAR defined as "the time derivative of the incremental energy (dW) absorbed by, or dissipated in an incremental mass (dm) contained in a volume element (dV) of a given density (ρ)".[1]

$$\text{SAR} = \frac{d}{dt}\left(\frac{dW}{dm}\right) = \frac{d}{dt}\left[\frac{dW}{\rho(dV)}\right] \tag{1}$$

Using the Poynting vector theorem for sinusoidally varying electromagnetic fields, Equation 1 can be presented as

$$\text{SAR} = \frac{\sigma}{2\rho} |E_i|^2 = \frac{\omega \epsilon_0 \epsilon''}{2\rho} |E_i|^2 \tag{2}$$

where σ is the tissue conductivity in siemens per meter, ϵ_0 is the dielectric constant of free space ($\epsilon_0 = 8.85 \times 10^{-12}$F/m), ϵ'' is the loss factor, $\omega = 2\pi f$, f is the frequency in hertz, and E_i is the peak value of the internal E field in volts per meter. The SAR is expressed in watts per kilogram (W/kg) or derived units (e.g., mW/g). The average SAR is defined as a ratio of the total power absorbed in the exposed body to its mass. The local SAR refers to the value within a defined unit volume or unit mass, which can be arbitrarily small.

It is very important to realize that even though Equation 2 expresses the rate at which the electromagnetic energy is converted into heat through well-established interaction mechanisms described in Part I, it provides a valid quantitative measure of all the interaction mechanisms that are dependent on the intensity of the internal E field.[2] At this point, some additional information may be relevant. For instance, since some effects of radiowaves modulated in amplitude at ELF (extremely low frequencies) are dependent on the E field intensity[3], once specific interaction mechanisms are better understood, they could be expressed in terms of the SAR and modulation characteristics, even though the interaction mechanism may not necessarily be thermal. However, should the direction of the internal E field with respect to the biological structures be of importance for a given interaction mechanism, the SAR would not provide full quantitative information. Similarly, for the interactions directly through the H field, the present SAR concept would not be sufficient. Bearing these limitations in mind, the SAR concept has proven to be a simple and useful tool in quantifying the interactions of RF radiation with living systems. It enables comparison of experimentally observed biological effects for various species under various exposure conditions, and it provides the only means, however imperfect, of extrapolating the animal data into potential hazards to humans exposed to RF radiation. It also facilitates planning and effective execution of therapeutic hyperthermic treatment.

B. Experimental Dosimetry and Units

Dosimetry in bioelectromagnetic research has been developing in two parallel but interacting and complementary streams: theoretical and experimental. Theoretical dosimetry deals with the analysis of simplified models as reviewed in Part III, Chapter 2, while experimental dosimetry concerns the development of methods and instruments suitable for measurements of the internal E fields, SARs, or related variables. Measurements of the external fields as relevant, for the evaluation of the exposure conditions are usually considered a part of experimental dosimetry (only occasionally but not appropriately named "densitometry"). To place the measurements of the external fields vs. the internal fields in proper perspective, one has to realize that the instruments and methods for measurements of internal fields are not suitable for use on humans except in hyperthermic treatment for cancer therapy. Therefore, to prevent potentially hazardous exposure to man, the external fields are measured. Data on the intensities of external fields that potentially constitute a hazard to man are obtained through experimental evaluation of the intensities of the internal fields (SARs) that cause damage to animals and determination of external fields which produce similar internal fields (SARs) in man.

Experimental dosimetry includes the following areas: measurements of the total SAR and its distribution in live or dead experimental animals, measurements of the average SAR and SAR distribution in models of animals and humans, and development of exposure devices for animals and in vitro preparations in which quantification of the exposure conditions and internal fields is facilitated.

Table 1
QUANTITIES, COMMON SYMBOLS, AND UNITS
USED IN EXPERIMENTAL DOSIMETRY

Symbol	Quantity	Unit
	External Fields	
$\overline{E}$	E field (vector)	V/m
$\overline{H}$	H field (vector)	A/m
S	Power density, energy flux density	W/m²
	(scalar)	mW/cm²
$\overline{S}$	Poyinting vector, $\overline{S} = \overline{E} \times \overline{H}$	W/m²
W	Energy density (scalar)	J/m³
$\overline{k}$	Propagation vector	
Z	Wave impedance, $Z = E/H$ (scalar)	Ω
	Internal Fields	
$\overline{E}_i$	E field (vector)	V/m
SAR	SAR	W/kg
	(av and local)	mW/g
ϵ	Complex relative permittivity, $\epsilon = \epsilon' - j\epsilon''$	—
σ	Conductivity, $\sigma = \omega\epsilon_0\epsilon''$	S/m

Experimental dosimetry plays a crucial role in the control of animal exposures and quantification of the data from animal experimentation. It enables comparison of experiments conducted in various laboratories, under different exposure conditions, and for different species. It also plays an important role in verifying the predictions of theoretical dosimetry. While theoretical methods, with appropriate simplifying assumptions, can in a straightforward manner predict the total or average SAR, the SAR distributions in complex models that are close approximations of the actual animal, and in the animal, can usually be more easily determined experimentally.

The fundamental quantities and units in the SI system used in experimental dosimetry are summarized in Table 1.

The dosimetric data can be used to extrapolate the results of animal experiments to human exposures. The physical scaling principle is then employed. For instance, to determine the equivalent human exposure at a frequency f_n (i.e., where roughly similar, but by no means identical SAR distribution exists) for a man of a height l_n and an animal of a height l_a exposed at a frequency f_a (with both man and animal having the same orientation in respect to the exposure field), the following relationship can be used:[4]

$$f_n = f_a \frac{l_a}{l_n} \tag{3}$$

It should be emphasized that the SAR distribution is only similar, not identical.[5]

In this chapter the basic knowledge on experimental dosimetry is summarized. The material is divided into three parts. The first deals with measurements of the fields both external and internal, in bodies exposed to RF fields. The second part covers dosimetric measurements based on temperature measurements. The concept of the SAR is used extensively in these two parts. The last part describes basic principles and some designs of exposure systems that are suitable for exposures with well-defined dosimetry. Effects related to the use of minitoring devices (e.g., electrodes) are also briefly outlined. A listing of dosimetry methods and their applications as described in this chapter are given in Table 2.

Table 2
EXPERIMENTAL DOSIMETRY METHODS

Av SAR	SAR distribution	E-field vector
Absorbed power measurements	E-field probes	E-field probes
Calorimetry	Nonperturbing thermometers	
	Thermography	

II. FIELD MEASUREMENTS

A. External Fields

1. Basic Definitions

RFs occupy a relatively wide segment of the electromagnetic spectrum with frequencies ranging from a few hundred kilohertz to about 300 GHz. The exposure fields can be produced intentionally by radiators such as those used in telecommunications and radar, or may be a result of imperfections in the design or manufacture of a given device, resulting in undesirable energy leakage, e.g., in RF dielectric heaters. The nature of these two types of fields and their characteristics that are important from the dosimetry viewpoint, are significantly different. One of the most important descriptors of an exposure field is whether it is the far field or the near field. The far-field region extends from a certain minimum distance from the antenna to infinity. This minimum distance is usually selected as $l \cong \dfrac{2D^2}{\lambda}$, where D is the larger dimension of the antenna and λ is the wavelength. This is equivalent to the fact that the power per unit solid angle is almost constant.[1] In this region the field has a predominantly plane-wave character, i.e., the E-field vector is perpendicular to the H-field vector, and they are both perpendicular to the direction of propagation. The ratio of the E-field intensity to the H-field intensity is constant at any location, and for free space is equal to

$$Z = \frac{E}{H} \cong 377 \ \Omega \tag{4}$$

In the far field, the field intensity is therefore fully defined by either the E-field intensity or the H-field intensity. Most frequently, for a plane-wave the power density is used to describe the exposure field.

$$S = \frac{E^2}{Z} = H^2 Z \tag{5}$$

The near-field region close to the radiator (antenna) comprises the radiating near-field region (Fresnel region) and the reactive (nonradiating) near-field regions. The reactive region is adjacent to the antenna and extends to about one wavelength from the antenna. In this region the E and H fields vary rapidly with distance. A portion of the electromagnetic energy is stored in this region. In the radiating near field the energy propagates away from the antenna, and the intensity of the E and H field decreases with distance, but the radiation still lacks the plane-wave character. The power density usually varies with the distance in an oscillatory fashion, rather than inversely proportionally with the square of the distance, as for the plane-wave (far field). For illustration, the variations of power density vs. distance are given in Figure 1.

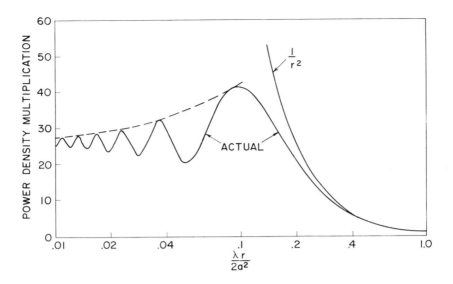

FIGURE 1.   Relative power density vs. distance along the axis of an aperture antenna. (r) The distance from the antenna; (a) the diameter of the antenna; (λ) the wavelength. (From Microwave Engineering Handbook, Horizon House.)

Table 3
ESSENTIAL PARAMETERS OF THE EXPOSURE FIELD

Frequency	Intensity	Orientation
Carrier frequency	Far field: power density	Far field: polarization, direction with respect to object
Modulation type (e.g., amplitude, sinusoidal, pulse)	Near field: E-field intensity	
Modulation characteristics (e.g., pulse duration, shape, and repetition rate)	H-field intensity	Near-field: directions of $\overline{E}$ and $\overline{H}$ with respect to the object

From the definitions and character of the fields it is obvious that while in the far field a measurement of only one field component (the E or the H field) is sufficient for a full description of the field, in the near field both fields should be measured. Furthermore, in the reactive zone, a measuring instrument is likely to significantly perturb the field and modify the radiation characteristics. Leakage fields, particularly at longer radiowaves, are in most cases of the near-field type. Essential parameters of an exposure field that have to be defined are summarized in Table 3.

2. Measurement Errors

Measurements of external fields, for evaluation of a potential hazard of exposure or for definition of exposure conditions in animal experimentation, are subject to several possible sources of interference and error.[6,7] Exposure may occur at more than one frequency either because of multiple sources of emission or emissions at more than one frequency from a single source (e.g., harmonics). Exposure fields may have various polarizations, as illustrated in Figure 2, which may not be known prior to measurements. Nearby objects may cause partial or total reflection or scattering of the waves, creating interference patterns with maxima and minima interspaced a quarter of the wavelength. The modulation characteristics of the

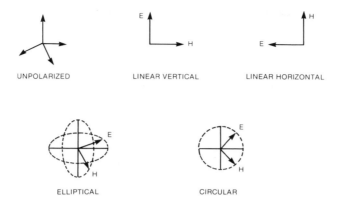

FIGURE 2.   Types of polarization of electromagnetic waves.

Table 4
DESIRABLE CHARACTERISTICS OF BROADBAND
INSTRUMENTS FOR MEASUREMENTS OF EXPOSURE FIELDS

Sensor	Meter
Response to one field only (E or H)	Response to modulated fields
Nonperturbing	Accuracy
Suitable for near and far field	Sensitivity
Small-good spatial resolution	Wide dynamic range
Broad frequency range, flat response characteristic	RFI resistance
Isotropic or three directions	Peak/av value option
Multiple signal integration	Short response time
Temperature insensitivity	Portability
Stability	Temperature insensitivity
Overload insensitivity	Stability
Linear response	Overload sensitivity
Low noise	Linear response

radiation are also important. Pulse shape and duration, repetition rate, and frequency of modulation for amplitude or frequency modulation can all very significantly affect the accuracy of measurements. Previously mentioned presence of the measuring instrument or any other object in the reactive field of the antenna can cause considerable change in the exposure field. Finally, other factors such as RF interference and reflecting surfaces in the vicinity may cause errors in measurements of exposure fields.

3. Broadband Instruments — Survey Meters
Two general categories of instruments are suitable, within certain restrictions, for measurements of RF exposure fields. They are broadband and narrowband type instruments. The broadband instruments, which have been developed specifically for hazard field surveys, have frequency-independent sensitivity, measure integrated intensity of the fields, and do not provide any information on the frequency spectrum of the field. The narrowband instruments, on the other hand, are selective and provide information about the field intensity at a selected frequency or rather within a very narrow frequency band, e.g., 1% around the nominal frequency.

The broadband instruments are more suitable for measurements in the near field as they usually cause only minimal perturbation of the field; however, their sensitivity is usually much less than that of the narrowband instruments. The salient characteristics of broadband survey instruments are listed in Table 4.

Various design concepts have been employed in E-field probes and basically one concept in H-field probes. Principles of operation of the broadband probes are outlined below.

One of the simplest E-field sensors consists of a short dipole connected to a diode. Such a sensor responds only to the field that is polarized parallel to the dipole, however when three dipoles are arranged in a mutually orthogonal configuration a nearly isotropic directional radiation characteristic can be obtained. For an electrically short dipole, i.e., kh < 1, where k is the propagation constant (in free space $k = 2\pi/\lambda$, λ is the wavelength) and 2h is the physical length of the dipole, the effective length h_e, and the input capacitance C_a, are independent of frequency and equal to:[9,10]

$$h_e = \frac{h(\Omega - 1)}{2(\Omega - 2 + 2\ln 2)} \tag{6}$$

$$C_a = \frac{4\pi\epsilon_0 h}{\Omega - 2 - 2\ln 2} \tag{7}$$

where

$$\Omega = 2 \ln (2h/a) \tag{8}$$

a is the antenna radius and ϵ_0 is the permittivity of free space. By selecting a proper detector diode and load impedance, a broadband frequency response can be obtained.[10]

The diode is connected to the signal processing circuitry by a pair of high-resistance leads. The high-resistance leads are necessary, because conductive leads act as antennas. The high-resistance leads provide an additional advantage of causing only negligible perturbation to the measured electromagnetic field. A description of an early design of a meter based on this principle is given elsewhere.[8] In some designs the dipole is also made of a resistive strip to achieve broadband operation.[11]

Several commercially available survey meters (Table 5) are based on this principle (e.g., Aeritalia, General Microwave, Holaday, Instruments for Industry). One of the major problems of the sensors with a diode detector is the nonlinearity of the diode characteristic. While for low-intensity E fields the output voltage of the diode is proportional to the square of the field intensity, at higher intensities the relationship becomes almost linear. This limits the dynamic range for accurate measurements.

The E-field intensity can also be measured by a thermocouple-based sensor.[12] The sensor consists of several miniature thermocouples connected in series, forming a thermopile as shown in Figure 3. The output signal of the thermocouple is proportional to the power dissipated in it, which in turn is proportional to the square of the current induced in the thermocouple elements acting as miniature dipoles. As the induced current is proportional to the E-field intensity polarized parallel to the dipole, the thermocouple sensor provides an output signal proportional to the square of the E-field intensity. A wide dynamic range can therefore be obtained. An arrangement of three mutually perpendicular thermocouple elements gives a practically isotropic spatial characteristic. Commercially available instruments based on this principle are manufactured by Narda (Table 5), and utilize overlapping thin films of antimony and bismuth (Figure 3).[12] The thermocouple sensor is connected with the external circuitry by high-resistance leads.

Thermocouple probes are true square-law detectors and are inherently accurate indicators of the time-averaged intensity for modulated exposure fields. They are more sensitive to overloading than the diode probes and their dynamic range, although theoretically unlimited, is restricted by the thermocouple sensitivity at one end and by the maximum power dissipated in the junction at the other.

Table 5
SALIENT CHARACTERISTICS OF REPRESENTATIVE COMMERCIALLY AVAILABLE BROADBAND SURVEY METERS

Manufacturer and model	Frequency	Intensity ranges	Type
Aeritalia (Italy), 307, RV13, RV14, RV15	1—500 MHz	10—1000 V/m	E field
Aeritalia (Italy), 307, RV16, RV17	2—1000 MHz	1—10 A/m	H field
General Microwave, Raham 12	0.01—18 GHz	2—200 mW/cm^2	E field
Holaday Industries, 3001	0.5—6 GHz	10^2—10^7 V^2/m^2	E field
Holaday Industries, 3002	0.5—6 GHz	10^4—10^7 V^2/m^2	E field
	5—300 MHz	1—100 A^2/m^2	H field
Narda Microwave, 8608	10—300 MHz	0.2—20 mW/cm^2 or	H field
	0.3—26 GHz	1—100 mW/cm^2	E field
Instruments for Industry RHM-1	0.01—220 MHz	10—300 V/m	E field

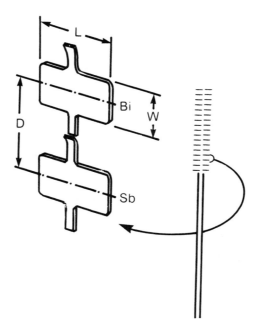

FIGURE 3.   Thermocouple E-field probe.[12]

A spherical, thin-film bolometer sensor, shown in Figure 4, was also developed[8,13] but not produced commercially. The sensor consists of two connected, gas-filled spheres. One of the spheres is coated with a thin, resistive film. This sphere, when exposed to an electromagnetic field, is heated by the currents induced by the E field. The second uncoated sphere is used as a reference. Since gas expansion in the coated sphere is proportional to temperature, the differential pressure between the spheres is proportional to the square of the E field intensity. Broadband operation can be obtained by proper selection of the film resistivity.

A H-field sensor consists of a small loop with a diode or a thermocouple connected by high-resistance leads to the external circuitry. The diameter of the loop, d, should be small compared with the wavelength, i.e., d < 0.1 λ_{min}, where λ_{min} is the wavelength at the

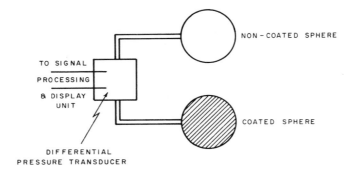

FIGURE 4.   Spherical bolometer E-field monitor.

highest frequency of operation of the sensor. Under this condition the effective length h_H of the loop antenna is

$$h_H = \frac{\pi^2 d^2 n}{2\lambda} \tag{9}$$

where n is the number of turns. The induced voltage is

$$V_H = -j\,\omega\mu_0 HnS \tag{10}$$

where $\omega = 2\pi f$, f is the frequency in hertz, μ_0 is the permeability of free space ($\mu_0 = 4\pi$ 10^{-7}H/m), H is the intensity of the H field perpendicular to the loop in ampere per meter, and S is the surface area of the loop in square meters. Through a proper resistive loading of the diode it is possible to obtain a relatively flat response of the probe over a relatively wide range of frequencies.[14,15] Three mutually perpendicular loops produce a nearly isotropic response of an H-field probe.

One of the problems that has to be considered for a loop, H-field sensor is its undesirable response to the E field. While a dipole does not respond to the H field, the situation is not that straightforward for the loop. For a plane-wave the ratio of the voltage induced by the E field to that induced by the H field is[16]

$$\frac{V_E}{V_H} \simeq -j2\pi\left(\frac{d}{\lambda}\right) \tag{11}$$

for a circular loop, and

$$\frac{V_E}{V_H} \simeq -j3\pi\left(\frac{w}{\lambda}\right)\frac{\Omega - 4.32}{\Omega - 3.17} \tag{12}$$

for a square loop of side length w and wire radius a, where $\Omega = 2 \ln (4w/a)$ is a wire thickness parameter. Equations 11 and 12 indicate that a circular loop is less sensitive than a square loop, and that the E-field sensitvity can be minimized by decreasing the size of the loop. The other possibility for minimizing the sensitivity of the H-field sensors to E fields is offered by the Moebius design.[17]

A probe equipped with three dipoles and three loops with zero-bias Schottky diode detectors was also developed.[15] Three orthogonal vector components of the H and E field can be

Table 6
SALIENT CHARACTERISTICS OF REPRESENTATIVE COMMERCIALLY AVAILABLE SPECTRUM ANALYZERS AND FIELD INTENSITY METERS

Manufacturer and model	Frequency Range	Frequency Resolution	Amplitude range
Spectrum Analyzers			
Hewlett-Packard, 8566A	100 Hz—40 GHz	10 Hz—3 MHz	−134 to +30 dBm
Hewlett-Packard, 8555A	10 MHz—40 GHz	100 Hz—3000 Hz	−127 to +10 dBm
Tektronix, 492	50 kHz—21 GHz	100 Hz—1 MHz	−123 to +40 dBm
Anritsu, MS62C	50 Hz—1700 MHz	30 Hz—100 kHz	−130 to +20 dBm
Polarad, 640B	3 MHz—40 GHz	300 Hz—1 MHz	−110 to +30 dBm
Field Strength Meters — Test Receivers			
Rohde & Schwarz, MSU	25—1000 MHz	100 Hz	−130 to 0 dBV
Rohde & Schwarz, HFH	0.1—30 MHz	100 Hz—4 kHz	−130 to −10 dBV
Scientific Atlanta, 1640APZ	20 MHz—32 GHz	1%	−100 to −40 dBm
Electro-Metrics, EMS-25	20 kHz—1 GHz	50 Hz—50 kHz	−160 to 0 dBV
Anritsu, M262E	500 kHz—30 MHz	4 kHz	−138 to 10 dBV/m
Anritsu, M518A	25 MHz—520 MHz	15 kHz	−130 to 10 dBV/m
Ailtech, NM-37/57A	30 MHz—1 GHz		−100 to 10 dBm

measured simultaneously and independently. The antenna elements are arranged to minimize the interactions between individual elements.

All types of nonperturbing field probes can be connected with the external circuitry by an optical-fiber telemetry system. This further minimizes perturbation of the surveyed field and is particularly useful in near-field measurements. The use of the optical-fibers rather than cables with metal conductors limits perturbation of the test field.[18,19] An additional advantage resulting from the use of optical fibers is the elimination of RF interference (RFI). Single, optical fiber kits, which include a transmitter and a receiver, are commercially available (Hewlett-Packard, Burr-Brown) and a simple end reliable system can be built at low cost.

4. Narrowband Instruments

The second category of instruments for measurment of external fields comprises narrowband, selectively tuned instruments such as spectrum analyzers and field intensity meters or panoramic receivers. Salient characteristics of a few commercially available instruments are given in Table 6, while a summary comparison between the broadband and narrowband instruments is outlined in Table 7. A typical narrowband instrument consists of two elements, a calibrated, usually broadband or tunable antenna and a measurement system capable of rather sophisticated processing of the received signals. The size of the linearly polarized antennas is usually comparable with the wavelength. An ominidirectional spatial response can be obtained by positioning the antennas along three mutually orthogonal directions and summing up the results, e.g., for the E field in cartesian coordinates

$$E = [E_x^2 + E_y^2 + E_z^2]^{1/2} \tag{13}$$

Table 7
COMPARISON BETWEEN BROADBAND (SURVEY METERS) AND
NARROWBAND (SPECTRUM ANALYZERS, TEST RECEIVERS) INSTRUMENTS

Characteristic	Broadband instruments	Narrowband instruments
Usable frequency range	Broad	Broad
Frequency resolution	None	Excellent
Modulation response	Limited	Excellent
Sensitivity	Moderate	Very high
Dynamic range	Moderate ($\sim$30 dB)	Very wide ($>$ 100 dB)
Spatial resolution	Excellent possible (usually $\sim$5 cm)	Usually poor (large antennas)
Near-field response	Accurate (ca. $\pm$ 1 dB)	Inaccurate, unless special small sensors used
Directional response	Isotropic	Linear or planar
Field perturbation	Minimal	Considerable, depends on sensor
Antenna size	Small	Comparable with wavelength
Portability	Excellent	Limited or poor

where E_x, E_y, and E_z are the field intensities measured with the antenna positioned along the x, y, and z directions, respectively. Directional antennas can also be assembled to form an isotropic response system. Several types of commercially available antennas, e.g., log-periodic antennas, provide a planar radiation pattern, i.e., giving genuine response for all polarizations in the plane of the antenna.

Special purpose antennas can also be designed and calibrated to provide desired characteristics for a specific task, e.g., a small-diameter, multiple-turn loop to measure the H-field emissions from a device.

Calibrated antennas can be used in conjunction with a spectrum analyzer or a field intensity meter. A spectrum analyzer provides visual, quantitative information on the spectral composition of the signal received by the antenna. The display is usually given on a cathode ray tube (CRT). Spectrum analyzers are capable of making absolute frequency and amplitude measurements, operate over a wide-amplitude dynamic range, have high frequency and amplitude resolution, and high sensitivity. They also provide means for observation, storage, and recording the output in a convenient form. More sophisticated signal analyzers capable of digital signal processing, i.e., Fourier analyzers, can be used in place of spectrum analyzers. Field strength meters or test receivers are tuneable, narrowband voltmeters. They do not provide information about modulation characteristics of the signal but give only the rms or peak value. However, information about the modulation characteristics can be retrieved through the analysis of the fequency spectrum.

Attempts have also been made to develop a personal dosimeter. The main goal of such a device is to record, as fully as possible, the exposure conditions of a person occupationally exposed to radiowaves. A fundamental difficulty in developing a useful dosimeter, apart from technical problems, is conceptual. A dosimeter, unless very complex, is only able of recording cumulative exposure, i.e., the total dose equal to the power density integrated over time, and cannot distinguish short-term, high-level from long-term, low-level exposures. However, biological effects and hazards are not a direct function of the total dose but the dose rate, i.e., exposure to 200 mW/cm² for 10 min is not equal to exposure to 1 mW/cm² for 2000 min. In all attempts to develop a personal dosimeter this fundamental problem has not been adequately addressed. Additional technical difficulties are posed by the frequency response of the dosimeter, the perturbation and partial shielding of the exposure field by the dosimeter wearer. Efforts were made to design a personal dosimeter whose sensor was to contain isotropic probes for E and H fields at frequencies from 2 to 300 MHz, and E-field probes at frequencies from 300 MHz to 23 GHz,[20] however due to technical difficulties this effort was discontinued. A description of other earlier, unsuccessful attempts is given elsewhere.[7]

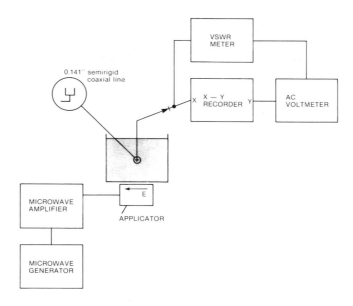

FIGURE 5.   Block diagram of the E-field mapping system using a monopole as a microwave probe.

B. Internal Fields

Since the present state of knowledge indicates that the thermal and nonthermal interactions at higher frequencies of electromagnetic fields with mammalian tissues are related to the intensity of the internal E field, only E-field sensors have been developed. There are two possibilities of measuring the intensities of the E field by direct means. The simplest way is to utilize a monopole or dipole antenna at the end of a coaxial line which is connected to a conventional detection and measuring system. This technique is effective when the direction of the E field is known, so that the coaxial line can be placed perpendicularly to the direction of the field, ensuring minimal perturbation to the field. This is frequently the case for applicators used in hyperthermia systems. A simple system for measuring the E field intensity produced by a waveguide applicator in a liquid tissue phantom is shown in Figure 5.[21] However, such probes are inadequate when the direction of the E field is not known, since an improperly oriented coaxial line introduces a large perturbation of the measured field.

A nonperturbing, implantable E field probe consists of a very short dipole and a miniature Schottky-barrier diobe connected by high-resistance leads with the external circuitry. The dipoles and the high-resistance leads are deposited on a dielectric substrate. Amplitude and frequency responses of such a probe placed in a lossy, dielectric medium have been investigated.[22-26] Two types of probes can be employed, a bare probe and an insulated probe, as illustrated in Figure 6. The bare probe lenght is 2h, and the diameter is 2a. The insulated probe of the same dimensions is surrounded by a cylindrical, dielectric sheath having the dielectric permittivity ϵ_i. The dielectric medium around the probe is assumed homogenous within the probe-sampling volume.[22] For a short probe the sampling volume is approximately equal to a sphere of a radius h centered on the probe.[22] The equivalent circuits of the probes are shown in Figure 7.[22] The input admittance of the bare probe in any dissipative medium is[22]

$$Y = G + j\omega C \frac{\pi h}{\ln(h/a - 1)} (\sigma + j\omega\epsilon) \tag{14}$$

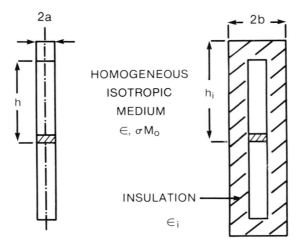

FIGURE 6. Bare and insulated cylindrical, dipole E-field probes.

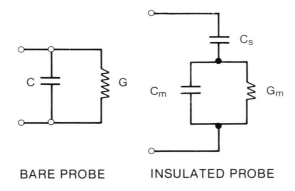

FIGURE 7. Equivalent circuits of electrically short bare and insulated E-field probes.[22]

where ϵ is the dielectric permittivity of the biological tissue surrounding the probe and σ is the conductivity of the tissue ($\sigma = \omega\epsilon_0\epsilon''$). Equation 14 is valid for a thin (a $\ll$ h) and electrically short probe ($\alpha h \ll 1$ and $\beta h \ll 1$, where α and β are the attenuation and propagation constants, respectively). The input admittance is therefore dependent on the permittivity of the surrounding medium.

The input admittance of the insulated probe is[22]

$$Y = \frac{\sigma\pi h}{\ln(h/b) - 1} \frac{\gamma^2}{p^2 + (1 + \gamma)^2} + j\omega \frac{\epsilon_i\pi h}{\ln(b/a)} \frac{p^2 + (1 + \gamma)}{p^2 + (1 + \gamma)^2} \quad (15)$$

where

$$\gamma = \frac{\epsilon_i}{\epsilon} \frac{\ln(h/b) - 1}{\ln(b/a)} \quad (16)$$

and

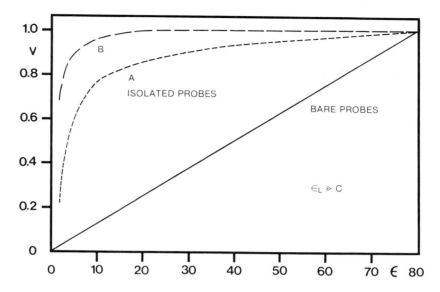

FIGURE 8. Normalized output voltage of bare and insulated probes as a function of relative, dielectric constant of the medium. The probes dimensions are (Probe A) a = 0.10 mm, b = 0.34 mm; (Probe B) a = 0.14 mm, b = 1.34 mm. For both probes h = 5.56 mm, h_i = 7.14 mm, and ϵ_i = 2.1 (Teflon®).[22]

$$p = \frac{\sigma}{\omega\epsilon} = \frac{\epsilon''}{\epsilon'} \tag{17}$$

For ϵ_i small compared with ϵ, and a sufficient thickness of the dielectric sheath, the admittance can be approximated by[22]

$$Y = j\omega \frac{\epsilon_i \pi h}{\ln(b/a)} \tag{18}$$

and is therefore, independent of the dielectric properties of the medium in which the probe is immersed, but only on the dielectric permittivity ϵ_i of the probe insulation.

The voltage across the load admittance for the bare probe is[22]

$$V = -hE \frac{Y}{Y + Y_L} \tag{19}$$

where E is the intensity of the E field parallel to the dipole and Y_L is the load admittance. For the voltage to be independent of the medium permittivity the load admittance must be much less than the probe input admittance, i.e.,

$$|Y_L| \ll |Y| \tag{20}$$

for the permittivities of all measured media. However, the input admittance can be very small,[22] making the condition (Equation 20) difficult to satisfy. Therefore, only a properly insulated probe is capable of providing a signal proportional to the E field and independent of the dielectric properties of the test medium. Experimental results for a bare probe and the same probe with two dielectric sheaths of various thicknesses are shown in Figure 8.[22]

Table 8
SALIENT CHARACTERISTICS OF COMMERCIALLY AVAILABLE, TISSUE-IMPLANTABLE E-FIELD PROBES

Manufacturer	Narda	EIT	Holaday
Model	26088	979	IME-01
Recommended instrument	—	—	HI-3003
Frequency range (MHz)	Not specified	100—12,400 MHz	5—1,000 MHz
Isotropic/single axis	Isotropic	Isotropic ±2 dB	Isotropic ±0.5 dB
Isotropic response (dB)			
Output connector	Microtech EP-75-1 minicon	Microtech EP-75-1 minicon	Supplied with cable
Output cable	—	—	Up to 6 m
Sensitivity in free space	1 mV/mW/cm^{2a}	4 mV/mW/cm^{2a}	0.2 mW/cm^{2b}
Dimensions (diameter, length)	DIA = 3.3 mm^c	DIA = 10 mm	DIA = 20.3 mm
	L = 276 mm	L = 300 mm	L = 432 mm

[a]   At 2,450 MHz.
[b]   Full scale.
[c]   The tip only (25-mm long).

The normalized voltage increases linearly with the medium, dielectric constant for the bare probe, but is practically independent of the medium, dielectric constant (for values > 10) for the probe in the thick insulation. In all cases the load admittance is much greater than the probe input addmittance ($Y_L \gg Y$).

Insulation also reduces the error due to proximity of an interface. For instance, the error is < 5% for the interface at a distance of h/3 from the probe.[24]

The reception of the incident signal by a parallel wire transmission line formed by the resistive leads was compared with that of the dipole.[26] The leads distort the field pattern somewhat, but for high-resistance leads the distortion is small. High resistance also reduces scattering of the incident wave. The highly resistive transmission line behaves as a low-pass filter, limiting applications of the probe to the fields modulated at frequencies much lower than f_m[26]

$$f_m = \frac{1}{8\pi crs^2} \qquad (21)$$

where c is the capacitance per unit length, r is the resistance per unit length of the thin-film leads, and s is the length of the leads.

Single- and three-dipole (isotropic) probes were designed and tested[18,27] and a few models are commercially available. Essential characteristics of commercially available probes are listed in Table 8.

A passive, remote-sensing method for measurements of the internal fields has also been proposed.[28] In this method, miniature, insulated diodes are imbedded in the tissue. When irradiated with two signals at different frequencies, diode nonlinearities generate intermodulation products whose amplitudes are related to the intensities of the irradiating fields. This method, the feasibiltiy of which was established,[28] does not seem to have any advantages over implantable E-field probes.

A summary of E-field probes and associated detector circuits for biological applications between 0.2 MHz and 26 GHz and field strengths from < 1 to 1000 V/m has been published recently.[28A]

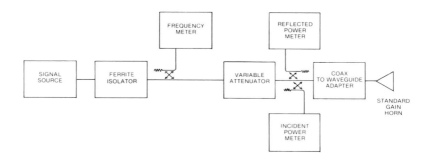

FIGURE 9.   Experimental arrangement for the free-space, standard field method of calibration.

C. Calibration Techniques

Probes for measurements of external and internal fields have to be calibrated over the whole frequency and amplitude range of operation. The calibration consists of placing the probe in a well-defined field of known intensity and taking output readings. While external field probes are calibrated in air, internal field probes should be calibrated in the tissue-equivalent dielectric, i.e., a dielectric having the complex permittivity close to that of tissues at the frequencies of interest.

To calibrate external field probes, three types of techniques can be used: free space, standard field methods, guided wave methods, and transfer probe methods.[7,29]

1. Free Space Methods

In the free space methods, a plane-wave field of a known power density is produced in free space either in an open range or in an anechoic chamber. A basic block diagram of the arrangement is shown in Figure 9. The power density S at the axis at a distance r from a standard gain horn in the far-field zone is given by

$$S = \frac{P_T G}{4\pi r^2} \tag{22}$$

where P_T is the power transmitted by the standard horn and G is the horn gain. The transmitted power is determined from measuring the forward and reflected powers. The dual directional coupler (Figure 9) has to be calibrated. The standard horn gain can be determined experimentally[30] or calcualted.[31-33] Since the distance between the horn and the calibrated probe is frequently too short for far-field operation, near-field corrections have to be included.[32,33] Operation in the far field is frequently precluded because of the limited size of the anechoic chamber or limited power of the signal sources to produce the required power density. There are several methods of calculating the gain of pyramidal horns in the far and near field.[29,31,32] The simplest is given here.[33] For a horn having dimensions as shown in Figure 10, the gain is equal to[33]

$$G = 10 \log (AB) + 10.08 - R_H - R_E \tag{23}$$

where G is gain in decibels R_H and R_E are the near-field reduction factors due to the H- and E-plane flare of the horn, respectively, and

$$A = \frac{a}{\lambda} , \ B = \frac{b}{\lambda} , \ L_H = \frac{\ell_H}{\lambda} , \ L_E = \frac{\ell_E}{\lambda} , \ R = \frac{r}{\lambda} \tag{24}$$

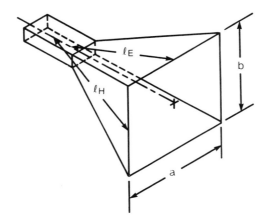

FIGURE 10. Pyramidal horn, designation of dimensions.

where a, b, ℓ_H, and ℓ_E are horn dimensions shown in Figure 10, λ is the wavelength, and r the distance from the horn.

$$R_H = (0.01\alpha) (1 + 10.19\alpha + 0.51\alpha^2 - 0.097\alpha^3) \tag{25}$$

$$R_E = (0.1\beta^2) (2.31 + 0.053\beta) \tag{26}$$

where

$$\alpha = A^2 \left(\frac{1}{L_H} + \frac{1}{R} \right) \tag{27}$$

$$\beta = B^2 \left(\frac{1}{L_E} + \frac{1}{R} \right) \tag{28}$$

The gains calculated from these simple equations are within 0.05 dB of the accurate computations[31] for R_H and $R_E < 5$ dB, $R > L_H$, and $R > L_E$.[33]

In addition to the uncertainty in the gain and the transmitted power, multipath interference is another source of error in the determination of the standard, free-space power density (Equation 22). More information on the free-space method can be found in References 29 and 33 to 35. At lower frequencies an open-ended waveguide radiator can be used instead of a standard gain horn.

2. Guided Wave Methods

The E-field sensors can be calibrated in guided structures such as rectangular waveguides and transverse electromagnetic (TEM) cells. These methods are only useful when the dimensions of the sensor are small compared with the cross-section of the structure. Furthermore, the calibration guide has to be terminated by a matched load, so that the standing waves are minimized.

In a rectangular waveguide operating in the fundamental TE_{10} mode the E-field intensity (rms value) at the center is equal to

$$E = \left(\frac{2PZ}{ab}\right)^{1/2} \tag{29}$$

where P is the power delivered to the waveguide, a and b are the dimensions of the waveguide, and

$$Z = \frac{Z_0}{[1 - (\lambda/2a)^2]^{1/2}} \tag{30}$$

($Z_0 = 377\Omega$ and λ is the free space wavelength). The power density at the center of the waveguide is equal to

$$S = \frac{E^2}{Z} = \frac{2P}{ab} \tag{31}$$

The E-field intensity varies cosinusoidally along the wider dimension a of the rectangular waveguide.

A TEM cell can also be conveniently used for calibration of the exteranl field probes. The TEM cell consists of a section of the rectangular, coaxial transmission line tapered on both ends to converge into a standard, coaxial line. The TEM cell can support various propagation modes,[36-38] the lowest being the TEM mode. When only the TEM mode is propagating (for the conditions that have to be satisfied see References 36 to 38), the intensity of the E field midway between the center and the outer conductor of the cell is equal to [33]

$$E = \frac{(ZP)^{1/2}}{D} \tag{32}$$

where Z is the characteristic impedance of the cell (usually 50 Ω), P is the power transmitted through the cell, and the D is the separation distance between the center conductor and the outer conductor of the cell. A standing wave in the cell can cause considerable calibration errors.

3. The Standard Probe Method

In the standard probe method, a probe to be calibrated is compared with an accurately calibrated "transfer standard".[29] The only requirement is that the field used for calibration be of a desired magnitude, sufficiently uniform over the calibration region and that it remain constant during the time required for calibration.

4. Implantable Probes

Since as indicated earlier, the output voltage of an implantable probe depends on the dielectric properties of the medium surrounding the probe, calibration has to be performed with the test probe in the environment simulating the actual conditions of use. Most implantable probes have a sufficiently thick dielectric insulation around the dipole, so that calibration in one medium having a dielectric constant between 10 and 80 is usually sufficient. Of course, calibration has to be performed at all frequencies of interest, or at a sufficient number of frequencies to obtain the full frequency characteristic of the probe through extrapolation. Various mixtures of electric materials (phantom materials) have been developed to simulate the electrical and thermal properties of biological tissues, as described in Section IV. These mixtures are utilized in calibrating, implantable probes. Basically, all three calibration methods that are used for free-space probes can be utilized for implantable probes when free space is replaced by the phantom material.

The simplest method of calibration utilizes an anechoic chamber and a semi-infinite slab of a tissue phantom material.[39] The E-field intensity in a sufficiently large slab (such that it can be considered infinite in all three dimensions) exposed to a plane-wave of known intensity can be calculated. If E_0 is the tangential E-field intensity at the interface between air and the dielectric slab, the field intensity along the direction of wave propagation is

$$E_z = E_0 e^{-\alpha z} \tag{33}$$

where z is the distance from the interface and α the attenuation constant.

$$\alpha = \frac{2\pi f}{c} (\epsilon_r/2)^{1/2} \left\{ [1 + (\sigma/2\pi f \epsilon_r \epsilon_0)^2]^{1/2} - 1 \right\}^{1/2} \quad (N/m) \tag{34}$$

where f is the frequency, c is the velocity of light in vacuum, ϵ_r is the relative dielectric constant, σ is the conductivity of the dielectric slab, and ϵ_0 is the permittivity of free space. This method is suitable at relatively high frequencies (above 2 GHz), so that the size of the slab is easily manageable. Also a relatively high power source is needed to provide sufficient power density.

In another calibration technique the probe is moved along the axis of a sphere made of a phantom material and irradiated by a plane-wave of known intensity.[27] The E-field intensity in the sphere can be accurately calculated. To avoid interference from the reflected and scattered fields, the calibration should be performed in an anechoic chamber.

A waveguide method can also be used to calibrate implantable probes.[40] A section of a rectangular waveguide is filled with a liquid phantom material and separated by a very thin ($< 0.01\lambda$) plastic window from the air-filled part of the waveguide. The dielectric constant of the window material should be low, so that the reflections from the window are minimal. The length of the dielectric-filled waveguide should be sufficient to simulate an infinitely long sample. The waveguide operates in the fundamental TE_{10} mode. The probe being calibrated is moved along the center of the waveguide axis in such a way that the dipole is parallel to the E-field direction or remains at a constant known angle (the angle must not be 90°). For the probe in the air section, the output voltage is related to the field intensity as[40]

$$V(air) = B_A E^2(air) \tag{35}$$

and in the dielectric

$$V(diel) = B_E E^2(diel) \tag{36}$$

At the interface

$$E(diel) = E(air) \tag{37}$$

and

$$\frac{B_E}{B_A} = \frac{V(diel)}{V(air)} \tag{38}$$

where B_E is the probe calibration factor in the test dielectric phantom material and B_A the probe calibration factor in air. By measuring V as a function of position on each side of the

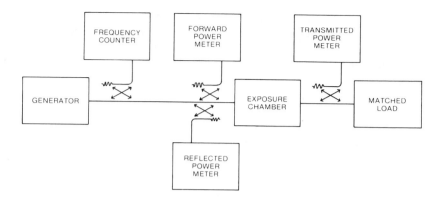

FIGURE 11. Experimental arrangement for measurement of the average SAR.

thin window and extrapolating the results to the interface, the ratio of the calibration coefficients can be determined. The calibration factor in the dielectric material can be then calculated by measuring the calibration factor in air as outlined in Section II.C.2. The dielectric window should be very thin and of low-permittivity material ($\epsilon \simeq 2$ to 3). A window thickness of 1 mm was found to be acceptable at 2.45 GHz.[40]

D. Absorbed Power Measurements

The average SAR can be determined by measurement of the total absorbed power in an exposed object, e.g., a model or an experimental animal. This technique is suitable only for exposures in electrically enclosed exposure chambers such as a TEM cell, waveguide, or resonant cavity. A simplified block diagram of the measurement system is shown in Figure 11.

The absorbed power can be determined as

$$P_A = P_F - P_R - P_T \tag{39}$$

where P_A is the absorbed power and P_F, P_R, and P_T are the measured forward, reflected, and transmitted powers, respectively. If the exposure chamber is not of a transmission type, e.g., a one-port cavity, Equation 39 simplifies to

$$P_A = P_F - P_R \tag{40}$$

As the absorption by an exposed animal changes as the animal moves, it is advisable to monitor and record the power levels during the whole exposure period, then to calculate the total absorbed energy from which the average absorbed power and the average SAR can be calculated. Several systems for this purpose have been described elsewhere.[41-46]

Equations 39 and 40 are valid provided that the losses (power absorbed) in the empty exposure chamber are negligible compared with the losses in the exposed object. If this condition is not satisfied, the power absorbed in the empty exposure chamber or the exposure chamber containing the animal cage has to be measured, e.g., using the same method of measuring the forward, reflected, and transmitted powers.

The corrected absorbed power is equal to

$$P_A = P_A - P_E \tag{41}$$

where P_E is the power absorbed in the exposure chamber, animal-restraining devices, etc.

Some errors can be associated with use of animal-restraining devices, since the power absorbed in them may be different in the presence of the animal due to the perturbation of the exposure field.

A correction to Equation 39 may also have to be introduced, if the load is not matched and significant reflections occur[46]

$$P_A = P_F - P_R - (1 - |\Gamma_L|^2)P_T \qquad (42)$$

where $|\Gamma_L|$ is the modulus of the reflection coefficient of the load.

III. THERMAL MEASUREMENTS

A. Basic Principles

The rate of temperature change in the subcutaneous tissue in vivo exposed to RF or microwave energy is related to the SAR, as[1]

$$\frac{\Delta T}{\Delta t} = (SAR + P_m - P_c - P_b)/C \qquad (43)$$

where ΔT is the temperature increase, Δt is the exposure duration, P_m is the metabolic heating rate, P_c is the rate of heat loss per unit volume due to thermal conduction, P_b is the rate of heat loss per unit volume due to blood flow, and C is the specific heat. If before the exposure a steady-state condition exists, i.e.,[1]

$$P_m = P_c + P_b \qquad (44)$$

during the initial period of exposure

$$\frac{\Delta T}{\Delta t} = \frac{SAR}{C} \qquad (45)$$

and the SAR can be determined from measurements of an increase in the tissue temperature over a short period of time following the exposure. For tissue phantoms and tissues in vitro, Equation 45 can be used as long as the thermal conductivity can be neglected, i.e., for short durations of exposure. Several methods of SAR determination are based on thermal measurements and utilization of Equation 45. The specific heat and density of various tissues is summarized in Table 9.[47]

B. Calorimetry

All calorimetric methods are suitable for determination of average SAR, but do not allow determination of the SAR distribution (i.e., variation in SAR from position to position in the irradiated body).

1. Analysis of Heating and Cooling Data

Heating and cooling data can be analyzed to estimate the energy absorbed by an irradiated sample.[48,49] For a system of mass m, specific heat C, and initial temperature T_0 heated at a constant rate from time t = 0 to t = τ, the temperature at any time reflects a balance between input and loss of heat in the heating period[48]

$$\frac{\partial T}{\partial t} = A - k\Delta T \qquad for \qquad 0 \leq t \leq \tau \qquad (46)$$

Table 9
TISSUE-SPECIFIC HEAT CAPACITY
AND DENSITY[47]

Tissue	Specific heat capacity (J/kg°C)	Density (kg/m³)
Skeletal muscle	3470	1070
Fat	2260	940
Bone, cortical	1260	1790
Bone, spongy	2970	1250
Blood	3890	1060

where $\partial T/\partial t$ is the heating rate, A is the rate of temperature increase in the absence of heat losses, and k is the proportionality constant $\Delta T = T - T_0$. In the cooling period

$$\frac{\partial T}{\partial t} = -k\Delta T \qquad t \geq \tau \tag{47}$$

If the system is absorbing energy at a constant rate W then

$$W = mCA \tag{48}$$

Integration of Equations 46 and 47 yields

$$\Delta T = A/k(1 - e^{-kt}) \qquad 0 \leq t \leq \tau \tag{49}$$

and

$$\Delta T = \Delta T_\tau e^{-k(t-\tau)} \qquad t \geq \tau \tag{50}$$

where ΔT_τ is the temperature difference at the termination of heating. To calculate W (Equation 48), the constant k is first determined from the slope of the cooling curve (Equation 49). This is done by finding the slope of ln ΔT vs. time for $t > \tau$ (e.g., by a linear regression analysis). Then A can be found from the slope of ΔT vs. e^{-kt} (Equation 49). Alternatively,[47] if the temperature of the system reaches a steady state before the heating is terminated, the rate of energy input equals the rate of energy loss, and Equation 49 becomes

$$\lim_{t \to \infty} \Delta T = A/k = \Delta T_{ss} \tag{51}$$

where ΔT_{ss} denotes the steady-state temperature increase. From Equations 48 and 51,

$$W = mCk\Delta T_{ss} \tag{52}$$

In this case k and ΔT_{ss} can be obtained directly from the data, i.e., the temperature vs. time.
Another method of calculation is based on combining Equations 49 and 51 to obtain

$$\ln(\Delta T_{ss} - \Delta T) = -kt + \ln \Delta T_{ss} \qquad t \leq \tau \tag{53}$$

where k is the slope and ΔT_{ss} is the intercept of $\ln(\Delta T_{ss} - \Delta T)$ vs. time.
One important advantage of the heating and cooling curve analysis is that this method

does not require calibration. The method is particularly suited for in vitro measurements. The analysis is difficult to apply to in vivo experiments because of the animal thermo-regulatory response to electromagnetic heating. A major limitation for both in vivo and in vitro experiments arises when the energy deposition in the exposed object is nonuniform while the temperature is measured only in few (frequently only one) locations. To limit the amount of power required to achieve a ΔT of a few degrees, the sample may be enclosed in Styrofoam® insulation.[48] In summary, the heating/cooling curve method is simple and can provide accurate results when carefully applied to samples/animals that are exposed under such conditions that energy deposition through the whole volume is relatively uniform and no thermoregulatory response is evoked by the exposure. To obtain reliable temperature measurements, only probes suitable for operation in RF fields should be used.

2. Dewar-Flask Calorimeter

A typical calorimeter contains a vacuum (Dewar) flask containing a measured amount of water and sealed with a cork stopper.[50] The water is stirred by a magnetic stirrer to minimize time to reach thermal equilibrium. Optionally, insulation may be added on the outside of the flask.[51] The temperature of water is measured with a thermistor and recorded with a resolution of 0.01°C.[50]

The SAR (W/kg) is determined as[52]

$$SAR = C_S[T_e \text{ (exposed)} - T_e \text{ (control)}]/t \tag{54}$$

where t is the exposure time, C_S is the specific heat capacity of the exposed sample (e.g., for a rat or mouse $C_S = 3448$ J/kg °C), and

$$T_e = [(Z_p + M_eC_e)(T_F - T_i)]/(M_SC_S) + T_F \tag{55}$$

where T_e is the temperature (°C) of the exposed sample upon insertion in the calorimeter, M_S is the mass of the sample (kg), M_e is the mass of water in the calorimeter (kg), C_e is the specific heat capacity of water ($C_e = 4185$ J/kg °C), T_i is the temperature of the calorimeter just before insertion of the sample (°C), T_F is the final temperature of the calorimeter (°C), and Z_p is the heat capacity of the calorimeter (J/°C).

Several precautions should be followed to obtain accurate results,[52] namely, there should be enough water just to allow total immersion of the exposed sample (animal cadaver) and the water temperature should be adjusted to approximately 0.5°C below the ambient air temperature. Irradiation time should be adjusted so that the resultant temperature following the insertion of the test sample is about 1°C higher. This limits the calorimeter drift. It is assumed that the thermal equilibrium is reached when the change in temperature is not > 0.01°C.

The heat capacity of the Dewar flask can be found from Equation 55 by adding a known amount of water at a known temperature.[50] Similarly, the specific heat capacity of an unknown sample can be obtained from Equation 55 when T_e is known.[50]

The Dewar calorimeter is a relatively simple and inexpensive instrument which can easily be calibrated and requires a relatively short time for one measurement (about 1 hr for a rat).[52] When the experiment is carefully performed, accuracy of a few percent can be achieved. Nonuniformity of energy deposition does not affect the accuracy of the average SAR obtained by this method. Liquids other than water can also be used, e.g., dodecane ($C_S = 2100$ J/kg °C).[50]

3. Twin-Well Calorimeter

A twin-well or differential calorimeter[2,53] consists of two identical metal cylinders large

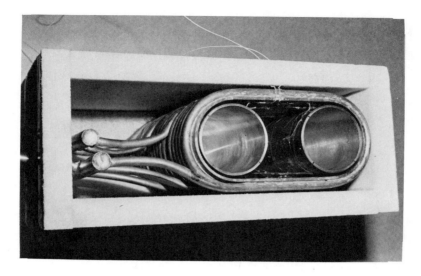

FIGURE 12.   Twin-well calorimeter (photo courtesy of Bioelectromagnetics Laboratory, University of Washington, Seattle, Wash.).

enough to hold a particular animal of interest (different sizes are used for different animals). Each cylinder is surrounded by a thermopile, an array of thermocouples connected in series (voltages additive).[2] The thermopiles of the two cylinders are connected in opposite polarities so that the voltages are subtracted. Therefore, when both cylinders are at the same temperature, the resulting output voltage from the two thermopiles is zero. The twin cylindrical wells are surrounded by a cylindrical heat exchanger carrying a circulating fluid. The exchanger is covered by an insulating jacket, as shown in Figure 12. The measurement procedure[53] consists of placing a pair of freshly killed animals of the same body weight into the calorimeter and equilibrating their temperatures — this usually takes about 2.5 hr.[2] One animal is then exposed briefly to RF energy, and the second serves as a reference. Immediately after treatment the animals are returned to the calorimeter and the heat added to the animal's body by irradiation is measured. The heat is determined by integrating the output voltage from the thermopiles over time (usually about 10 to 12 hr[2]). The system can be calibrated with a known amount of ice at 0°C or by adding a known amount of heat to one of the two identical thermal bodies in the calorimeter.[2] The use of a microprocessor greatly improves the accuracy of the integration over the long period of time (several hours) required to equilibrate the two wells.

The twin-well calorimeter is more accurate than the Dewar-flask calorimeter, its major drawback being the relatively long time required for one measurement.

C. Thermometry
1. Basic Concepts

Various physical principles can form a basis for temperature measurements. One of the major conditions for a thermometer to be suitable for temperature measurements under certain environmental conditions, e.g., RF exposure, is that it should give a true response to temperature but not other environmental conditions, e.g., the intensity of the exposure field. Some of the most popular thermometers, e.g., mercury-in-glass and platinum resistance, are not suitable for use in electromagnetic fields, while others, e.g., thermocouple or thermistor, can be used after special modifications. The salient characteristics of thermometers that should be considered in selecting one for a specific purpose are summarized in Table 10. Various types of popular thermometers were reviewed elsewhere.[54,55]

Table 10
SALIENT CHARACTERISTICS OF THERMOMETERS

Capability	Description
Sensitivity	Minimum ΔT detectable
Accuracy	Difference between the indicated and true temperature
Stability	Drift of accuracy over time
Response time	Time to reach 67% of the final reading when subjected to a step change in temperature
Size	Size determines the spatial resolution
Indication	Type of temperature indication: direct-indirect (e.g., tables); analog-digital; continuous-intermittent

2. Conventional Thermometer Probes

The difficulty of measuring temperature in electromagnetic fields with many conventional thermometers stems from three types of interaction between the thermometer and the field.[54] They are electromagnetic interference (EMI) (basically pick-up by the leads), direct heating of the temperature sensor, and perturbation of the field by the thermometer. Several techniques can be used to minimize these effects.[54] The interference and pick-up can be minimized by placing the leads of the sensor, e.g., a thermocouple or thermistor, perpendicular to the E field. Magnetic induction pick-up is reduced when the leads are tightly twisted. Several designs have been proposed for thermocouple and thermistor thermometers for use in electromagnetic fields.[56-61]

A stainless steel hypodermic needle containing a thermistor[57] can be used to measure temperature in saline solutions simulating tissue. All leads and connectors not immersed in the solution should be shielded from microwave radiation. The thermistor probe should be located beneath the surface of the solution. Ultraminiature (125-μm diameter), glass-coated thermistors can also be used for temperature monitoring in vivo, e.g., during heating.[60] The thermistor beads are bound to high-resistance ($>$ 164 kΩ/m) carbon-impregnated, flexible Teflon® leads.[60]

In some applications even a carefully designed thermocouple thermometer can provide satisfactory results. A small size of the thermocouple wires (0.05 mm in diameter) is a minimum requirement.[59] The error related to the current flow through the thermocouple junction due to the unequal resistance of the two wires can be minimized by a filter.[58] A 0.1-mm diameter thermocouple in a metal shield and a Teflon® jacket connected through two choke-coils with a shielded digital thermometer was successfully used for monitoring hyperthermia induced by a RF field (at 13.5 MHz).[61] However, use of thermocouples and thermistors in RF fields is subject to interference from many sources and is possible only in certain situations. Unless a sensor is carefully designed and thoroughly tested, serious errors in measurements can arise. As an alternative, thermocouples and thermistors can be used and inserted into the test site while the RF field is turned off for the duration of the temperature measurement. Before the field is turned on the thermometer has to be withdrawn. Usually special insertion guides are employed.[2]

3. Nonperturbing Temperature Probes

Several types of nonperturbing temperature probes have been developed,[62-75] however only few are available commercially.

A small, high-resistance thermistor with high-resistance leads[65] was developed into a commercial product manufactured by Vitek, Inc. The thermistor resistance is sensed by injecting a constant current and measuring the voltage across the thermistor by means of a high-impedance amplifier.[65]

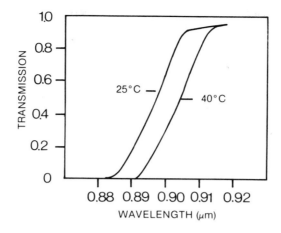

FIGURE 13.   Optical transmission through GaAs at two temperatures.[72]

Several optical temperature sensors take advantage of the fact that optical fibers, being low-loss dielectrics, are neither directly heated by an electromagnetic field, nor do they perturb the field. In the liquid-crystal, optical-fiber probe a strong temperature-dependence of the reflectance of red light by a liquid crystal is utilized.[62-64,71] The liquid crystal medium is sealed between a glass lens and a fiberoptic catheter. The catheter contains eight optical fibers in Teflon® tubing. Light (670 nm) emitted by a light-emitting diode (LED) is transmitted by four fibers to the sensor tip, and the remaining fibers receive light scattered by the sensor and carry it to a photodetector. The photodetector converts light into voltage. The actual system is somewhat more complicated as pulse modulation is employed. An increase in the sensor temperature results in an increased reflectance from the crystal and subsequently an increase in the detected voltage. By adjusting the crystal composition its useful range of operation (about 10°C) can be changed. Some problems were encountered with the time stability of the probe.[71]

Reflection of a light beam from a thermodilatable liquid can also be utilized to measure the temperature.[66-68] The shape and position of the liquid meniscus vary with temperature. The light beam is carried by optical fibers.

In the birefringent crystal thermometer a solid, single crystal of lithium tantalate is used as a sensor.[70] The principle of operation is based on the temperature-dependence of the difference in the indices of refraction along the two optical axes. The optical and electronic systems can then be similar to those used in other optical-fiber systems. To further improve the performance, a reference signal from the LED can be applied to the photodetector. The solid crystal sensor is more time-stable than liquid crystal sensors.

A very small (0.25 mm) and inherently stable sensor is based on the semiconductor band edge shift.[72] In semiconductor materials a forbidden zone separates the allowed electron energy levels with a typical energy spacing that corresponds to the quantum energy of infrared or visible light, e.g., for GaAs $\Delta E = 1.43$ eV, which corresponds to a wavelength of 0.87 μm at room temperature. At temperatures of interest in biological applications, the thermal energy is very small compared with the gap energy, so practically all electrons occupy the lower (valence) band, and the upper (conduction) band is empty. Light of a wavelength corresponding to the gap energy excites the electrons and is absorbed (Planck law). A typical dependence of the light transmission vs. wavelength is shown in Figure 13. It is evident that for a properly selected wavelength the absorption is strongly dependent on the temperature.

Table 11
CHARACTERISTICS OF NONPERTURBING THERMOMETERS

Type	Dynamic range (°C)	Accuracy (°C)	Diameter (mm)	Ref.
Thermistor[a]	30	0.1	0.94	65
Thermistor[b]	25	0.15	0.91	76
Liquid crystal	15	0.1	2	62—64,71
Thermodilable liquid	40	0.5	2	66—68
Birefringent crystal	30	0.3	1.5	70
GaAs semiconductor	25	0.1	0.25	72
Phosphor[c]	250	0.1	0.5	73, 74

[a] Available from Vitec Inc.
[b] Available from Narda Corp.
[c] Available from Luxtron Corp.

Properties of rare earth phosphors and fiber optics are utilized in another temperature monitoring system[73,74] that is manufactured by Luxtron Corp. The principle of operation is based on emissions by the phosphors of well-defined fluorescent lines, whose ratio of intensities depends on the temperature of the phosphor only. The ratio of intensities does not depend on amount of phosphor, level of illumination, or nature of optical coupling, and is not in any way affected by electromagnetic fields. Therefore, the temperature of the phosphor can be determined by measuring relative intensities of two colors from the glow of a very small amount of phosphor. The probes can be very small (0.2 mm) and have a wide range of temperature sensitivity (100°C), and modern electronics utilizing a microprocessor provide for accurate and convenient manual or automatic measurements.

Various methods[71,76,77] can be employed for evaluating the performance of a temperature probe for use in microwave fields, and some probes have been evaluated. Basic characteristics of nonperturbing thermometers are summarized in Table 11.[1,76]

D. Thermography

Infrared thermography technique for determination of temperature and SAR in models of biological bodies and animal cadavers exposed to strong electromagnetic fields was first developed by Guy[78] and has gained wide acceptance.[1,75,77,79-83] Thermographic cameras are commercially available and are used in various fields. A thermographic camera gives temperature distribution on a surface with a capability of spatially resolving the temperature gradients over any selected line within the limits of the camera resolution. A typical camera consists of an infrared radiation detector, usually a liquid-nitrogen-cooled chip of indium antimonide or mercury (cadmium) telluride, an optical scanning system, and a synchronized display unit.[81] Many cameras are also computer-compatible.

The response I(T) of a thermographic camera to the incident radiation can be represented by[81]

$$I(T) = \int_{\lambda_{min}}^{\lambda_{max}} p(\lambda,T)r(\lambda)d\lambda \qquad (56)$$

where T is the temperature, λ is the light wavelength, $p(\lambda,T)$ is a function describing the spectral radiation characteristics of the source, and $r(\lambda)$ is a function characterizing the spectral response of the instrument, including the detector characteristic, attenuation by optical elements, and electronic gain factors. In practice, I(T) is determined from a calibration of the system against black-body sources at a known temperature. For a gray source, the emittance characteristics of the source must be determined experimentally.[81]

To calibrate the camera, the differential characteristic, i.e., dI/dT as a function of temperature is determined by measuring finite differences $[I(T_1) - I(T_2)] / (T_1 - T_2)$ for various temperatures. Then, the average temperature $T_{av} = (T_1 + T_2)/2$ is used as the independent variable in the calibration.[81] Most thermographic cameras have calibration sources at various reference temperatures.

The optical properties of the material on which temperature profile is being measured with a thermographic camera affect the measurement accuracy. These are the emittance, reflectance, transmittance, and penetration depth.[81] The reflectance and emittance are interrelated, and for opaque objects their sum is = 1. The transmittance and penetration depth are similarly related. The penetration depth is defined as the depth in the medium from which 1/e of the radiation emitted by an elemental volume in the medium escapes to the surface. The penetration depth is of importance for materials with nonuniform temperature distribution ''in depth''. A nonuniformly heated transparent material which is hotter inside than near the surface appears warmer than a similarly heated opaque material. Tissue phantom materials developed for bioelectromagnetics studies were tested,[81] and for thicknesses > 0.4 mm no differences within the noise of 0.2°C were found (it is interesting to note that the muscle equivalent material 0.4 mm thick was visually, but not thermally, translucent). On the other hand, the transmittance of 0.05-mm thick polyethylene is 0.9, but its effect on determining the temperature distribution of the material underneath is negligible.[81]

The emittance of a surface is the ratio of the radiation flux per unit area of the emitter to that of a black-body radiator at the same temperature and under the same conditions. The emittance depends on the material, the surface smoothness, and the presence of oils, moisture, and dust on the surface. The emittance of 2.45-GHz bone equivalent material is about 0.8 to 0.9, and for muscle phantom very close to that of water, i.e., 0.98.[81] the emittance has to be taken into account when information about absolute temperature is needed.

The major limitations in thermographic measurements are the spatial resolution of the camera (the source has to be larger than the minimum surface area resolved) and the oblique angles of viewing (more than 30°).

When a thermographic camera is used to determine the temperature distribution or SAR in tissue-equivalent models or animal cadavers exposed to RF radiation fields, the following procedure is employed.[78] A model of an animal or other object is made of tissue-equivalent materials and cut along planes whose two-dimensional SAR (temperature) distribution is to be determined. To avoid water loss from the material, the open surfaces are covered with a very thin (0.05 mm or less) polyethylene film or silk screen. An animal is killed, frozen in dry ice, cast in a Styrofoam® block, bisected, covered with polyethylene film, reassembled, and equilibrated to room temperature.[2] The model or cadaver is then exposed under specific test conditions for a few seconds (preferably < 20) to a high-intensity field. The parts are separated and thermographic scans are taken. To obtain temperature or SAR information undistorted by thermal conduction the exposure time and the delay between taking thermographic scans must be minimized. The delay and the recording time should not be longer than 10 sec. High-power RF sources allow exposure times of < 20 sec. Large temperature gradients should also be avoided, as they increase thermal conduction. Gradients of 5 to 10°C appear reasonable. If the thermal conduction is neglected, the SAR can be calculated from the measured temperature using Equation 45.

E. Other Techniques

Microwave radiometry, ultrasonic radiometry, and ultrasonic time-of-flight temperature tomography can also be considered for measurements of temperature in RF fields.[83,84] These techniques are still in early stages of development, but may prove to be useful, particularly in medical applications where hyperthermia is produced by RF energy.

IV. MATERIALS FOR DIELECTRIC MODELS

Several dielectric mixtures whose electrical properties are close to those of various tissues have been developed. They find extensive use in experimental dosimetry involving construction of models of human and animal tissues. For various applications different consistencies of the material are desirable, e.g., for thermography measurements solid or semisolid materials are necessary, but for measurements with an implantable E-field probe liquid or semiliquid materials are highly desirable.

V. EXPOSURE SYSTEMS

A. General Requirements
Depending on the object and the purpose of the experiment, different characteristics of an exposure system are required. For instance, chronic exposure of a large number of rodents requires different features of the exposure system than exposure of a small in vitro preparation. Exposures to investigate potentially hazardous effects of electromagnetic fields usually call for uniform, far-field conditions, while exposures for therapeutic purposes require a specific profile of the energy deposition. Certain requirements are common for all exposure systems. These include (1) a well-defined exposure field; (2) controllable, ambient conditions (temperature, humidity); (3) ability to determine, preferably to measure, the SAR and/or temperature of the specimen.

The exposure characteristics that are important only in some situations are (1) possibility of providing the exposed subject with nutrients; (2) ability to monitor various functions or conditions, electrocardiogram e.g., (ECG), electroencephalogram (EEG), respiration; (3) simultaneous exposure of a number of test subjects or samples. Another underlying factor in selecting an exposure system is its cost. To limit the cost, enclosed systems, such as a TEM cell or cavities, are often used to simulate free space fields.

B. Free Field — Anechoic Chambers
This exposure system simulates the far-field, free-space exposure. The exposure takes place in a large room lined with a special material that is only minimally reflective for radiowaves over a frequency range of interest. Such enclosures are called "anechoic chambers" and have been extensively used in antenna research. The anechoic materials are available commercially in the form of pyramids or wedges, and the larger their size the lower the frequency of operation with small reflections. An exposure field is produced by an antenna, and the far-field conditions prevail at distances from the antenna greater than

$$R \geqslant \frac{2a^2}{\lambda} \tag{57}$$

where a is the greatest dimension of the antenna, and λ is the wavelength. The far-field power density on the antenna axis at a distance r is equal to

$$W = \frac{PG}{4\pi r^2} \tag{58}$$

where P is the power delivered to the antenna and G is the antenna gain. The area within

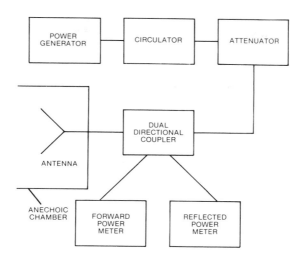

FIGURE 14. Anechoic chamber exposure system.

which the power density is relatively constant (e.g., within ± 0.5 dB) depends on the distance from the antenna and the antenna pattern.

A schematic diagram of an anechoic chamber exposure system is shown in Figure 14. To determine the power density, the system has to be calibrated (see Section II.C.1). Anechoic chambers are used for exposures of experimental animals usually at frequencies above 500 MHz. At lower frequencies enclosures become exceedingly large and costly. Temperature and humidity in anechoic chambers can be closely controlled. One of the main disadvantages is that when a large number of animals are exposed simultaneously, a relatively high power is required. To ensure the same exposure conditions of each test animal they have to be separated from each other by a sufficient distance to avoid multiple reflections.

By proper separation of specially designed animal cages, nonuniformities in the exposure field can be minimized.[91] The animal cages should be constructed from materials having electrical properties as close to those of free space as possible. Low-density Styrofoam® is an excellent choice, and low-dielectric, constant plastics (e.g., Plexiglas®, Teflon®, and acrylic) and dry wood can be used when mechanical strength is required.

To increase the power density within the exposure area for a given input power a special slot array[92] can be utilized.

An improved space utilization and a simulation of the free-field exposure is provided by "compact ranges". Two structures have been described[93] and used.[94] A prolate-spheroidal reflector having two foci, when illuminated by an open-ended waveguide placed in one focus, produces a uniform wavefront over a diameter of 1.5 wavelength in the other focus.[93] Another design uses absorber-lined horns.[93,94] The absorber-lined horn (rectangular or cylindrical) acts as a small-aperture radiating source with a radiation pattern that is approximately uniform at a plane near to the horn aperture. An absorber-lined horn operating at 2.45 GHz is shown in Figure 15.[93] The absorber is thin and of the resonant type.

C. Enclosed Exposure Systems
1. TEM Cell

Exposure conditions simulating those of free space can be obtained in a TEM cell utilizing a modified, shielded stripline configuration as shown in Figure 16. A simplified sketch of a TEM transmission cell is shown in Figure 17. The tapered sections are designed to match the rectangular, cross-section strip line to standard, 50-Ω coaxial connectors. The charac-

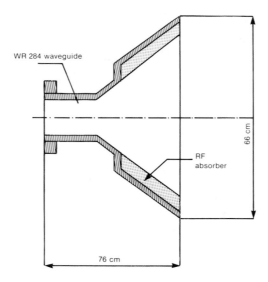

FIGURE 15.   Compact range — an absorber-lined horn at 2.45 GHz.[93]

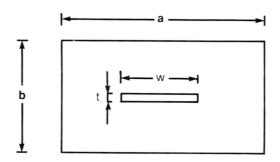

FIGURE 16.   Cross-section of a TEM cell.

teristic impedance of the shielded strip line (Figure 16) can be calculated from the following equation[95]

$$Z_0 = \frac{94.51}{\sqrt{\epsilon_r}\left[\dfrac{w}{b(1 - t/b)} + \dfrac{C_f}{0.0885\epsilon_r}\right]} \tag{59}$$

where ϵ_r is the relative dielectric constant of the medium between the conductors (for air lines $\epsilon_r = 1$), w, b, and t are the line dimensions as shown in Figure 16 and C_f is the fringing capacitance of the line in picofarads per centimeter. The fringing capacitance can be determined experimentally[95] or by various analytical methods.[96] For strip lines having very thin septum (t/b $\ll$ 1 and w/b $\geq$ 0.5), the fringing capacitance per unit length can be calculated within an accuracy better than 2% from[96]

$$C_f = \frac{2\epsilon_0}{\pi} \ln\left[1 + \coth\pi\left(\frac{a - w}{2b}\right)\right] \tag{60}$$

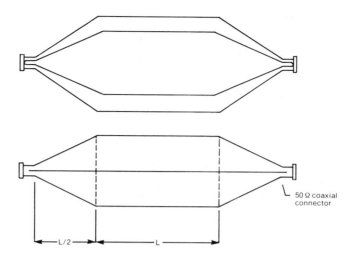

FIGURE 17.   Top and side view of a TEM cell.

where $\epsilon_0 = 8.854 \; 10^{-12}$ F/m. For thin-septum strip lines with $0.1 \leq w/b < 0.5$, a correction factor ΔC has to be added to C_f.[96]

$$\Delta C \simeq \frac{2\epsilon_0}{\pi} \left[\frac{8}{(1 + \sqrt{\lambda})^2 \, (1 + \lambda)} \right] \tag{61}$$

where

$$\lambda \simeq [1 - \exp(-2\pi w/b)]^{1/2} \tag{62}$$

For a finite thickness line the fringing field capacitance can be obtained from the following equation[96]

$$C_f = \frac{\epsilon_0}{\pi \ln 2} \left\{ \frac{b}{b - t} \ln\left(\frac{2b - t}{t}\right) + \ln\left[\frac{t(2b - t)}{(b - t)^2}\right] \right\}$$
$$\cdot \ln\left[1 + \coth\pi\left(\frac{a - w}{2b}\right) \right] \tag{63}$$

The fields in a TEM cell resemble those of a plane-wave as long as only the fundamental TEM mode propagates in the cell. The approximate cut-off frequencies for the higher order modes are given by[95]

$$(f_c)_{n,m} = \frac{c(n^2a^2 + m^2b^2)^{1/2}}{2ab} \tag{64}$$

where c is the velocity of light ($c \simeq 3 \times 10^8$ m/sec). The E-field distribution for the first few modes is shown in Figure 18. More accurate and detailed data on higher-order modes in TEM cells are given elsewhere.[97] The cut-off frequency of the TE_{10} mode is equal to

$$(f_c)_{1,0} = \frac{c}{2a} \tag{65}$$

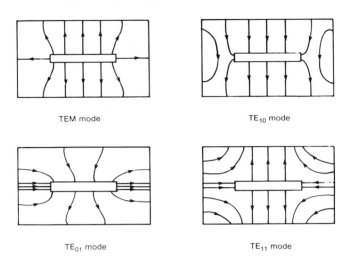

FIGURE 18.   The E-field distribution in a TEM cell.

TEM transmission cells can be used for exposures even at frequencies higher than that given by Equation 65.[97,98] The exposed object however, should not occupy more than one third of the volume between the septum and outer conductor.[95]

The E-field intensity (when operating in the TEM mode) is[95]

$$E = \sqrt{PZ_0}/d \tag{66}$$

where P is the net power flowing through the cell, Z_0 is the cell characteristic impedance, and d is the distance between the septum and the wall. Sources of error in determination of the field intensity from Equation 66 include[95] (1) the uncertainties in P (uncertainties in coupler calibration and RF power measurements, mismatch between the cell, RF source, and the cell termination); (2) the uncertainties in Z_0; (3) the nonuniformity of the E field inside the cell resulting from field perturbation by the test object.

2. Resonant Cavity

Single- and multiple-mode resonant cavities can be used as exposure chambers. One of their major advantages is that the total SAR can be easily determined from measurements of the net input power. Other advantages are low cost and limited space requirements. The main disadvantage is that usually only one object (animal) can be exposed in a well-defined field. However, a system consisting of many cavities connected to a single power source offers a viable alternative to exposures in an anechoic chamber.[99]

A cylindrical cavity can be excited in the TE_{11} mode circularly polarized to provide relatively constant and easily quantifiable coupling of the energy to the test animal. Such cavities have been designed and extensively used in studies of rodents at 918 MHz[99] and 2.45 GHz.[100] The cavities can be made of wire screen to facilitate ventilation and visibility of the test object. The test animal is placed in a plastic cage equipped with a food pellet dispenser and a water bottle. The food pellets are dry and absorb very little microwave energy. The water bottle made of glass is electrically decoupled from the animal by two concentric, one quarter-wavelength coaxial chokes.[99] This is essential for eliminating the marked increases in power absorption in the animal when it comes in contact with the highly conductive water spout.

A rectangular cavity operating in two modes can be employed to produce desired intensities

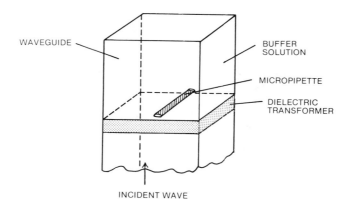

FIGURE 19.   Waveguide exposure system for cell cultures.[105,106]

of the E and H fields within an exposure volume occupying a relatively small part of the center of the cavity.[101] This type of a cavity can be used at frequencies between 100 MHz and 1 GHz.[101]

A multimode rectangular cavity can also be used for exposures.[102] A mode stirrer is employed to increase the number of modes excited in the cavity and to improve the uniformity of the exposure field.

3. Near-Field Synthesizer

Exposures at lower RFs (a few tens of megahertz) can be conveniently accomplished in a near-field synthesizer.[103,104] The synthesizer consists of two parallel plates. A uniform E field can be generated between the plates and a loop inductor placed between the plates can produce a H field of any desired orientation. Since the intensities of the two fields can be controlled independently, any desired exposure conditions simulating actual near-field exposure can be obtained.

D. Cell Culture Exposure Systems

In exposing cell cultures, blood samples, and other solutions containing microorganisms, it is extremely important to ensure that the E field intensity and temperature are well-defined (and uniform if possible).[104] The situation in which the sample holder may significantly modify the exposure field should be avoided.

Various waveguide[105-107] and cavity[108] exposure systems for cell cultures have been developed. Most of the systems have provisions for temperature control of the culture. A waveguide exposure system[105,106] is shown in Figure 19. The sample is contained in a micropipette immersed in a buffer solution. To couple the energy to the sample effectively, a layer of a dielectric material is used (the thickness of the dielectric is one quarter guide wavelength in the dielectric and the dielectric constant is approximately $= \sqrt{2\epsilon}$, where ϵ is the relative dielectric constant of the buffer). The buffer solution can be circulated to maintain a constant temperature.

A specially designed, coaxial-transmission-line exposure cell[104,109] contains a 5-mℓ sample, which can be exposed at frequencies up to 1 GHz. A simplified diagram is shown in Figure 20. The temperature of the sample can be controlled, and 37°C or less can be maintained at field strength in excess of 2.5 kV/m. E- and H-field probes are used to monitor the reflection coefficient from the sample. The field intensity in the sample can easily be calculated from an analysis of the equivalent circuit.[104,109]

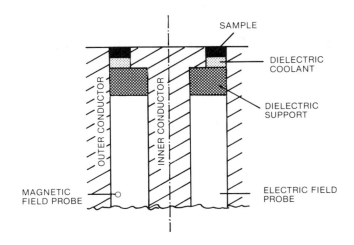

FIGURE 20. Simplified diagram of cell culture transmission line exposure system.[104,109]

E. Monitoring Electrodes

Metal electrodes and connecting wires conventionally used to record electrophysiological processes when used in electromagnetic fields significantly perturb the exposure field and cause highly localized heating at electrode/tissue boundaries, due to the field enhancement.[110] Several types of electromagnetic-field-compatible electrodes have been designed.[110-114] Even nonmetallic electrodes when made of a material of high conductivity can enhance the field.[114]

A nonperturbing electrode system utilizing low conductivity plastics for electrodes and leads[114] is satisfactory for recordings of EEGs and ECGs. The system consists of a pair of electrodes and leads prefabricated as a single unit. The electrode is 10 cm long, 2 mm wide, 0.25 mm thick, and made of carbon-loaded Teflon® (CLT) having a conductivity of 45 S/m (Technical Fluorocarbons Engineering, Warwick, R.I., Polymer Corp., Reading, Pa.) The lead is a strip of 72-S/m CLT of the same width and thickness as the electrode. The lead and the electrode are taped one atop the other with double-sided cellophane adhesive tape. The field enhancement due to these electrodes is very low.[114]

VI. SUMMARY OF EXPERIMENTAL RESULTS

The various experimental dosimetry methods described in this chapter have been used to verify theoretical calculations for models and to determine the SAR in experimental animals. A representative summary of the data is given in Table 12.

VII. CONCLUDING REMARKS

Experimental dosimetry is essential in determining the exposure field and the whole-body or the body-part average SARs in experiments with animals exposed to RF fields. Furthermore, carefully performed experiments are crucial in verification of theoretical predictions and delineation of their limitations. For instance, experimentally obtained whole-body, average SAR for humans are three to four times greater than those calculated.[128] The differences in the spatial distribution of the SAR between the theoretical estimations and the measured values are even greater.[133]

The two approaches to dosimetry at RFs, namely the theoretical and the experimental, are complementary, and as both have certain advantages and limitations, these should be considered in selecting one or the other or both approaches to solve a given problem.

Table 12
REPRESENTATIVE RESULTS OF EXPERIMENTAL DOSIMETRY STUDIES

Experiment	Technique	Frequency (MHz)	Field	Polarization[a]	SAR Average	Distribution[b]	Ref.
Spheres, prolate spheroids	Near-field synthesizer and anechoic chamber, thermography	3—1600		Various		+	5
Phantom-scale models of man	Anechoic chamber, thermography	80—250 (scaled)	Far	E, H, k	Given	+	117
Prolate spheroidal models	Two-plate strip line, thermography	285—4000	TEM	E, H, k	Approx.	Approx.	116
Mice	Anechoic chamber, flask calorimeter	2600	Far	E	1.15 W/kg/mW/cm²	–	51
Mice, 24—25 g	Free-space, calorimeter	2450	Far	E	1.2—1.4 W/kg/mW/cm²	–	118
Mice, 30—50 g	Anechoic chamber, twin-well calorimeter	2450	Far		0.32—0.42 W/kg/mW/cm²	–	119
Mice, 27 g	Anechoic chamber, flask calorimeter	2450	Far	E, H, k	0.58—1.21 W/kg/mW/cm²	–	122
Voles, 22—24 g	Free-space calorimeter	2450	Far	E	1.41—1.5 W/kg/mW/cm²	–	118
Whiptail lizards, 19—24 g	Free-space, calorimeter	2450	Far	E	1.85—2.68 W/kg/mW/cm²	–	118
Ring doves, 120—144 g	Free-space, calorimeter	2450	Far	E	0.57—0.75 W/kg/mW/cm²	–	118
Rats, 190—380 g and cylindrical models	Multimode cavity, twin-well calorimeter	2450	Cavity	–	1.4—2.3 W/kg/W	–	53
	Anechoic chamber, twin-cell calorimeter	2880	Far		~0.3—0.29 W/kg/mW/cm²	–	53

Table 12 (continued)
REPRESENTATIVE RESULTS OF EXPERIMENTAL DOSIMETRY STUDIES

Experiment	Technique	Exposure conditions			SAR		Ref.
		Frequency (MHz)	Field	Polarization[a]	Average	Distribution[b]	
Rats, 132—490 g	TE_{11} cavity, differential power	918	Cavity	Variable	1.16—3.15 W/kg/W	—	99
Spheroidal models of rats	TE_{11} cavity, thermography	2450	Cavity	E/H, k	2.64, 1.99 W/kg/W	+	100
Rat, 324 g	TE_{11} cavity, thermography	2450	Cavity	k	2.34—247 W/kg/W	+	100
Rats, 96—390 g	Two-plate stripline, thermography	285—4000	TEM	k	Approx.	Approx.	116
Spheroidal models of rats, 300—330 g	TE_{11} cavity, thermography	918	Cavity	E/H, k	Given	+	99
Rats, 388 g	TE_{11} cavity, thermography	918	Cavity	E/H	0.98—1.3 W/kg/W	+	99
Rats, 355—550 g	Free-space, calorimeter	2450	Far	E	Enhance- ment factor given	—	118
Rats, 100—250 g proximity enhancement	Free-space, calorimeter	2450	Far	E	0.15—0.23 W/kg/mW/cm²	—	119
Rats, 6—440 g	Anechoic chamber, twin-well calorimeter	2450	Far	E	0.1—1.25 W/kg/mW/cm²	—	118
Rats	Anechoic chamber, calorimeter, and optional fiber probe	2450	Far	E, k	Head and whole-body	—	127
Rabbits, 2—2.25 kg	Free-space, calorimeter	2450	Far	E	0.076—0.085 W/kg/mW/ cm²	—	118
Speroid rhesus monkey phantom, 3.5 kg	TEM cell, differential power	10—50	TEM	E/H, k	Given	—	121
Rhesus monkeys, 3.2—4.2 kg	TEM cell, differential power	10—50	TEM	E, H, k	Given	—	121

Model	Apparatus	Frequency (MHz)	Exposure	Field[a]	Dosimetry	Distribution[b]	Ref.
Full-scale model of Rhesus monkey	Anechoic chamber, calorimeter, and optical fiber probe	225, 1.29	Far	E	Given	+	129, 130
Rectangular (146 × 79 × 20 cm) human phantom	TEM cell, net power absorbed	10, 20, 30	TEM	k	10, 3 × 10⁻³, 5 × 10⁻³ mW/cm³	−	115
Spheroidal human phantom, 70 kg	TEM cell, differential power	10—50	TEM	E, H, k	Given	−	121
Phantom-scale models of man	Near-field synthesizer anechoic chamber and resonant cavity, thermography	3—1600		Various	Given	+	2, 80
	TM_{110} and TE_{102} cavity, thermography	24, 31 (scaled)	Cavity	Various	Given	+	5
Men	TEM cell (70 Ω) differential power	8.5, 23.25	TEM	E, H, k	Given	−	123, 128
Sealed spheroidal models of man	Monopole in anechoic chamber, fiber-optic thermometer	75 (scaled)	Far, near		Given	−	124
Phantom-scale models of man	Anechoic chamber, parallel plate, digital thermometer	70—280 (scaled)	Far	E, H, k	Approx.	Approx.	120, 125, 126
	Anechoic chamber, flash calorimeter	300—470 (scaled)	Far	E, k	Head, whole-body	−	127
Full-scale models of man	Anechoic chamber, implantable E-field probe	350, 915	Far	E, H, k	Given	+	131—133
	Anechoic chamber, implantable E-field probe	350, 915	Near (dipole, dipole above ground, slot)	E, H, k	Given	+	133, 134

[a]   E designates the E field parallel to the longest axis, H designates the H field parallel to the longest axis, and k designates the waves propagating along the longest axis.

[b]   − Distribution not measured, + distribution measured.

ACKNOWLEDGMENTS

Our thanks to the Editors for their constructive and helpful criticism. Special thanks are due to Mrs. Judy Smith of the Radiation Protection Bureau, Health and Welfare Canada, for the editorial assistance.

REFERENCES

1. Radiofrequency electromagnetic fields. Properties, quantities and units, biophysical interaction, and measurements, National Council on Radiation Protection and Measurements, Report No. 67, Washington, D.C., March 1981.
2. **Guy, A. W.,** Non-ionizing radiation: dosimetry and interaction, *Proc. ACGIH Top. Symp.,* 1979, 75.
3. **Adey, W. R.,** Tissue interactions with nonionizing electromagnetic fields, *Physiol. Rev.,* 61, 435, 1981.
4. **Durney, C. H., Iskander, M. F., Massoudi, H., Allen, S. J., and Mitchell, J. C.,** Radiofrequency Radiation Dosimetry Handbook, 3rd ed., Report SAM-TR-80-32, U.S. Air Force School of Aerospace Medicine, Brooks Air Force Base, Tex., 1980.
5. **Guy, A. W., Webb, M. D., and Sorensen, C. C.,** Determination of power absorption in non exposed to high frequency electromagnetic fields by thermographic measurements on scale models, *IEEE Trans. Biomed. Eng.,* 23, 361, 1976.
6. **Wacker, P. F. and Bowman, R. R.,** Quantifying hazardous electromagnetic fields: scientific basis and practical considerations, *IEEE Trans. Microwave Theor. Tech.,* 19, 178, 1971.
7. **Tell, R. A.,** Instrumentation for measurement of radiofrequency electromagnetic fields: equipment, calibrations and selected applications, in *Proc. Course Adv. Biol. Eff. Dosimetry Low Energy. Electromagnetic Fields,* Grandolfo, M. and Michaelson, S. M., Eds., Plenum Press, New York, 1982.
8. **Bowman, R. R.,** Some recent developments in the characterization and measurement of hazardous electromagnetic fields, in *Proc. Int. Symp. Biol. Eff. Health Hazards Microwave Radiat.,* Polish Medical Publishers, Warsaw, 1974, 217.
9. **King, R. W. P.,** *Theory of Linear Antennas,* Harvard University Press, Cambridge, 1956, 184.
10. **Kanda, M.,** Analytical and numerical techniques for analyzing an electrically short dipole with a nonlinear load, *IEEE Trans. Antennas Propag.,* 28, 71, 1980.
11. **Hopfer, S. and Adler, Z.,** An ultra broad-band (200 kHz—26 GHz) high-sensitivity probe, *IEEE Trans. Instrum. Meas.,* 29, 445, 1980.
12. **Aslan, E.,** Broad-band isotropic electromagnetic radiation monitor, *IEEE Trans. Instrum. Meas.,* 21, 421, 1972.
13. **Fletcher, K. and Woods, D.,** Thin-film spherical bolometer for measurement of hazardous field intensities from 400 MHz to 40 GHz, *Non-Ioniz. Radiat.,* 57, September, 1969.
14. **Greene, F. M.,** Development of magnetic near-field probes, Publ. No. 75-127, J.S. Department of Health, Education and Welfare, Cincinnati, 1975, 75.
15. **Babij, T. M. and Bassen, H. I.,** Development of E/H near-field probe, *IEEE Trans. Electromagn. Compat.,* in press.
16. **Whiteside, H. and King, R. W. P.,** The loop antenna as a probe, *IEEE Trans. Antennas Propag.,* 12, 71, 1964.
17. **Duncan, P. H.,** Analysis of the Moebius loop magnetic field sensor, *IEEE Trans. Electromagn. Compat.,* 16, 83, 1974.
18. **Bassen, H., Herman, W., and Hoss, R.,** EM probe with fiber optic telemetry system, *Microwave J.,* 20, 35, 1977.
19. **Bassen, H. I. and Hoss, R. J.,** An optically linked telemetry system for use with electromagnetic-field measurement probes, *IEEE Trans. Electromagn. Compat.,* 20, 483, 1978.
20. **Postow, E.,** Plans to develop a radiofrequency personnel dosimeter, *Proc. Workshop Prot. Personnel Against Radiofrequency Electromagnetic Radiat.,* Rev. 3-81, Mitchell, J. C., Ed., U.S. Air Force School of Aerospace Medicine, Brooks Air Force Base, Tex., 1981, 93.
21. **Gajda, G., Stuchly, M. A., and Stuchly, S. S.,** Mapping of the nearfield pattern in simulated biological tissues, *Electron. Lett.,* 15, 120, 1979.
22. **Smith, G. S.,** A comparison of electrically short bare and insulated probes for measuring the local radio frequency electric field in biological systems, *IEEE Trans. Biomed. Eng.,* 22, 477, 1975.
23. **Mousavinezhad, S. H., Chen, K.-M., and Nyquist, D. P.,** Response of insulated electric field probes in finite heterogeneous biological bodies, *IEEE Trans. Microwave Theor. Tech.,* 26, 599, 1978.

24. **Smith, G. S.,** The electric-field probe near a material interface with application to the probing of fields in biological bodies, *IEEE Trans. Microwave Theor. Tech.,* 27, 270, 1979.

25. **King, R. W. P., Prasad, S., and Sandler, B. H.,** Transponder antenna in and near a three-layered body, *IEEE Trans. Microwave Theor. Tech.,* 28, 586, 1980.

26. **Smith, G. S.,** Analysis of miniature electric field probes with resistive transmission lines, *IEEE Trans. Microwave Theor. Tech.,* 29, 1213, 1981.

27. **Stuchly, M. A., Kraszewski, A., and Stuchly, S. S.,** Implantable electric field probes — some performance characteristics, *IEEE Trans. Biomed. Eng.,* 31, 526, 1984.

28. **Brodwin, M. E., Taflove, A., and Matz, J. E.,** A passive electrodeless method for determining the interior field of biological materials, *IEEE Trans. Microwave Theor. Tech.,* 24, 514, 1976.

28A. **Bassen, H. I. and Smith, G. S.,** Electric field probes — a review, *IEEE Trans. Antennas Propag.,* 31(5), 710, 1983.

29. **Baird, R. C.,** Methods of calibrating microwave hazard meters, in *Proc. Int. Symp. Biol. Health Hazards Microwave Radiat.,* Polish Medical Publishers, Warsaw, 1974, 228.

30. **Hickman, T. G. and Heaton, R. A.,** Measurement of gain, in *Microwave Antenna Measurements,* Hollis, J. S., Lyon, T. J., and Clayton, L., Eds., Scientific Atlanta, 1970, chap. 8.

31. **Jull, E. V.,** Finite-range gain of sectoral and pyramidal horns, *Electron. Lett.,* 6, 680, 1970.

32. **Hunter, J. D. and Morgan, I. G.,** Near-field gain correction for transmission between horn antennas, *IEEE Trans. Instrum. Meas.,* 26, 58, 1977.

33. **Larsen, E. B.,** Techniques for producing standard EM fields from 10 kHz to 10 GHz for evaluating radiation monitors, in *Proc. Symp. Electromagnetic Fields Biol. Syst.,* Stuchly, S. S., Ed., Intern. Microwave Power Inst., Edmonton, Alberta, Canada, 1979, 96.

34. **Bowman, R. R.,** Field strength above 1 GHz: measurement procedures for standard antennas, *Proc. IEEE,* 55, 981, 1967.

35. **Bassen, H. I. and Herman, W. A.,** Precise calibration of plane-wave microwave power density using power equation technique, *IEEE Trans. Microwave Theor. Tech.,* 25, 701, 1977.

36. **Crawford, M. L.,** Generation of standard EM fields using TEM transmission cells, *IEEE Trans. Electromagn. Compat.,* 16, 189, 1974.

37. **Weil, C. M., Joines, W. T., and Kinn, J. M.,** Frequency range of large-scale TEM mode rectangular strip lines, *Microwave J.,* 24, 93, 1981.

38. **Hill, D. A.,** Bandwidth, limitations of TEM cells due to resonances, *J. Microwave Power,* 18, 181, 1983.

39. **Cheung, A. Y.,** Electric field measurements within biological media, in *Proc. Symp. Biol. Eff. Meas. Radio Publ. 77-8026, Frequency/Microwaves,* Food and Drug Administration, U.S. Department of Health, Education and Welfare, Rockville, Md., 1977, 117.

40. **Hill, D. A.,** Waveguide technique for the calibration of miniature implantable electric-field probes for use in microwave-bioeffects studies, *IEEE Trans. Microwave Theor. Tech.,* 30, 92, 1982.

41. **Christman, C. L., Ho, H. S., and Yarrow, S.,** A microwave dosimetry system for measured sampled integral-dose rate, *IEEE Trans. Microwave Theor. Tech.,* 22, 1267, 1974.

42. **Greene, F. M.,** Measurement of RF power-absorption in biological specimens (10 to 100 MHz), Publ. 77-146, U.S. Department of Health, Education and Welfare, Washington, D. C., 1977.

43. **Edwards, W. P. and Ho, H. S.,** RF cavity irradiation dosimetry, *IEEE Trans. Microwave Theor. Tech.,* 23, 311, 1975.

44. **Allen, S. J.,** Measurement of power absorption by human phantoms immersed in radio-frequency fields, *Ann. N.Y. Acad. Sci.,* 247, 494, 1975.

45. **Hill, D. A.,** Human microwave absorption from 7 to 40 mHz, in Proc. Workshop Prot. Personnel Against Radiofrequency Electromagnetic Radiat., Rev. 3-81, Mitchell, J. C., Ed., U.S. Air Force School of Aerospace Medicine, Brooks Air Force Base, Tex., 1981, 170.

46. **Hill, D. A.,** Human whole-body radiofrequency absorption studies using a TEM-cell exposure system, *IEEE Trans. Microwave Theor. Tech.,* 30, 1847, 1982.

47. **Guy, A. W., Lehmann, J. F., and Stonebridge, J. B.,** Therapeutic applications of electromagnetic power, *Power IEEE,* 62, 55, 1974.

48. **Allis, J. W., Blackman, C. F., Fromme, M. L., and Benane, S. G.,** Measurement of microwave radiation absorbed by biological systems. I. Analysis of heating and cooling data, *Radio Sci.,* 12(6S), 1977.

49. **McRee, D. J.,** Determination of energy absorption of microwave radiation using the cooling curve technique, *J. Microwave Power,* 9, 263, 1974.

50. **Blackman, C. F. and Black, J. A.,** Measurement of microwave radiation absorbed by biological systems, 2 analysis by Dewar-flask calorimetry, *Radio Sci.,* 12(6S), 9, 1977.

51. **Allen, S. J. and Hunt, W. D.,** Calorimetric measurements of microwave energy absorption by mice after simultaneous exposure of 18 animals, *Radio Sci.,* 14(6S), 1, 1979.

52. **Durney, C. H., Iskander, M. F., Massoudi, H., Allen, S. J., and Mitchell, J. C.,** Radiofrequency Radiation Dosimetry Handbook, 3rd ed., Report SAM-TR-80-32, U.S. Air Force School of Aerospace Medicine, Brooks Air Force Base, Tex., 1980, 29.

53. **Phillips, R. D., Hunt, E. L., and King, N. W.,** Field measurement, absorbed dose, and biologic dosimetry of microwaves, *Ann. N.Y. Acad. Sci.,* 247, 499, 1975.
54. **Cetas, T. C.,** Thermometry in strong electromagnetic fields, Publ. 77-8055, *U.S. Department of Health, Education and Welfare, Washington, D.C.,* 1977, 261.
55. **Cetas, T. C.,** Invasive thermometry, in *Proc. AAPM Summer Sch.,* Phys. Aspects Hyperthermia, Dartmouth College, Hanover, N.H., 1981.
56. **McRee, D. J. and Pendergrass, F. T.,** Interaction of 2450-MHz microwave field with thermocouples and thermistors, *Health Phys.,* 25, 180, 1973.
57. **McAfee, R. D., Cazenavette, L. L., and Shubert, H. A.,** Thermistor probe error in an X-band microwave field, *J. Microwave Power,* 9, 177, 1974.
58. **Chakraborty, D. P. and Brezovich, I. A.,** A source of themrocouple error in radiofrequency electric fields, *Electron. Lett.,* 16, 853, 1980.
59. **Szwarnowski, S., Sheppard, R. J., Grant, E. H., and Bleehen, N. M.,** A thermocouple for measuring temperature in biological material heated by microwaves at 2.45 GHz, *Br. J. Radiol.,* 53, 711, 1980.
60. **Samaras, G. M., Harrison, G. H., and Blaumanis, O. R.,** In vivo temperature/heat flow sensing in microwave fields, in Proc. 3rd Int. Symp. Cancer Ther. Hyperthermia, Fort Collins, Colo., 1980, 192.
61. **Eno, K., Kato, H., Nishida, T., Kano, E., Sugahara, T., Tanaka, H., and Ishida, T.,** Physical basis of RF hyperthermia for cancer therapy, a non-perturbed and non-perturbing thermometer at RF heating, *J. Radiat. Res. (Tokyo),* 22, 265, 1981.
62. **Johnson, C. C. and Rozzell, T. C.,** Liquid crystal fiber optic RF probes, temperature probe for microwave fields, *Microwave J.,* 18, 55, 1975.
63. **Johnson, C. C., Durney, C. H., Lords, J. L., Rozzell, T. C., and Livingston, G. K.,** Fiberoptic liquid crystal probe for absorbed radio-frequency power and temperature measurement in tissue during irradiation, *Ann. N.Y. Acad. Sci.,* 247, 527, 1975.
64. **Rozzell, T. C., Johnson, C. C., Durney, C. H., Lords, J. L., and Olsen, R. G.,** A nonperturbing temperature sensor for measurements in electromagnetic fields, *J. Microwave Power,* 9, 241, 1974.
65. **Bowman, R. R.,** A probe for measuring temperature in radiofrequency-heated material, *IEEE Trans. Microwave Theor. Tech.,* 24, 43, 1976.
66. **Deficis, A. and Priou, A.,** Nonperturbing microprobe for measurement in electromagentic fields, *Microwave J.,* 20, 55, 1977.
67. **Deficis, A.,** Fiber optic microprobes for microwave electromagnetic field measurements, Digest Microwave Theor. Tech. Int. Symp., Palo Alto, Calif., 1975, 300.
68. **Deficis, A. and Priou, A.,** Nonperturbing microprobe for measurements in electromagnetic fields, Publ. 77-8055, Food and Drug Administration, U.S. Department of Health, Education and Welfare, 1977, 283.
69. **Cain, C. A. and Chen, M. M.,** The viscometric thermometer, Publ. 77-8055, Food and Drug Administration, Department of Health, Education and Welfare, 1977, 295.
70. **Cetas, T. C.,** A birefringent crystal optical thermometer for measurements of electromagnetically induced heating, Symp. Proc. Biol. Eff. Meas. Radiofrequency/Microwaves, Publ. 77-8026, Food and Drug Administration, Department of Health, Education and Welfare, Rockville, Md., 1977, 338.
71. **Livingstone, G. K.,** Thermometry and dosimetry of heat with specific reference to the liquid crystal optical fiber temperature probe, *Radiat. Environ. Biophys.,* 17, 233, 1980.
72. **Christensen, D. A.,** A new non-perturbing temperature probe using semiconductor band edge shift, in Dig. IEEE/MTTS Int. Microwave Symp., San Diego, Calif., 1977.
73. **Wickesheim, K. A. and Alves, R. V.,** Ratioing fluoroptic temperature sensor for induced hyperthermia, presented at the 3rd Int. Symp. Cancer Ther. Hyperthermia, Drugs, Radiat., Fort Collins, Colo., 1980.
74. **Wickersheim, K. A. and Alves, R. B.,** Recent advances in optical temperature measurement, *Ind. Res. Dev.,* December, 1979.
75. **Cetas, T. C.,** Temperature measurement in microwave diathermy fields: principles and probes, Proc. Int. Symp. Cancer Ther. Hyperthermia Radiat., Washington, D.C., 1975, 193.
76. **Hochuli, C. U.,** Procedure for evaluating nonperturbing temperature probes in microwave field, Publ. 81-8143, Food and Drug Administration, U.S. Department of Health and Human Services, 1981.
77. **Hochuli, C. U. and Kantor, G.,** An analysis of minimally perturbing temperature probe and thermographic measurements in microwave diathermy, *IEEE Trans. Microwave Theor. Tech.,* 29, 1285, 1981.
78. **Guy, A. W.,** Analyses of electromagnetic fields induced in biological tissues by thermographic studies on equivalent phantom, *IEEE Trans. Microwave Theor. Tech.,* 19, 205, 1971.
79. **Guy, A. W.,** History and state of the art on the quantitation of the interaction of electromagnetic fields with biological structures, NATO ASI Interaction Electromagnetic Fields with Structures, Norwich, England, 1979.
80. **Guy, A. W., Webb, M. D., and McDougall, J. A.,** RF radiation absorption patterns: human and animal modeling data, Publ. 77-183, U.S. Department of Health, Education and Welfare, Washington, D. C., 1977.

81. **Cetas, T. C.,** Practical thermometry with a thermographic camera — calibration, transmittance, and emittance, measurements, *Rev. Sci. Instrum.,* 49, 245, 1978.
82. **Ho, H. S. and Faden, J.,** Experimental and theoretical determination of absorbed microwave dose rate distributions in phantom heads irradiated by and aperture source, *Health Phys.,* 33, 13, 1977.
83. **Cristensen, D. A.,** Current techniques for noninvasive thermometry, *Proc. AAPM Summer Sch.,* Phys. Aspects Hyperthermia, Dartmouth College, Hanover, N.H., 1981.
84. **Christensen, D. A.,** Thermal dosimetry and temperature measurements, *Cancer Res.,* 39, 2325, 1979.
85. **Bolzano, Q., Garay, O., and Steel, F. R.,** An attempt to evaluate exposure of operators of portable radios at 30 MHz, Conf. Proc. 29th IEEE Vehicular Technol. Soc., Arlington Heights, Fla., 1979, 187.
86. **Cheung, A. Y. and Koopman, D. W.,** Experimental development of simulated biomaterials for dosimetry studies of hazardous microwave radiation, *IEEE Trans. Microwave Theor. Tech.,* 24, 669, 1976.
87. **Thansandoti, A.,** Monitoring Time-varying Biological Impedance at Microwave Frequencies, Ph.D. thesis, Carleton University, Ottawa, Canada, 1981.
88. **Paulsson, L. E.,** Measurements of 0.915, 2.45 and 9 GHz absorption in the human eye, Proc. 6th Eur. Microwave Conf., Rome, 1976.
89. **Toler, J. and Seals, J.,** RF dielectric properties measurement system: human and animal data, Tech. Rep. A-1862, Biomedical Research Group, Engineering Experimental Station, Georgia Institute of Technology, Atlanta, 1977.
90. **Hand, J. W., Robinson, J. E., Szwarnowski, S., Sheppard, R. J., and Grant, E. H.,** A physiologically compatible tissue-equivalent liquid bolus for microwave heating of tissues, *Phys. Med. Biol.,* 24, 426, 1979.
91. **Oliva, S. A. and Catravas, G. N.,** A multiple-animal array for equal power density microwave irradiation, *IEEE Trans. Microwave Theor. Tech.,* 25, 433, 1977.
92. **Hagmann, M. J. and Gandhi, O. P.,** Radiators for microwave biological effects research — waveguide slot array with constant radiation intensity, *Radio Sci.,* 12(6S), 97, 1977.
93. **Bassett, H. L., Ecker, H. A., Johnson, R. C., and Sheppard, A. P.,** New techniques for implementing microwave biological-exposure systems, *IEEE Trans. Microwave Theor. Tech.,* 19, 197, 1971.
94. **McRee, D. and Walsh, P.,** Microwave exposure system for biological specimens, *Rev. Sci. Instrum.,* 42, 1860, 1971.
95. **Crawford, M. L.,** Generation of standard EM fields using TEM transmission cells, *IEEE Trans. Electromagn. Compat.,* 16, 189, 1974.
96. **Weil, C. M.,** The characteristic impedence of rectangular transmission lines with thin center conductor and air dielectric, *IEEE Trans. Microwave Theor. Tech.,* 26, 238, 1978.
97. **Weil, C. M., Joines, W. T., and Kinn, J. B.,** Frequency range of large-scale TEM mode rectangular strip lines, *Microwave J.,* 24, 93, 1981.
98. **Hill, D. A.,** Bandwidth limitations of TEM cells due to resonances, *J. Microwave Power,* 18, 1983.
99. **Guy, A. W. and Chou, C. K.,** System for quantitative chronic exposure of a population of rodents to UHF fields, Biological Effects of Electromagnetic Waves, Publ. 77-8011, Bureau of Radiological Health, U.S. Department of Health, Education and Welfare, Rockville, Md., 1976, 389.
100. **Guy, A. W., Wallace, J., and McDougall, J. A.,** Circularly polarized 2450-MHz waveguide system for chronic exposure of small animals to microwaves, *Radio Sci.,* 14(6S), 63, 1979.
101. **Spiegel, R. J., Oakey, W. E., and Bronaugh, E. L.,** A variable-volume cavity electromagnetic near-field simulator, IEEE Trans. Electromagn. Compat., 22, 289, 1980.
102. **Heynick, L. N., Polson, P., and Karp, A.,** A microwave exposure system for primates, *Radio Sci.,* 12(6S), 103, 1977.
103. **Greene, F. M.,** Development of an RF near-field exposure synthesizer (10 to 40 MHz), Publ. 76-160, National Institute for Occupational Safety and Health, U.S. Department of Health, Education and Welfare, Cincinnati, 1976.
104. **Guy, A. W., Webb, M. D., and McDougall, J. A.,** RF radiation absorption patterns: human and animal modeling data, Publ. 77-183, National Institute for Occupational Safety and Health, U.S. Department of Health, Education and Welfare, Cincinnati, 1977.
105. **Chen, K. C. and Lin, C. J.,** A system for studying effects of microwaves on cells in culture, *J. Microwave Power,* 13, 251, 1978.
106. **Lin, J. C.,** A new system for investigating nonthermal effect of microwaves on cells, in Biological Effects of Electromagnetic Waves, Publ. 77-8011, Bureau of Radiological Health, Department of Health, Education and Welfare, Rockville, Md., 1976, 350.
107. **Partlow, L. M., Bush, L. G., Stensaas, L. J., Hill, D. W., Riazi, A., and Gandhi, O. P.,** Effects of millimeter-wave radiation on monolayer cell cultures. I. Design and validation of a novel exposure system, *Bioelectromagnetics,* 2, 123, 1981.
108. **McEwen, A., Tanabe, E., Vaguine, V., Williams, N., Li, G. C., and Hahn, G. M.,** Temperature controlled cavity applicator and radiofrequency wave irradiation of mammalian cells in vitro, Proc. Symp. Electromagnetic Fields Biol. Syst., Stuchly, S. S., Ed., International Microwave Power Institute, Edmonton, Alberta, Canada, 1979, 113.

109. **Guy, A. W.,** A method of exposing cell cultures to electromagnetic fields under controlled conditions of temperature and field strengths, *Radio Sci.,* 12(6S), 87, 1977.

110. **Johnson, C. C. and Guy, A. W.,** Nonionizing electromagnetic wave effects in biological materials and systems, *Proc. IEEE,* 60, 692, 1972.

111. **Larsen, L. E., Moore, R. A., and Acevedo, J.,** A microwave decoupled brain-temperature transducer, *IEEE Trans. Microwave Theory. Tech.,* 22, 438, 1974.

112. **Lords, J. L., Durney, C. H., Borg, A. M., and Tinney, C. E.,** Rate effects in isolated hearts inducted by microwave irradiation, *IEEE Trans. Microwave Theor. Tech.,* 21, 834, 1973.

113. **Tyazhelov, V. V., Tigranian, R. E., and Khizhniak, E. P.,** New artifact-free electrodes for recording of biological potentials in strong electromagnetic fields, *Radio Sci.,* 12(6S), 121, 1977.

114. **Flanigan, W. F., Bowman, R. R., and Lowell, W. R.,** Nonmetallic electrode system for recording EEG and ECG in electromagnetic fields, *Physiol. Behav.,* 18, 531, 1977.

115. **Allen, S. J.,** Measurements of power absorption by human phantoms immersed in radiofrequency fields, *Ann. N.Y. Acad. Sci.,* 247, 494, 1975.

116. **Gandhi, O. P.,** Strong dependance of whole animal absorption on polarization and frequency of radio-frequency energy, *Ann. N.Y. Acad. Sci.,* 247, 532, 1975.

117. **Chou, C. K. and Guy, A. W.,** Quantitation of microwave biological effects, in Proc. Symp. Biol. Eff. Meas. Radio Frequency/Microwaves, Hazzard, D. G., Ed., Publ. 77-8026, Bureau of Radiological Health, U.S. Department of Health, Education and Welfare, Rockville, Md., 1977, 81.

118. **Gandhi, O. P., Hagmann, M. J., and D'Andrea, J. A.,** Part-body and multi-body effects on absorption of radio-frequency electromagnetic energy by animals and models of man, *Radio Sci.,* 14(6S), 15, 1979.

119. **Kinn, J. B.,** Whole-body dosimetry of microwave radiation in small animals: the effects of body mass and exposure geometry, *Radio Sci.,* 12(6S), 61, 1977.

120. **Gandhi, O. P., Sedigh, K., Beck, G. S., and Hunt, E. L.,** Distribution of electromagnetic energy deposition in models of man with frequencies near resonance, in Biological Effects of Electromagnetic Waves, Publ. 77-8011, Bureau of Radiological Health, Food and Drug Administration, U.S. Department of Health, Education and Welfare, Rockville, Md., 1976, 44.

121. **Allen, S. J., Hurt, W. D., Krupp, J. H., Ratliff, J. A., Durney, C. H., and Johnson, C. C.,** Measurement of radiofrequency power absorption in monkeys, monkey phantoms, and human phantoms exposed to 10—50 MHz fields, in Biological Effects of Electromagnetic Waves, Publ. 77-8011, Bureau of Radiological Health, Food and Drug Administration, U.S. Department of Health, Education and Welfare, Rockville, Md., 1976, 83.

122. **Carnie, A. B., Hill, D. A., and Assenheim, H. M.,** Dosimetry for a study of effects of 2.45-GHz microwaves on mouse testis, *Bioelectromagnetics,* 1, 325, 1980.

123. **Hill, D. A.,** Human whole-body radiofrequency absorption studies using a TEM-cell exposure system, *IEEE Trans. Microwave Theor. Tech.,* 30, 1847, 1982.

124. **Iskander, M. F., Massoudi, H., Durney, C. H., and Allen, S. J.,** Measurements of the RF power absorption in speroidal human and animal phantoms exposed to the near field of a dipole source, *IEEE Trans. Biomed. Eng.,* 28, 258, 1981.

125. **Gandhi, O. P.,** Conditions of strongest electromagnetic power in man and animals, *IEEE Trans. Microwave Theor. Tech.,* 233, 1021, 1975.

126. **Gandhi, O. P.,** State of the knowledge for electromagnetic absorbed dose in man and animals, *Proc. IEEE,* 68, 24, 1980.

127. **Hagmann, M. J., Gandhi, O. P., D'Andrea, J. A., and Chatterjee, I.,** Head resonance: numerical solutions and experimental results, *IEEE Trans. Microwave Theor. Tech.,* 27, 809, 1979.

128. **Hill, D. A.,** The effect of frequency and grounding on whole-body absorption of humans in E-polarized radiofrequency field, *Bioelectromagnetics,* 5, 131, 1984.

129. **Olsen, R. G.,** Distribution of absorbed energy in primate models, in Proc. Workshop Prot. Personnel Against Radiofrequency Electromagnetic Radiat., Rev. 3-81, Mitchell, J. C., Ed., U.S.A.F School of Aerospace Medicine, Brooks Air Force Base, Tex., 1981, 149.

130. **Olsen, R. G., Griner, T. A., and Pettyman, G. D.,** Far-field dosimetry in a Rhesus monkey model, *Bioelectromagnetics,* 1, 149, 1980.

131. **Stuchly, S. S., Barski, M., Tam, B., Hartsgrove, G., and Symons, S.,** Computer-based scanning system for electromagnetic dosimetry, *Rev. Sci. Instrum.,* 54, 1547, 1983.

132. **Kraszewski, A., Stuchly, M. A., Stuchly, S. S., Hartsgrove, G., and Adamski, D.,** Specific absorption rate distribution in a full-scale model of man at 350 MHz, *IEEE Trans. Microwave Theor. Tech.,* 32, 779, 1984.

133. **Stuchly, M. A., Kraszewski, A., and Stuchly, S. S.,** Exposure of human models in the near and far field — a comparison, *IEEE Trans. Biomed. Eng.,* 32, 609, 1985.

134. **Stuchly, S. S., Kraszewski, A., Stuchly, M. A., Hartsgrove, G., and Adamski, A.,** Energy deposition in a model of man in the near-field, *Bioelectromagnetics,* 6, 1985.

Chapter 2

COMPUTER METHODS FOR FIELD INTENSITY PREDICTIONS

James C. Lin

TABLE OF CONTENTS

I. INTRODUCTION

The propagation of electromagnetic waves in biological materials is governed by dielectric constant, conductivity, source configuration, and the geometrical factors that describe the tissue structure. These parameters also determine the quantity of energy a given biological body extracts from the radiation. When the radius of curvature of the body surface is large compared to the wavelength and beam width of the impinging radiation, planar tissue models may be used to estimate the absorbed energy and its distribution inside the body. Otherwise, the absorbed energy will be dictated by the size of the body, curvature of its surface, the ratio of body size to wavelength, and the source characteristics.

The purpose of this chapter is to present a selective account of electromagnetic wave propagation in biological media, with special emphasis on the energy coupling and distribution characteristics in models of biological structures. Such information is essential for analyzing the interrelationships among various observed biological effects, for separating known and substantiated effects from those that are speculative and unsubstantiated, for assessing therapeutic effectiveness of electromagnetic waves, and for extracting diagnostic information from field effects.

There exists a wide variety of methods for quantifying fields in biological bodies. The extent of computer usage varies, depending on specific information sought and complexity of tissue geometry. The whole-body absorption of electromagnetic energy by humans and laboratory animals is of interest because it is related to the energy required to alter the thermoregulatory system of the exposed subject, and because it may serve as an index for extrapolating experimental results to human exposures. The distribution of absorbed electromagnetic waves within an irradiated body is important because (1) it relates to specific responses of the body, (2) it facilitates understanding of phenomena, and (3) it contributes to definition of mechanisms of interaction.

This chapter outlines a number of techniques that have been successfully employed to analyze the propagation and absorption characteristics of electromagnetic radiation in tissue structures. There are two general approaches: one involves extensive use of analytical development and the other relies more heavily on numerical formulation. Analytical procedures are most suited for calculation of distribution of absorbed energy in simplified tissue geometries such as plane slabs, cylinders, and spheroids, whereas numerical methods offer the prospect of analyzing the coupling of electromagnetic energy to animal and human bodies which is difficult, if not impossible, to approach analytically. The advantages and limitations of various methods for field computations, along with representative results, are considered in the following sections.

II. PLANAR TISSUE GEOMETRIES

A. Semi-Infinite Layers

The reflection and transmission of a plane wave at a planar tissue interface depend on the frequency, polarization and angle of incidence of the wave, and on the dielectric constant and conductivity of the tissue. For a linearly polarized plane wave impinging normally on a boundary separating two semi-infinite media, the reflection and transmission coefficients are respectively given by

$$R = \frac{\eta_2 - \eta_1}{\eta_2 + \eta_1} \tag{1}$$

and

Table 1
REFLECTION COEFFICIENT (MAGNITUDE IN PERCENT) BETWEEN BIOLOGICAL TISSUES AT 37°C

	Frequency (MHz)	Air	Fat (bone)	Lung	Muscle (skin)	Blood	Saline
Air	433	0	46	76	82	81	83
	915	0	43	73	78	79	80
	2,450	0	41	71	76	77	79
	5,800	0	39	70	75	76	78
	10,000	0	37	70	74	76	78
Fat (bone)	433		0	46	56	56	60
	915		0	43	52	54	57
	2,450		0	42	50	53	57
	5,800		0	42	50	53	56
	10,000		0	45	52	54	58
Lung	433			0	14	13	19
	915			0	12	14	18
	2,450			0	10	15	19
	5,800			0	10	14	19
	10,000			0	10	13	18
Muscle (skin)	433				0	4	6
	915				0	4	7
	2,450				0	5	10
	5,800				0	4	9
	10,000				0	3	9
Blood	433					0	6
	915					0	4
	2,450					0	5
	5,800					0	5
	10,000					0	6
Saline	433						0
	915						0
	2,450						0
	5,800						0
	10,000						0

$$T = \frac{2\eta_2}{\eta_2 + \eta_1} \tag{2}$$

where η_1 and η_2 are the intrinsic impedance of the media. When $\eta_1 \cong \eta_2$, e.g., the electromagnetic properties of the media are approximately equal, there is no reflection and the transmission is maximum. In contrast, there is complete reflection when η_2 is zero. Note that these equations are sufficiently simple to be easily programmed on a hand-held calculator to assess electromagnetic reflection and transmission behavior at planar tissue interfaces.

Table 1 summarizes the magnitude of the reflection coefficient at the boundary separating various tissues. The fraction of normally incident power reflected by the discontinuity is given by R^2. Clearly, about one half of the incident power is reflected at these boundaries. Further, the reflection coefficient for tissue-tissue interfaces generally is smaller than air-tissue interfaces. The values of percent reflected power for tissue-tissue interfaces range

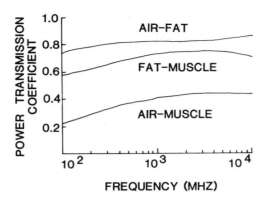

FIGURE 1.   Power transmission coefficients at three tissue interfaces as functions of frequency.

Table 2
DEPTH OF PENETRATION OF AN ELECTROMAGNETIC WAVE IN BIOLOGICAL TISSUES AS A FUNCTION OF FREQUENCY

Frequency (MHz)	Saline	Blood	Muscle (skin)	Lung	Fat (bone)
			Tissue		
		Depth of Penetration (cm)			
433	2.8	3.7	3.0	4.7	16.3
915	2.5	3.0	2.5	4.5	12.8
2,450	1.3	1.9	1.7	2.3	7.9
5,800	0.7	0.7	0.8	0.7	4.7
10,000	0.2	0.3	0.3	0.3	2.5

from a low of 5 for muscle-blood to a high of 50 for bone-biological fluid interfaces. This suggests that the closer are the dielectric properties on both sides of the interface, the smaller is the power reflection.

The fraction of transmitted power is related to the power transmission coefficient $(1 - R^2)$. It is readily apparent from Table 1 that the transmitted power at air-tissue interfaces is quite substantial at radio and microwave frequencies. Moreover, Figure 1 shows that the power transmission coefficient is highly frequency dependent, especially at lower frequencies.

As the transmitted wave propagates in the tissue medium, energy is extracted from the wave and absorbed by the medium. This absorption will result in a progressive reduction of the power density of the wave as it advances in the tissue. This reduction is quantified by the depth of penetration δ, which is the distance in which the power density decreases by a factor of e^{-2}. Table 2 presents the calculated depth of penetration in selected tissues using typical dielectric constants and conductivities provided in Part I. A graphical representation of penetration depth vs. frequency for blood, muscle, and fat is given in Figure 2. It is seen that δ is frequency dependent and takes on different values for different tissues. In particular, the penetration depth for fat and bone is nearly five times greater than for higher-water-content tissues.

A wave of general polarization usually is decomposed into its orthogonal linearly polarized

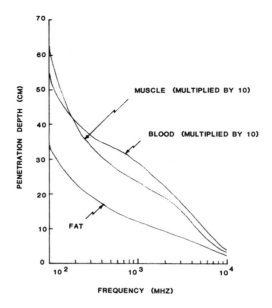

FIGURE 2. Depth of penetration for blood, muscle, and fat as functions of frequency.

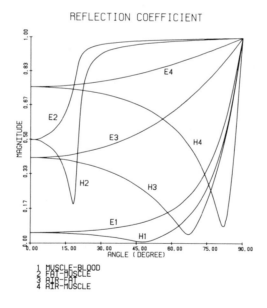

FIGURE 3. Magnitudes of reflection coefficient for E and H polarized plane waves at 2450 MHz.

components whose electric or magnetic field parallels to the interface. These components can be treated separately and combined afterward. Figures 3 and 4 illustrate the magnitude and phase of the reflection coefficients of representative tissue interfaces at a temperature of 37°C for irradiation at 2450 MHz. The figures clearly show the difference between E and H polarization. E polarization, also called perpendicular polarization, and H polarization,

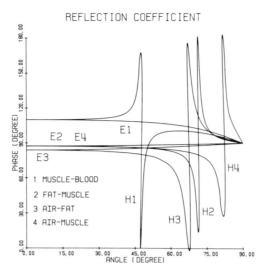

FIGURE 4.   Phase of reflection coefficient for E and H polarized 2450-MHz plane waves.

also referred to as parallel polarization, are defined in Figure 13 of the "Introduction" section of this book. For E polarization, there is only a slight variation in magnitude and phase of the reflection coefficient with incidence angle. For H polarization, however, there is a pronounced dependence on incidence angle. The reflection coefficient reaches a minimum magnitude and has a phase angle 90° at the Brewster angle. Thus, the H-polarized wave is totally transmitted into the muscle medium at the Brewster angle.

B. Multiple Layers

When there are several layers of different tissues, the reflection and transmission characteristics become more complicated. Multiple reflections can occur between the skin and s.c. tissue boundaries, with a resulting modification of the reflection and transmission coefficients.[1,2] In general, the transmitted wave will combine with the reflected wave to form standing waves in each layer. This phenomenon becomes especially pronounced if the thickness of each layer is less than the penetration depth for that tissue medium. Plane waves impinging on the human body considered as consisting of parallel layers of s.c. fat and more deeply lying muscle have been studied in detail by Schwan and Li.[3]

For the situation depicted in Figure 5, the electric field strength in the fat layer is given by

$$E_f = F_\ell E_0 \left[e^{-(\alpha_2 + j\beta_2)z} + R_{32} \, e^{(\alpha_2 + j\beta_2)z} \right] \tag{3}$$

and the electric field in the underlying muscular tissue is given by

$$E_m = F_t E_0 e^{-(\alpha_3 + j\beta_3)z} \tag{4}$$

where α_2, β_2 and α_3, β_3 are the attenuation and propagation coefficients in fat and muscle, respectively. The layer function $F\ell$ and the transmission function F_t are given by

$$F_\ell = T_{12} / \left[e^{(\alpha_2 + j\beta_2)\ell} + R_{21} R_{32} e^{-(\alpha_2 + j\beta_2)\ell} \right] \tag{5}$$

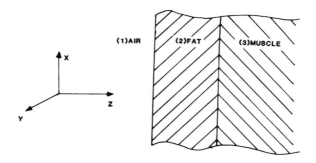

FIGURE 5. Plane wave impinging on a composite fat-muscle layer.

$$F_t = (T_{12}T_{23})/[e^{(\alpha_2 + j\beta_2)\ell} + R_{21}R_{32}e^{-(\alpha_2 + j\beta_2)\ell}] \tag{6}$$

where T_{12} and T_{23} are the transmission coefficients at the air-fat and fat-muscle boundaries, respectively. R_{21} and R_{32} denote, respectively, the reflection coefficients at these boundaries; ℓ is the thickness of the fat layer. The power absorption in a given layer can be obtained from the equation

$$P_a = \frac{1}{2} \sigma E_f^2 \tag{7}$$

where σ is the electrical conductivity.

Figure 6 shows the numerical results obtained using the above equations and the dielectric data given in Part I. The values are normalized to the specific absorption rate (SAR) (see Equations 1 and 2 in Part III, Chapter 1) in muscle at the fat-muscle boundary. Note the absorbed energy is much lower in fat than in muscle. The standing wave maximum becomes bigger in fat and the penetration into muscle becomes less as the frequency increases.

The absorbed microwave energy in models composed of planar layers of skin, fat, and muscle can be analyzed in a similar manner,[3,4] except the distribution of absorbed energy becomes more complex. Figure 7 shows that in addition to frequency dependence, the peak SARs exhibit considerable fluctuation with thickness of the s.c. fatty layer. The incident power density is, in this case, 10 W/m² and the skin layer is 0.2 cm thick. Note that the peak SARs are always higher in the skin layer for planar models at microwave frequencies. The depth of penetration for 10 GHz radiation in skin is less than 0.05 cm; thus, the transmitted energy is almost completely absorbed in the skin and the SAR is rather unaffected by changing fatty layer thickness. The fact that SAR is highest in the skin is significant, since skin is populated with thermosensitive free nerve endings which may be excited along with cutaneous pain receptors when the absorbed energy exceeds the normal range that can be handled by thermoregulation.

Figure 8 shows the distribution of induced electric field strength in a layer of muscle beneath layers of fat, muscle, and bone for two frequencies.[2] It is seen that in addition to frequency dependence, the electric fields exhibit considerable fluctuation within each tissue layer. While the standing-wave oscillations become bigger at 2450 than 915 MHz, microwave energy at both frequencies can penetrate into more deeply situated tissues. This result, together with Figures 6 and 7, implies that at frequencies between 300 and 3000 MHz, sufficient energy may be transmitted and reflected to allow interrogation of organs within the body. Furthermore, at these frequencies, electromagnetic energy can penetrate into more deeply situated tissues, making it most hazardous to humans in an uncontrolled situation.

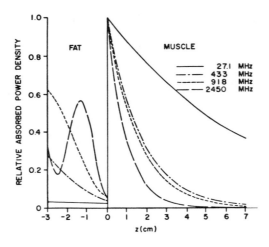

FIGURE 6.   SAR distributions (absorbed power density) in planar fat-muscle layers. (From Johnson, C. C. and Guy, A. W., *Proc. IEEE,* 60, 692, 1972. With permission. © 1972 IEEE.)

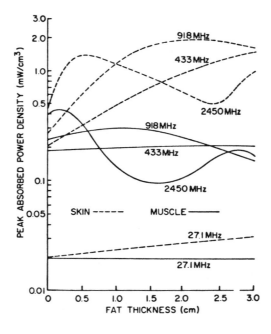

FIGURE 7.   Peak SAR (absorbed power density) in models composed of skin-fat-muscle layers. (From Johnson, C. C. and Guy, A. W., *Proc. IEEE,* 60, 692, 1972. With permission. © 1972 IEEE.)

III. BODIES OF REVOLUTION

Although depth of penetration and reflection and transmission characteristics in planar tissue structures provide considerable physical insight into coupling and distribution of radio and microwave energy, biological bodies generally are more complex in form and exhibit

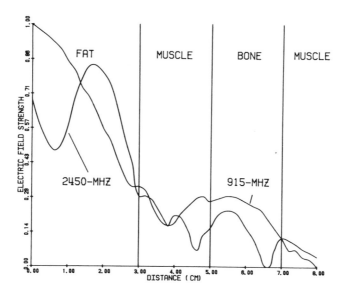

FIGURE 8. Distribution of electric field strength in planar layers of fat-muscle-bone-muscle tissue model.

substantial curvature that can modify radio and microwave energy transmission and reflection. For bodies with complex shape, the propagation characteristics depend critically on polarization and on orientation of the incident wave with respect to the body, as well as the ratio of body size to wavelength. These complications place severe limitations on reflection and transmission calculations for bodies of arbitrary shape and complex permittivity. This section presents a summary of analytic/numeric results that have been obtained for homogeneous and multilayered models based on bodies of revolution that approximate mammalian tissue structures.

A. Spherical Models

The distribution of the electric field induced inside a homogeneous dielectric sphere by a plane wave propagating in the positive z direction and polarized along the x axis is given by

$$\underline{E} = E_0 \, e^{j\omega t} \sum_{n=1}^{\infty} (-j)^n \frac{2n + 1}{n(n + 1)} [a_n^t \, \underline{m}_{oln}^{(1)} + jb_n^t \, \underline{n}_{eln}^{(1)}] \tag{8}$$

where a_n^t and b_n^t denote the magnetic and electric modes of oscillations inside the sphere, respectively, and $\underline{m}_{oln}$ and $\underline{n}_{eln}$ are two independent solutions called vector spherical wave functions.[5] They are all functions of the wavelength, dielectric constant, conductivity, and size of the sphere. The infinite series of Equation 8 is readily adaptable for computer calculation using the numerical values of dielectric permittivity and conductivity given in Part I. The power absorption is once again given by Equation 7 and the total time rate of energy absorption is obtained from

$$W_a = W_t - W_s \tag{9}$$

with the time rate of total energy derived from the incident waves and the energy scattered by the sphere given by

$$W_t = - \frac{E_0^2}{120k_0^2} \text{Re} \sum_{n=1}^{\infty} (2n + 1) (a_n^r + b_n^r) \tag{10}$$

$$W_s = \frac{E_0^2}{120k_0^2} \sum_{n=1}^{\infty} (2n + 1) (|a_n^r|^2 + |b_n^r|^2) \tag{11}$$

and

$$a_n^r = a_n^t \left[\frac{j_n (Nk_0a)}{h_n (k_0a)} - \frac{j_n (k_0a)}{h_n (k_a)} \right] \tag{12}$$

$$b_n^r = \left[b_n^t \frac{N j_n (Nk_0a)}{h_n (k_0a)} - \frac{j_n (k_0a)}{h_n (k_0a)} \right] \tag{13}$$

Where $N^2 = \epsilon$ is the relative dielectric permittivity, $k_0 = 2\pi/\lambda_0$, and μ_0, ϵ_0, and λ_0 are the permeability, permittivity, and wavelength in vacuum, respectively; j_n and h_n are the spherical Bessel functions and a is the radius of the sphere.

Some representative calculations of the SAR are shown in Figure 9 for two models at 918 and 2450 MHz.[2,6] The 6-cm-diameter sphere approximates a cat or rhesus monkey-size brain, while the 14-cm-diameter sphere is more typical of a human adult-size brain. The figures illustrate the SAR distributions along the three rectangular coordinate axes whose origin coincides with the center of the sphere. An incident plane wave power density of 10 W/m² is assumed. It is seen that for 918 MHz, maximum absorption occurs near the center of both the cat- and human-size brain sphere. When the frequency is increased to 2450 MHz, the location of peak SAR for the cat-size brain sphere remains near the center, whereas that for a human-size brain sphere is moved to an anterior location.

In general, standing wave patterns with many oscillations are observed. Note that while peak and average SAR in the cat brain are larger by a factor of two than in the human brain at 918 MHz, the peak absorption is four times and the average absorption is three times greater in the cat brain than in the human brain. A comparison of available studies[6-9] indicates that the peak absorption may be as much as five times greater than the average, and the enhanced absorptions near the center of these brain models may be two to three orders of magnitude greater than that expected from the planar tissue models. The increased absorptions are due to a combination of high dielectric constant and curvature of the model, which produces a strong focusing of energy toward the interior of the sphere that more than compensates for the transmission losses through the tissue.[10]

The peak absorption per unit volume, average absorption per unit volume, and average absorption per unit surface area as functions of frequency and radius of spherical brain model are illustrated in Figure 10. It can be seen that the absorbed energy varies widely with sphere size and frequency. In general, the absorption initially increases rapidly with an increasing radius and is then followed by some resonant behavior. The peaks of these resonant oscillations are related to the maxima, or hot spots, in the distribution of absorbed energy inside the head model, as shown in Figure 9. Therefore, $(2\pi a/\lambda_0) < 0.4$, where a is the sphere radius and λ_0 is the wavelength in vacuum. In general, hot spots do not occur inside the sphere. However, for some combinations of irradiation frequency and radius hot spots will occur, e.g., in spheres with radii between 2 and 8 cm at 918 MHz and between 0.9 and 5 cm at 2450 MHz. For spheres whose radii exceed the size ranges mentioned above, the maximum absorption appears at the anterior portion (exposed surface) of the sphere, and the penetration depth at the surface becomes a dominating factor. The planar model discussed in Section II may be applied to obtain a theoretical estimation of the absorbed energy in this case.

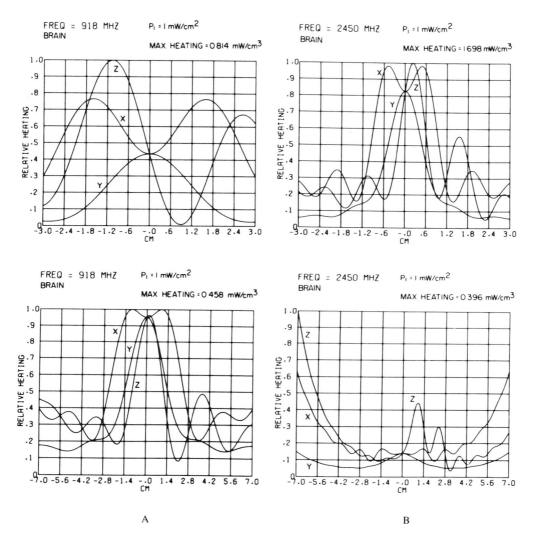

FIGURE 9. Predicted SAR distribution (heating pattern) along the three rectangular axes of spherical models of brain exposed to plane waves. (From Lin, J. C., Guy, A. W., and Kraft, G. H., *J. Microwave Power*, 8, 275, 1973. With permission.)

The frequency dependence of energy absorption is illustrated in the upper graphs in Figure 10 for the head of a small animal, such as a cat or rhesus monkey, and a human-head-size sphere. Besides the occurence of resonant peaks at these frequencies, with increasing frequency, energy is absorbed in a decreasingly smaller volume as result of shortened penetration depth.

The effects of skin, fat, bone, dura, and cerebrospinal fluid on the absorption of radio frequency radiation by the brain have been investigated in several laboratories[7,11-14] using more complex spherical structures where the spherical core of brain is surrounded by five concentric shells of tissues (see Table 3). It is interesting to note that if brain sizes remain unchanged, but the overall sphere diameter is increased to account for the outer tissue layers, absorption in brain tissues may be increased by 25% for human- and cat-size heads at 918 MHz or decreased by 70% or more in the case of 2450 MHz (see Figures 11 and 12).

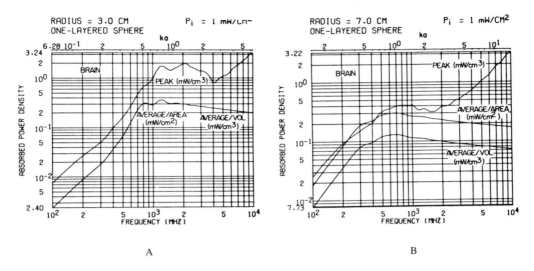

FIGURE 10. Electromagnetic energy absorption characteristics for spherical models of brain.

Table 3
COMPOSITION OF A SIX-LAYER MODEL
OF A MAMMALIAN HEAD

	Dielectric constant (MHz)		Thickness (cm)	
	918	2450	3.5	10
Brain	34.42-j 15.49	32.78-j 15.37	2.88	6.98
CSF	80.85-j 14.05	77.0-j 13.94	0.20	1.10
Dura	51.40-j 25.08	47.52-j 11.42	0.05	0.8
Bone	5.56-j 0.856	5.0-j 0.857	0.20	0.7
Fat	5.56-j 0.856	5.0-j 0.857	0.07	0.27
Skin	51.40-j 25.08	47.52-j 11.42	0.10	0.15

Moreover, surface absorption is greatly increased in the case of layered models, while fat and bone always absorb the least amount of energy.

If the outer diameter of the sphere remains the same, while the composition of the tissue layers is allowed to be either layered or homogeneous, the peak and average SARs show very little change except when the radius of the spherical head is between 0.1 and 1.0 times the wavelength in air. The peak and average SARs for layered models may be several times as great as for homogeneous models. Enhancement is apparently the result of resonant coupling of energy into the sphere by the outer tissue layers.

It should be mentioned that the computed absorption characteristics are highly sensitive to the dielectric constant and conductivity assumed as a function of frequency. There are two divergent sets of brain dielectric constant and conductivity in the literature.[2,8,13,15,16] Depending on the particular values chosen for these quantities, the interpretation of the relative hazard at 2450 MHz has varied from least to most hazardous.[7,11,13,14] The conclusion that 2450 MHz represents an extremely hazardous frequency for humans most likely stems from the inordinately high (by a factor of 2) dielectric constant used in these computations.[17-19]

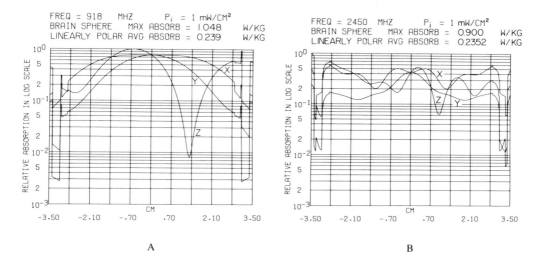

FIGURE 11. Distribution of absorbed energy in a six-layered spherical model of the cat head. (From Lin, J. C., *IEEE Trans. Biomed. Eng.*, 23, 371, 1976. With permission. © 1976 IEEE.)

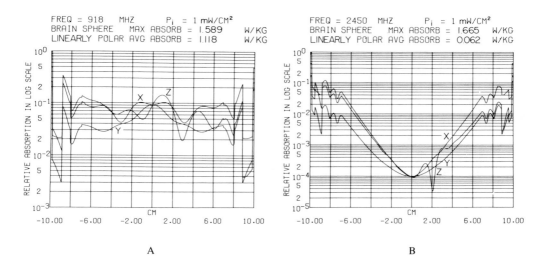

FIGURE 12. SAR distribution in a six-layered spherical model of an adult human head. (From Lin, J. C., *IEEE Trans. Biomed. Eng.*, 23, 371, 1976. With permission. © 1976 IEEE.)

A study also has been made of the interaction of circularly polarized plane electromagnetic waves with six-layered spherical models of the mammalian head.[14] The approach is a classic one; Mie equations as given in Equations 8 to 13 are modified to account for the two polarizations that are orthogonal in space. Two sets of representative calculations are shown in Figures 13 and 14 for 918 and 2450 MHz, respectively. The 7-cm-diameter sphere represents a cat- or monkey-size head, while the 20-cm-diameter sphere is typical of a human adult-size head. In each case, the absorbed power distribution is normalized to the maximum along the z axis. An incident power density of 10 W/m² is again assumed.

As for the case of linearly polarized plane waves, the maximum absorption for 918 MHz occurs near the center of a cat-size head, whereas the maximum absorption for a human-

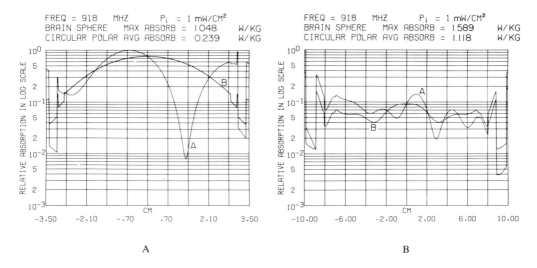

FIGURE 13.   Relative absorption of circularly polarized 918-MHz radiation in spherical models of mammalian head. (From Lin, J. C., *IEEE Trans. Biomed. Eng.*, 23, 371, 1976. With permission. © 1976 IEEE.)

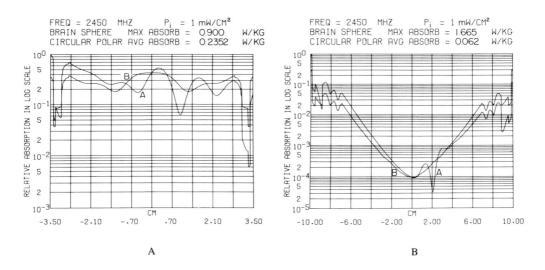

FIGURE 14.   Circularly polarized 2450-MHz microwave-induced SAR distribution in spherical models of mammalian head. (From Lin, J. C., *IEEE Trans. Biomed. Eng.*, 23, 371, 1976. With permission. © 1976 IEEE.)

size head is at the surface. When the frequency is increased to 2450 MHz the location of maximum absorption for both the smaller and larger spheres shifts to the leading surface. The distribution of absorbed energy for circularly polarized waves is more uniform compared with the linearly polarized case. In fact, the absorbed energy distribution in the planes transverse to the direction of propagation is rotationally symmetric, i.e., it is independent of angular variation (see the B curves). Note also that the maximum energy absorbed in the spherical head models varies only slightly between the two frequencies studied. However, a greater quantity of energy is deposited in the inner sphere (representing the brain of a human head) for 918 than 2450 MHz radiation.

A number of investigations[2,20-22] have used muscle spheres as a first-order approximation

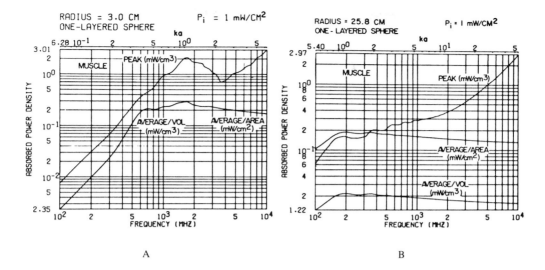

FIGURE 15.   Absorption characteristics of spherical body models.

for the extrapolation to human beings of results obtained from laboratory animals and as an index to whole-body absorption of electromagnetic energy as a function of frequency. The spherical model is attractive since exact solutions for absorbed energy can be obtained for all frequencies and body sizes. The absorptions for homogeneous muscle spheres, whose volumes correspond to small animals, such as a rat and standard man, computed as functions of frequency, are shown in Figure 15. While in this case the peak absorption is of very limited utility, the average absorption per unit surface area is related to the time and power required to overload the thermoregulatory capacity of an exposed object. Note that the average absorption for the rat model is at least ten times higher than for a muscle sphere representing a human body at frequencies greater than 500 MHz. Absorptions increase, in both cases, rapidly with frequency until the sphere diameter approaches the free space wavelength of impinging radiation. A number of resonant oscillations appear which tend to increase the amount and nonuniformity of absorbed energy. Above this range the absorption falls off slowly, indicating that details of body surface curvature are of little significance.

We have, thus far, dealt mainly with situations where the diameter of the sphere is comparable to or larger than a wavelength in air. When the sphere is small compared with a wavelength, the absorbed energy distribution varies almost as the square of the radius or distance from the magnetic axis (the axis parallel to the direction of magnetic field vector) as shown in Figure 16. If the sphere is extremely small compared with a wavelength, the absorbed energy distribution becomes nearly uniform in the x and y directions, but decreases continuously with distance from the exposed surface (see Figure 17). This behavior can be explained by a quasi-static field theory.[21] Accordingly,, for a plane wave polarized in the x direction which propagates in the z direction, the induced electric field inside a dielectric sphere is given by

$$\underline{E}_t = E_0 e^{j\omega t} \left[\frac{3}{\epsilon} \hat{x} - j \frac{kr}{2} (\cos\phi \ \hat{\theta} - \cos\theta \ \sin\phi \ \hat{\phi}) \right] \qquad (14)$$

where E is the strength of the incident electric field and r is the radial variable. The whole-body absorption rate is given by

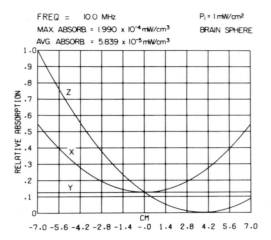

FIGURE 16.   SAR distribution along the x, y, and z axes of a small brain sphere. (From Lin, J. C., Guy, A. W., and Johnson, C. C., *IEEE Trans. Microwave Theor. Tech.*, 21, 791, 1973. With permission. © 1973 IEEE.)

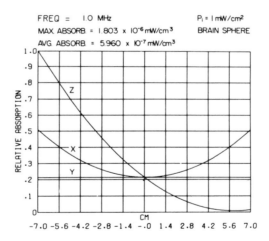

FIGURE 17.   Predicted SAR distribution in a sphere whose size is extremely small compared to the wavelength. X, Y, and Z are the orthogonal coordinates of a rectangular system. (From Lin, J. C., Guy, A. W., and Johnson, C. C., *IEEE Trans. Microwave Theor. Tech.*, 21, 791, 1973. With permission. © 1973 IEEE.)

$$P_a = \frac{1}{2} \sigma E_0^2 V \left[\frac{9}{|\epsilon|^2} + \frac{1}{10} (ka)^2 \right] \tag{15}$$

where V and a are, respectively, the volume and radius of the spherical model.

The electric field component of the incident plane wave couples to the object in the same fashion as an electrostatic field. This gives rise to a constant induced electric field inside the sphere which has the same direction but is reduced by $3/\epsilon$ from the applied electric field

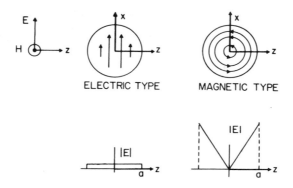

FIGURE 18. A sketch illustrating the behavior of electric and magnetic fields under quasi-static conditions of irradiation. (From Lin, J. C., Guy, A. W., and Johnson, C. C., *IEEE Trans. Microwave Theor. Tech.*, 21, 791, 1973. With permission. ©1973 IEEE.)

for biological materials and is independent of sphere size. Similarly, the magnetically induced electric field inside the body is identical to the quasi-static solution whose magnitude is given by $E = \pi f \mu r H$, where f is the frequency, μ is the permeability, r is the radius, and H is the magnetic field component. Thus, the magnetic field component of the incident plane wave produces an internal electric field that varies directly with distance away from the axis and in proportion to the frequency.

A sketch depicting how the induced electric fields combined inside the sphere is shown in Figure 18. It is seen that the magnetically induced electric field encircles the y axis (magnetic axis) and gives rise to an eddy current whose magnitude increases with distance away from the y axis. This indicates that while the H-induced energy absorption in a mouse or larger animal is much greater than the E-induced component, electrically and magnetically induced absorption may be equally significant in even smaller animals at lower radio frequencies (RFs) (below 30 MHz). Moreover, for a small insect or pupae the electric field will be the predominant factor.

The variation of average and maximum energy absorption with frequency for a human-size sphere is illustrated in Figure 19. In the frequency range from 1 to 20 MHz, the maximum absorption rate is only 10^{-6} to 10^{-3} W/kg/W/m² of incident power. Inspection of the maximum absorption rate induced by a plane wave, a quasi-static electric field, and a quasi-static magnetic field shows that absorption in the frequency range between 1 and 20 MHz is primarily due to the magnetic induction and is characterized by a square-of-frequency dependence. The approximate frequency dependence of average or total energy absorption throughout the frequency range from 1 MHz to 10 GHz is indicated by the dashed line. For frequencies between 1 and 20 MHz the average absorption varies as the square of the frequency. In the frequency range of 20 to 200 MHz, the average absorption increases directly in proportion to frequency and attains a maximum of about 2×10^{-3} W/kg/W/m² of incident power at 200 MHz. The average absorption rate remains fairly constant with increasing frequency. (It actually is inversely proportional to frequency for higher frequencies.) There is thus little doubt that electromagnetic energy absorption varies both with frequency and body size, and in a predictable manner.

B. Prolate Spheriodal Models

Since the bodies of humans and experimental animals are seldom spherical in shape, we need a more appropriate model to analytically and numerically describe the induced fields

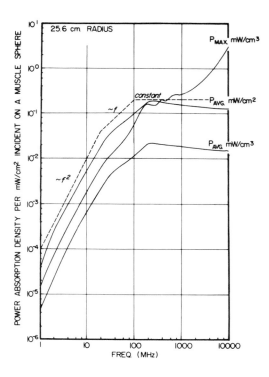

FIGURE 19.   Frequency dependence of absorption in
a spherical model of the human body. (From Lin, J. C.,
Guy, A. W., and Johnson, C. C., *IEEE Trans. Microwave Theor. Tech.*, 21, 791, 1973. With permission.
© 1973 IEEE.)

and absorbed energy inside experimental subjects. A prolate spheroid emulates more closely
the shape of mammalian bodies, but most analyses have been restricted to homogeneous
models for humans and experimental animals.[23-29] The basic approach[30] is to expand the
incident, scattered, and transmitted electric fields in terms of vector wave functions in
spheroidal coordinates. The expansion coefficients are determined from the boundary conditions demanding that the tangential components of the fields must be continuous across
the surface of the spheroid. The absorbed power is obtained from Equation 7 and the average
absorptions are found from relationships similar to those of Equations 9 to 13.

In addition, for frequencies below resonance (resonance is defined as the condition of
maximum absorption), long-wavelength formulations[24,25] and quasi-static approximations[31]
have been used to obtain absorption information. More recently, geometric-optics approximations have been developed for computation of absorption characteristics of prolate spheroidal models of humans at frequencies whose wavelengths are short compared with body
size.[19]

Three orientations of the impinging plane wave with respect to the body must be distinguished (see Figure 20): E-polarization in which the electric field is parallel to the major
axis of the spheroid, H-polarization in which the magnetic field vector is parallel to the
major axis, and K-polarization in which both electric and magnetic field vectors are perpendicular to the major axis of the spheroid. In general, E-polarization produces the highest
energy absorption for frequencies up to and slightly beyond the resonance region.

The induced electric fields inside a dielectric prolate spheroid (semimajor axis, a; semi-
minor axis, b) in a plane wave electromagnetic field with long wavelength ($\lambda > a$) may be
represented by

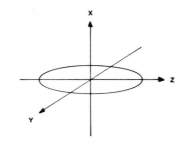

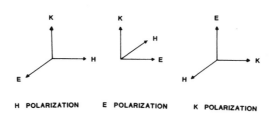

H POLARIZATION   E POLARIZATION   K POLARIZATION

FIGURE 20. Polarization states of a plane wave impinging on a prolate spheroidal body model.

$$\underline{E}_e = \frac{E_0}{C_1} \hat{z} - j^{1/2}E_0 \frac{\omega\mu_0}{377} (z \cos\phi \, \hat{\rho} - z \sin\phi \, \hat{\phi} - \rho \cos\phi \, z) \qquad (16)$$

for E-polarization and

$$\underline{E}_n = \frac{-E_0\hat{y}}{C_2} + \frac{1}{2} E_0 \frac{\omega\mu_0\rho\hat{\phi}}{377} \qquad (17)$$

for H-polarization. The whole-body energy absorption rate is generally given by

$$P_a = \frac{1}{2} \sigma \int_{Volume} E \cdot E^* \, dv \qquad (18)$$

where $\sigma = \omega\epsilon_0\epsilon''$ is the electrical conductivity and E^* denotes the complex conjugate of the induced field E inside the body. The whole-body rates of absorbed energy, by substituting Equations 16 and 17 into 18, are

$$P_{ae} = \frac{1}{2} \sigma E_0^2 \left(\frac{4}{3} \pi ab^2\right) \left[\frac{1}{C_1^2} + \frac{1}{20} \left(\frac{\omega\mu_0}{377}\right)^2 (a^2 + b^2)\right] \qquad (19)$$

for E-polarization and

$$P_{ah} = \frac{1}{2} \sigma E_0^2 \left(\frac{4}{3} \pi ab^2\right) \left[\frac{1}{C_2^2} + \frac{1}{10} \left(\frac{\omega\mu_0}{377}\right)^2 b^2\right] \qquad (20)$$

for H-polarization.[31] The constants C_1 and C_2 are given by

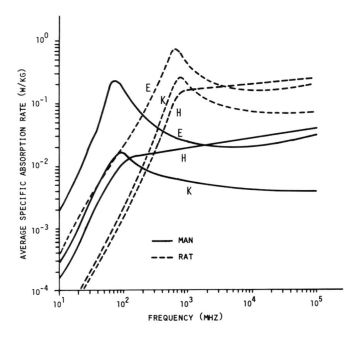

FIGURE 21. Predicted frequency dependence of absorbed energy in spheroidal models of biological bodies.

$$C_1 = (1 - \epsilon)\eta_0[(1 - \eta_0^2)\coth^{-1}\eta_0 + \eta_0] + \epsilon$$

$$C_2 = 1 + \frac{1}{2}(\epsilon - 1)\eta_0[(1 - \eta_0^2)]\coth^{-1}\eta_0 + \eta_0]$$

where $\epsilon = \epsilon' - j\epsilon''$ is the complex relative permittivity and η_0 is the eccentricity given by

$$\eta_0 = \frac{a}{(a^2 - b^2)^{1/2}}$$

Since C_1 and C_2 do not depend on spatial variables, the induced fields within the spheroid are uniform and independent of size when the external field is uniform. For $\epsilon > 1$, the field inside the spheroid is weaker than the applied field. Moreover, the whole-body energy absorption depends not only on the strength of impressed fields, but also on the orientation of the field with respect to the major axis of the body.

As in the case of spherical models, the absorption is produced by an electrically induced current in the direction of the applied E-field vector, combined with a circulating current induced by the incident H field. One would therefore expect the electrically induced absorption to be uniform, whereas the absorption due to the circulating eddy current would be zero at the center and increase as square of the distance from the center.

Computations for the absorbed energy as a function of frequency and body size have been made using homogeneous muscle material to serve as an index of SARs in mammals and to serve as a guide for extrapolating data from experimental animals to human beings, particularly with regard to averages of absorbed energy.[32] Figure 21 shows the theoretically projected frequency dependence of absorbed energy for humans and for laboratory rats. Note that for a given incident field orientation, the average SAR for humans may either be higher or lower than for rats, depending on the frequency. For example, at 70 MHz, the average

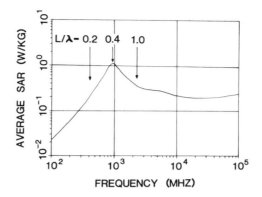

FIGURE 22.   The three distinct regions of absorption
as a function of frequency.

SAR is the highest for humans, having a value of 0.25 W/kg for an incident power density
of 10 W/m^2; the average SAR for a rat is only 0.0125 W/kg. In contrast, the average SAR
of 0.8 W/kg at 700 MHz is the highest for rats; the corresponding value for humans is less
than 1/25. It is thus extremely important to take into account the body size and operating
frequency to draw any relationship between the biological effects that arise in the laboratory
and corresponding effects that might occur in humans at a given incident power density.

The frequency for maximal absorption (resonance frequency) depends on the subject and
its orientation with respect to the incident field. In general, the shorter the subject, the higher
the resonance frequency and vice versa. Further, the frequency dependence of whole-body
or average absorption may be partitioned into three regions. This may be illustrated using
the orientation that is most efficient in energy coupling, E-polarization. Figure 22 gives the
whole-body absorption data for a prolate spheroidal model of the rat irradiated with plane
waves in free space. For frequencies well below resonance such that the ratio of the longest
body dimension (L) to free space wavelength (λ) is less than 0.2, the average SAR is
characterized by a f^2 dependence. The average absorption goes through a resonance in the
region where $0.2 < L/\lambda < 1.0$. In this case, the average SAR rapidly increases to a
maximum near $L/\lambda = 0.4$ and then falls off as $1/f$. At frequencies for which $L/\lambda > 1.0$,
the whole-body absorption decreases slightly but approaches asymptotically the geometrical
optics limit of about one half of the incident power (1 − power reflection coefficient).

It should be noted that the resonant absorption length of 0.4 λ is in good agreement with
results from antenna analysis. In addition, whole-body absorptions for H- and K-polarizations
are totally different. The resonances are not nearly as well defined as for E-polarization. In
fact, the whole-body absorption curve for H-polarization gradually reaches a plateau and
stays at that plateau for higher frequencies.

The distribution of absorbed energy in a prolate spheroidal model whose size is small
compared with wavelength is shown in Figure 23 for the three distinct polarizations. The
height of the prolate spheroid equals 1.75 m with a major-to-minor axis ratio of 6.34 and
a 70-kg mass corresponding to a human body. The dielectric constant and conductivity are
those for muscular tissue at 10 MHz. These graphs are qualitatively similar to those for
spherical models. As expected, the absorbed energy is highest for E-polarization. In fact,
there is approximately an order of magnitude difference in the peak rate of absorbed energy,
depending on the polarization.

For E-polarization the eddy current is zero on the x axis, thus a low relative absorption
is seen along the x axis which is only due to the electrically induced current. The absorption
rate is elevated considerably, however, indicating a strong coupling of the impressed electric

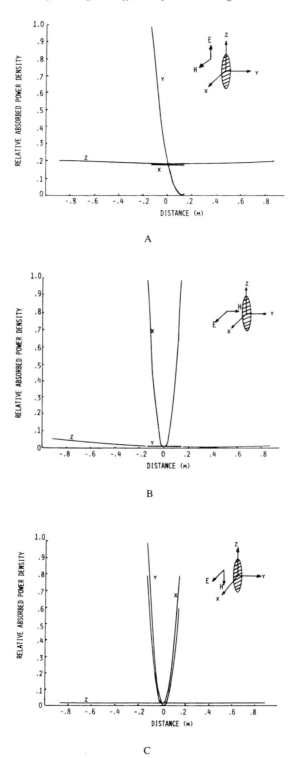

FIGURE 23.   SAR distributions along the x, y, and z axes of
a prolate spheroidal model of man. The maximum absorptions
are 6.14, 1.93, and 0.64 mW/kg for E, K, and H polarizations,
respectively. The incident power density is 1 W/m².

field into the interior of the prolate spheroid. The slender spheroid forces the internal electric field to more closely correspond to the incident field because of the boundary condition that requires continuity of tangential fields. The relatively flat energy distribution along the z axis again shows the strong coupling of impressed electric field and relatively weak eddy-current contribution due to a smaller cross-section for intercepting the magnetic flux. The distribution, along the y axis, indicates that the electrically and magnetically induced field components are nearly equal. The electric polarization field and the circulating eddy current add at the front side and subtract on the back side of the spheroid to render an absorption pattern that peaks at the front surface and is reduced to almost zero deeper inside the spheroid.

For H-polarization, the electrically induced current flows along the x axis (direction of incident E field) and the eddy-current field encircles the z axis (direction of incident H field). The relatively low power on the z axis comes solely from the incident electric field. The combination of E- and H-induced components generates a displaced parabolic energy absorption pattern along both the x and y axes. Clearly, magnetically induced eddy current predominates in this case and the absorption is highest along the transverse circumference at the middle of the prolate spheroid. For K-polarization, both the electric and magnetic components of the incident field are again along the minor axes of the spheroid. The electrically induced current flows along the x axis. The incident magnetic field induces an eddy-current electric field that encircles the y axis and the absorption is lowest at the axis. Whereas in both E- and H-polarization cases, the peak absorption occurs at the front surface of the spheroid (the surface of the spheroid first seen by the incident plane wave), this is not the case for K-polarization. Maximum absorption appears at $x = \pm b$ and the absorbed energy varies parabolically along the x axis. This results from the large quantity of magnetic flux intercepted by the broad cross-section (and resulting concentration of eddy current). It should be noted that the results illustrated in Figure 23 match very well with experimental measurements.[33] Moreover, the peak absorptions may be two orders of magnitude higher than those for dielectric spheres of equal mass.

Contours of absorbed energy distribution in prolate spheroidal models close to resonance are illustrated in Figures 24 and 25 for small animals. In this case, the square of induced electric fields are shown; the SARs can be easily calculated using Equation 7 and the dielectric properties for muscle. The top graphs show the distribution over the plane parallel to the electric field vector through the center. The lower graphs give the distribution over the plane parallel to the magnetic field vector through the middle of the spheroid. The nonuniform nature of the distribution is apparent in all cases; each graph shows several absorption peaks. The intense fields in the prolate spheroid models are about 30% higher than those predicted using a sphere. The corresponding curve for a spherical object usually falls just above the curves for H- and K-polarizations and well below the curve for E-polarizations, as one would expect from symmetry considerations.

IV. COMPLEX TISSUE MODELS

We have summarized above some of the analytical approaches to quantify the absorbed energy in biological objects. It should be recognized that while spheres and spheroids are good models of some animal bodies and certain body parts, they may not always be adequate for humans and experimental animals under a variety of exposure situations. More realistic models, such as models of human bodies formed from cubes, have been developed to account for the irregular shapes.[34-40] These models, based on numerical techniques, hold great promise for accurately predicting energy absorption and its distribution in biological objects exposed to radio and microwave radiation. In what follows we shall summarize a number of computer techniques that have been applied with some success in solving electromagnetic energy absorption and distribution problems. We shall also describe some results obtained using

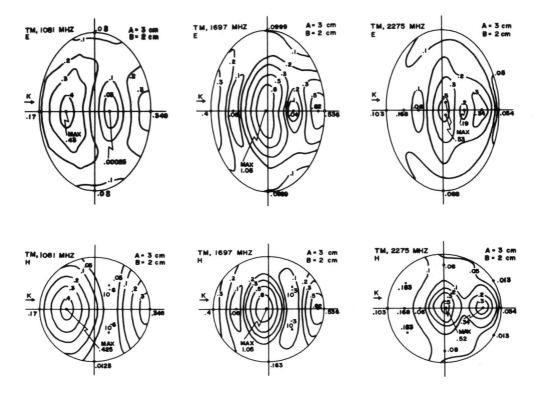

FIGURE 24.   SAR (square of induced electric field) contours in a prolate spheroidal model of small animal body (mouse) in an E-polarized plane wave field of 1 V/M at the three resonance frequencies.

the volume integral equation method for human bodies exposed to plane waves in free space. Lastly, we shall present two techniques that promise to provide highly efficient procedures, for calculating field intensity using models with very small step volumes.

A. A Brief Survey

The methods used to treat biological bodies of realistic shapes include the volume integral equation method[34-36,38] and the surface integral equation method.[41-43] The method of moments[44] is used in both approaches for finding solutions to the unknown fields inside the body. The approaches differ, however, in specifics, in that the surface integral equation method finds the unknown currents on the body surface and calculates the interior fields from the surface currents, the reciprocity theorem, and a "measurement matrix". In contrast, the volume integral equation requires determination of unknown fields throughout the volume of the body using the volume equivalence principle and the method of moments.

1. Volume Integral Equation Method

The numerical technique that has been adopted for most of the field intensity computations is the volume integral equation method employing the volume equivalence principle.[36,38,45,46] The method of moments is used to transform the integral equation into a matrix equation by subdividing the body into N simply shaped cells. This is accomplished with the aid of an appropriate set of expansion functions chosen to satisfy the boundary conditions and a set of weighting (testing) functions to reduce the matrix fill-in time. The total electric field at each of the N cells is given by matrix inversion. A more detailed description of the volume integral equation method is included in the next section.

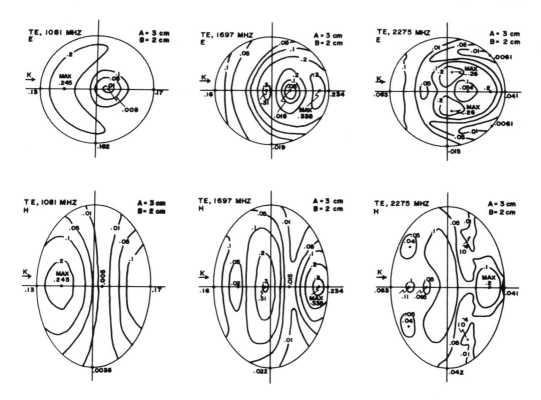

FIGURE 25. Distribution contours for induced electric field inside a prolate spheroidal model of small animal body (mouse) in an H-polarized plane wave field of 1 V/M. The frequencies are for the first, second, and third resonances.

However, it should be noted that a fundamental limitation of this method is the use of full or nearly full matrices and therefore, the requirement of extensive computer storage and long running time. Even with the availability of larger and faster computers, this difficulty may not be completely resolved. The need for excessively large numbers of mathematical cells to render a more accurate representation of the body will give rise to an equally large and full matrix. The inversion of large, full matrices often leads to numerical instabilities in the solution. Nevertheless, the method does allow the use of inhomogeneous models with up to 180 cells and is the only method available at present for computing field intensity distributions. In fact, this method has been employed, successfully, to calculate whole-body averaged absorption and to obtain regional distribution of absorbed radio frequency energy using inhomogeneous block models composed of rectangular cells.[36-38,40,47] Recently, this method also has been used to study the interaction of the near-zone field of an antenna with biological bodies.[37,48,49]

2. Surface Integral Equation Method

Another approach for predicting the distribution of absorbed electromagnetic energy is the surface integral equation method.[41,42,50,51] This method makes use of two coupled integral equations, i.e., the electric field and magnetic field integral equations for the tangential components of the field on the surface separating the biological body from air. The unknown surface currents are found by Fourier decomposition and the moment method. The fields inside the biological body are calculated using the previously computed surface currents, the reciprocity theorem, and the concept of measurement matrix.[42,43,52,53]

The method begins with the matrix representation of the coupled integral equations. If the body is assumed to be rotationally symmetric, the incident wave and the induced current could then be expanded in a Fourier series expansion in the angle of rotation. This reduces the problem to that of solving a system of orthogonal modes. The method further expands the surface components in terms of triangular expansion or basis functions and allows the testing functions to be the complex conjugate of the basis functions taking advantage of the orthogonality property. Thus, the major advantage of introducing the Fourier series is to enable each mode to be treated completely independently of all other modes. This results in a much smaller size, manageable matrix equation to be evaluated for the unknown expansion coefficients which determine the surface currents. It should be noted that for biological bodies, triangular expansion and testing functions are preferred over flat pulse expansion functions.[42] In fact, an expansion function with a continuous first derivation may constitute an even better choice for the expansion basis function. In any event, once the surface currents are obtained, the fields everywhere, or SAR at each point inside the body, can be calculated using the reciprocity theorem.[52,53] The total absorption can be found by integrating the surface Poynting vector.

The validity of this surface integral equation method has been substantiated by using a dielectric sphere.[42] Calculations for a human torso modeled by a homogeneous muscle body of revolution with a height of 1.78 m at 30, 80, and 300 MHz showed enhanced absorption in the neck region for all three frequencies and both vertical and horizontal polarization.[43] Note that the vertical direction is aligned with the long dimension of the torso and serves as the axis of symmetry. The strongest absorption in the torso model was found to occur with vertical polarization and near the first resonance frequency of the torso (80 MHz). In general, the surface integral equation method is applicable to any arbitrarily shaped homogeneous body of revolution. The method can be used not only with incident plane waves, but also with a wide variety of other field exposure conditions, including direct contact situations and near-zone sources.

Since both the surface and volume integral equation methods for field intensity prediction rely on the method of moments for implementation, it is instructive to compare the relative advantages of these two techniques. For simplicity, consider a homogeneous cube with N samples on each side: the computer storage requirements are $24 N^2$ and $3 N^3$ for the surface and volume integral methods, respectively.[42] For sufficient sampling to ensure accurate description of field variations, N is usually a large number. Thus, the surface integral equation method requires significantly less unknowns for homogeneous models. Moreover, in cases where permittivity and conductivity values are large, such as in biological bodies, the wavelength becomes contracted inside the body, and a much larger number of cells than that indicated above may actually be needed. If the model is inhomogeneous, then the volume integral equation would prove to be more suitable. It is possible, however, to generalize the surface integral equation technique to account for inhomogeneities by employing the invariant imbedding procedure.[54]

3. Extended Boundary Condition Method

An alternate surface integral equation method for computing the distribution of absorbed electromagnetic energy in realistic models of biological body is the extended boundary condition method.[55] This method employs analytic continuation, spherical harmonics, and the equivalence theorem to calculate fields inside the body. Specifically, the fields induced inside the body are replaced by equivalent surface currents. These surface currents are such that they reduce the total fields inside the body to zero. Thus, upon applying the boundary conditions at the surface of the body, an integral equation results which relates the incident field to the surface currents. This equation is then cast in a form suitable for numerical computation by expansion of the various field quantities in vector spherical harmonics.[56] It should be noted that the expansion coefficients for the incident field are known; only the coefficients for the surface currents need to be determined. This is accomplished by applying the orthogonality properties of the vector spherical harmonics, giving rise to a set of simultaneous equations that can be solved for the unknown coefficients. The induced fields are then determined in terms of the surface currents through the equivalence theorem.

This technique, thus far, has been applied to treat axi-symmetric models such as spherical, spheroidal, and finite cylindrical bodies[56,57] with limited success. Because the complex arguments associated with the vector spherical harmonics, especially the spherical Bessel functions, tend to be singular for large conductivities, the application of the technique is severely restricted. It appears that this method is amenable predominantly for predicting fields at high frequencies (i.e., frequencies well above the resonance region). The technique fails to yield convergent induced field distributions in lossy biological bodies in the resonance region for models of large aspect ratio in which the long dimension is very large compared with the short dimension of the body. In this case, an ill-conditioned set of matrix equations results from the large number of terms required to fit the vector spherical harmonic expansion to a nonspherical body.[58]

A number of attempts have been made to formulate computational schemes by which the ill-conditioned matrix obtained in the resonance region could be avoided. Among them the iterative techniques[59] proposed to overcome the convergence-related stability problem appear to have merit. This iterative procedure has the unique feature of approximating the interior of the body with several overlapping subregions, each represented by an expansion appropriate to its geometry. These subregional expansions are linked to each other by explicit matching in the appropriate overlapping zones. Since all the subregional field expansions are simultaneously solved, continuous and convergent field values are assured throughout the interior of the body.

In this method a noniterative procedure is used first to obtain the initial estimate of the tangential fields on the body surface. This is done by replacing the biological model with a perfectly conducting one of the same size and shape. Therefore, for subsequent iterations, the incident field and the results of the previous iteration for the surface fields are used to obtain the unknown coefficients which determine the surface fields for the current iteration. This is done by applying the orthogonality properties of the vector spherical harmonics and solving the resultant system of simultaneous equations. To determine the unknown internal field expansion coefficients, the boundary condition is satisfied at an appropriate number of points on each of the subregional surfaces, and, in addition, the continuity of internal fields is enforced at an appropriate number of points in each of the overlapping zones. The set of equations developed through these manipulations is simultaneously solved for the unknown coefficients of the subregional internal field expansions. The iterative procedure outlined above is continued until a preset convergence criterion is fulfilled. This may be satisfied by requiring the incremental surface electric current density to approach zero. Once the internal field distribution is known, and thus the distribution of SAR, the whole-body-averaged SAR can be obtained by averaging the SAR distribution over the entire model volume.

This method has been applied to calculate the distribution of absorbed energy in two body models exposed to plane wave irradiation in the resonance and postresonance frequency ranges.[59] The calculated results for homogeneous prolate spheroidal and capped cylindrical models of the human body showed applicability for frequencies up to 300 MHz, whereas the simple extended boundary condition method was restricted to frequencies less than 70 MHz for similar models. It was found that the number of iterations required to obtain a convergent solution did not change with various schemes for subdividing the volume. From the perspective of computational efficiency, however, subregional geometries and expansion functions with appropriate geometric conformation are found to be advantageous.

B. Cubic Cell Models

In this section we shall briefly summarize the volume integral equation method for field competition and present some results obtained for plane wave exposures.

The basic equation for the total induced electric field $\underline{E}(\underline{r})$ inside a biological body of volume V with conductivity $\tau(\underline{r}')$, and absolute dielectric permittivity $\epsilon(\underline{r}')$, and an incident electric field $\underline{E}^i(\underline{r})$ is given by the equation

$$\left[1 + \frac{\tau(\underline{r})}{3j\omega_0\epsilon_0}\right] \underline{E}(\underline{r}) - PV \int_v \tau(\underline{r}')\underline{E}(\underline{r}') \cdot G(\underline{r},\underline{r}')dv' = \underline{E}^i(\underline{r}) \qquad (21)$$

with the Green's function,

$$G(\underline{r},\underline{r}') = j\omega\mu_0 \left(\underline{I} + \frac{1}{k_0^2} \nabla\nabla\right) \frac{\exp(jk_0|\underline{r} - \underline{r}'|)}{4\pi|\underline{r} - \underline{r}'|}$$

where $\tau(\underline{r}) = \sigma(\underline{r}) + j\omega[\epsilon(\underline{r}) - \epsilon_0]$, $\underline{r}$ is the vector from the origin of coordinates to a given point, $\underline{r}'$ is the vector from the origin to the differential volume element dv', $\underline{I}$ is the unity dyad, k_0 is the free space wave number, ∇ signifies the gradient operation, and $P\overline{V}$ symbolizes the principal value of the integral.

According to the method of moments, the body may be partitioned into N cubic subvolumes or cells which are sufficiently small for the electric field and dielectric permittivity to be constant within each cell. The integral equation is then transformed into a system of 3N simultaneous linear equations for the three orthogonal components of the electric field at the center of each cell. The simultaneous equations may be written in matrix form as

$$[G] [E] = -[E^i] \qquad (22)$$

where [G] is a 3N × 3N matrix and [E^i] and [E] are column matrices representing incident and induced electric field at the center of each cell. The elements of [G] can be evaluated as shown in the References 34 and 37. In particular, the diagonal elements of the [G] matrix may be evaluated exactly by approximating each subvolume with a sphere of equal volume centered at the position of an interior point. If the actual shape of the subvolume differs appreciably from that of a sphere, this approximation may lead to unsatisfactory numerical results.[37,60] In such cases, a small cylindrical volume may be created around an interior point. It may also be necessary to evaluate these terms by numerical integration throughout the cubic subvolume for increased accuracy. The evaluation of off-diagonal elements of the [G] matrix is considerably simplified since it does not involve principal value operations. Therefore, for a given applied field configuration, the induced electric fields inside the body are obtained by matrix inversion. That is,

$$[E] = -[G]^{-1} [E^i] \qquad (23)$$

Several computer programs have been developed to implement this numerical procedure. Factors that influence the computational accuracy include frequency, body size, cell dimensions, and computer memory. It has been found that reliable numerical results can be obtained if the linear dimensions of the cell do not exceed a quarter free-space wavelength.[34] For a computer with sufficient capacity to invert a 120 × 120 matrix, the maximum number of cells is limited to 40. If we assume, for simplicity, symmetries between the right and the left half and the front and the back of 1.7-m-tall adult human body, this computer would handle approximately a cell size around 10^{-5} m^3. Once the 10^{-5}-m^3 cell size is adopted,

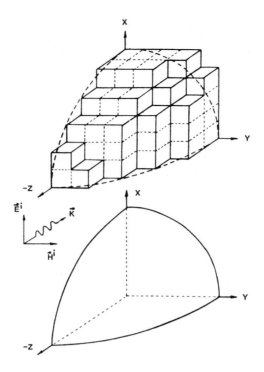

FIGURE 26. The approximation of one eighth of a sphere by an equivalent cubic-cell-formed structure. (From Rukspollmuang, S. and Chen, K. M., *Radio Sci.*, 14, 51, 1979. With permission.)

750 MHz would be the highest frequency that can be considered for field intensity calculation without violating the criterion that the linear dimension of the cell not exceed a quarter free-space wavelength.

The computational resources necessary to obtain even a regional SAR using this method of moment approach is quite extensive. A relatively full complex matrix, $3N \times 3N$ in dimensions, is required for a model with N cells. The computation time required for a noniterative solution of the matrix equation is therefore proportional to between N^2 and N^3, which increases rapidly as N increases. The faithfulness with which a cubic cell model approximates the detailed structure of a biological body and the maximum usable frequency increases with the number of cells. In fact, substantial errors will occur if

$$N \leq \frac{L}{\lambda'} \frac{2\pi}{\sqrt{6}} \tag{24}$$

where L is the linear dimension of the body and λ' is the wavelength in the body (see Part III, Chapter 1).

The accuracy of the numerical method can be verified by comparison with known results from exact analytic solutions based on well-characterized geometrical bodies such as spheres. It should be noted that perfect agreement between the exact solution based on Mie theory and the numerical method based on the volume integral equation is not expected unless a large number of cubic cells are used to simulate the sphere. Figure 26 shows one eighth of a sphere approximated by one eighth of a "cubic model of a sphere", which is constructed

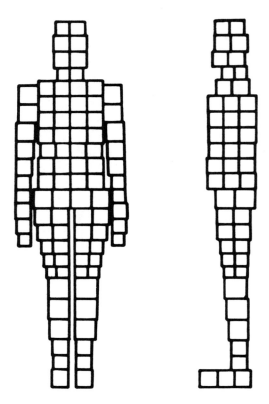

FIGURE 27.   A cubic-cell representation of human body. (From Hagman, M. J., Gandhi, O. P., D'Andrea, J. A., and Chatterjee, I., *IEEE Trans. Microwave Theor. Tech.*, 27, 809, 1979. With permission. © 1979 IEEE.)

from 73 cubic cells. Clearly, a better approximation can be achieved by a larger number of smaller cubic cells. Nevertheless, for a brain sphere constructed from 40 cubic cells at a frequency of 918 MHz, the computed maximum field intensity deviated from the exact solution by less than 9%.[61]

A model of a human body consisting of 180 cubic cells that accounts for the anatomic and biometric characteristics of human beings is shown in Figure 27. The model is 1.75 m tall and can be made either homogeneous or inhomogeneous by using an equivalent or a volume-weighted complex permittivity for each cell.[62] The distributions of absorbed energy inside the model in free space at 80 MHz (near resonance) for homogeneous and inhomogeneous tissue properties are illustrated in Figure 28 and 29, respectively. The electric field vector is oriented along the height of the model and the plane wave propagates from front to back of the body. It is interesting to observe that the absorbed energy in the upper torso is higher in the middle and back layers than the front layer on which the plane wave impinges; the highest absorption in the transverse dimension occurs generally near the center of the torso. Energy deposition along the height of the body shows several peaks which appear in the neck, the thoracic region, the pelvic/thigh region, and the calf, for the homogeneous model. The SARs for 10 W/m² of incident power density are 86 W/kg for the pelvic/thigh region; 79 W/kg for the calf; 47 W/kg for the thorax; and 34 W/kg for the neck. It should be noted that the distribution of absorbed energy in the homogeneous model is close to that measured using scaled models.[37,38]

303

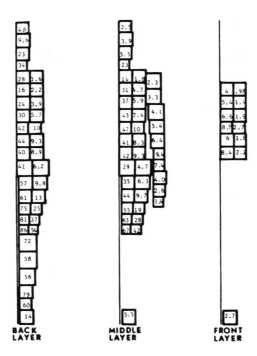

FIGURE 28. SARs for a homogeneous 180-cell model of man exposed to vertically polarized, 80-MHz plane wave. The values are expressed as W/kg/mW/cm² × 100. (From Hagman, M. J., Gandhi, O. P., D'Andrea, J. A., and Chatterjee, I., *IEEE Trans. Microwave Theor. Tech.*, 27, 809, 1979. With permission. © 1979 IEEE.)

The effect of inhomogeneity produced by volume-weighted cell complex permittivity is mostly revealed in the relative strength of absorbed energy in different regions of the body. The general distribution of the absorbed energy shows very little difference between homogeneous and inhomogeneous models. That is, enhanced absorptions occur in the same region, but the SARs in the pelvic/thigh and thoracic regions are now, respectively, 12 and 5.1 W/kg/W/m² incident power density. The SARs in the neck and the calf, however, are reduced to 2.8 and 6.8 W/kg, respectively. Since the heart and the pelvic/thigh regions are characterized by tissues with high water content, whereas the neck and the calf are composed largely of tissues with low water content, this suggests that energy absorption in the body is increased in regions with larger quantity of muscular tissue and is decreased in regions that consist mainly of bony tissues. Changes in coupling characteristics that stem from the inhomogeneity, however, may also have contributed to the difference in absorbed energy distribution.

The average absorption or whole-body SAR for the model of the human body shown in Figure 27 as a function of frequency is illustrated in Figure 30. Again, the electric field vector is along the height of the body and the plane wave propagates from front to back of the model with a power density of 10 W/m². A homogeneous complex permittivity approximately two thirds of that for muscle is used in the calculations. Note that the whole-body SAR increases with frequency until it reaches a maximum of about 0.23 W/kg at 77 MHz (resonance frequency) and then decreases according to 1/frequency. The experimental data shown in Figure 30 are obtained from a saline-filled scale model of the human body. It is seen that the calculated absorption is in good agreement with that found experimen-

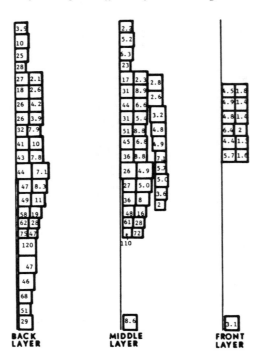

FIGURE 29. SARs for an inhomogeneous 180-cell model of human body exposed to vertically polarized, 80-MHz plane wave. The values are given in W/kg/ mW/cm² × 100. (From Hagman, M. J., Gandhi, O. P., D'Andrea, J. A., and Chatterjee, I., *IEEE Trans. Microwave Theor. Tech.*, 27, 809, 1979. With permission. © 1979 IEEE.)

tally,[38,63] except for the resonant frequency which is somewhat lower (70 MHz) in the experimental case. It should be mentioned that the whole-body SAR given in Figure 30 is typically within 10% of that estimated from prolate spheroidal models of the same height and dielectric property. Further, when inhomogeneous complex permittivities are used with the model, the whole-body SAR changes less than 2% from that depicted in Figure 30. Thus, if one is primarily concerned with average absorption over the body, a homogeneous prolate spheroidal model may be quite adequate.

While moment methods based on the volume integral equation remain the most successful numerical procedure for computation of SAR distribution in complex tissue geometries, the requirement of a full $3N \times 3N$ matrix presents severe limitations. The computation times required to provide even regional SAR distribution of sufficient resolution to delineate the resonant frequency for the head region are enormous. A minimum of 340 cells was needed, increasing the computation time by a factor of four over the 180-cell model[63,64] (see also Table 4).

Matrix inversion operations consume the largest block of time in moment-method solutions for the cubic cell models. The computer time required is proportional to the cube of the number of cells. However, the matrix generated is usually diagonally dominant and well conditioned. For a human-size body, iterative procedures for matrix inversion are practical at frequencies below about 60 MHz. The convergence rate decreases with increasing frequency and fails above 90 MHz. This is most likely caused by the decrease in degree of diagonal dominance with increasing frequency. A number of approaches have been inves-

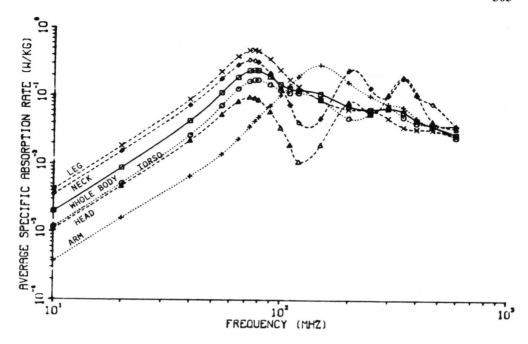

FIGURE 30. Average SARs for a homogeneous 180-cell model of human body exposed to vertically polarized plane wave in free space. The incident power density in 10 W/m². (From Hagman, M. J., Gandhi, O. P., D'Andrea, J. A., and Chatterjee, I., *IEEE Trans. Microwave Theor. Tech.*, 27, 809, 1979. With permission. © 1979 IEEE.)

Table 4
SOME TYPICAL COMPUTATION TIMES ON
THE DEC-20 FOR THE VARIOUS MAN MODELS

No. of cells	Frequency (MHz)	Half-bandwidth	No. of iterations	Computation time (hr, min)
180	27.12	50	20	0, 5
180	350.00	100	20	0, 7
340	27.12	100	20	0, 22
340	350.00	180	20	0, 33
580	27.12	250	30	1, 50
626	27.12	250	15	1, 21
626	77.00	250	38	2, 31
760	27.12	300	40	5, 9
1132	27.12	350	40	11, 36

tigated to alleviate this difficulty. A semi-iterative procedure called Band Approximation Method appears to be an efficient algorithm that can be profitably used to invert large matrices generated by arbitrarily high frequencies for man models and converges significantly faster than standard iterative algorithms.[64]

For a finite biological body of arbitrary shape, the total induced electric field inside the body is given by Equation 22 such that

$$[G] \, [E] \; = \; - [E^i] \qquad (25)$$

where [G] is the 3N × 3N matrix involving the size, position, frequency, and complex permittivity of the individual cells as well as the Green's function. The column vectors [E] and [E^i] are the unknown induced and the known incident electric field within each cell inside the body. The matrix [G] may be rendered to have rapidly decreasing magnitudes in elements with increasing distance from the diagonal by proper ordering of cells in the model. Let [B] be a band-section about the diagonal of a matrix [G] and [C] the remainder of the matrix [G] such that

$$([B] + [C]) [E] = -[E^i] \tag{26}$$

This equation may be written as

$$[E] = -[B]^{-1} [E^i] - [B]^{-1} [C] [E] \tag{27}$$

Iterations for the Band Approximation Method are defined by

$$[E]_{j+1} = -[B]^{-1} [E^i] - [B]^{-1} [C] [E]_j \tag{28}$$

The computer memory space requirement could be drastically reduced if the matrix [G] is replaced with its LU decomposition, where L and U are the lower and upper triangular matrices, respectively.

Let [e_j] denote the error of the jth iteration from the exact solution [E], then

$$[e_{j+1}] = -\{[B]^{-1} \cdot [C]\} \cdot [e_j] \tag{29}$$

Since the magnitude of elements with increasing distance from the diagonal rapidly decreases, all eigen values of $\{[B]^{-1} \cdot [C]\}$ have magnitudes less than unity. This assures that the iterations will converge for sufficiently large bandwidths.

The required bandwidth for convergence for the 180-cell block model of man is shown in Figure 31. Since a plane of symmetry was assumed, the matrix was 270×270 in dimension. It can be seen that the required bandwidth increases with frequency. This is attributable to slower decrease in the magnitudes of the matrix elements with distance from the diagonal at higher frequencies. At these frequencies, the radiation term 1/distance becomes an increasingly significant contributor to field intensity. It should be noted that the computational time at 915 MHz was about one third that of noniterative solutions, and was considerably less at lower frequencies due to reduced bandwidth requirements.

C. Models with Very Small Cell Sizes

Several studies have shown that the calculated whole-body-averaged SAR (Figure 32) increases with the number of cells employed in the moment method computation.[37,64,65] This is attributable to the insufficient number of cells used to represent field variations inside the body. In general, moment method algorithms using a pulse function basis require that the field be approximately constant throughout each cell. According to Equation 24, the upper bound on usable cell size is given by

$$\ell/\lambda' < 0.39 \text{ and } 0.27 \tag{30}$$

for high and low conductivity dielectrics, respectively. In this case, for tissues with high and low conductivities, ℓ is the length of the side of a square cell and λ' is the wavelength in tissues. Thus, there must be at least three cells per wavelength, if not four. Otherwise, the magnitude and phase resolutions would be such that even with convergence, the reliability of the computed SAR would be questionable.

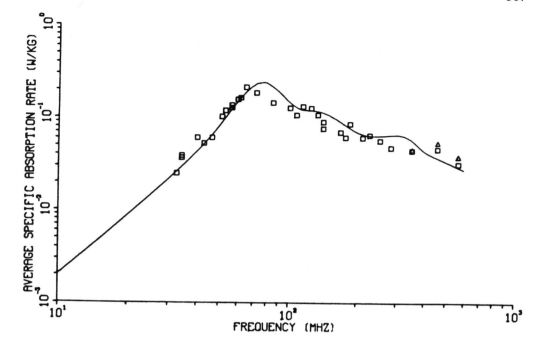

FIGURE 31. Whole-body-averaged absorption for a homogeneous cubic-cell model of man exposed to vertically polarized plane wave in free space. Squares and triangles represent measured values for normal saline and phantom mixture-filled figurines. The incident power density is 10 W/m². (From Hagman, M. J., Gandhi, O. P., and Durney, C. H., *IEEE Trans. Microwave Theor. Tech.*, 27, 804, 1979. With permission. © 1979 IEEE.)

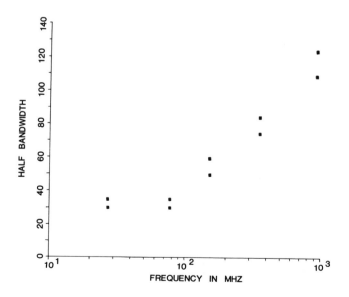

FIGURE 32. Bandwidth for convergence of iterations in a 180-cell model.

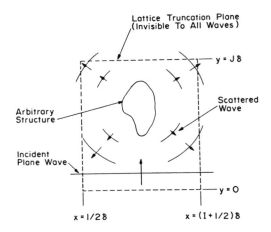

FIGURE 33.   The concept of time domain wave track-
ing in FD-TD method of solution.

To achieve sufficiently fine sampling throughout the entire volume of the model and to
permit more accurate geometric representation of the model, a reasonable cell size would
be $\ell < 0.1\lambda'$. Such fine discretization implies man models with 10,000 to 100,000 cells at
microwave frequencies. Since, for moment method algorithms, the required computer storage
is proportional to N^2 and the computation time rises as N^3, moment method solutions would
be impractical with presently available computational resources. Clearly, new approaches
based on different principles are needed to give detailed information on the distribution of
absorbed microwave energy inside biological bodies.

Presently, there exist two alternative methods that promise to provide highly efficient
procedures for field intensity calculations using a large number of cells in a finely discretized
model. One technique involves the use of finite difference algorithms for the time-dependent
Maxwell's equations[65-69] and the other employs a fast-Fourier-transform (FFT) procedure[70]
to solve the electric field integral equation.

1. The FD-TD Method

The finite difference, time domain (FD-TD) approach is an attempt to solve Maxwell's
curl equations by directly modeling the propagation of waves into a volume of space containing the biological body (Figure 2). By repeatedly implementing a finite difference representation of the curl equations at each cell of the corresponding space lattice, the incident
wave is tracked as it first propagates to the body and then interacts with it through surface
current excitation, transmission, and diffraction. This wave-tracking process is completed
when the steady-state behavior is observed at each lattice cell. Considerable simplification
is achieved by analyzing the interaction of the wavefront with a part of the body surface at
a time, rather than attempting a simultaneous solution of the entire problem.

Time-stepping for the FD-TD method is accomplished by an explicit finite difference
procedure.[71] For a cubic cell lattice space, this procedure involves positioning the electric
and magnetic field components about a unit cell of the lattice (Figure 33) and then evaluating
the components at alternate half-time steps. In this manner, centered difference expressions
can be used for both the space and time increments without solving simultaneous equations
to compute the fields at the latest time step.

The body of interest is mapped into the lattice space by first choosing the lattice increment
and then assigning values of permittivity and conductivity to each cell. The boundary
conditions at media interfaces are naturally generated by the curl equations. Thus, once a

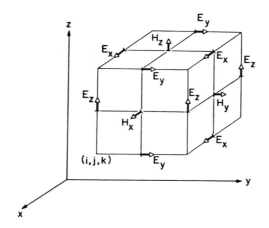

FIGURE 34. Positions of field components about a
unit cell of the lattice space.

computer program is developed, the basic routines need not be changed for different model
geometries. In fact, inhomogeneities and fine structural details could be modeled with a
maximum resolution of one unit cell (Figure 34).

The explicit formulation of the FD-TD method is particularly suited for execution with
minimum computer storage and run-time using current array-processing computers. The
required computer storage and run-time increase only linearly with N, the number of cells.
In fact, it has been shown that the FD-TD method is capable of solving for more than 1
million unknown field components within a few minutes on an array-processing computer.[69]
Field intensities have been predicted to within 2.5% accuracy relative to known analytical
and experimental bench marks. It is anticipated that in the near future this FD-TD technique
could be improved to allow solutions for field penetration and absorption in body sizes
greater than ten wavelengths in three dimensions with a spatial resolution better than 0.1
wavelength (Figure 35). With the exception of a few early attempts, a majority of current
efforts have been directed toward application of the FD-TD method to electromagnetic
interaction in time-varying inhomogeneous media,[66,72] with metallic bodies of revolution,[67,69]
and with lossy dielectric structures.[68,72]

2. The FFT Method

The FFT method makes use of the fact that the integral equation for the electric field is
in the form of a convolution integral. This integral, when Fourier transformed, becomes a
simple algebraic equation in the frequency domain which can be solved, iteratively, to obtain
the unknown electric field. Specifically, the integral such as that in Equation 21 is first
replaced by a double discrete summation to preserve the convolutional nature of the equation.
A FFT algorithm is used to efficiently evaluate the kernel and current distributions, which
are then algebraically multiplied to give the transform of the internal electric field. The
inverse FFT then yields the internal electric field in the spatial domain. Since the FFT method
computes a circular rather than a linear convolution, the FFT lengths must be chosen to
ensure that the convolution is linear inside the biological body. This choice of FFT lengths
decouples the fields due to the periodic scattering, introduced by the discrete Fourier transform. This also obviates the need for large intervals which would otherwise have restricted
the number of cells that could be ascribed to the biological body.

The interative solution begins by assuming almost any value for the field. A sequence of
steps is performed that provides better and better estimates of the unknown field until it

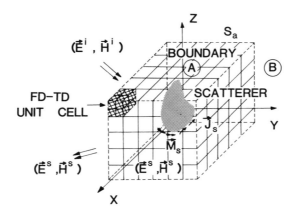

FIGURE 35.   A three-dimensional body embedded in a
FD-TD lattice space.

converges on the correct answer. It is important for the success of the FFT approach that a
fast convergence is achieved so that the solution is reached within a few iterations. There
are several iteractive techniques which may be employed with the FFT method. These include
the stationary iterative scheme associated with the K-space formulation,[73] the steepest descent
scheme,[74] and the conjugate gradient method.[75,76] In general, the number of iterations required
for convergence could decrease by as much as an order of magnitude for these three schemes,
in the same order as the presentation.

It is noteworthy that the numerical implementation of the FFT method is iterative. Therefore, the computer storage requirement is proportional to N and the computation time is
proportional to 4N log 8N per iteration, where N is the number of cells used to represent
the body. This may allow use of 10,000 to 100,000 inhomogeneous cells to model three-
dimensional bodies at higher frequencies, as well as with greater anatomic details, for
distributive SAR calculations. Efforts in this direction have just begun. Results obtained for
two-dimensional problems demonstrate accuracy comparable to known analytic solutions
for homogeneous and layered circular cylindrical bodies.[70]

A major difficulty with both the FD-TD and FFT methods is that it is necessary to allot
a large number of cells to model the space exterior to the body. However, the enormous
number of cells possible with these methods permit many more to be assigned to the body
than is possible with the method of moments. Also, in the FD-TD method additional cells
are needed, in comparison with FFT, to model the exterior region to a distance where the
scattered radiation can be successfully truncated. This disadvantage must be balanced with
the slightly longer computation time needed by the FFT method as compared with the FD-
TD method, which is proportional to N. It is reasonable to expect these relatively recent
developments in methodologies to provide detailed information on the distribution of absorbed radio and microwave energy in realistic models of humans and animals.

REFERENCES

1. **Schwan, H. P. and Piersol, G. M.,** Absorption of electromagnetic energy in body tissues, *Am. J. Phys. Med.,* 33, 371, 1954.
2. **Johnson, C. C. and Guy, A. W.,** Nonionizing electromagnetic wave effects in biological materials and systems, *Proc. IEEE,* 60, 692, 1972.
3. **Schwan, H. P. and Li, K.,** Hazards due to total body irradiation by radar, *Proc. IRE,* 44, 1572, 1956.
4. **Schwan, H. P. and Li, K.,** The mechanism of absorption of ultrahigh frequency electromagnetic energy in tissue as related to the problem of tolerance dosage, *IRE Trans. Med. Electron.,* 4, 45, 1956.
5. **Stratton, J. A.,** *Electromagnetic Theory,* McGraw-Hill, New York, 1941.
6. **Lin, J. C., Guy, A. W., and Kraft, G. H.,** Microwave selective brain heating, *J. Microwave Power,* 8, 275, 1973.
7. **Shapiro, A. R., Lutomirski, R. F., and Yura, H. T.,** Induced fields and heating within a cranial structure irradiated by an electromagnetic plane wave, *IEEE Trans. Microwave Theory Tech.,* 19, 187, 1971.
8. **Kritikos, H. N. and Schwan, H. P.,** Hot spot generated in conduction spheres by EM waves and biological implications, *IEEE Trans. Biomed. Eng.,* 19, 53, 1972.
9. **Ho, H. S. and Guy, A. W.,** Development of dosimetry for RF and microwave radiation, *Health Phys.,* 29, 317, 1975.
10. **Lin, J. C.,** Microwave biophysics, in *Microwave Bioeffects and Radiation Safety,* Stuchly, M. A., Ed., IMPI, Edmonton, Alberta, Canada, 1978, 15.
11. **Weil, C. M.,** Absorption characteristics of multi-layered sphere models exposed to UHF/microwave radiation, *IEEE Trans. Biomed. Eng.,* 22, 468, 1975.
12. **Guy, A. W., Lin, J. C., and Chou, C. K.,** Electrophysiologic effects of electromagnetic fields on animals, in *Fundamental and Applied Aspects of Nonionizing Radiation,* Michaelson, S. M. et al., Eds., Plenum Press, New York, 1975, 167.
13. **Joines, W. T. and Spiegel, R. J.,** Resonance absorption of microwaves by the human skull, *IEEE Trans. Biomed. Eng.,* 21, 46, 1975.
14. **Lin, J. C.,** Interaction of two cross-polarized electromagnetic waves with mammalian cranial structures, *IEEE Trans. Biomed. Eng.,* 23, 371, 1976.
15. **Schwan, H. P.,** Electrical properties of tissues and cell suspensions, *Adv. Biol. Med. Phys.,* 4, 147, 1957.
16. **Schwan, H. P.,** Survey of microwave absorption characteristics of body tissues, in *Proc. 2nd Tri-Service Conf. Biol. Effects of Microwave Energy,* Pattishall, E. G. and Banghart, F. W., Eds., 1958, 126.
17. **Lin, J. C.,** Microwave properties of fresh mammalian brain tissues at body temperature, *IEEE Trans. Biomed. Eng.,* 22, 74, 1975.
18. **Lin, J. C., Grove, H. M., and Sharp, J. C.,** Comparative measurement of dielectric properties of fresh mammalian tissues, in *Proc. Conf. Proc. Electromagnetic Meas.,* 1975, 246.
19. **Schwan, H. P. and Foster, K. R.,** RF-field interaction with biological systems: electrical properties and biophysical mechanisms, *Proc. IEEE,* 68, 104, 1980.
20. **Schwan, H. P.,** Microwave biophysics, in *Microwave Power Engineering,* Okress, E. C., Ed., Academic Press, New York, 1968, 213.
21. **Lin, J. C., Guy, A. W., and Johnson, C. C.,** Power deposition in a spherical model of man exposed to 1—20 MHz electromagnetic fields, *IEEE Trans. Microwave Theory Tech.,* 21, 791, 1973.
22. **Kritikos, H. N. and Schwan, H. P.,** The distribution of heating potential inside lossy spheres, *IEEE Trans. Biomed. Eng.,* 22, 457, 1975.
23. **Johnson, C. C., Durney, C. H., and Massoudi, H.,** Long-wavelength electromagnetic power absorption in prolate spheroidal models of man and animals, *IEEE Trans. Microwave Theory Tech.,* 23, 739, 1975.
24. **Durney, C. H., Johnson, C. C., and Massaudi, A.,** Long wave-length analysis of plane wave irradiation of a prolate spheroidal model of man, *IEEE Trans. Microwave Theory Tech.,* 23, 246, 1975.
25. **Massoudi, H., Durney, C. H., and Johnson, C. C.,** Long wavelength electromagnetic power absorption in ellipsoidal models of man and animals, *IEEE Trans. Microwave Theory Tech.,* 24, 41, 1977.
26. **Durney, C. H., Iskander, M. F., Massoudi, H., and Johnson, C. C.,** An empirical formula for broadband SAR calculations of prolate spheroidal models of humans and animals, *IEEE Trans. Microwave Theory Tech.,* 27, 758, 1979.
27. **Wu, C. L. and Lin, J. C.,** Absorption and scattering of EM waves by prolate spheroidal models of biological structures, in *IEEE AP-S Int. Symp.,* Stanford, Calif., 1977, 142.
28. **Barber, P.,** Electromagnetic power absorption in prolate spheroidal models of man and animals at resonance, *IEEE Trans. Biomed. Eng.,* 24, 513, 1977.
29. **Rowlandson, G. I. and Barber, P. W.,** Absorption of high frequency RF energy by biological models: calculations based on geometrical optics, *Radio Sci.,* 14, 43S, 1979.
30. **Asano, A. and Yamamoto, G.,** Light scattering by spheroidal particles, *Appl. Opt.,* 14, 29, 1975.
31. **Lin, J. C.,** Whole-body exposure in the near zone of HF electromagnetic fields, in *Int. Electromagnetic Waves and Biology Symp.,* Jouy en Josas, France, 1980.

32. **Durney, C. H., Iskander, M. F., Massoudi, H., Allen, S. J., and Mitchell, J. C.,** *Radio Frequency Radiation Dosimetry Handbook,* 3rd ed., Brooks AFB, Tex., 1980.

33. **Guy, A. W., Webb, M. D., and Sorenson, C. C.,** Determination of power absorption in man exposed to HF electromagnetic fields by thermographic measurements on scale models, *IEEE Trans. Biomed. Eng.,* 23, 361, 1976.

34. **Liversay, D. E. and Chen, K. M.,** Electromagnetic fields induced inside arbitrary shaped biological bodies, *IEEE Trans. Microwave Theory Tech.,* 22, 1273, 1974.

35. **Guru, B. S. and Chen, K. M.,** Experimental and theoretical studies in electromagnetic field induced inside finite biological bodies, *IEEE Trans. Microwave Theory Tech.,* 24, 433, 1976.

36. **Chen, K. M. and Guru, B. S.,** Internal EM field and absorbed power density in human torsos induced by 1—500 MHz EM waves, *IEEE Trans. Microwave Theory Tech.,* 25, 746, 1977.

37. **Chen, K. M.,** Interaction of electromagnetic fields with biological bodies, in *Research Topics in Electromagnetic Theory,* Kong, J. A., Ed., John Wiley & Sons, New York, 1980, 290.

38. **Hagman, M. J., Gandhi, O. P., and Durney, C. H.,** Numerical calculation of electromagnetic energy deposition for a realistic model of man, *IEEE Trans. Microwave Theory Tech.,* 27, 804, 1979.

39. **Durney, C. H.,** Electromagnetic dosimetry for models of humans and animals: a review of theoretical and numerical techniques, *Proc. IEEE,* 68, 33, 1980.

40. **Gandhi, O. P.,** Electromagnetic absorption in inhomogeneous model of man for realistic exposure conditions, *Bioelectromagnetics,* 3, 81, 1982.

41. **Wu, T. K. and Tsai, L. L.,** Electromagnetic fields induced inside arbitrary cylinders of biological tissue, *IEEE Trans. Microwave Theory Tech.,* 25, 61, 1977.

42. **Wu, T. K. and Tsai, L. L.,** Scattering from arbitrary-shaped lossy dielectric bodies of revolution, *Radio Sci.,* 12, 709, 1977.

43. **Wu, T. K.,** Electromagnetic fields and power deposition in body of revolution models of man, *IEEE Trans. Microwave Theory Tech.,* 27, 279, 1979.

44. **Harrington, R. F.,** *Field Computation by Moment Methods,* McGraw-Hill, New York, 1968.

45. **Schelkunoff, S. A.,** Field equivalence theorems, *Comm. Pure Appl. Math.,* 4, 43, 1951.

46. **Lin, J. C. and Wu, C. L.,** Scattering of microwaves by dielectric materials used in laboratory animal restrainers, *IEEE Trans. Microwave Theory Tech.,* 24, 219, 1976.

47. **Gandhi, O. P., Hagman, M. J., and D'Andrea, J. A.,** Part-body and multi-body effects on absorption of radio frequency electromagnetic energy by animals and by models of man, *Radio Sci.,* 14, 155, 1979.

48. **Karimullah, K., Chen, K. M., and Nyquist, D. P.,** Electromagnetic coupling between a thin-wire antenna and a neighboring biological body, *IEEE Trans. Microwave Theory Tech.,* 28, 1218, 1980.

49. **Chatterjee, I., Hagman, M. J., and Gandhi, O. P.,** Electromagnetic energy deposition in an inhomogeneous block model for near-field irradiation conditions, *IEEE Trans. Microwave Theory Tech.,* 28, 1452, 1980.

50. **Poggio, A. J. and Miller, E. K.,** Integral equation solutions of three-dimensional scattering problems, in *Computer Techniques for Electromagnetics,* Mittra, R., Ed., Pergamon Press, Elmsford, N.Y., 1973, 159.

51. **Massoudi, H., Durney, C. H., Barber, P. W., and Iskander, M. F.,** Post resonance EM absorption by man and animals, *Bioelectromagnetics,* 3, 333, 1982.

52. **Mautz, J. R. and Harrington, R. F.,** Radiation and scattering from bodies of revolution, *Appl. Sci. Res.,* 20, 405, 1969.

53. **Harrington, R. F. and Mautz, J. R.,** Green's functions for surfaces of revolution, *Radio Sci.,* 7, 603, 1972.

54. **Pogorzelski, R. J. and Wu, T. K.,** Computations of scattering from inhomogeneous penetrable elliptic cylinders by means of invariant imbedding, in *URSI Symp. Electromagnetic Wave Theory,* Stanford, 1977, 323.

55. **Waterman, P. C.,** Matrix formulation of electromagnetic scattering, *Proc. IEEE,* 53, 805, 1965.

56. **Barber, P. W. and Yeh, C.,** Scattering of electromagnetic waves by arbitrary shaped dielectric bodies, *Appl. Opt.,* 14, 2864, 1975.

57. **Barber, P. W.,** Resonance electromagnetic absorption by nonspherical dielectric objects, *IEEE Trans. Microwave Theory Tech.,* 25, 373, 1977.

58. **Wall, D. J. N.,** Method of overcoming numerical instabilities associated with the T-matrix method, in *Acoustic, Electromagnetic and Elastic Wave Scattering,* Varadan, V. V. and Varadan, V. K., Eds., Pergamon Press, Elmsford, N.Y., 1980.

59. **Lakhtakia, A., Iskander, M. F., and Durney, C. H.,** An iterative extended boundary condition method for solving the absorption characteristics of lossy dielectric objects of large aspect ratios, *IEEE Trans. Microwave Theory Tech.,* 31, 640, 1983.

60. **Graglia, R. D.,** Integral Equations for Anisotropic Scatterers, Ph.D. dissertation, Department of Electrical Engineering and Computer Sciences, University of Illinois, Chicago, 1983.

61. **Rukspollmuang, S. and Chen, K. M.,** Heating of spherical vs. realistic models of human and infrahuman heads by electromagnetic waves, *Radio Sci.,* 14, 51, 1979.

62. **Hagman, M. J., Gandhi, O. P., D'Andrea, J. A., and Chatterjee, I.,** Head resonance: numerical solutions and experimental results, *IEEE Trans. Microwave Theory Tech.,* 27, 809, 1979.
63. **Gandhi, O. P., Hunt, E. L., and D'Andrea, J. A.,** Deposition of EM energy in animals and in models of man with and without grounding and reflector effects, *Radio Sci.,* 12, 39S, 1977.
64. **Deford, J. F., Gandhi, O. P., and Hagman, M. J.,** Moment-method solutions and SAR calculations for inhomogeneous models of man with large number of cells, *IEEE Trans. Microwave Theory Tech.,* 31, 848, 1983.
65. **Hagman, M. J.,** Numerical Studies of Absorption of Electromagnetic Energy by Man, Ph.D. dissertation, Department of Electrical Engineering, University of Utah, Salt Lake City, 1978.
66. **Taflove, A. and Brodwin, M. E.,** Numerical solution of steady-state EM scattering problems using the time dependent Maxwell's equation, *IEEE Trans. Microwave Theory Tech.,* 23, 623, 1975.
67. **Taflove, A.,** Application of the finite-difference time domain method to sinusoidal steady-state electromagnetic-penetration problems, *IEEE Trans. EM Compatibility,* 22, 191, 1980.
68. **Taflove, A. and Lin, J. C.,** Finite difference time domain computation of microwave absorption in models of biological bodies, in *Abstr. Bioelectromagnetics Soc. Annu. Meet.,* Washington, D.C., 1981, 62.
69. **Taflove, A. and Umashankar, K.,** A hybrid moment method/finite difference time domain approach to electromagnetic coupling and aperture penetration into complex geometries, *IEEE Trans. Ant. Propagation,* 30, 617, 1982.
70. **Borup, D. T. and Gandhi, O. P.,** Fast-Fourier transform method for calculation of SAR distributions in finely discretized inhomogeneous models of biological bodies, in *Abstr. Bioelectromagnetics Soc. Meet.,* Boulder, Colo., 1983, 66.
71. **Yee, K. S.,** Numerical solutions of initial boundary value problems involving Maxwell's equations in isotropic media, *IEEE Trans. Ant. Propagation,* 14, 303, 1966.
72. **Taflove, A. and Brodwin, M. E.,** Computation of the electromagnetic fields and induced temperatures within a model of the microwave-irradiated human eye, *IEEE Trans. Microwave Theory Tech.,* 23, 888, 1975.
73. **Bojarski, N. N.,** N-Dimensional fast Fourier transform tomography for incomplete information, and its application to electromagnetic inverse scattering theory, in *Mathematical Methods and Applications of Scattering Theory,* Desanto, J. A. et al., Eds., Springer-Verlag, Berlin, 1980, 277.
74. **Sarkar, T. and Rao, S. M.,** An iterative method for solving electrostatic problems, *IEEE Trans. Ant. Propagation,* 30, 611, 1982.
75. **Hestenes, M. and Stiefel, E.,** Method of conjugate gradients for solving linear systems, *J. Res. Natl. Bur. Standards,* 49, 409, 1952.
76. **Sarker, T. K., Siarkiewicz, K. R., and Stratton, R. F.,** Survey of numerical methods for solution of large systems of linear equations for electromagnetic field problems, *IEEE Trans. Ant. Propagation,* 29, 847, 1981.

Chapter 3

THERMOREGULATION IN THE PRESENCE OF MICROWAVE FIELDS

Eleanor R. Adair

TABLE OF CONTENTS

I. INTRODUCTION

Thermoregulation is the maintenance of the body temperature within a prescribed range under conditions in which the thermal load on the body may vary. Thermal loads result from changes in metabolic heat production and from alterations in ambient conditions (temperature, ambient vapor pressure, air velocity, insulation, and other environmental variables that may alter the temperature of the integument). The deposition of thermalizing energy in the body by exposure to radio frequency electromagnetic (RF/EM) radiation is a unique exception to the energy flows normally encountered by earthly organisms, although metabolic activity in skeletal muscle can also deposit large amounts of thermal energy directly into deeper tissues.

All animals exhibit substantial response to changes in body temperature. Each species is characterized by a preferred level of body temperature at which its functioning and well being is optimal. While cold-blooded animals (ectotherms) regulate their body temperatures largely through behavioral selection of a preferred microclimate, warm-blooded animals (endotherms) have available, in addition, an involuntary physiological system that functions to produce and dissipate body heat. In general, endotherms rely largely on behavioral thermoregulation even though physiological responses can maintain a narrower range of body temperatures. When behavioral responses, which include the thermostatic control of the immediate microenvironment, are rendered difficult or inoperative, the body temperature of endotherms is regulated by the physiological mechanisms that control heat production, the distribution of heat in the body, and the avenues and the rate of heat loss from the body to the environment.[1-5]

The goal of much research into the biological consequences of exposure to RF/EM radiation is the understanding of how such exposure may compromise the normal biological functioning of *human beings*. Since, for ethical reasons, many experimental maneuvers cannot be performed on human subjects, studies of animal subjects must be substituted. The investigation of thermoregulation in the presence of microwave fields is no exception in this regard. While human beings have been exposed inadvertently to ultralow, "background" levels of both naturally occurring and man-made sources of RF/EM radiation, few individuals have voluntarily exposed themselves to significant levels for the scientific assessment of any responses, thermoregulatory or otherwise. The few extant data[6] quantify the thermal sensations of human subjects derived from stimulation of restricted body areas at a few discrete microwave frequencies as compared with similar sensations derived from stimulation at frequencies in the infrared. Since thermal sensation is the initial event that often underlies changes in thermoregulatory responses, these data are extremely valuable.

The physiological mechanisms the human body uses to maintain a nearly constant internal temperature are well described and well understood quantitatively, far better, indeed, than the responses of laboratory animals. Therefore, the following discussion of thermoregulation in the presence of RF/EM fields will be based upon our knowledge of how human thermoregulatory responses change (1) in the presence of conventional thermal stimuli in the environment, (2) during exercise-induced hyperthermia, and (3) on predictions made by sophisticated computer models of the human thermoregulatory system. The best quantitative data derived from animal subjects are cited whenever applicable.

II. FUNDAMENTALS OF THERMOREGULATION

Most of the vital internal organs of human beings function most efficiently when maintained at a relatively constant temperature near 37°C. While the temperature of individual body parts may vary characteristically from this norm, significant departures are associated with disease states or possibly lethal conditions. The usual range of "normal" body temperatures

(35.5 to 40°C) encompasses circadian variation, vigorous exercise, variations in ambient temperature, sequelae of food intake, age factors, cyclical variation in females, and emotional factors. Temperatures outside of this range must be related to disease states, unusual activity, or extraordinary environmental conditions.

Body tissues are extremely vulnerable to changes in temperature, particularly to over-heating. The intricate system of mechanisms developed by humans for regulating internal body temperature is therefore not surprising. In man, as in all endotherms, two distinct control systems are available to accomplish thermoregulation: a behavioral system involving conscious, voluntary acts that adjust the characteristics of the air-skin interface, and a physiological system involving the involuntary responses of the body that generate and dissipate body heat. It is the physiological system to which we first turn our attention.

A. Body Heat Balance

The most important principle involved in the study of physiological thermoregulation is the first law of thermodynamics or the law of conservation of energy.[7] In the steady state, the heat produced in the body is balanced by the heat lost to the environment such that storage of heat is minimal. This can be expressed by a generalized heat balance equation:

$$\underline{M} \pm \underline{W} = \pm \underline{R} \pm \underline{C} \pm \underline{E} \pm \underline{S} \text{ [W] or [W/m}^2] \tag{1}$$

in which $\underline{M}$ = the rate at which thermal energy is produced through metabolic processes, $\underline{W}$ = power or the rate at which work is produced by or on the body, $\underline{R}$ = heat exchange with the environment via radiation, $\underline{C}$ = heat exchange with the environment via convection, $\underline{E}$ = rate of heat loss due to the evaporation of body water, and $\underline{S}$ = rate of heat storage in the body.

It is important to note that all terms in Equation 1 must be in the same units, e.g., watts. As the equation is written, negative values of $\underline{R}$, $\underline{C}$, and $\underline{E}$ all may cause a rise in the body temperature; positive values may cause a fall. Work $(\underline{W})$ is positive when accomplished by the body (e.g., riding a bicycle), and this potential energy must be subtracted from metabolic energy $(\underline{M})$ to find the net heat developed within the body. When $\underline{W}$ is negative (e.g., walking downhill), this heat is added to the metabolic heat. Usually, $\underline{E}$, evaporative heat loss, is positive; when $\underline{E}$ is negative, condensation occurs and thermal injury is possible.

Because heat exchange by radiation $(\underline{R})$, convection $(\underline{C})$, or evaporation $(\underline{E})$ is always related in some way to the surface area of the body, it is useful to describe each term in Equation 1 in terms of energy per unit surface area, e.g., W/m². The most commonly used measure of body surface area in humans is that proposed by DuBois,[8]

$$A_D = 0.202 \ \underline{w}^{0.425} \ \underline{h}^{0.725} \tag{2}$$

where A_D = DuBois surface [m²], $\underline{w}$ = body mass [kg], and $\underline{h}$ = height [m].

As noted by Kleiber,[9] $\underline{h}$ for similar body shapes is proportional to a mean linear dimension equal to $\underline{w}^{1/3}$. Therefore, Equation 2 may be generalized to compare humans of different sizes:

$$A_D = k_1 \ \underline{w}^{0.425} \ k_2 \ \underline{w}^{1/3 \times 0.725}$$

$$= k_1 \ k_2 \ \underline{w}^{(0.425 + 0.242)}$$

$$= k_1 \ k_2 \ \underline{w}^{2/3} \tag{3}$$

The surface area-to-body mass ratio varies between species so that it is difficult to give a general rule for the determination of surface area. Indeed, many methods have been devised for the direct measurement of the surface area of experimental animals, a discussion of which would be out of place here. In most cases, the surface area is some function of $\underline{w}^{2/3}$ and the resting $\underline{M}$ of most species varies as $\underline{w}^{3/4}$. Since the latter relationship is most generally accurate, metabolic heat production is often expressed in units of W/kg.

No term appears in Equation 1 for heat transfer through *conduction* which is usually insignificant in man under normal conditions. However, conduction combined with mass transfer forms the mode of heat transfer called *convection*, a significant form of heat loss in man. Convective heat transfer in air ($\underline{C}$) is a linear function of body surface area (A) and the convective heat transfer coefficient (h_c) is a function of ambient air motion to the 0.6 power ($V^{0.6}$). The amount of heat lost by the body through convection depends upon the difference between the surface temperature of the skin (T_{sk}) and the air temperature, usually taken as the dry bulb temperature (T_{db}). The expression for heat loss via convection is

$$\underline{C} = kV^{0.6}A(\overline{T}_{sk} - T_{db}) \quad [W/m^2] \qquad (4)$$

where the heat transfer coefficient (h_c) equals $kV^{0.6}$ and k depends upon certain properties of the surrounding medium such as density, viscosity, etc., as well as a shape/dimension factor for the body. Clothing complicates the analysis and is often evaluated in terms of insulation (clo) units.[10-12] This variable is discussed in more detail below (cf. Section III.A).

Heat transfer by *radiation* ($\underline{R}$) is independent of T_{db}. The wavelengths of the radiation exchanged between two objects are related to their respective surface temperatures and the net heat transfer by radiation is proportional to the difference between their absolute temperatures to the fourth power and to the relative absorptive and reflective properties of the two surfaces. In general, the net radiant heat exchange (where h_r = radiant heat transfer coefficient) between a nude man and the environment involves estimation of the mean radiant temperature (MRT). MRT (alternate symbol T_r) can be derived from the temperature (T_g) of a blackened hollow sphere of thin copper (usually 0.15 m diameter) having similar heat transfer characteristics to those for the human body (e.g., $h_c/h_r = 0.178$).[13] Thus,

$$MRT = (1 + 0.222 \, V^{0.5}) (T_g - T_{db}) + T_{db} \quad [°C] \qquad (5)$$

Again, clothing complicates the analysis as it does for heat transfer by other modes. Solar heat load poses other special problems.[14] Heating by RF/EM radiation may provide further complications to the analysis of radiant heat exchange between a man and his environment, although Berglund[15] has demonstrated that the latter can be accomplished using conventional methods.

The fourth avenue of heat loss available to man is that due to the *evaporation* of water. The latent heat of vaporization of water at normal body temperature is 0.58 kcal/g; thus, the body loses this amount of heat when water is evaporated from its surfaces. Water in the expired air is continually being lost from the respiratory surfaces as is water that continuously diffuses through the skin (insensible water loss). These two avenues make equal contribution to a heat loss totalling $\sim$25% of the resting metabolic heat production ($\underline{M}$) in a thermoneutral environment. However, the major avenue of evaporative heat loss in man is sweating. Evaporation depends on the vapor pressures of the ambient air and the evaporating surface and is thus a direct function of both dry bulb (T_{db}) and wet bulb (T_{wb}) temperatures. For a body under conditions of $T_{db} = T_{wb}$, the air is at 100% RH and thus no water can be evaporated from the skin surface. When the ambient air is at less than 100% RH, evaporation can take place. The interrelationships between these variables, as well as tolerance limits for heat and cold by humans, can be determined from a standard psychrometric chart.

Evaporative heat transfer ($\underline{E}$) for a human body can be quantified as:

$$\underline{E} = 2.2 \, h_c A(P_{sk} \cdot \phi_a P_a)F_{pcl} \quad [W/m^2] \qquad (6)$$

where the evaporative area (A) is taken as 1.83 m² for an average or "standard" man, P_{sk} is the vapor pressure of water at skin temperature, $\phi_a P_a$ is the ambient vapor pressure at RH = ϕ_a, h_c is the convective heat transfer coefficient, and F_{pcl} is a dimensionless permeability factor. The reader is referred to the *ASHRAE Handbook*[16] for additional details related to Equation 6. This relationship applies to nude man and will be confounded by the insulation value and permeability of clothing. It should also be noted that E represents the evaporative cooling allowed by the environment (E_{max}) and is in no way related to the level of evaporative cooling required (E_{req}) by the man.

B. Endogenous Heat Production (M)

The basal metabolic rate (BMR) is defined as the heat production of a human in a thermoneutral environment (33°C), at rest mentally and physically, at a time exceeding 12 hr from the last meal. The standard BMR for man is ~250 ml O_2 per minute, or 84 W or 0.8 MET (where 1 MET = 58.2 W/m²). The BMR also corresponds to ~1.2 W/kg for a 70-kg "standard" man. The BMR is altered by changes in active body mass, diet, and endocrine levels, but probably not by living in the heat.[17] In resting man, most of the heat is generated in the core of the body — the trunk, viscera, and brain — despite the fact that these regions represent only about one third of the total body mass. This heat is conducted to the other body tissues and its elimination from the body is controlled by the peripheral vasomotor system. The range of M for humans, considering work performed and assorted physiological variables such as age, sex, and size, is roughly 40 to 800 W/m² (1 to 21 W/ kg for "standard" man) depending on physical fitness and level of activity. If deep body temperature is altered, either from heat storage in warm environments or febrile disease, there is a comparable change in M.[18] Similar changes occur when deep body temperature rises during exposure to RF/EM radiation.[19]

In cold environments, shivering is usually preceded by a generalized increase in muscle tone and piloerection which can increase M by about 35%.[20] Active shivering begins locally and over time recruits more and more of the body's skeletal musculature, eventually increasing heat production to as much as four to five times the resting level (i.e., 160 to 200 W/m²). However, if cold exposure lasts many hours, a doubling of the resting M is all that can be expected.[21] Further increases in M must be produced by exercise.

C. Avenues of Heat Loss

Both changes in vasomotor tonus and evaporation of body water through active sweating are mechanisms of body heat loss. In general, vasomotor control normally operates to achieve regulation of the body temperature when man is in thermoneutral environments; sweating is activated in warmer environments and during exercise.

1. Vasomotor Control

Convective heat transfer via the circulatory system is under the control of the sympathetic nervous system. In the cold, vasoconstriction of the peripheral vasculature in arm, leg, and trunk skin minimizes heat loss from the body core to the skin, leaving a residual conductive heat flow of 5 to 9 W/m²/°C temperature difference between body core and skin. In thermoneutral environments, when the peripheral vessels are vasodilated, each liter of blood at 37°C that flows to the skin and returns 1°C cooler allows the body to lose ~1 kcal or 1.16 W·hr of heat.[22] During vigorous exercise in the heat, peripheral blood flow can increase almost tenfold; this increase is essential to eliminate the increased metabolic heat produced in the working muscles.

Tissue "conductance" represents the combined effect of two channels of heat transfer in the body, conduction through layers of muscle and fat and convective heat transfer by the blood. Tissue thermal conductance is defined as the rate of heat transfer per unit area during steady state when a temperature difference of 1°C is maintained across a layer of tissue [W/ m²/°C].[7] Although this variable is not amenable to direct measurement in the living organism, tissue conductance can be estimated for resting humans if we assume that all the heat is produced in the core of the body and is transferred to the skin and thence to the environment. Thus, we may write

$$H_L = K (T_{re} - \overline{T}_{sk}) \quad [W/m^2] \tag{7}$$

in which H_L represents the heat loss (neglecting that lost through respiration), K is conductance, T_{re} is rectal (or core) temperature, and $\overline{T}_{sk}$ is the average skin temperature. In the cold (22 to 28°C) conductance is minimal for both men and women, ranging between 6 and 9 W/m²/°C. In warm environments, conductance increases markedly, with women showing a faster increase than men.[23] In general, measurements of heat transfer made at the skin surface yield only minimal rates of blood flow, since countercurrent heat exchange between arteries and veins in the peripheral tissues affects the observed heat flow.

2. Evaporative Water Loss through Sweating

Evaporation of sweat from the skin is an efficient way of losing heat even in environments warmer than the skin. Thus, evaporative heat loss must take care of both metabolic heat and that absorbed from the environment by radiation and convection. Thermalizing energy from RF/EM fields will be dealt with similarly to heat produced by normal metabolic processes in resting man or absorbed by exposure to warm environments.

Normal secretory functioning of the ~2.5 million sweat glands in the skin of a human being is essential to prevent dangerous hyperthermia. Secretion is under the control of the sympathetic nervous system and occurs when the ambient temperature rises above 30 to 31°C and/or the internal body temperature rises above 37°C. Local sweating rate has also been found to depend on the local skin temperature.[24] Individuals who are physically fit and/or acclimated to warm environments sweat at a lower internal body temperature and show an increased responsiveness of the sweating mechanism when challenged by exercise.[25] Dehydration or increased salt intake will alter plasma volume and decrease the efficiency of sweating during exercise.[26]

It is important to recognize that evaporative water loss through sweating occurs in man, the great apes, certain other primates (to a greater or lesser extent), and a few other species (e.g., horses). Many animals (e.g., dogs and cats) use panting (increased respiratory frequency coupled with copious saliva production) to accomplish evaporative heat loss. Other species (e.g., rodents) have no such physiological mechanisms and must rely on behavioral responses that include seeking shade, burrows, or aqueous environments and/or grooming their bodies with fluid (water, urine, saliva) to aid evaporative cooling.[27]

D. Neurophysiological Control of Thermoregulation

The operating characteristics of the thermoregulatory system have much in common with those of an automatic control system involving negative feedback. The body temperature appears to be regulated at a set or reference level. Temperature sensors located in various parts of the body detect thermal perturbations and feed this information to a central integrator and controller. The controller integrates sensory information from various parts of the body and controls the effector systems for heat production and heat loss. If a thermal disturbance shifts the body temperature away from the set or reference level, this change will be detected by the controller through signals from the temperature sensors, and appropriate effector responses will be mobilized to return the body temperature back to the set level.

1. Thermosensitive Neural Tissue

The type and vigor of thermoregulatory effector activity depends on the temperature of the skin; thus, specialized receptors must exist in the outer skin layers that detect changes in their own temperature. Two populations of cutaneous free nerve endings increase their activity to warming (warm receptors) or to cooling (cold receptors), respectively.[28] Hensel and Kenshalo[29] heated and cooled the facial skin of cats, while recording from single fibers in the infraorbital nerve, and determined response functions for both types of receptor. An increase in skin temperature (T_{sk}) above thermoneutrality ($\sim$33°C) produced a sharp increase in impulse frequency of warm fibers, while cold fibers decreased theirs. Warm fiber response ceased at a T_{sk} above 45°C indicating thermal death of the fibers. Skin temperatures below 33°C produced an increase in impulse frequency of cold fibers while suppressing activity in the warm fibers.

Extensive neurophysiological research, beginning a century ago with studies of lesioned animals,[30,31] has shown that certain central nervous system (CNS) tissue is either essential for or involved in the control of autonomic mechanisms of heat production and heat loss in endotherms. Barbour,[32] in 1912, showed that a primitive thermode implanted in the anterior brainstem of rabbits could be used to trigger many thermoregulatory responses. When this device was warmed, heat loss responses were initiated; when cooled, heat production and conservation responses were stimulated. These primitive experiments provided the impetus for more than 60 years of neurophysiological research, much of which has involved the search for temperature sensors in those areas of the CNS where lesions or thermal stimulation influence thermoregulatory effector responses.

Many CNS sites have been explored with microelectrodes in experimental animals to reveal the existence of single units whose firing rates depend upon the temperature of the tissue in which they are located.[33-35] In these experiments, local tissue temperature is controlled with implanted thermodes.[36] The activity patterns of CNS thermal sensors resemble those of the skin sensors in many respects. CNS sites that exhibit such thermosensitivity include the medial preoptic area of the hypothalamus, lateral hypothalamus, midbrain reticular formation, medulla, motor cortex, thalamus, spinal cord, and deep viscera.[37-40] Many preoptic cells respond to temperature changes in the skin, the spinal cord, and other CNS thermosensitive sites as well as in their own vicinity.[41-43] Many neurophysiologists hold this to be evidence that the preoptic area of the hypothalamus is the site of the central "thermostat" as well as the locus of the "integrator-controller". Unfortunately, there is still no direct evidence that these thermally responsive neurons are the actual sensors involved with thermoregulation. A further assumption is that the CNS of humans contains thermal sensors in sites analogous to those that have been found in experimental animals. The strongest evidence marshaled for this assumption is the uniformity of the neurophysiological findings between ectotherms and endotherms.[44]

Temperature changes localized in several thermosensitive sites trigger behavioral as well as physiological thermoregulatory responses.[45-49] Any means by which thermosensitive CNS structures may be thermally stimulated has the potential for altering the normal thermoregulatory responses of endotherms. RF/EM radiation is such a potential stimulus. Under appropriate conditions of frequency, intensity, and duration, RF/EM radiation can be absorbed by biological targets in such a way that tissues below the surface may be heated. Often, the patterns of absorbed energy are extremely complex, depending not only on the parameters of the radiation, but also on the size, shape, and electrical characteristics of the exposed tissue.[50] Complex absorption may give rise to nonuniform heating of tissue, a potentially unique challenge to the thermoregulatory system.[51]

2. Characteristics of the Thermoregulatory "Integrator-Controller"

The integration of neural signals from the various thermal sensors in the body by a "controller" located (perhaps) in the hypothalamus, the comparison of the integrated signal with a reference level, and the generation of neural effector "commands" are all hypothetical constructs that aid our understanding of thermoregulatory processes. There are nearly as many models of the thermoregulatory "controller" as there are researchers in the field. Many early models featured two thermoregulation centers rather than one — one controlling body heat loss, the other body heat production;[31,52] however, we now believe that one center or controller can accomplish both types of regulation.[36] All models have other features in common: the controlling center resides in the brain and contains a reference against which error signals of various types are compared.

Models have been of many types: verbal descriptions,[31,52,53] pictorial models of the neurophysiological system,[54-57] physical analogs of the thermoregulatory system or portions thereof,[58] electrical analogs,[59] mathematical models,[60-65] neuronal models,[66-68] and chemical or neural transmitter models.[69,70] Hardy[4] gives a comprehensive account of the development of modeling in thermal physiology. The use of such simulation models to predict the human response to RF/EM radiation is discussed in a later section of this chapter.

III. LIMITS OF HUMAN HEAT TOLERANCE TO RF RADIATION

The human thermoregulatory response to the deposition of significant amounts of thermalizing energy from RF/EM radiation (i.e., sufficient to elevate the temperature of tissues in all or part of the body) will not differ substantially from the response to heat generated in the body tissues by other means. The basic problem posed by excessive heating of the body, whatever the source, is whether the heat-loss capability of the thermoregulatory system is sufficient to prevent heat storage in the body. Thus, for a human being exposed to a RF/EM field, the generalized heat balance Equation 1 may be rewritten:

$$(\underline{M} \pm \underline{W}) + \underline{A}_{RF} = \pm \underline{R} \pm \underline{C} \pm \underline{E} \pm \underline{S} \quad [W] \text{ or } [W/m^2] \tag{8}$$

where $\underline{A}_{RF}$ represents the rate of energy absorption from the RF/EM field. Neglecting the work factor, the sum of the heat production, heat exchange by radiation and convection, and the absorbed RF energy will provide a useful estimate of the evaporative cooling required ($\underline{E}_{req}$):

$$\underline{E}_{req} = \underline{M} \pm \underline{C} \pm \underline{R} + \underline{A}_{RF} \quad [W] \text{ or } [W/m^2] \tag{9}$$

If the maximum available evaporative cooling ($\underline{E}_{max}$) is less than $\underline{E}_{req}$, $\underline{S}$ will be positive and body temperature will rise.

In general, the degree of heat stress placed on the body can be predicted by the simple ratio of $\underline{E}_{req}/\underline{E}_{max}$ which yields a measure of the percentage of the skin surface that is wet with sweat.[71] Values of $\underline{E}_{req}/\underline{E}_{max} < 20\%$ yield a state of thermal comfort, while higher percentages indicate tolerance limits. This ratio has also been called the Heat Stress Index (HSI).[72] HSI values in excess of 30% are judged to be uncomfortable and may interfere with concentration and fine motor performance, but are also judged to be tolerable; values up to 60% result in finite tolerance times and interfere with work; values between 60 and 100% represent severe, intolerable conditions.[72]

The amount of heat storage ($\underline{S}$) in the body, calculated from Equation 6, can also be used to predict heat tolerance times (i.e., when $\underline{E} < \underline{E}$ req).[73] Thus, $\underline{S} = 0$ to 25 kcal/hr may not even be noticed; $\underline{S} > 80$ kcal/hr represents the usual voluntary tolerance limit, double this value (160 kcal/hr) incurs a 50% risk of collapse, and triple this value (240 kcal/hr) is intolerable.

A. Environmental Variables: Temperature, Vapor Pressure, and Clothing

Tolerance limits for human beings can be determined from a standard psychrometric chart

Table 1
COMFORT-HEALTH-INDEX — HUMAN RESPONSES

C-H-I	Sensations	Physiological responses	Health hazard
(T_{db} at 50% RH) (°C)			
45	Unbearable	Rapid body heating; failure of body temperature regulation	Circulatory collapse
40	Very uncomfortable, hot	100% Wetted area	Temporary exposure dangerous
35	Uncomfortable, hot	50% Wetted area; physiological strain	"Danger line" for heat stroke
30	Slightly uncomfortable		
25	Warm; comfort zone; cool	25% Wetted area; zone of vasomotor regulation	No health hazard
20	Slightly uncomfortable; cold	Vasoconstriction of hands and feet	Muscular pains; peripheral circulation impaired

which diagrams the vapor pressure of the ambient air as a function of both T_{wb} and T_{db}. A comfort-health-index (C-H-I) for human responses, extracted from such a chart, is shown in Table 1 (adapted from Hardy[74]). The table gives the thermal sensations, physiological thermoregulatory responses, and associated health hazards as these variables relate to T_{db} at RH = 50%. This level of RH prevails in much of the U.S. (except for certain high-altitude and desert regions); therefore, the T_{db} given will be appropriate for a high percentage of the U.S. population and can be applied to sedentary man in the indoor environment. No data exist related to exposure to RF/EM radiation that can be added to this table, but it is most probable that intensities of exposure that provide significant energy deposition in the body core will alter the C-H-I in much the same way as would an equivalent metabolic load. This conclusion was reached by Nielsen and Nielsen[75] who demonstrated the equivalence of passive heating (through diathermy) and active heating (through exercise) on thermoregulation in two male subjects. Stolwijk[76] has predicted, using a mathematical model of the human thermoregulatory system, that at T_{db} = 30°C and RH = 50% deposition of 500 W of RF power for 30 min into the trunk core of a sedentary man with intact thermoregulation would elevate the temperature of the central blood by ~1.5°C. This rate of energy deposition is equivalent to moderate exercise (5 METS) such as pedaling a bicycle ergometer at 50 rpm with a load of 600 kpm/min.[77] (The kilopondmeter per minute, kpm/min, is used as the unit of work, $\underline{W}$, by exercise physiologists. A kilopondmeter is a kilogram mass under normal gravity raised through a distance of 1 m.) If the sweating response were disabled (as when the body is immersed in water or RH = 100%), deposition of 500 W for 30 min into the trunk core would elevate the temperature of the central blood by > 4°C. Stolwijk's model has also been used to predict shifts in many other thermoregulatory parameters (body temperatures, heat storage, metabolic rate, evaporative water loss, skin blood flow, cardiac output, and effective thermal conductance) that may occur when significant amounts of RF/EM energy are deposited in the head core or leg.[76]

A somewhat different approach has been taken in the *Radiofrequency Radiation Dosimetry Handbook, 2nd ed..*[78] Using a rectal temperature limit of 39.2°C as a criterion of heat-stress tolerance and an arbitrary 60-min exposure period, the radiation conditions (SAR_{60}) were determined that would fulfill these criteria under different environmental conditions (i.e., T_{db} from 30 to 68°C and RH = 80, 50, and 20%). Equation 8 was used to predict the maximum whole-body specific absorption rate (SAR) that a healthy man could tolerate for 1 hr in a specified environment. One such calculation yielded an SAR_{60} of 3.11 W/kg for T_{db} = 40°C and RH = 80%. Then the SAR curves for an average man[78] were used to calculate the power density for any given frequency and polarization that will produce this SAR[60]. These SAR curves, derived from a prolate spheroidal model of man at 10 W/m², are found in Figure 17 of the work cited. Thus, an SAR_{60} of 3.11 W/kg would represent a power density of roughly 130 W/m² for irradiation by an electromagnetic planewave (E-polarization) at the resonant frequency of 70 MHz. SAR_{60} values calculated in this manner are not to be interpreted as safe exposure limits, however; they are simply estimates of the upper tolerance levels for a healthy man with normal heat loss capabilities.

The above calculations assume that the subject is a healthy nude male in a resting state. The clothed human body presents a resistance to the flow of heat from the skin to the environment that is a direct function of the thickness of the air layer trapped by the clothing. The standard insulation unit, called the clo,[10] was derived empirically and equals 0.18°C·m²·hr/kcal or 0.155°C·m²/W. The insulation value of a standard business suit is defined as 1 clo. Typical indoor clothing worn in offices today ranges from 0.4 clo in summer (short, light dress; light slacks and short sleeved shirt) to ~1.0 clo in winter (heavy slacks, blouse, light sweater, and jacket; heavy trousers, sweater and shirt, jacket).[79] A rule of thumb[80] states that for each 0.1-clo deviation from the usual 0.6-clo baseline insulation for sedentary persons doing light office work (1 MET), the air temperature for comfort can be offset by 1°F (0.55°C). This temperature offset can be doubled when the workload is increased to 4 to 5 METS.

The limiting factor to lowering office temperatures for energy conservation appears to be related to cold hands and feet. It is difficult to design practical mittens that warm the hands and still allow the manual dexterity required for most jobs. The recent proposal[81] to provide RF/EM comfort heating of human beings in otherwise cool surroundings appears most attractive. On the other hand, avoidance of thermal discomfort in hot environments that may include a source of RF/EM radiation will be accomplished primarily by minimizing the percentage of the body surface area that is wetted by sweat. This can be accomplished by (1) increasing air movement, (2) reducing the ambient vapor pressure, and (3) removing as much clothing as possible consistent with acceptable social norms.

B. Exercise

During physical exercise, the human thermoregulatory system functions in the presence of an internal disturbance. When a proportional, negative-feedback-control system (the thermoregulatory system) encounters such a disturbance, it usually maintains an offset in its regulated variable (the regulated internal body temperature, in this case). This occurs during muscular exercise. At the initiation of activity, heat production in the skeletal muscles rises rapidly, producing a temporary large increase in heat storage within the body. The rising body temperature generates an error signal relative to the "set" body temperature, which drives the mechanisms of heat loss at an ever-increasing rate until heat loss equals heat production. When a balance is attained, the body temperature will equilibrate at a hyperthermic level that is characteristic of the level of exercise being performed.[82] Exercise represents the single condition that provides a maximal strain to the thermoregulatory system under normal circumstances and allows measurement of the various modes by which the body adjusts to this strain.[83]

Exercise changes neither the "set point" for body temperature nor the normal operating characteristics of the thermoregulatory system. Nielsen and Nielsen[75] used short-wave diathermy to deposit heat directly into the deep tissues of the trunk in human subjects. In other experiments, the same subjects exercised on a bicycle ergometer at a work rate that was adjusted so that the heat load during cycling and diathermic heating was the same (approximately 5 METS). In the steady state, the rectal temperature increased by the same amount during the two procedures as did thermal conductance as assessed by changes in skin blood flow. Sweating rates were also comparable. Thus, passive heating by diathermy and the heat generated during active exercise produce the same kind of thermal disturbance in the body as a whole, although the distribution of heat in individual tissue compartments of the body is very different in the two cases. These important findings also demonstrate that the thermoregulatory consequences of whole-body energy deposition by RF/EM radiation may be predicted by the consequences of equivalent heat loads produced by exercise. This rationale underlies predictions of changes in thermoregulatory responses to microwave power deposition made using mathematical models of the human thermoregulatory system.

C. Febrile States

It is important to differentiate between the hyperthermia produced by exercise and that occurring during febrile disease. Exercise produces an elevation in body temperature to a level above the normal, regulated (or "set") level. The magnitude of the heat dissipation response that is stimulated during exercise is directly related to the magnitude of this temperature offset.[84] On the other hand, during fever the increase in the regulated body temperature is defended in the same way as is the normal body temperature during euthermia. During the generation and maintenance of fever the thermoregulatory system behaves in a manner consistent with an elevation of the "set" temperature.[85] This was demonstrated in a human subject during daily bouts of exercise and rest.[86] When the subject contracted a febrile disease, his rectal temperature rose and fell during the periods of exercise and rest, but at an overall level about 2°C higher than the other (healthy) subjects. An elevated internal temperature at the threshold for vasodilation also occurs during fever.[83]

Fever is generated differently in different thermal environments. In warm environments, heat loss will be curtailed and vasoconstriction will occur; if these responses are inadequate to increase heat storage, heat production will rise. In cold environments where heat loss responses are already minimal, greatly increased heat production (vigorous shivering) is the only way to raise the body temperature. Once a pyrogenic substance introduced into the body displaces the set temperature, the controller for thermoregulation will mobilize any appropriate effector mechanism to increase the storage of body heat.[87] Behavioral responses that increase skin temperature (by selection of a warmer environment or by increasing body insulation) generate an identical fever to that produced by changes in physiological responses when a given dose of pyrogenic substance is administered.[88]

The evolution of a febrile episode is well understood, although the actual mechanism by which the set point is elevated by the pyrogen remains a mystery. When the set temperature is elevated by a circulating pyrogen, a large load error is generated because the body temperature is below the new set point. Heat loss is curtailed, vasoconstriction occurs, and metabolic heat production increases in an attempt to raise the body temperature to the new set level. This "chill" phase gives way to a "plateau" phase once the body temperature has risen to the febrile level. At this point, the error signal no longer exists and the effector responses return to the magnitude existing before fever onset, regulating in normal fashion around the new elevated set point. When the disease runs its course or an antipyretic is given, the normal set point is restored, giving rise to a new load error of opposite sign to that occurring at fever onset. During this "crisis" phase, vigorous heat loss responses are stimulated (vasodilation and sweating) which rapidly return the body temperature to normal. Thus, it is clear that the hyperthermia that may result from exposure to RF/EM radiation differs greatly from that occurring during febrile episodes, resembling much more closely the hyperthermia of exercise.

IV. THERMAL SENSATION AND THERMOREGULATORY BEHAVIOR

Since thermalization can occur in body tissues that are exposed to RF/EM radiation, such radiation must be considered as part of the thermal environment to which man and animals may potentially be exposed.[81] Although physiological responses (e.g., vasodilation, sweating) may be triggered automatically by thermal stimuli, the sensation of tissue heating is necessary for the initiation of appropriate behavioral action. Excitation of temperature-sensitive neural structures residing within the outermost 0.6 mm of mammalian skin[89] is believed to underlie the sensation of changes in skin temperature. Whether or not an exposure to RF/EM radiation produces a sensation of warmth will depend upon many parameters of the signal (e.g., frequency, modulation, intensity, duration, and the surface area of the body exposed). Many of these same parameters influence the magnitude of thermal sensations derived from IR radiation.[90-97] Infrared radiation is absorbed in the uppermost layers of the skin in close proximity to the temperature-sensitive structures; a similar absorption profile will be obtained for the highest microwave frequencies (10 GHz or higher). However, lower RF/EM frequencies will be absorbed in complex patterns at other depths, rendering prediction of thermal sensations difficult. It has been suggested recently that psychophysical investigations of thermal sensations aroused by microwaves should be guided by past experience with more conventional means of thermal stimulation, such as IR radiation.[6]

A. Thermal Sensations Produced by Microwaves

Absolute thresholds for the detection of RF irradiation by human observers have been determined in several studies,[98-104] all of which have involved brief exposures (10 sec or less) and restricted areas of forehead or forearm skin. In general, the shorter the wavelength of the irradiation, the less energy is required to provoke a just-detectible sensation of warmth.[105] When a 37-cm² area of the forehead was irradiated for 4 sec, the mean absolute threshold of warmth was 335 W/m² for 3-GHz irradiation, 126 W/m² for 10 GHz, and 42 W/m² for >1000 GHz (far-IR).[101] Irradiation of small areas of skin by 3- or 10-cm pulsed (2500 pps, 0.4 μsec pulse width) microwaves has to last at least 5 sec in order for the minimal intensity to evoke a thermal sensation, the exact intensity depending upon the area stimulated.[99,101] At shorter durations, stimulus intensity has to be greatly increased in order to evoke comparable sensations of warmth. An unusual aspect of the sensation derived from RF/EM irradiation is its persistence following cessation of the stimulus.[98,101] This property, together with longer-than-average latencies to detection, probably derives from the thermal inertia of the affected tissues[102,104] and the greater volume of tissue involved when deeply penetrating radiation is involved.[50]

A recent study has incorporated indirect assessment of the energy absorbed during 10-sec exposures of the forearm (average area = 107 cm²) of human subjects to 2450-MHz CW microwaves.[104] Sensations of warmth were experienced by the subjects when the energy dose rate was about 900 J/m² as opposed to about 200 J/m² when the same skin area was exposed to IR radiation. These threshold values correspond to incident powers of 270 and 17 W/m², respectively, and are similar to the results reported above.

B. Changes in Thermoregulatory Behavior in the Presence of Microwaves

Thermoregulatory behavior refers to voluntary actions by an organism that control the thermal characteristics of the air-skin interface, thereby facilitating regulation of the body temperature at a stable level. Few experimental data exist that relate changes in thermoregulatory behavior of human subjects to imposed RF/EM fields. However, experimental results from animal subjects afford meaningful insights with good predictive value. Low intensity microwave fields influence thermoregulatory behavior like other sources of heat. Lizards will bask in proximity to a 900-W/m² microwave source and regulate their body temperature thereby.[106] Rats, trained to press a lever for IR heat in the cold, will reduce the rate of lever pressing when a low-intensity field of 2450-MHz microwaves is present.[107] The higher the microwave intensity to which they are exposed (range = 50 to 200 W/m²), the less IR radiation is demanded by the rats. Squirrel monkeys will select a cooler environment when irradiated for 10-min periods by 2450-MHz planewaves at a power density of 60 to 80 W/m². [108] Higher power densities stimulate correspondingly greater reductions in selected air temperature, assuring regulation of skin and deep body temperatures at the normal level. Except during the first irradiation period of a series or during the early minutes of a single long (2½-hr) exposure, duration of irradiation has no significant effect upon the air temperature selected by an animal or the body temperatures achieved. [109]

Extensive training of the experimental animal in thermoregulatory behavior[108] is certainly advantageous and may be essential to survival when RF/EM fields are present. [110] The presence of visible or auditory cues correlated with RF/EM irradiation facilitates learning of appropriate responses leading to escape from potentially lethal fields, [111,112] while careful shaping of appropriate responses accelerates such learning dramatically. [113] Previous experience with similar stimulus-response contingencies has also been demonstrated to be helpful. [114] Extrapolation of these general findings to the human condition, especially in industrial settings that incorporate high-power sources of RF fields, leads to the conclusion that education of the worker, together with environmental signals (visual and auditory) that are associated with equipment function and malfunction, should aid recognition of the potential for untoward exposure and insure timely retreat from the field. [113]

V. ADDITIONAL SUPPORTING DATA FROM ANIMAL STUDIES

In addition to the behavioral experiments summarized above, several studies have assessed individual physiological responses of heat production and heat loss in the presence of RF/EM fields with varying degrees of success. Experimental animals must be restrained to afford the opportunity to monitor thermoregulatory variables, and restraint in and of itself often imposes a stress that interferes with normal thermoregulation. Many of these studies have involved exposure to 2450-MHz radiation. There is a clear and present need for definitive experiments of this type, using both higher primates and human subjects, and a wider range of frequencies.

A. Threshold Effects

Microwave exposure at 2450 MHz can lower the metabolic heat production of mice[115] and rats[116] in thermoneutral and cool environments. In the studies cited, the thermoregulatory consequences of the absorbed microwave energy could not be evaluated while the exposures were in progress because of the exposure geometry employed, the mice being held inside a waveguide and the rats in a multimodal cavity (microwave oven). Microwave dose was varied, but the environmental temperature during exposure was uncontrolled in the cavity and held at 24°C in the waveguide. The metabolic reduction produced by microwave exposure was found to be dose dependent, although postural adjustments by the mice (undocumented because the waveguide was opaque) may have reduced microwave absorption. [115] Bradycardia and arrhythmia were also exhibited by the rats, responses which may have resulted from significant localized heating in the brain. When the environmental temperature in the waveguide was varied (range = 20 to 35°C), [117] no difference in oxygen consumption from sham-exposed levels occurred when mice were in cool environments; however, an *increase* in metabolic rate was measured in mice exposed at an average dose rate of 10 W/kg and above in a warm waveguide temperature of 35°C. The latter result is comparable to the metabolic elevation which occurs during hyperthermias produced by high ambient temperatures: when the animal is unable to eliminate the heat it produces, body temperature rises and so does the metabolic rate because the cellular processes are now uncontrolled.

Adair and Adams[19] exposed squirrel monkeys to both brief (10-min) and prolonged (90-min) periods of 2450-MHz CW microwaves while T_{db} was held constant at 15, 20, or 25°C. Microwave power density ranged from 25 to 100 W/m² (SAR range = 0.4 to 1.5 W/kg).

Reliable reductions in metabolic heat production (M) were initiated at all T_{db} by 10-min whole-body exposures at power densities of 40 to 60 W/m² and above. The magnitude of the M reduction was linearly related to microwave intensity above the threshold level. Termination of the microwave field was followed by a rapid M rebound. The change in M produced by a given power density was the same at all T_{db}. During 90-min exposures, the vigorous M reduction to microwave onset adapted slowly, ensuring continued precise regulation of the internal body temperature. In the steady state, the reduction in metabolic heat production was nearly always the same as the whole-body SAR of the field to which the animal was exposed. These experiments demonstrated clearly that cold-exposed endotherms compensate readily for microwave-induced body heating by reducing endogenous heat production. Precise regulation of the internal body temperature is achieved thereby.

Vasomotor tonus in cutaneous tail veins of the squirrel monkey has been indexed by changes in local skin temperature during exposure of the whole body to 2450-MHz CW microwaves.[118] At an ambient temperature (26°C) just below that at which the vessels of the tail normally vasodilate, criterion vasodilation is initiated by 5-min exposures at a microwave power density of 80 W/m². This intensity deposits energy equivalent to ~20% of the monkey's resting metabolic heat production, but produces no observable change in deep body temperature. Intensity increments of 30 to 40 W/m² for 1°C reductions in T_{db} below 26°C produce identical threshold responses. No vasodilation occurs during IR exposures of equivalent power density, indicating that noncutaneous thermosensitive structures may mediate microwave activation of thermoregulatory responses in the peripheral vasomotor system.

The minimal power density of 10-min exposures to 2450-MHz CW microwaves has been determined that will reliably initiate thermoregulatory sweating from the foot of squirrel monkeys restrained in ambient temperatures just below the upper critical temperature (36°C) at which sweating occurs spontaneously in this species.[119] This threshold power density was 60 to 80 W/m² (SAR $\cong$ 1.1 W/kg), a value comparable to microwave thresholds for other thermoregulatory responses in this species. Above the threshold, the magnitude of the change in sweating rate at a given T_{db} was directly related to the power density of the imposed field. These responses are qualitatively similar to those of human subjects.[120] The local sweat rate of exercising humans increases linearly as the temperature of the body core rises during exercise at a given ambient temperature. But a higher core temperature is necessary to initiate sweating at lower skin temperatures and the rate of increase in sweating during exercise is lower and lower in cooler and cooler environments. This comparison indicates that, in the monkey, the rate of thermoregulatory sweating initiated by microwave exposure depends not only on the ambient (skin) temperature, but also upon the temperature of the body core as it is directly increased by absorbed microwave energy.

A recent report by Gordon[121] describes a threshold SAR of 29 W/kg for initiation by 2450-MHz microwaves of an increase in the evaporative water loss of mice held at T_{db} = 20°C inside a waveguide. However, mice and other rodents neither sweat nor pant when hyperthermic; thermoregulation is restricted to behavioral maneuvers such as spreading saliva on the pelt to aid evaporative cooling. Since neither metabolic rate nor body temperatures were measured in this study, and since the animals could not be observed inside the opaque waveguide, the published results are open to question.

B. Intense or Prolonged Exposure

The thermoregulatory consequences of exposure to intense microwave fields have been explored by Michaelson and co-workers.[122,123] Whole-body exposure of dogs to 2880-MHz pulsed (360 pps, 2 to 3 μsec pulse width) microwave fields at an average power density of 1 kW/m² (SAR = 3.7 W/kg) for 6 hr or 1.65 kW/m² (SAR = 6.1 W/kg) for 2 to 3 hr produces a characteristic triphasic change in internal body temperature. These phases are

(1) an initial increase in core temperature, (2) a plateau phase at the hyperthermic level, and (3) thermoregulatory failure. Presumably, specific heat loss mechanisms (e.g., panting) mobilized by these exposures are able to counterbalance partially the thermalizing energy absorbed by the animal, but only temporarily. The strain placed upon the thermoregulatory system ultimately exhausts the heat loss capabilities of the dog and death ensues due to hyperpyrexia unless the animal is removed from the field.

The environmental temperature at which the exposure occurs is also very important.[124] Thus, dogs can tolerate exposures at SARs of 3.7 and 6.1 W/kg when T_{db} = 11°C and exposures at 3.7 W/kg when T_{db} = 22°C without becoming hyperthermic. However, at T_{db} = 40.5°C, dangerous hyperthermia can occur within 20 min at SAR = 6.1 (a value roughly twice the resting metabolic heat production of the dog). Hydration during microwave exposure permits an extended tolerance at high SARs, presumably through an increased capacity for respiratory evaporative heat loss (panting). One of Michaelson's notable findings was the development of tolerance to RF/EM exposure at SAR = 6.1 W/kg as the number of such exposures increased; e.g., it took 60 min on the first day, but 220 min on the 34th day, for dogs to generate a rise in rectal temperature of 1.5°C. This phenomenon resembles acclimatization to hot environments (cf. Goldman[125]).

Thermographic studies on tissue-equivalent models of men and animals[49,126] have demonstrated regions of high local SAR during exposure of the whole body to planewave RF/EM fields. Wrists, ankles, and neck (also animal tails) are predicted to be foci of enhanced energy absorption where excessive elevations of tissue temperature may occur. Studies by Krupp[127] of anesthetized rhesus monkeys exposed to planewave 210-MHz RF/EM radiation at power densities from 50 to 270 W/m² demonstrate substantial increases in core temperature, but no evidence for tissue "hot spots" in wrist, ankle, thigh, or biceps. Increased convective heat transfer via blood flow during such exposure clearly results in the protection of individual tissues.[76,128] Similarly, mathematical models[129-132] have predicted regions of high local SAR within the brain of animals exposed to RF/EM fields. Should such "hot spots" occur in the vicinity of the hypothalamic thermoregulatory center, this would have great import for normal thermoregulation. Burr and Krupp[133] mapped the distribution of RF energy absorption in the heads of live rhesus monkeys and concluded that normal thermoregulatory processes (especially increased local blood flow) prevent inordinate increases in local temperature until a threshold SAR of 5 W/kg is reached or exceeded. Significant elevation of the brain temperature of anesthetised rats was measured following exposure to 1600-MHz microwaves at 200 to 800 Wm². [134] The deep brain temperature of conscious rats can rise by as much as 1.0°C during 60-min exposure to 2800-MHz pulsed fields at low power densities.[135] Recent studies of squirrel monkeys by Adair and colleagues[135a,135b] featured hypothalamic heating and cooling by an implanted thermode (tube perfused with temperature-controlled liquid) during microwave exposure to a head-resonant frequency. The experiments not only failed to find evidence for brain hot spots, but also led to the conclusion that the hypothalamic thermoregulatory center is no more important than other thermally sensitive sites in the body for the mobilization of autonomic thermoregulatory responses. Other reports[136,137] indicate that the anterior hypothalamic temperature of squirrel monkeys engaged in behavioral thermoregulation seldom rises more than 0.2 to 0.3°C above normal baseline levels when the animals are exposed to 2450-MHz CW fields; indeed, efficient thermoregulatory behavior acts to prevent such a temperature rise by increasing the skin-to-ambient temperature gradient, thereby facilitating heat loss from the body.

Whole-body exposure of experimental animals at frequencies near resonance for the species in question may pose special problems for the thermoregulatory system because of greatly enhanced SARs at any given field strength.[78] Investigation of this problem is just beginning; it deserves intensive experimental attention in the future. Two studies in which rhesus monkeys (anesthetized or awake) were exposed to near-resonant frequency (219 or 225 MHz, respectively) radiation at power densities near 100 W/m² reported rapid elevations in deep body temperature of animals exposed at T_{db} near thermoneutrality.[127,138] However, the failure to measure heat-production or heat loss responses in the exposed animals renders interpretation of these findings extremely difficult.

It is essential that quantitative studies of thermoregulatory function always be conducted at the same time of day, because a circadian rhythm of regulated changes in body temperature is characteristic of all endothermic species. Lotz[139] has demonstrated an influence of this circadian rhythm on the elevation of rectal temperature in rhesus monkeys exposed or sham-exposed to 1.29-GHz pulsed radiation at a whole-body SAR = 4.1 W/kg. The 8-hr exposures occurred either during the day or during the night. The change in body temperature resulting from these exposures was nearly identical (~1.6°C), but the peak temperature was ~1°C lower at night, a value equal to the normal nocturnal fall in regulated body temperature in this species. Plasma cortisol was elevated only during day exposure. Lotz suggested that this response may depend simply on the absolute level of body temperature rather than the magnitude of temperature change during RF/EM exposure.

Studies too numerous to mention corroborate the general finding in many species that absorbed doses of microwave energy equivalent to the resting metabolic heat production will elevate the deep body temperature of an experimental animal by 1°C or more. Of special note is the work of deLorge[140-143] who has correlated the rise in deep body temperature with the cessation of food-reinforced operant behavior (work stoppage). Estimates of whole-body SAR may not be the best predictors of work stoppage from one frequency of RF/EM radiation to another, however; deLorge suggests that energy deposition in the head or brain may be a more reliable predictor than is the increase in body temperature. Since the animals in deLorge's experiments are often tested in restraining devices, it should be possible to monitor local body temperatures (including in the brain) as well as individual thermoregulatory responses such as metabolic heat production, vasomotor tonus, and sweating rate in order to derive a more complete assessment of the effects of RF/EM radiation on changes in thermoregulatory processes during ongoing behavior.

C. Drug-Microwave Interactions

Michaelson reported that dogs display a greater susceptibility to microwave heating after administration of pentobarbital sodium, morphine sulfate, or chlorpromazine, indicating that mechanisms of heat loss may be compromised by treatment with certain drugs.[122] A similar effect was demonstrated[144] in rats rendered hypothermic by cortisone injections: following this treatment, RF/EM exposure at 40 W/kg produced greater increases in core temperature than occurred in nontreated control animals. Smialowicz[145-147] has utilized similar drug-microwave interactions to demonstrate subtle heating produced by low intensity (≤10 W/m²) exposure to 2450-MHz CW microwaves in both mice and rats. Mice injected i.p. with 5-HT creatine sulfate complex (20 mg/kg) and rats injected i.v. with *Salmonella typhimurium* LPS (100 μg/kg) become hypothermic when held at T_{db} = 22°C. Postinjection exposure to RF/EM radiation generates an increased colonic temperature that is a direct function of the power density of the imposed field. Saline-injected control animals, similarly exposed, do not show comparable elevations of core temperature. Thus, chemical impairment of thermoregulatory mechanisms permits detection of significant changes in the body temperature of experimental animals exposed to low-intensity RF/EM fields that were formerly believed to be nonthermogenic.

VI. APPLICATIONS OF SIMULATION MODELS OF HUMAN THERMOREGULATORY SYSTEM TO DEPOSITION OF RF/EM POWER WITHIN THE BODY

Since the thermal characteristics of the human body are reasonably well known, it is possible to develop mathematical simulation models of physiological thermoregulation in reasonable detail and validate them against experimental data.[59-62,64,65,148] Most models separate the thermoregulatory system into two major components: the regulated or passive system, and the regulatory or controlling system.

The passive system consists of a simplified representation of the thermal characteristics of all body tissues including metabolic heat production, heat capacitance, local tissue temperature, heat transfer by conduction and convection inside the body, and via evaporation, conduction, convection, and radiation between skin and environment. The controlling system consists of the response characteristics of the structures which sense body temperatures, the central neural integration, neural effector pathways, and the effector mechanisms themselves (e.g., shivering muscles, secreting sweat glands, and the tone of peripheral blood vessels in the skin and elsewhere which control the convective transfer between different parts of the body).

In order for such a model to have maximal utility, it must contain all the changes in physiological responses that may occur within the ranges of imposed environmental thermal stress, metabolic heat production, and body temperatures that encompass its anticipated uses. The model of Stolwijk and Hardy[64] accurately predicts these response changes to both sudden changes in environmental conditions[149] and to bouts of exercise.[65] It seems well suited to predict changes in thermoregulatory responses that may result from deposition of RF/EM energy into the body and has provided the basis for several such analyses.[128,150,151]

In recent years, considerable progress has been made in the quantification of the absorption of RF/EM energy by animals and humans.[49,126,132,152,153] Many studies have pointed out nonuniformities of energy deposition within exposed tissue volumes and the potential for hot spots, especially in the cranial cavity. Mathematical models embodying considerable detail[129-131,152,154-158] are now available to simulate the deposition of RF/EM energy in the human body. Thus, it is logical to combine the two types of models for the evaluation of physiological thermoregulatory responses to microwave power deposition.[128,159] However, the interweaving of two simulation models, each of which was developed for different purposes using different rationales and objectives, poses particular problems and should be approached with caution. Prior assumptions and simplifications of different types that may have been incorporated into the separate models may lead to internal incompatibilities which may not be readily apparent.

Stolwijk[159] has used a combined simulation model to demonstrate that during the localized deposition of RF/EM power the rise in local temperatures in the human body, especially in the brain, may be very much less than anticipated. Similar conclusions have been drawn by Way et al.,[128] who used a modified Stolwijk model that provided for focusing of RF/EM radiation in the hypothalamus. Greatly enhanced evaporative heat loss, skin blood flow, and core-to-skin conductance serve to protect individual tissues of the body during deposition of RF/EM energy. Thus, regions of high tissue temperature (i.e., thermal "hot spots") are rather unlikely, because the convective heat transfer via blood flow will dissipate deposited energy over the heat capacitance of the whole body.

It is possible that poorly perfused tissues such as the ocular lens or resting muscle may sustain substantial temperature elevation if high local specific absorption rates are created during RF/EM exposure. However, most body tissues are well vascularized and very large increases in local blood flow will occur as soon as a tissue temperature of 42 to 43°C is reached.[23,160] Thus, cases where a combined simulation model predicts local tissue temperatures in excess of this level should be viewed with great caution until the presence of an increased blood perfusion response has been verified.

REFERENCES

1. **Adair, E. R.,** Autonomic thermoregulation in squirrel monkey when behavioral regulation is limited, *J. Appl. Physiol.,* 40, 694, 1976.
2. **Adair, E. R. and Wright, B. A.,** Behavioral thermoregulation in the squirrel monkey when response effort is varied, *J. Comp. Physiol. Psychol.,* 90, 179, 1976.
3. **Corbit, J. D.,** Behavioral regulation of body temperature, in *Physiological and Behavioral Temperature Regulation,* Hardy, J. D., Gagge, A. P., and Stolwijk, J. A. J., Eds., Charles C Thomas, Springfield, Ill., 1970, 777.
4. **Hardy, J. D.,** Models of temperature regulation — a review, in *Essays on Temperature Regulation,* Bligh, J. and Moore, R., Eds., North-Holland, Amsterdam, 1972, 163.
5. **Stitt, J. T., Adair, E. R., Nadel, E. R., and Stolwijk, J. A. J.,** The relation between behavior and physiology in the thermoregulatory response of the squirrel monkey, *J. Physiol. (Paris),* 63, 424, 1971.
6. **Stevens, J. C.,** Thermal sensation, infrared and microwaves, in *Microwaves and Thermoregulation,* Adair, E. R., Ed., Academic Press, New York, 1983, 191.
7. **Bligh, J. and Johnson, K. G.,** Glossary of terms for thermal physiology, *J. Appl. Physiol.,* 35, 941, 1973.
8. **DuBois, D. and DuBois, E. F.,** A formula to estimate approximate surface area, if height and weight are known, *Arch. Int. Med.,* 17, 863, 1916.
9. **Kleiber, M.,** *The Fire of Life,* R. E. Kreiger, Huntington, N.Y., 1975.
10. **Gagge, A. P., Burton, A. C., and Bazett, H. C.,** A practical system of units for the description of the heat exchange of man with his environment, *Science,* 94, 428, 1941.
11. **Winslow, C.-E. A., Gagge, A. P., and Herrington, L. P.,** Heat exchange and regulation in radiant environments above and below air temperature, *Am. J. Physiol.,* 131, 79, 1940.
12. **Sprague, C. H. and Munson, D. M.,** A composite ensemble method for estimating thermal insulation values of clothing, *ASHRAE Trans.,* 80(1), 120, 1974.
13. **Woodcock, A. H., Pratt, R. L., and Breckenridge, J. R.,** Theory of the globe thermometer, Res. Study Rep. BP-7, Quartermaster R & E Command, U.S. Army Natick Laboratories, Natick, Mass., 1960.
14. **Roller, W. L. and Goldman, R. F.,** Estimation of solar radiation environment, *Int. J. Biometeorol.,* 11, 329, 1967.
15. **Berglund, L. G.,** Characterizing the thermal environment, in *Microwaves and Thermoregulation,* Adair, E. R., Ed., Academic Press, New York, 1983, 15.
16. American Society of Heating, Refrigerating and Air-Conditioning Engineers, Inc., Physiological principles, comfort, and health, in *ASHRAE Handbook, 1981 Fundamentals,* ASHRAE, Atlanta, Ga., 1981, 8.1.
17. **Goldman, R. F.,** Effect of environment on metabolism, in *Proc. 1st Ross Conf. on Metabolism, 26—28 June, 1978, Jackson Point, Maine,* Ross Laboratories, Columbus, Ohio, 1980, 117.
18. **Stitt, J. T., Hardy, J. D., and Stolwijk, J. A. J.,** PGE_1 fever: its effect on thermoregulation at different low ambient temperatures, *Am. J. Physiol.,* 227, 622, 1974.
19. **Adair, E. R. and Adams, B. W.,** Adjustments in metabolic heat production by squirrel monkeys exposed to microwaves, *J. Appl. Physiol. Respir. Environ. Exercise Physiol.,* 52, 1049, 1982.
20. **Swift, R. J.,** The effects of low environmental temperature upon metabolism. II. The influence of shivering, subcutaneous fat, and skin temperature on heat production, *J. Nutr.,* 5, 227, 1932.
21. **Iampietro, P. F., Vaughn, J. A., Goldman, R. F., Kreider, M. B., Masucci, F., and Bass, D. E.,** Heat production from shivering, *J. Appl. Physiol.,* 15, 632, 1960.
22. **Hardy, J. D.,** Regulation of body temperature in man — an overview, in *Energy Conservation Strategies in Buildings,* Stolwijk, J. A. J., Ed., Yale University Printing Service, New Haven, Conn., 1978, 14.
23. **Cunningham, D. J.,** An evaluation of heat transfer through the skin in the human extremity, in *Physiological and Behavioral Temperature Regulation,* Hardy, J. D., Gagge, A. P., and Stolwijk, J. A. J., Eds., Charles C Thomas, Springfield, Ill., 1970, 302.
24. **Nadel, E. R., Mitchell, J. W., Saltin, B., and Stolwijk, J. A. J.,** Peripheral modifications to the central drive for sweating, *J. Appl. Physiol.,* 31, 828, 1971.
25. **Nadel, E. R., Pandolf, K. B., Roberts, M. F., and Stolwijk, J. A. J.,** Mechanisms of thermal acclimation to exercise and heat, *J. Appl. Physiol.,* 37, 515, 1974.
26. **Greenleaf, J. E.,** Blood electrolytes and exercise in relation to temperature regulation in man, in *The Pharmacology of Thermoregulation,* Schoenbaum, E. and Lomax, P., Eds., S. Karger, Basel, 1973, 72.
27. **Cabanac, M.,** Thermoregulatory behavioral responses, in *Microwaves and Thermoregulation,* Adair, E. R., Ed., Academic Press, New York, 1983, 307.
28. **Hensel, H., Iggo, I., and Witt, I.,** A quantitative study of sensitive cutaneous thermoreceptors with C afferent fibers, *J. Physiol. (London),* 153, 113, 1960.
29. **Hensel, H. and Kenshalo, D. R.,** Warm receptors in the nasal region of cats, *J. Physiol. (London),* 204, 99, 1969.

30. **Aronsohn, E. and Sachs, J.,** Die Beziehungen des Gehirns zur Körperwärme und zum Fieber, *Pflügers Arch. Ges. Physiol.,* 37, 232, 1885.
31. **Ott, I.,** The heat center in the brain, *J. Nerv. Ment. Dis.,* 14, 152, 1887.
32. **Barbour, H. G.,** Die Wirkung unmittelbarer Erwärmung und Abkühlung der Wärmezentra auf die Körpertemperatur, *Arch. Exp. Pathol. Pharmakol.,* 70, 1, 1912.
33. **Nakayama, T., Eisenman, J. S., and Hardy, J. D.,** Single unit activity of anterior hypothalamus during local heating, *Science,* 134, 560, 1961.
34. **Nakayama, T., Hammel, H. T., Hardy, J. D., and Eisenman, J. S.,** Thermal stimulation of electrical activity of single units of the preoptic region, *Am. J. Physiol.,* 204, 1122, 1963.
35. **Guieu, J. D. and Hardy, J. D.,** Integrative activity of preoptic units. I. Response to local and peripheral temperature changes, *J. Physiol. (Paris),* 63, 253, 1971.
36. **Hammel, H. T.,** Regulation of internal body temperature, *Annu. Rev. Physiol.,* 30, 641, 1968.
37. **Hellon, R. F.,** Thermal stimulation of hypothalamic neurones in unanaesthetized rabbits, *J. Physiol. (London),* 193, 381, 1967.
38. **Cabanac, M., Stolwijk, J. A. J., and Hardy, J. D.,** Effect of temperature and pyrogens on single-unit activity in the rabbit's brain stem, *J. Appl. Physiol.,* 24, 645, 1968.
39. **Simon, E.,** Temperature regulation: the spinal cord as a site of extrahypothalamic thermoregulatory functions, *Rev. Physiol. Biochem. Pharmacol.,* 71, 1, 1974.
40. **Rawson, R. O. and Quick, K. P.,** Evidence of deep-body thermoreceptor response to intra-abdominal heating of the ewe, *J. Appl. Physiol.,* 28, 813, 1970.
41. **Wit, A. and Wang, S. C.,** Temperature-sensitive neurones in preoptic/anterior hypothalamic region: effects of increasing ambient temperature, *Am. J. Physiol.,* 215, 1151, 1968.
42. **Hellon, R. F.,** Environmental temperature and firing rate of hypothalamic neurones, *Experientia,* 25, 610, 1969.
43. **Wünnenberg, W. and Hardy, J. D.,** Response of single units of the posterior hypothalamus to thermal stimulation, *J. Appl. Physiol.,* 33, 547, 1972.
44. **Cabanac, M., Hammel, H. T., and Hardy, J. D.,** *Tiliqua scincoides:* temperature sensitive units in lizard brain, *Science,* 158, 1050, 1967.
45. **Satinoff, E.,** Behavioral thermoregulation in response to local cooling of the rat brain, *Am. J. Physiol.,* 206, 1389, 1964.
46. **Adair, E. R.,** Hypothalamic control of thermoregulatory behavior: preoptic-posterior hypothalamic interaction, in *Recent Studies of Hypothalamic Function,* Lederis, K. and Cooper, K. E., Eds., S. Karger, Basel, 1974, 341.
47. **Lipton, J. M.,** Behavioral temperature regulation in the rat: effects of thermal stimulation of the medulla, *J. Physiol. (Paris),* 63, 325, 1971.
48. **Cabanac, M.,** Thermoregulatory behavior, in *Essays on Temperature Regulation,* Bligh, J. and Moore, R., Eds., North-Holland, Amsterdam, 1972, 19.
49. **Carlisle, H. J. and Ingram, D. L.,** The effects of heating and cooling the spinal cord and hypothalamus on thermoregulatory behaviour in the pig, *J. Physiol. (London),* 231, 353, 1973.
50. **Guy, A. W.,** Analysis of electromagnetic fields induced in biological tissues by thermographic studies on equivalent phantom models, *IEEE Trans. Microwave Theory Tech.,* MTT-19, 205, 1971.
51. **Adair, E. R.,** Sensation, subtleties, and standards: synopsis of a panel discussion, in *Microwaves and Thermoregulation,* Adair, E. R., Ed., Academic Press, New York, 1983, 231.
52. **Meyer, H. H.,** Theorie des Fiebers und seiner Behandlung, *Verh. Dtsch. Bes. Inn. Med.,* 30, 15, 1913.
53. **Rubner, M.,** *Die Gesetze des Energieverbrauchs bie der ernahrung,* Franz Deutiche, Leipzig, 1902.
54. **DuBois, E. F.,** *Fever and the Regulation of Body Temperature,* Charles C Thomas, Springfield, Ill., 1948.
55. **Benzinger, T. H.,** On physical heat regulation and the sense of temperature in man, *Proc. Natl. Acad. Sci. U.S.A.,* 45, 645, 1959.
56. **Hensel, H.,** Physiologie der thermoreception, *Ergeb. Physiol.,* 47, 166, 1952.
57. **Chatonnet, J. and Cabanac, M.,** The perception of thermal comfort, *Int. J. Biometeorol.,* 9, 183, 1965.
58. **Aschoff, J. and Wever, R.,** Modellversuche zum gegenstrome-Wärmeaustausch in der Extremität, *Z. Gesamte Exp. Med.,* 130, 385, 1958.
59. **MacDonald, D. K. C. and Wyndham, C. H.,** Heat transfer in man, *J. Appl. Physiol.,* 3, 342, 1950.
60. **Burton, A. C.,** The application of the theory of heat flow to the study of energy metabolism, *J. Nutr.,* 7, 497, 1934.
61. **Wissler, E. H.,** Steady state temperature distribution in man, *J. Appl. Physiol.,* 16, 734, 1961.
62. **Wissler, E. H.,** A mathematical model of the human thermal system, *Bull. Math. Biophys.,* 26, 147, 1964.
63. **Smith, P. E. and James, E. W.,** Human responses to heat stress, *Arch. Environ. Health,* 9, 332, 1964.
64. **Stolwijk, J. A. J. and Hardy, J. D.,** Temperature regulation in man — a theoretical study, *Pflügers Arch. Gesamte Physiol.,* 291, 129, 1966.
65. **Stolwijk, J. A. J. and Hardy, J. D.,** Control of body temperature, in *Handbook of Physiology,* Sec. 9, Lee, D. H. K., Ed., American Physiological Society, Bethesda, Md., 1977, 45.

66. **Hammel, H. T.,** Neurons and temperature regulation, in *Physiological Controls and Regulation,* Yamamoto, W. S. and Brobeck, J. R., Eds., W. B. Saunders, Philadelphia, 1965, 71.
67. **Hardy, J. D. and Guieu, J. D.,** Integrative activity of preoptic units. II. Hypothetical network, *J. Physiol. (Paris),* 63, 264, 1971.
68. **Bligh, J.,** Neuronal models of mammalian temperature regulation, in *Essays on Temperature Regulation,* Bligh, J. and Moore, R., Eds., North-Holland, Amsterdam, 1972, 105.
69. **Feldberg, W. and Myers, R. D.,** Effects on temperature of amines injected into the cerebral ventricles. A new concept of temperature regulation, *J. Physiol. (London),* 173, 226, 1964.
70. **Myers, R. D.,** Hypothalamic mechanisms of pyrogen action in the cat and monkey, in *CIBA Symp. on Pyrogens and Fever,* Churchill, London, 1970.
71. **Gagge, A. P.,** A new physiological variable associated with sensible and insensible perspiration, *Am. J. Physiol.,* 120, 277, 1937.
72. **Belding, H. S. and Hatch, T. F.,** Index for evaluating heat stress in terms of resulting physiological strain, *Heat./Piping/Air Cond.,* 27, 129, 1955.
73. **Goldman, R. F.,** Prediction of human heat tolerance, in *Environmental Stress,* Follinsbee, L. J. et al., Eds., Academic Press, New York, 1978, 53.
74. **Hardy, J. D.,** Thermal comfort and health, *ASHRAE J.,* 77, 43, 1970.
75. **Nielsen, B. and Nielsen, M.,** Influence of passive and active heating on the temperature regulation of man, *Acta Physiol. Scand.,* 64, 323, 1965.
76. **Stolwijk, J. A. J.,** Whole body heating — thermoregulation and modeling, in *Physical Aspects of Hyperthermia,* Nussbaum, G. H., Ed., American Institute of Physics, New York, 1982, 565.
77. **Gagge, A. P., Stolwijk, J. A. J., and Nishi, Y.,** An effective temperature scale based on a simple model of human physiological regulatory response, *ASHRAE Trans.,* 77, 247, 1971.
78. **Durney, C. H., Johnson, C. C., Barber, P. W., Massoudi, H., Iskander, M. F., Lords, J. L., Ryser, D. K., Allen, S. J., and Mitchell, J. C.,** *Radiofrequency Radiation Dosimetry Handbook,* 2nd ed., Report SAM-TR-78-22, prepared for U.S. Air Force School of Aerospace Medicine, Brooks AFB, Tex., 1978.
79. **Goldman, R. F.,** The role of clothing in achieving acceptability of environmental temperatures between 65F and 85F (18C and 30C), in *Energy Conservation Strategies in Buildings. Comfort, Acceptability and Health,* Stolwijk, J. A. J., Ed., John B. Pierce Foundation of Connecticut, New Haven, Conn., 1978.
80. ASHRAE monographs for practical application of ASHRAE research: thermal comfort conditions, *ASHRAE J.,* 16, 90, 1974.
81. **Pound, R. V.,** Radiant heat for energy conservation, *Science,* 208, 494, 1980.
82. **Nielsen, M.,** Die regulation der Körpertemperatur bei Muskelarbeit, *Skand. Arch. Physiol.,* 79, 193, 1938.
83. **Nadel, E. R., Ed.,** *Problems with Temperature Regulation During Exercise,* Academic Press, New York, 1975.
84. **Robinson, S.,** Physiological adjustments to heat, in *Physiology of Heat Regulation and the Science of Clothing,* Newburg, C. H., Ed., W. B. Saunders, Philadelphia, 1949, 193.
85. **von Liebermeister, C.,** *Handbuch der Pathologie und Therapie des Fiebers,* Vogel, Leipzig, 1875.
86. **Macpherson, R. K.,** The effect of fever on temperature regulation in man, *Clin. Sci.,* 18, 281, 1959.
87. **Stitt, J. T.,** Fever versus hyperthermia, *Fed. Proc., Fed. Am. Soc. Exp. Biol.,* 38, 39, 1979.
88. **Crawshaw, L. I. and Stitt, J. T.,** Behavioural and autonomic induction of prostaglandin E_1 fever in squirrel monkeys, *J. Physiol. (London),* 244, 197, 1975.
89. **Hensel, H.,** Neural processes in thermoregulation, *Physiol. Rev.,* 53, 948, 1973.
90. **Cain, W. S.,** Spatial discrimination of cutaneous warmth, *Am. J. Psychol.,* 86, 169, 1973.
91. **Hardy, J. D. and Oppel, T. W.,** Studies in temperature sensation. III. The sensitivity of the body to heat and the spatial summation of the end organ responses, *J. Clin. Invest.,* 16, 533, 1937.
92. **Herget, C. M., Granath, L. P., and Hardy, J. D.,** Warmth sense in relation to the area of skin stimulated, *Am. J. Physiol.,* 135, 20, 1941.
93. **Kenshalo, D. R., Decker, T., and Hamilton, A.,** Spatial summation on the forehead, forearm, and back produced by radiant and conducted heat, *J. Comp. Physiol. Psychol.,* 63, 510, 1967.
94. **Marks, L. E. and Stevens, J. C.,** Temporal summation related to the nature of the proximal stimulus for the warmth sense, *Percept. Psychophys.,* 14, 570, 1973.
95. **Stevens, J. C. and Marks, L. E.,** Spatial summation and the dynamics of warmth sensation, *Percept. Psychophys.,* 9, 391, 1971.
96. **Stevens, J. C., Marks, L. E., and Simonson, D. C.,** Regional sensitivity and spatial summation in the warmth sense, *Physiol. Behav.,* 13, 825, 1974.
97. **Stevens, J. C., Oculicz, W. C., and Marks, L. E.,** Temporal summation at the warmth threshold, *Percept. Psychophys.,* 14, 307, 1973.
98. **Schwan, H. P., Anne, A., and Sher, L.,** *Heating of Living Tissues,* Report NAEC-ACEL-534, U.S. Naval Air Engineering Center, Philadelphia, 1966.
99. **Eijkman, E. and Vendrik, A. J. H.,** Dynamic behavior of the warmth sense organ, *J. Exp. Psychol.,* 62, 403, 1961.

100. **Vendrik, A. J. H. and Vos, J. J.,** Comparison of the stimulation of the warmth sense organ by microwave and infrared, *J. Appl. Phys.*, 13, 435, 1958.

101. **Hendler, E., Hardy, J. D., and Murgatroyd, D.,** Skin heating and temperature sensation produced by infrared and microwave irradiation, in *Temperature: Its Measurement and Control in Science and Industry*, Vol. 3, Hardy, J. D., Ed., Reinhold, New York, 1963, 211.

102. **Hendler, E. and Hardy, J. D.,** Infrared and microwave effects on skin heating and temperature sensation, *IRE Trans. Med. Electron.*, ME-7, 143, 1960.

103. **Hendler, E.,** Cutaneous receptor response to microwave irradiation, in *Thermal Problems in Aerospace Medicine*, Hardy, J. D., Ed., Unwin, Surrey, 1968, 211.

104. **Justesen, D. R., Adair, E. R., Stevens, J. C., and Bruce-Wolfe, V.,** A comparative study of human sensory thresholds: 2450-MHz microwaves vs. far-infrared radiation, *Bioelectromagnetics*, 3, 117, 1982.

105. **Michaelson, S. M.,** Cutaneous perception of microwaves, *J. Microwave Power*, 7, 67, 1972.

106. **D'Andrea, J. A., Cuellar, O., Gandhi, O. P., Lords, J. L., and Nielson, H. C.,** Behavioral thermoregulation in the whiptail lizard (*Cnemidophorus tigris*) under 2450 MHz CW microwaves, in *Biological Effects of Electromagnetic Waves*, Abstr. URSI General Assembly, Helsinki, 1978, 88.

107. **Stern, S., Margolin, L., Weiss, B., Lu, S.-T., and Michaelson, S. M.,** Microwaves: effect on thermoregulatory behavior of rats, *Science*, 206, 1198, 1979.

108. **Adair, E. R. and Adams, B. W.,** Microwaves modify thermoregulatory behavior in squirrel monkey, *Bioelectromagnetics*, 1, 1, 1980.

109. **Adair, E. R. and Adams, B. W.,** Behavioral thermoregulation in the squirrel monkey: adaptation processes during prolonged microwave exposure, *Behav. Neurosci.*, 97, 49, 1983.

110. **Carroll, D. R., Levinson, D. M., Justesen, D. R., and Clarke, R. L.,** Failure of rats to escape from a potentially lethal microwave field, *Bioelectromagnetics*, 1, 101, 1980.

111. **Levinson, D. M., Riffle, D. W., Justesen, D. R., Carroll, D. R., Bruce-Wolfe, V., and Clarke, R. L.,** Escape behavior by mice and rats in 2450- and 918-MHz fields: evaluation of sonic stimulation as a cue, in Abstr. 3rd Annu. Meet. Bioelectromagnetics Society, BEMS, Gaithersburg, Md., 1981, 40.

112. **Levinson, D. M., Grove, A. M., Clarke, R. L., and Justesen, D. R.,** Photic cuing of escape by rats from an intense microwave field, *Bioelectromagnetics*, 3, 105, 1982.

113. **Justesen, D. R.,** Sensory dynamics of intense microwave irradiation: a comparative study of aversive behaviors by mice and rats, in *Microwaves and Thermoregulation*, Adair, E. R., Ed., Academic Press, New York, 1983, 203.

114. **Bruce-Wolfe, V. and Adair, E. R.,** Operant control of microwave radiation by squirrel monkeys, in Abstr. 3rd Annu. Meet. Bioelectromagnetics Society, BEMS, Gaithersburg, Md., 1981, 42.

115. **Ho, H. S. and Edwards, W. P.,** Oxygen-consumption rate of mice under differing dose rates of microwave radiation, *Radio Sci.*, 12 (6S), 131, 1977.

116. **Phillips, R. D., Hunt, E. L., Castro, R. D., and King, N. W.,** Thermoregulatory, metabolic, and cardiovascular responses of rats to microwaves, *J. Appl. Physiol.*, 38, 630, 1975.

117. **Ho, H. S. and Edwards, W. P.,** The effect of environmental temperature and average dose rate of microwave radiation on the oxygen-consumption rate of mice, *Radiat. Environ. Biophys.*, 16, 325, 1979.

118. **Adair, E. R. and Adams, B. W.,** Microwaves induce peripheral vasodilation in squirrel monkey, *Science*, 207, 1381, 1980.

119. **Adair, E. R.,** Microwaves and thermoregulation, in *USAF Radiofrequency Radiation Bioeffects Research Program — A Review*, Mitchell, J. C., Ed., Report No. Aeromedical Review 4-81, USAFSAM/RZP, Aerospace Medical Division (AFSC) Brooks AFB, Tex., 1981.

120. **Nadel, E. R., Bullard, R. W., and Stolwijk, J. A. J.,** Importance of skin temperature in the regulation of sweating, *J. Appl. Physiol.*, 31, 80, 1971.

121. **Gordon, C. J.,** Effects of ambient temperature and exposure to 2450-MHz microwave radiation on evaporative heat loss in the mouse, *J. Microwave Power*, 17, 145, 1982.

122. **Michaelson, S. M.,** Thermal effects of single and repeated exposures to microwaves — a review, in *Biological Effects and Health Hazards of Microwave Radiation*, Czerski, P. et al., Eds., Polish Medical Publishers, Warsaw, 1974, 1.

123. **Michaelson, S. M., Thomson, R. A. E., and Howland, J. W.,** Physiologic aspects of microwave irradiation of mammals, *Am. J. Physiol.*, 201, 351, 1961.

124. **Michaelson, S. M.,** Thermoregulation in intense microwave fields, in *Microwaves and Thermoregulation*, Adair, E. R., Ed., Academic Press, New York, 1983, 283.

125. **Goldman, R. F.,** Acclimation to heat and suggestions, by inference, for microwave radiation, in *Microwaves and Thermoregulation*, Adair, E. R., Ed., Academic Press, New York, 1983, 275.

126. **Guy, A. W., Webb, M. D., and Sorensen, C. C.,** Determination of power absorption in man exposed to high frequency electromagnetic fields by thermographic measurements in scale models, *IEEE Trans. Biomed. Eng.*, 23, 361, 1974.

127. **Krupp, J. H.,** *In vivo* temperature measurements during whole-body exposure of *Macaca mulatta* to resonant and non-resonant frequencies, in *Microwaves and Thermoregulation,* Adair, E. R., Ed., Academic Press, New York, 1983, 95.

128. **Way, W. I., Kritikos, H., and Schwan, H. P.,** Thermoregulatory physiologic responses in the human body exposed to microwave radiation, *Bioelectromagnetics,* 2, 341, 1981.

129. **Kritikos, H. and Schwan, H. P.,** Hot spots generated by EM waves in lossy spheres and biological implications, *IEEE Trans. Biomed. Eng.,* 19, 53, 1972.

130. **Kritikos, H. and Schwan, H. P.,** The distribution of heating potential inside lossy spheres, *IEEE Trans. Biomed. Eng.,* 22, 457, 1975.

131. **Kritikos, H. and Schwan, H. P.,** Formation of hot spots in multilayer spheres, *IEEE Trans. Biomed. Eng.,* 23, 168, 1976.

132. **Kritikos, H. and Schwan, H. P.,** Potential temperature rise induced by electromagnetic field in brain tissues, *IEEE Trans. Biomed. Eng.,* 26, 29, 1979.

133. **Burr, J. G. and Krupp, J. H.,** Real-time measurement of RFR energy distribution in the *Macaca mulatta* head, *Bioelectromagnetics,* 1, 21, 1980.

134. **Merritt, J. H., Chamness, A. F., Hartzell, R. H., and Allen, S. J.,** Orientation effects on microwave-induced hyperthermia and neurochemical correlates, *J. Microwave Power,* 12, 167, 1977.

135. **Brainard, G., Albert, E., DeSantis, M., Postow, E., and Parker, J.,** The effects of low power density microwaves on rat hypothalamic temperatures, *Proc. 1983 Symp. on Electromagnetic Fields in Biological Systems,* IMPI, Edmonton, Alberta, Canada, 1979, 14.

135a. **Adair, E. R., Adams, B. W., and Rawson, R. O.,** Role of the hypothalamus in thermoregulatory responses to microwaves (Abstr. Bull.), in *6th Annu. Meet. Bioelectromagnetics Soc.,* Atlanta, 1984, 37.

135b. **Adair, E. R. and Adams, B. W.,** Brain 'thermostat' plays minor role in metabolic response to microwaves (Abstr.), in *Int. Symp. Interaction of Electromagnetic Fields with Biological Systems,* International Union of Radio Science (U.R.S.I), Florence, Italy, 1984, 56.

136. **Adair, E. R.,** Microwave irradiation and thermoregulatory behavior, in Behavioral Effects of Microwave Radiation, Monahan, J. C. and D'Andrea, J. A., Eds., U.S. Department of Health and Human Services Publ. FDA 85-8238, PHS-FDS, Rockville, Md., 1985, 84.

137. **Adair, E. R., Adams, B. W., and Akel, G. M.,** Minimal changes in hypothalamic temperature accompany microwave-induced alteration of thermoregulatory behavior, *Bioelectromagnetics,* 5, 13, 1984.

138. **Lotz, W. G.,** Hyperthermia in rhesus monkeys exposed to a frequency (225 MHz) near whole-body resonance, in *NAMRL Tech. Report No. 1284,* Naval Aerospace Medical Research Laboratory, Pensacola, Fla., 1982.

139. **Lotz, W. G.,** Influence of the circadian rhythm of body temperature on the physiological response to microwaves: day *vs* night exposure, in *Microwaves and Thermoregulation,* Adair, E. R., Ed., Academic Press, New York, 1983, 445.

140. **deLorge, J. O.,** Operant behavior and colonic temperature of squirrel monkeys *(Saimiri sciureus)* during microwave irradiation, NAMRL-1222, Naval Aerospace Medical Research Laboratory, Pensacola, Fla., 1976.

141. **deLorge, J. O.,** The effects of microwave radiation on behavior and temperature regulation in rhesus monkeys, in Biological Effects of Electromagnetic Waves, Selected Papers of the USNC/URSI Annual Meeting, Boulder, Colo., October 20 to 23, 1975, Johnson, C. C. and Shore, M. L., Eds., HEW Publ. (FDA) 77-8010, U.S. Government Printing Office, Washington, D.C., 1977, 158.

142. **deLorge, J. O. and Ezell, C. S.,** Observing-responses of rats exposed to 1.28 and 5.62-GHz microwaves, *Bioelectromagnetics,* 1, 183, 1980.

143. **deLorge, J. O.,** The thermal basis for disruption of operant behavior by microwaves in three animal species, in *Microwaves and Thermoregulation,* Adair, E. R., Ed., Academic Press, New York, 1983, 379.

144. **Putthoff, D. L., Justesen, D. R., Ward, L. B., and Levinson, D. M.,** Drug-induced ectothermia in small mammals, the quest for a biological microwave dosimeter, *Radio Sci.,* 12 (6S), 73, 1977.

145. **Smialowicz, R. J., Compton, K. L., Riddle, M. M., Rodgers, R. R., and Brugnolotti, P. L.,** Microwave radiation (2450 MHz) alters the endotoxin-induced hypothermic response of rats, *Bioelectromagnetics,* 1, 353, 1980.

146. **Smialowicz, R. J., Riddle, M. M., Brugnolotti, P. L., Rogers, R. R., and Compton, K. L.,** Detection of microwave heating in 5-hydroxytryptomine-induced hypothermic mice, *Radiat. Res.,* 88, 108, 1981.

147. **Smialowicz, R. J.,** Detection of radiofrequency radiation-induced whole body heating following chemical impairment of thermoregulation, in *Microwaves and Thermoregulation,* Adair, E. R., Ed., Academic Press, New York, 1983, 431.

148. **Crosbie, R. J., Hardy, J. D., and Fessenden, E.,** Electrical analog simulation of temperature regulation in man, *IEEE Trans. Biol. Med. Electron.,* 8, 245, 1961.

149. **Stolwijk, J. A. J.,** A Mathematical Model of Physiological Temperature Regulation in Man, Report CR-1855 NASA, Washington, D.C., 1971.

150. **Emery, A. F., Short, R. E., Guy, A. W., Kraning, K. K., and Lin, J. C.,** The numerical thermal simulation of the human body when undergoing exercise or nonionizing electromagnetic irradiation, *Trans. Am. Soc. Mech. Eng.,* 284, 1976.

151. **Spiegel, R. J., Deffenbaugh, D. M., and Mann, J. E.,** A thermal model of the human body exposed to an electromagnetic field, *Bioelectromagnetics,* 1, 253, 1980.

152. **Gandhi, O. P.,** Electromagnetic absorption in an inhomogeneous model of man for realistic exposure conditions, *Bioelectromagnetics,* 3, 81, 1982.

153. **Guy, A. W. and Chou, C.-K.,** Electromagnetic heating for therapy, in *Microwaves and Thermoregulation,* Adair, E. R., Ed., Academic Press, New York, 1983, 57.

154. **Burr, J. G., Cohoon, D. K., Bell, E. L., and Penn, J. W.,** Thermal response model of a simulated cranial structure exposed to radiofrequency radiation, *IEEE Trans. Biomed. Eng.,* 27, 452, 1980.

155. **Johnson, C. C. and Guy, A. W.,** Nonionizing EM effects in biological materials and systems, *Proc. IEEE,* 60, 692, 1972.

156. **Lin, J. C., Guy, A. W., and Johnson, C. C.,** Power deposition in a spherical model of man exposed to 1—20 MHz electromagnetic fields, *IEEE Trans. Microwave Theory Tech.,* 21, 791, 1973.

157. **Shapiro, A. R., Lutomirski, R. F., and Yura, H. T.,** Induced fields and heating within the cranial structure irradiated by an electromagnetic wave, *IEEE Trans. Microwave Theory Tech.,* 19, 187, 1971.

158. **Weil, C. M.,** Absorption characteristics of multilayered sphere models exposed to UHF/microwave radiation, *IEEE Trans. Biomed. Eng.,* 22, 468, 1975.

159. **Stolwijk, J. A. J.,** Mathematical models of thermal regulation, *Ann. N.Y. Acad. Sci.,* 335, 98, 1980.

160. **Sekins, K. M.,** Microwave Hyperthermia in Human Muscle: An Experimental and Numerical Investigation of the Temperature and Blood Flow Fields Occurring during 9.5-MHz Diathermy, Ph.D. dissertation, University of Washington, Seattle, 1981.

Chapter 4

INTERACTION OF NONMODULATED RADIO FREQUENCY FIELDS WITH LIVING MATTER: EXPERIMENTAL RESULTS

Sol M. Michaelson

TABLE OF CONTENTS

I. INTRODUCTION

The elucidation of the biological effects of exposure to radio frequency (RF) (including microwave [mw]) energies requires a careful review and critical analysis of the literature. This entails separating well-documented effects and mechanisms from speculative and unsubstantiated reports.

The organs and organ systems affected by exposure to RF energies respond by means of functional (physiological) reactions and/or structural alterations (pathologic). Some reactions to RF exposure may lead to measurable biological effects that remain within the range of normal (physiological) compensation and are not hazardous. Some reactions, on the other hand, may lead to effects which may be potential or actual hazards to health. Other uses of RF may improve the efficiency of certain physiological processes and can thus be used for therapeutic purposes.

Investigations have shown that exposure at certain power densities, when absorbed for several minutes or hours, can result in pathophysiological manifestations in laboratory animals. Such effects may or may not be characterized by a measurable rise in temperature, that is a function of thermoregulatory processes and active adaptation of the animal. The end result is either reversible or irreversible change, depending on the irradiation conditions and the physiological state of the animal. At lower power densities, evidence of pathological changes or physiological alteration is nonexistent or equivocal. Much discussion, nevertheless, has taken place on the relative importance of thermal or nonthermal effects of RF energy absorption.

II. PROBLEMS OF EXTRAPOLATION

Proper investigation of the biological effects of electromagnetic fields requires an understanding and appreciation of biophysical principles and "comparative medicine". Such studies require interspecies "scaling", the selection of biomedical parameters that reflect basic physiological functions, identification of specific and nonspecific reactions, and differentiation of adaptational or compensatory changes from pathological manifestations. In comparing results of experiments performed in the same or different laboratories, standardization of conditions is mandatory.

Even by using approaches where absorbed energy patterns in a test animal are set to approximate as closely as possible the patterns that may exist in man under certain exposure conditions, the intrinsic physical and physiological dissimilarities between species further confound the problem of extrapolating between animals and to man. In addition to the obvious external geometric differences, the differences in internal vascular anatomy and mechanisms of heat dissipation in fur-bearing animals compared to man must be taken into consideration.

Experimental animal models are used extensively for the study of physical factors in the environment to assure human health and safety. The best we can do experimentally is to create an arbitrary set of conditions which we consider to be as relevant as possible for the purpose of the study. Many factors, such as methods of animal care, the role of seasonal and circadian rhythms, temperature and humidity, etc., as well as psychosocial interactions, must be considered in experimental design and analysis of results.

Meticulous care must be exerted in defining experimental conditions. One of the problems in studying biological effects generally is the selection of the most appropriate animal species and extrapolation to man. Animals are quite often selected only on the basis of convenience, economy, or familiarity and without regard to their suitability for the problem under study. One should not extrapolate to larger animals or man results obtained in small laboratory animals without consideration of size and energy distributions as well as metabolic and physiological differences.

Reliability of laboratory studies with animals depends on the following:

1. Selection of an animal model with consideration of its physiologic responses
2. Methods applied for investigation of biological processes
3. Extrapolation of data from animals to man

The parameters to be specified when reporting biological effects of MW/RF exposures are noted in Tables 1 and 2.

The investigator has to determine whether an observed difference between exposed and nonexposed animals can be attributed to the causal factor in question, or whether it might simply have occurred spontaneously, "by chance." In cases in which no difference is observed, the question is whether any conclusion at all is warranted and if so, what confidence we should place on it. For guidance the investigator turns to statistical analysis. He or she must appreciate, however, that statistics cannot decide what is truth and what is not. Statistical analysis should be considered only as a powerful adjunct to, and an integral part of, experiment and observation. But as with any tool, its usefulness will be enhanced only if its inherent limitations are properly recognized.

A. Physical and Physiologic Scaling

Proper dosimetry in experimental procedures and the importance of realistic scaling factors for extrapolation of data obtained with small laboratory animals to man are clear. Detailed discussions that serve as bases for scaling have been presented by several authors.[1-8]

For the most part, research on biological effects of MWs has dealt with the problem of hazard assessment for which two complex, integrative tasks must be systematically accomplished to reach an objective, scientifically valid solution. One of these tasks is the deter-

Table 1
FACTORS THAT AFFECT MICROWAVE/RF ABSORPTION

Physical Parameters
  Frequency
  Polarization
  Modulation (AM, FM, pulse, CW)
  Power density (peak and average)
  Field pattern (near or far)
  Field uniformity
  Type of transmitting and
    radiating equipment
  Chamber material
  Chamber dimensions

Biological parameters
  Tissue dielectric properties
  Size; geometry
  Animal orientation relative
    to polarization
  Spatial relations
    among animals

Artifacts
  Ground or conductor plate
  Container (material, size)
  Metal implants
  Shielding materials
  Metal or nonmetallic
    objects in the field

Environmental factors
  Temperature
  Humidity

Table 2
FACTORS THAT INFLUENCE BIOLOGICAL
RESPONSES TO THE SAME SAR

Subject variables
  Species: sex; age; weight
  Sensitivity
  Number of subjects
  Interventions (anesthetics;
    drugs; electrodes; lesions)
  Animal husbandry

Environmental variables
  Temperature; humidity
  air flow
  Lighting
  Noise
  Odor

Concomitant variables
  Genetic predisposition
  Base line of the response
  Functional and metabolic
    disorders

Experimental variables
  Acclimation procedures
  Duration of exposure
  Number and schedule of
    exposures
  Mode of exposure (partial-
    or whole-body)
  Sampling technique
  Time between exposure and
    sampling
  Time of day of exposure
  Restraint devices
  Investigator-animal
    interaction

mination of the exposure required for deleterious changes in the body functions of experimental animals. This determination requires a quantitative evaluation and comparison of the many experiments that have been conducted. The second complex task is the extrapolation of the results of animal experiments to man, a process referred to as "scaling".

There are many factors which complicate these two tasks. Various body functions may be affected at different levels of RF exposure. The most sensitive function could be taken as the determinant of a hazard level, but the possible difference between an effect and a deleterious change in function is an important question in hazard assessment. Different conditions of exposure, including MW frequency, waveform, method of exposure (cavity, free space, waveguide, etc.), and orientation of the exposed subject with respect to the E and H vectors of the field may all lead to differences in the effect of the exposure of biological systems. A major consideration is the need for the reporting of complete information about experiments to permit the quantitative comparisons of experiments and effects.

Many of the above-mentioned physical parameters have been extensively studied by Schwan and associates[9-18] using a "model" approach. The experimental work of Guy and Lehmann and associates[19-28] has demonstrated that the model approach of Schwan and associates[9-18] agrees qualitatively and predicts temperature values obtained by direct measurement.

Schaffer[29] has reviewed experimental data involving MW heating of small animals and proposed experiments to relate temperature-time responses to MW exposure. These data reveal wavelength, animal species, efficiency of the thermoregulatory mechanisms, thermal peculiarities, and environmental conditions to be important factors in the species-specific response to exposure to RF.

Hoeft[30] has pointed out that extrapolating the results of MW heating experiments from various species of animals to man has been done frequently without regard for the interspecies differences in mass and size. His analysis provides a theoretical basis for extrapolation and suggests ways to improve experiments. The exposure times required to produce a 5°C temperature rise in man and experimental animals were calculated as a function of the MW intensity, using a simplified model. These calculations showed that specific RF intensities will elevate temperatures more quickly in small animals than larger ones.

The relationship between metabolic activity and body mass (W) is the same for a variety of mammals. Heat production is proportional to $W^{0.75}$ for animals ranging in size from mice to elephants.[31] Many other constant proportionalities in physiological processes have also been documented by Adolph.[32] He has analyzed the interrelations of physiological processes within and among species, using the heterogonic equation $Y = aX^k$. The interrelations found imply quantitative orderliness among many diverse characteristics. These interrelations apply not only to rates of physiological processes, but also to sizes of organs, numbers of reduplicated structures, and biochemical compositions.

The average life span of the rat (approximately 2.5 years) corresponds to 15 to 17 years of a human life. The differences in the life spans of animals and humans must be considered in the experimental study of the long-term effects of substances and in the extrapolation of the results to man.

Johnson[33] has listed recommendations for specifying electromagnetic wave exposure conditions in bioeffects research wherein exposure of an animal (including man) to an electromagnetic field of given magnitude will produce a particular rate of energy absorption (W/kg) in tissue. This absorption is dependent on the size and geometry of the animal relative to the wavelength. This wavelength-to-animal size relationship (λ-to-a, where a is one half the longest axis dimension of the body) is also a critical factor in the relative absorption cross-section, the ratio of the absorbed energy per second to the power incident on the geometrical cross-sectional area of the animal.[10] At longer wavelengths, the orientation of the animal with respect to the E and H vectors also affects the relative absorption cross-section[2] (cf. Part III, Chapters 1 and 2).

The process of scaling in RF bioeffects research is not restricted to extrapolation of the physical absorption of energy, but includes the problems of quantitatively comparing the relative efficiency of various biological processes in different animals to each other and to those processes in man. The same biological processes in different animals are not necessarily affected by the same specific absorption rate (SAR). It is also possible that the absorbed energy in a particular area or tissue may not be the most useful indicator of disturbances of some biological functions, if those functions are perturbed by a systemic stimulus instead of a localized stimulus (a distant manifestation of local absorption at another site). The need for proper dosimetry in experimental procedures and the importance of using appropriate scaling factors for extrapolation of data obtained with small laboratory animals to man are thus clearly indicated.

From the results of theoretical analysis of a simple physical model and comparisons of size, metabolism, and thermal tolerance, some general conclusions can be drawn concerning the validity of scaling results from one species of animal to another, particularly man. Energy absorption from a plane wave depends on frequency and object geometry, orientation, and electrical properties. For a given object and orientation, there is a typical frequency for maximum energy absorption. Since this relationship varies greatly with the size of the animal, one cannot directly extrapolate results at the same frequency to animals greatly different in size. It is of practical importance to realize that experiments on biological effects at 2.45 GHz on small animals, like mice and rats, do not scale to man at 2.45 GHz, but rather approximately to effects on man at ~VHF frequencies (~0.1 GHz), an important region of the broadcast spectrum. Approximating the bodies of all animals and man as prolate spheroids, an attempt has been made to ascertain what maximum absorption of lower frequency energy is required for larger animals if the total absorbed dose rate (at the same plane wave exposure field intensity) is to be the same as that obtained at a higher frequency for a smaller animal[1] (cf. Part III, Chapters 1 and 2).

When absorbed energy patterns in a test animal are used to approximate the pattern that may exist in man under certain conditions of exposure, the intrinsic physical and physiological dissimilarities between species must be considered. In addition to geometric differences, the differences in internal vascular anatomy and mechanisms of heat dissipation in fur-bearing animals compared to man are extremely important.

Responses induced by exposure to RF depend on the amount of energy absorbed which is a function of wavelength, orientation, and electrical properties of the object(s) being irradiated.[1] Maximum absorption during whole-body irradiation of small animals apparently occurs at frequencies between approximately 0.5 and 3 GHz, and for man, at around 0.06 to 0.1 GHz, with a peak at about 0.08 GHz. At frequencies below 0.03 GHz, absorption drops off rapidly and is also much less at frequencies above 0.5 GHz.

B. SAR — Specific Absorption Rate

RF energy absorption which is dependent on polarization, frequency, and the immediate physical environment is rapidly converted to thermal energy. This thermal energy is then rapidly redistributed by conduction, convection (e.g., blood flow) within and to a lesser extent, radiation from the biological target. The tissue heat capacity and heat transfer processes influence the dose and dose distribution within the body (cf. Part III, Chapter 3).

An effort is being made to standardize dosimetric measures of RF exposure by employing a quantity called the SAR. The unit-mass, time-averaged rate of RF energy absorption is specified in SI units of watts per kilogram (W kg^{-1}).[34] The amount of energy absorbed by a given mass of material, which is termed specific absorption (SA), i.e., Joules per kilogram (J/kg), is the product of SAR times the duration of exposure (in seconds). Thus, the *specific absorption* rate (SAR) is the time rate at which RF electromagnetic energy is imparted to a component or mass of a biological body. The SAR is applicable to any tissue or organ of interest, or is expressed as a whole-body average.

Whole-body SAR is maximal when the long axis of a body is parallel to the E-field vector and is four tenths the wavelength of the incident field. At 2.45 GHz (λ = 12.5 cm), for example, Standard Man (long axis 175 cm) will absorb about half of the incident energy. If the human whole-body SAR is divided by the basal metabolic rate (BMR) for man, a ratio is obtained that provides a measure of the thermal load incurred due to a known incident power density.[35] Table 3 illustrates the variation of this ratio with frequency at two incident

Table 3
RATIO OF SAR TO BASAL
METABOLIC RATE FOR AN
AVERAGE MAN EXPOSED TO
FAR-FIELD INCIDENT POWER
DENSITIES OF 10 and 50 W/m^2

Frequency (GHz)	Av SAR/BMR (%)	
	10 W/m^2	50 W/m^2
0.01	0.13	0.65
0.02	0.60	3.00
0.05	5.80	29.00
0.06	10.00	50.00
0.08	16.00	80.00
0.10	12.00	60.00
0.20	5.20	26.00
0.50	3.70	18.50
1.00	2.90	14.50
2.00	2.50	12.50
5.00	2.50	12.50
10.00	2.50	12.50
20.00	2.50	12.50

Adapted from Stuchly, M. A., Health Aspects of Radiofrequency and Microwave Radiation Exposure, Part 2, Department of National Health and Welfare, Ottawa, 1978.

power densities. In the region of human whole-body resonance (0.06 to 0.08 GHz), this ratio reaches a maximum value (about 0.16 for an incident far-field power density of 10 W/m^2). The ratio drops off rapidly on either side of this peak, and at 0.01 GHz and below, the ratio would be less than 0.001.[36]

At frequencies that result in maximal absorption, which defines whole-body resonance, the electrical cross-section of an exposed body increases. This increase occurs at a frequency near 0.07 GHz for Standard Man and results, as shown in Table 4, in an approximately eightfold increase in absorption relative to that in a 2.45-GHz field.

III. CELLULAR AND MOLECULAR BIOLOGY

A. Chromosome-Genetic Effects
Some investigators have reported chromosome changes in various plant and animal cells in tissue culture.[37-41] Huang et al.[42] have reported no changes. Reported chromosomal changes include translocations, structural aberrations,[39,40] polyploidy,[43] and stickiness.[40] Exposures ranged from 70 W/m^2 to more than 2 kW/m^2. In a study of meiotic cells removed from exposed male mice,[44] translocations were observed at meiosis I at 1 and 5 W/cm^2. The problems inherent in the use of chromosomal aberrations as indicators of genetic damage have been noted by Savage.[45]

There is considerable interest in the interaction of millimeter waves with biological media in order to understand the mechanisms of biological effects of RF energies. A number of investigators have reported sharp, distinct resonances in the absorption of millimeter wave RF energy by various biochemical and biological preparations[46-50] or that exposure to millimeter waves produces biological effects which exhibit a sharp frequency depend-

Table 4
SAR FOR ANIMALS AND MAN (W/kg for 10 W/m² incident PD)

Species	Max absorption (GHz)	Frequency (GHz)						
		0.02—0.03	0.07	0.30	1.00	2.45	3.00	10.00
Mouse	2.00	8×10^{-4} (0.05)	0.008 (0.04)	0.06 (1.50)	0.04 (13)	1.00 (36)	0.965 (36.60)	0.322 (12.40)
Rat	0.60	1.8×10^{-3} (0.12)	0.0125 (0.06)	0.3 (7.50)	0.6 (20)	0.23 (8)	0.26 (9.60)	0.25 (9.60)
Rabbit	0.32	5×10^{-3} (0.33)	0.050 (0.22)	0.80 (20)	0.250 (8.30)	0.15 (5.40)	0.08 (2.96)	0.07 (2.69)
Rhesus	0.30	1.7×10^{-3} (0.01)	0.0125 (0.06)	0.195 (5.00)	0.10 (3.33)	0.07 (2.50)	0.065 (2.41)	0.060 (2.30)
Dog	0.20	1.5×10^{-3} (0.10)	0.010 (0.04)	0.100 (2.50)	0.050 (1.67)	0.040 (1.40)	0.037 (1.40)	0.030 (1.15)
Human (1 year)	0.15	0.004	0.040	0.15	0.065	0.055	0.050	0.042
Man (av)	0.07	0.015	0.225	0.04	0.03	0.028	0.027	0.026

Note: ( ) SAR is relative to average man at same frequency.

ence.[49,51-55] These studies suggest sharp millimeter wave frequency-dependent lethal and mutagenic effects on microorganisms, on metabolic control of growth of cells, on oncogenic viruses, viability properties of cells, protective effects on X-irradiated and cytotoxic-drug-treated animals, etc. It has been suggested that these differences might provide the basis for frequency-specific health hazards and possibilities for new forms of cancer therapy. Millimeter waves and resonance effects will be discussed in Part III, Chapter 5.

The mutagenic potential of microwave energy has been evaluated by various techniques including point mutations in bacterial assays,[56,58] the dominant lethal test in mammalian systems,[57] or genetic transmission in *Drosophila*,[59,60] with inconsistent results. Some of these reports are summarized in Table 5.[61]

Anderstam et al.[58] found no change in mutation induction in *Escherichia coli* or *Salmonella typhimurium* exposure to 0.027 and 2.45 GHz CW and 3.07 pW at SARs from 4 to 100 W/kg. Baranski et al.[78] were not able to attribute mutagenic effects or metabolic changes in *Physarum polycephalum* or *Aspergillus nidulans* to specific effects of 2.45 GHz at 100-W/m² continuous wave or pulsed microwaves. Correlli et al.[70] also reported no mutagenesis after exposure to 2.6 to 4.0 GHz at 20 W/kg. They investigated the effects of RF on colony-forming ability (CFA) and molecular structure (determined by IR spectroscpy) of *E. coli B* bacterial cells in aqueous suspension. Cells were exposed for 10 hr at SARs of 20 W/kg (equivalent to 5 kW/m²). No RF-induced effects on either CFA or molecular structure were observed.

Other studies of mutagenic effects of RF in bacteria and yeasts have given negative results. A study by Blackman et al.[56] involved exposure of *E. coli* WWU to 1.70- or 2.45-GHz RF at 20 to 500 W/m² for 3 to 4 hr. Dutta et al.[80] exposed *Saccharomyces cerevisiae* D4 to 2.45-GHz continuous wave RF at 400 W/m² or to 8.5- to 9.6-GHz pulsed RF at 10 to 450 W/m² for 120 min with negative results.

Two studies report effects of RF on sister chromatid exchange.[81,82] In the first study, Chinese hamster ovary cells were exposed in vitro to 2.45-GHz RF at unstated power density levels and durations. Sister chromatid exchanges were observed in RF-exposed cells; however, the same level of exchanges was produced in control cells by heating them to the same temperature as that produced by RF exposure. The authors concluded that the production

Table 5

GENETIC AND MUTAGENIC EFFECTS OF RF EXPOSURE

Effects	Source of test specimen	Exposure conditions				Ref.
		Frequency (GHz)	Intensity (W/m²)	SAR (W/kg)	Duration (days × min)	
Change in thermal denaturation profile and hyperchronicity of DNA extracted from testes following exposure	Mouse	1.7(CW) 0.985 (CW)	≤500 100	≤2.4 (est) ≤0.26 (testis)	1 × 80	57, 62
No chromosome aberrations in white blood cells	Chinese hamster	2.45 (CW)	50—450	21	5 × 15	42
No sister chromatid exchange in bone marrow cells	Mouse	2.45 (CW)	200	15	28—480	63
No chromosome aberrations in CHO-K1 cell line if temperature maintained	Chinese hamster	2.45 (CW)	≤2,000	≤360 (est)	1 × 30	64
No chromosome aberrations or change in mitotic activity in regenerating liver cells in rat	Rat	0.013 (CW)	$(4.45\ kV_{p-p}/m)$ $(44.1\ kV_{p-p}/m)$	1.3 (est)	1 × 1,680 2,640	65
No mutation induction	*Escherichia coli* *E. coli*	2.45 (CW) 1.7 (CW)	100 or 500 $(250\ V_{p-p}/m)$	15 or 70 3	1 × 180 — 240 1 × 120	56
No mutation induction observed in Ames tester strains	*Salmonella typhimurium*	2.45 (CW) 8.6—9.6 (PW)	200 100, 450	40 18, 80 (est)	1 × 90 1 × 90	66
Reduction in survival concomitant with rise in sample temperature	*E. coli*	8.6—9.0 (PW)	10—200	≥50	1 × 90	67
	S. typhimurium *Saccharomyces cerevisiae*	8.6—9.0 (PW) 8.6—9.0 (PW)	≤450 ≤450	≤80 ≤80	1 × 90 1 × 120	
No reduction in survival or mutational events	*S. cerevisiae*	70 (CW) 73 (CW)	≤600	≤17 (est)	1 × 180	68
	S. cerevisiae	9.4 (CW) 17 (CW)		≤28 (est)	1 × 300 1 × 1,440	69

Table 5 (continued)
GENETIC AND MUTAGENIC EFFECTS OF RF EXPOSURE

Effects	Source of test specimen	Exposure conditions				Ref.
		Frequency (GHz)	Intensity (W/m²)	SAR (W/kg)	Duration (days × min)	
No detectable lethal events due to no change in CFUs	E. coli	2.6—4.0 (CW)		19	1 × 480	70
No observable change in molecular structures because no change in infrared spectrum	E. coli	3.2 (CW)		21 or 16	1 × 660—720	70
No repairable DNA damage	E. coli	8.6 (PW)		12	1 × 60—420	66
No change in growth pattern; enhanced colony-forming activity	E. coli	2.45 (CW)	0.05—500	0.008—75	1 × 240	71
No change in mutation frequencies at either of two loci controlling requirements for adenine and tryptophan	S. cerevisiae	2.45 (CW) / 8.5—9.6 (PW)	200 / 10—450	40 / ≤80 (est)	1 × 120 / 1 × 120	66
No mutagenic effects in exposed embryos	Drosophila melanogaster	2.45 (CW)	46,000—65,000	100	1 × 360	72
No changes in generation time, sex ratio, or sex-linked lethal mutations in offspring	D. melanogaster	2.45 (CW)		150—210 (est)	1 × 45	60
No mutations in adult males as evidenced by chromosome loss; nondisjunction; or sex-linked recessive lethals	D. melanogaster	0.029 (CW) / 0.146 (CW)	(600 V$_{rms}$/m) / (62.5 V$_{rms}$/m)	0.024 (est) / 0.015 (est)	1 × 720 1 × 720	59
No change in mutation induction	E. coli	0.027 (CW) / 2.45 (CW) / 3.07 (PW)		≤4 / 35—100 / 35—100	1 × 60—400 / 1 × 60—400 / 1 × 60—400	58
	Salmonella typhimurium	0.027 (CW) / 2.45 (CW) / 3.07 (PW)		≤4 / 35—100 / 35—100	1 × 60—400 / 1 × 60—400 / 1 × 60—400	
No mutagenic changes (recessive lethals) in adult females	D. melanogaster	0.098 (CW)	(0.3 V$_{rms}$/m, [FM] at audio)	0.0004 (est)	224 × 1,140	73
No significant germ-cell mutagenesis in weekly breedings	Rat	2.45 (CW)	50	4.7—0.9	106 × 240	74
	Rat	2.45 (CW)	100	2	5 × 300	74

Effect	Organism	Frequency (MHz)	Power density	SAR	Exposure	Ref.
Same, except decrease in pregnancies, indicating temporary sterility caused by elevated testicular temperatures	Rat	2.45 (CW)	280	5.6	20 × 240	74
Induction of a repressed protein, colicin, indicating a change in the genetic processes	E. coli	37 (CW)	0.0l—10		1 × 30—120	75
Change in growth rate that was very frequency specific, indicating an alteration in the processes of the cell	Saccharomyces cerevisiae	41.650 — 41.825 (CW)	10—30 ≤100		1 × 180	54
Chromosome aberrations in lung cells in vitro at two frequencies, but not at two closely related frequencies, 0.015 or 0.025 MHz	Chinese hamster	0.019 (PW) 0.021 (PW)	(Up to 300 kV_{p-p}/m		1 × 30	37 76
Increase in chromosome translocations in sperm cells	Mouse	9.4 (PW)	1—100	0.05—5 (est)	10 × 60	44
Increased mutations and lethality	Salmonella typhimurium	2.45 (CW)	51,000		1 × 0.03—0.48	77
No change in dominant lethality	Mouse	2.45 (CW)		43	1 × 30	79

Note: est = Estimated; av = average.

Adapted from U.S. Environmental Protection Agency, Biological Effects of Radiofrequency Radiation, Elder, J. A. and Cahill, D. F., Eds., Rep. No. EPA-600/8-83-026F, Research Triangle Park, N. C., 1984 of sister chromatid exchanges is not related to RF. In the second study, mice were exposed to 2.45-GHz RF at 200 W/m² (SAR 21 W/kg) for 8 hr/day for 28 days. Incidences of sister chromatid exchange in bone marrow cells of irradiated mice, sham-irradiated control mice, and standard control mice were compared. No statistically significant differences were detected.

Meltz and Walker[83] investigated whether there were any RF-induced alterations in DNA repair in normal human fibroblasts maintained in vitro after the DNA was damaged by a selected dose of UV light. Power densities of 10 or 100 W/m² (0.350 and 1.2 GHz) caused no perturbation of the DNA repair process. Brown et al.[84] treated mice with streptozooocin, a mutagenic/carcinogenic agent known to damage DNA in the rodent liver, and exposed the mice to 400-MHz RF to determine if excision repair of the DNA would be inhibited. Power densities of 16 and 160 W/m² (SAR 0.29 and 2.9 W/kg) did not alter the level of excision repair.

Although Mickey[85] reported increased mutagenesis in *Drosophila* exposed to a 0.02-GHz field, no effects on mutagenesis were observed by Pay et al.[60] using 2.45 GHz or by Mittler[59] with 0.003- and 0.150-GHz RF energy.

Varma and co-workers[57] reported increased mutagenesis, using the dominant lethal test with male mice exposed to 1.7 GHz continuous wave at 100 and 500 W/m² for 90 and 30 min, respectively. Mice exposed to 2.45 GHz CW at 1 kW/m² for 10 min and at 500 W/m² three times, 10 min each, within 1 day also showed increased mutagenesis. Mice subjected to four 10-min exposures at 500 W/m², over a period of 2 weeks, showed no increase in dominant lethality above control levels. Saunders et al.[79] found no evidence of change in dominant lethality in mice exposed to 2.45 GHz CW, 43 W/kg.

Berman et al.[74] exposed male rats daily to 0.425 GHz CW (day 12 of gestation to 90 days of age, 100 W/m² 4 hr/day) or 2.45 GHz CW at 50, 100, or 280 W/m² from day 6 of gestation to 90 days of age, 4 or 5 hr/day. No significant evidence of germ cell mutagenesis or alteration in reproductive efficiency was detected.

It is known that the rate of induction of mutations will increase with increasing temperature. It is possible, therefore, that artifacts or thermal stress could be factors in some of the reported studies. Baranski and Czerski[86] note there is no satisfactory evidence of microwave-induced genetic effects at low to modest power densities. Recent experimental studies of primitive organisms and rodents[57,87,88] have confirmed earlier findings that microwave exposure at power densities below 100 W/m² is not mutagenic in these organisms. Other studies[89] have failed to find effects that differ from those resulting from RF heating. Janiak and Szmigielski[90] likewise reported no significant differences in the sequence and time-course of cell membrane injury between cells treated in a water bath and those heated with 2.45-GHz microwaves.

The principal technical problem in studying RF-induced effects on cells is that the studies are often conducted using conventional apparatus designed for cell studies such as flasks, dishes, holders, agitators, water baths, incubators, etc. Various elements of such apparatus may distort the RF fields in such a way that the SAR of energy in the entire cell culture or, more likely, in small regions of the culture media may be considerably higher or lower than field measurements would indicate. Some progress has been made in designing cell culture apparatus that will provide accurate, calibrated exposure to RF fields, but results of much of the earlier work on cell and tissue cultures must be questioned with relation to the actual absorbed RF energy in the cell culture media.[91-93]

In summary, there is no solid evidence that exposure to RF induces mutations in bacteria, yeasts, or fruit flies. The results of some studies suggest that RF induces mutations in mammals. Critical review has cast doubt on these findings. Other studies have shown no mutagenic effects of RF on mammals. Evidence for cytogenetic effects of RF is mixed. The lowest power density at which cytogenetic effects have been reported is 200 W/m²,[94] but these results are contradicted by Chen et al.[40] who failed to find cytogenetic effects at 2 to 5 kW/m².

B. Hyperthermia and Cell Kinetics

Investigations of the effects of MW-induced hyperthermia on cell kinetics have provided background information on the potential application of hyperthermia as an adjunct in cancer therapy. The use of MW-induced heating as an adjunct to other methods of cancer therapy is an area of investigation which warrants more comprehensive review than is possible for this chapter.

C. Carcinogenesis

There is no reliable evidence that RF initiates carcinogenesis. There are, nevertheless, sporadic reports of such relationships. Szmigielski et al.[95,96] described accelerated development of spontaneous and benzopyrene-induced skin cancer in mice exposed to 2.45 GHz, 150 W/m² (6 to 8 W/kg). At 50 W/m² (2 to 3 W/kg) the cancer induction was comparable to that which occurs with chronic stress resulting from confinement. Preskorn et al.,[97] however, reported decreased tumor growth and greater longevity in mice exposed to intense levels of 2.45-GHz energy, i.e., sufficient to make the pregnant mice hyperthermic.

A report by Prausnitz and Susskind[98] has been misinterpreted as supporting a link between exposure to MWs and the development of neoplasia. In this study male Swiss albino mice were exposed to 9.27-GHz microwaves at 1000 W/m² for 4.5 min daily, 5 days/week for 59 weeks (estimated SAR = 40 to 50 W/kg, half the LD_{50} for the mice) which resulted in a 3.3°C average increase in colonic temperature. The authors reported an increased incidence of ''leukosis'', defined as a noncirculatory neoplasia of white blood cells, which has been misinterpreted to be synonymous with leukemia. This report, however, is deficient in several respects.[99] No histological or biological criteria were provided to characterize or define a ''leukosis''; no pathological material was illustrated. The historical evidence of ''leukosis'' in the mouse strain used in the studies was not provided. Finally, there was an endemic infection in the colony which contributed to the lethality. In fact, the data suggest that the mice exposed to MWs showed enhanced survival.

IV. BIOCHEMICAL CHANGES

Biochemical alterations have been reported to result from exposure to RF energies. Such effects generally appear to be reversible and no well-defined characteristic response pattern has been determined, nor is it known whether the changes are direct or indirect effects of exposure.

Effects on mitochondria isolated from exposed animals have been reported,[100] but there was no effect on rat liver mitochondria exposed in vitro to 2.4 GHz, 1 to 4 W/Kg,[101] or 10 to 12 GHz, at a maximum of 20 W/m².[102] No effect of MW exposure has been found on a number of enzymes and proteins irradiated in vitro.[103-106] Albert et al.[107] exposed Chinese hamsters at 2.45 GHz, 500 W/m², for 0.5 to 4.5 hr over a period of 1 to 21 days and found no change in liver ATP.

The limited number of studies on oxidative enzyme systems has yielded mixed results. Exposure of suspensions of the membrane-bound enzyme cytochrome oxidase to sinusoidally modulated 2.45-GHz MW energy at an SAR of 26 W/kg did not significantly affect its activity during exposure.[108]

Dumansky et al.[109] reported a decrease in liver glycogen content together with increased lactic acid levels and phosphorylase activity which was interpreted as evidence of impairment of glycogen synthesis in the livers of rats chronically exposed to 2.45- or 10-GHz MWs at 500 to 2000 W/m². Exposure to 2.45- or 10-GHz MWs at 0.25 to 10 W/m² decreased the proteolytic activity of the mucous membrane of the small intestines of experimental animals, whereas the invertase and adenosine triphosphatase activity increased. Altered proteolytic activity was suggested to be due to alteration in the structure and physicochemical characteristics of the mucous membrane. Reduced synthesis of macroglobulin and macroglobulin antibodies resulted from exposure of experimental animals to 0.5 W/m[2].[110]

Dose-dependent transient elevations in serum glucose, blood urea nitrogen, and uric acid were noted following far-field exposure of rabbits to 2.45 GHz for 2 hr at intensities of 50, 100, and 250 W/m[2].[111] There were detectable differences between continuous wave and pulse-modulated exposures of equivalent average power density. There was an increase in colonic temperature of 1.7 to 3.0°C at 100 and 250 W/m[2]. Exposure of rats for 15 min to pulsed 2.86 GHz at 50, 100, 200, 500, or 1000 W/m[2] resulted in statistically significant changes in serum albumin and phosphorous levels only at 1000 W/m[2];[112] there was no change in serum glucose levels in rats exposed to 2.86- and 0.43-GHz pulse-modulated fields at an average power density of 50 W/m[2]. Single or repeated exposures of rabbits to 3 or 10 GHz at 50 to 250 W/cm[2] resulted in alterations in serum albumin/globulin ratio, an effect attributable to effects on the liver or adrenals.[113]

Baranski[115] found inconsistent changes in cholinesterase activity in both rabbit and guinea pig brains following 3 months of exposure, 1 hr/day, with pulsed fields of 250 W/m[2]. He also found a decrease in cholinesterase activity in the brains of guinea pigs after a single 3-hr exposure to 35 W/m[2] of pulsed 2.45-GHz MWs. Increase to 250 W/m[2] caused a further decrease in activity. Pulsed energy was found to produce a more severe effect than continuous wave (CW) exposures of the same average power density, suggesting that these effects are due to peak fields. Nikogosyan[116] found an increase in blood cholinesterase activity after a single 90-min exposure to 10-cm waves at 400 W/m[2]. Revutsky and Edelman[117] also report an increase in specific cholinesterase activity in rabbit blood exposed in vitro to 2.45-GHz MWs. It should be noted that what is measured in blood is "pseudocholinesterase" which has no neural-related activity. On the other hand, Olcerst and Rabinowitz[118] found no effect on aqueous cholinesterase exposed to 2.45-GHz CW up to 1.25 kW/m[2] for ¹/₂ hr or 250 W/m[2] for 3 hr. No effect was found on cholinesterase activity in defibrinated rabbit blood exposed for 3 hr to 210, 350, or 640 W/m[2], 2.45 GHz, CW or pulsed. Under similar exposure conditions, there was no effect on release of bound calcium or magnesium from rabbit red blood cells.

Ho and Edwards[119] studied oxygen consumption in mice exposed to 2.45 GHz at various SARs in a waveguide exposure system and found that the animals compensated homeostatically to SAR of 10.4 W/kg or greater by a decrease in metabolic rate to compensate for thermal loading. Normal metabolic activity was resumed following cessation of exposure.

Some of the reports on effects of RF on subcellular systems and biochemical changes are summarized in Tables 6 to 8.[61]

V. RESPONSE TO LOCAL RF EXPOSURE

Some consideration has been given in animal and human experiments to the temperature rise in various organs and tissues as functions of incident energy wavelength, thickness of the s.c. fatty layer, blood-circulation rate, and other factors. The increase in temperature of tissues during local exposure is linear for short periods (1 to 3 min) and proportional to the magnitude of the MW energy absorbed. With exposures in excess of 3 min, the extent of the thermal effect and distribution of heat in tissues is determined by heat-regulating mechanisms[137,138] (see Part III, Chapter 3). The thermal effect depends on exposure duration. Deep-lying muscles are heated to a greater extent only during the first 20 min of exposure. When the thigh region is exposed to microwaves, there is a greater temperature rise in the muscles than in the skin and s.c. fatty layer.[139-142] The distribution of temperatures among

Table 6
BIOCHEMICAL EFFECTS OF RF

End point measured or effects	Experimental system or model	Exposure conditions				Ref.
		Frequency (GHz)	Duration (min)	SAR (W/kg)	Modality	
No change in activity of membrane-bound enzymes measured spectrophotometrically	Red blood cell membrane; mitochondrial inner membrane	2.45 (SW)[a]	10	26	Waveguide	108
	Endoplasmic reticulum	2.45 (CW)	5	42	Waveguide	103
No difference in respiratory activity	Mitochondria	2.45 (CW)	30—120	17.5, 87.5	Anechoic chamber Far field	120
	Mitochondria	2—4 (Swept) 3.4 (CW)	10	1.6—2.3 41	Coaxial airline	101
No change in formation of microtubules	Tubulin (rabbit brain)	3.1 (PW)	15	112—430	Far field	121
No change in migration of proteins within axonal membrane	Vagus nerve cell	3.1 (PW)	24 hr	~10—100	Far field	121
No changes in infrared spectra of proteins and nucleic acids in Escherichia coli exposed before drying	Dried film of E. coli cells	3.2 (CW)	8, 10, 11 hr	20	Waveguide	70

[a] Sine-wave modulated.

Adapted from U.S. Environmental Protection Agency, Biological Effects of Radiofrequency Radiation, Elder, J. A. and Cahill, D. F., Eds., Rep. No. EPA-600/8-83-026F, Research Triangle Park, N.C., 1984.

Table 7
EFFECTS OF RF ON MOLECULAR SYSTEMS

End point measured or effects	Experimental system or model	Exposure conditions				Ref.
		Frequency (GHz)	Duration (min)	SAR (W/kg)	Modality	
No change in UV difference spectra measured over pH range 2.5—5.5	Bovine serum albumin	1.70 (CW) 2.45 (CW)	30	30—100	Waveguide	104
UV spectra and binding constants for mononucleotides showed no difference from controls	Ribonuclease	1.70 (CW) 2.45 (CW)	30	39	Waveguide	122
No change in enzyme activity	Glucose-6-phosphate dehydrogenase; adenylate kinase; NADPH cytochrome c reductase	2.45 (CW)	5	42	Waveguide	103
No difference in melting curves	DNA	2.45 (CW)	60, 960	67, 160	Far field	114
Inactivation of enzyme; probably temperature inhomogeneity effect at very high doses	Horseradish peroxidase	2.45 (CW)	5, 10, 20, 30, 40	62,500—375,000	Waveguide	124
Heat inactivation of enzymes found at highest SAR (T = 50°C) corresponded closely to heat-treated controls	Glucose-6-phosphate dehydrogenase; lactate dehydrogenase; acid phosphatase; alkaline phosphatase	2.8 (PW)	4.5, 18.5	~200 — 500	Waveguide	105
Heat inactivation of enzyme found at SARs > 165 W/kg	Lactate dehydrogenase	3.0 (CW)	20	33—960	Waveguide	106

Adapted from U. S. Environmental Protection Agency, Biological Effects of Radiofrequency Radiation, Elder, J. A. and Cahill, D. F., Eds., Rep. No. EPA-600/8-83-026F, Research Triangle Park, N.C., 1984.

Table 8a

EFFECTS OF RF ON CLINICAL CHEMISTRY AND METABOLISM

Effects	Test specimen	Frequency (GHz)	Intensity (W/m²)	SAR (W/kg)	Duration (days × min)	Δt (°C)	Ref.
No effect on serum chemistry values	Rat	0.918 (CW)	25	1.0	91 × 600	0	125
Increase in serum glucose	Rabbit	2.45 (CW and PW)	50, 100, 250	0.8—4.0 (est)	1 × 120	0, 10, 1.7 (PW) 0, 0, 2.9 (PW)	111
Increase in blood urea nitrogen	Rabbit	2.45 (CW)	250	4.0 (est)	1 × 120	2.9	
No increase in blood urea nitrogen	Rabbit	2.45 (CW)	50 and	0.8, 1.6 (est)	1 × 120	0	
	Rabbit	2.45 (PW)	100, 50, 100, 250	0.8—4.0 (est)	1 × 120	0, 0, 1.7	
Increase in uric acid values	Rabbit	2.45 (CW and PW)	100, 250	1.6, 4.0 (est)	1 × 120	0, 1.7 (PW) 0, 2.9 (CW)	
No change in uric acid	Rabbit	2.45 (CW and PW)	50	0.8	1 × 120	0	
No effect on other serum chemistry values	Rabbit	2.45 CW and PW)	50, 100, 250	0.8—4.0 (est)	1 × 120	0, 1.7 (PW) 0, 2.9 (CW)	
Increased iron and manganese levels in brain	Rat	1.6 (CW)	800	48 (est)	1 × 10	4.5	126
Decrease in specific metabolic rate (ambient T = 24°C)	Mouse	2.45 (CW)		10.4	1 × 30		119

Table 8b

End point measured or effects	Experiment system or model	Exposure conditions				Ref.
		Frequency (GHz)	Modality	SAR (W/kg)	Duration (min)	
Increase in red blood electrophoretic mobility 30 min postexposure (SAR ≥ 10 W/kg)	Red blood cell	1.0 (CW)	Stripline	5—45	4, 8, 15, 30	132
Increase in K^+ efflux and Na^+ influx	Red blood cell	1.0 (CW)	Stripline	45	30	133
K^+ transport no different from heat-treated controls; no change in osmotic fragility	Red blood cell	2.45 (CW)	Monopole far field	3—57	60, 120, 180, 240	134
K^+ transport not different from controls at corresponding temperatures; no difference in hemoglobin release	Red blood cell	2.45 (CW)	Anechoic chamber far field	200	45	135
Passive transport of Na^+ and Rb^+ increased at transition temperature	Red blood cell	2.45 (CW)	Waveguide	100, 190, 390	60	136
No significant changes in K^+ efflux, hemoglobin release, or osmotic fragility	Red blood cell	2.45 (CW) 3.00 (CW) 3.95 (CW)	Waveguide	22—200	20, 180	89

Table 8c

Effects	Test specimen	Exposure conditions					
		Frequency (GHz)	Intensity (W/m²)	SAR (W/kg)	Duration (days × min)	Δt (°C)	Ref.
Increase in specific metabolic rate (ambient T = 35°C)	Mouse	2.45 (CW)		8.6	1 × 30	0	127
Increased NADH fluorescence	Rat (exposed brain)	0.591 (CW)	138	0.36—2.2	1 × 0.5	0	128
Decreased ATP							
Decreased CP							
Decreased ATP	Rat (exposed brain)	0.591 (CW)	50	0.13—0.8	1 × 0.5	0	128
Decreased CP							
Increase in oxygen consumption	Rat	2.45 (120 Hz PW)		6.5, 11.1	1 × 30	0.9, 1.8	129
Decrease in metabolic heat production	Monkey	2.45 (CW)	60	0.9	1 × 10	0	130
No effect on blood coagulation	Human plasma	2.45 (CW)	100—2800	1.3—38 (est)	1 × 30	Not reported	131

Note: Δt = Colonic temperature increase.

Adapted from U.S. Environmental Protection Agency, Biological Effects of Radiofrequency Radiation, Elder, J. A. and Cahill, D. F., Eds., Rep. No. EPA-/8-83-026F, Research Triangle Park, N.C., 1984.

357

the various tissues depends on the rate of blood circulation. Thus, rapid heating of all tissues was observed in rabbit thighs during MW exposure when the local blood supply was occluded. When the paws of rabbits were exposed to 2.45 GHz, the temperature rise in the deep tissues of the leg could vary substantially with repeated local exposures. While the temperature in the thigh muscles rose 4°C after the first 10-min exposure at 1.2 to 1.5 kW/m², the rise was only about 2°C after the sixth or seventh treatment, suggesting that adaptive reactions had come into play. Adaptive reactions were also noted on the opposite unexposed leg after daily conditioning exposures to MWs. After denervation of the rear extremity, the temperature rise in the thigh muscles remained the same regardless of the number of exposures.[143]

Engle et al.,[144] Gersten et al.,[145] and Rae et al.[142] exposed the hind legs of dogs and the forearms of human volunteers and found that increments in temperature of muscle were slightly higher than those of s.c. fat. Lehmann et al.[24,25] note that variations in the thickness of the s.c. tissues involved may explain apparent differences in results of various investigations. Examination of the hindleg of a dog reveals that the s.c. layer is usually only a few millimeters thick and in many cases contains little fat. Similarly, the human forearm has a comparatively thin layer of s.c. fat in most cases. Thus, one explanation for the greater increases in temperature of muscle than of fat in some studies might be that, under the particular conditions of the experiments, relatively small amounts of energy were absorbed in the thin layer of s.c. fat. For a quantitative discussion of such effects, see the ''Introduction'' and Part III, Chapters 1 and 2.

A. Comparison of Exposure to Microwaves and Infrared

According to some investigators,[146-149] hyperthermia develops much more efficiently with MWs than with IR exposure. With 2.45 GHz, 500 W/m², 30-min exposure of the spinal region in rabbits, a 1 to 1.5°C colonic temperature rise was observed, while 3.5 kW/m² was necessary to produce the same effect with IR.

Several studies, notably those of Lehmann and associates,[26-28] have compared the heating patterns and temperature distributions resulting from 2.45- and 0.915-GHz MW exposure to those resulting from IR exposure, using human volunteers. The highest temperature distributions with infrared application were obtained in the most superficial tissues, while the highest temperatures in MW application were obtained close to the s.c. fat-muscle interface. MW application produced somewhat higher temperatures in all cases in the deep tissues. A marked difference was found between temperature distributions produced by the two MW frequencies prior to the blood flow changes resulting from temperature increases in the tissues. The 0.9-GHz exposure produced a greater increase in the temperature of the deeper musculature. This difference was much less pronounced after blood flow changed, about 10 min after start of exposure, since the increased flow cooled the skin and s.c. tissues specifically. Thus, after blood flow increased, the difference in the deep heating properties between the two frequencies diminished. A more detailed discussion of thermoregulation in the presence of RF fields is found in Part III, Chapter 3.

B. Therapeutic Application of RF (Microwave) Energies (Diathermy)

Diathermy is the therapeutic induction of heat in the tissue beneath skin and s.c. fat. The temperature elevation results in increased metabolic activity and in the dilation of blood vessels, causing increased blood flow. The increased blood flow increases the oxygen supply to the treated areas and accelerates the removal of carbon dioxide and metabolic products. In addition, microscopic thrombi may be unclogged or removed when constricted vessels are opened.

VI. REPRODUCTION, GROWTH, AND DEVELOPMENT

A. The Gonads

The effect of MWs on the testes has been studied fairly extensively.[150-152] Exposure of the scrotal area at high power densities (>500 W/m²) results in varying degrees of testicular damage such as edema, enlargement of the testis, atrophy, fibrosis, and coagulation necrosis of seminiferous tubules in rats and rabbits, exposed to 2.45, 3, and 10 GHz. Exposure to 3 GHz, 80 W/m² did not affect mating of mice or rats.[153] Pituitary gonadotropic function was preserved in female mice exposed to 3 GHz, 100 W/m², twice daily for 5 months.[154]

Saunders and Kowalczuk[155] exposed the caudal area of anesthesized mature male mice to 2.45-GHz RF in a waveguide system for 30 min. Half-body SARs ranging from 18 to 75 W/kg were estimated from measurements of forward, reflected, and transmitted powers. The corresponding colonic temperatures at the end of exposure ranged from 35.3 to 42.2°C. Other anesthetized mice were sham-exposed. Their mean colonic temperature was 32.6°C (4 to 5°C lower than unanesthetized mice). For comparison, the caudal area of other anes-thestized mice were inserted for 30 min in a copper well heated by a water bath to 37, 41, 43, or 45°C, which resulted in colonic temperatures from 36.4 to 40.7°C. Six days after treatment, sections of testes were scored for cell damage and enumeration of sperm. Extensive degeneration of the spermatogenic epithelium was evident for RF exposure at 75 W/kg and for direct heating to 45°C. At SARs of 57 and 46 W/kg (T 43 and 41°C) marked depletion of spermatids and spermatocytes, but not spermatogonia, was observed. At the lower SARs (37, 30, 18, and 0 W/kg) or a temperature of 37°C, no effects were seen.

Temperature-sensing probes were also implanted in the testes of other groups of mice, and testicular temperatures were related to SAR values. Such measurements of testicular temperature indicated the existence of a threshold of about 39°C for depletion of sperma-tocytes and of about 41°C for 50% cell death after 6 days of RF exposure or direct heating. The corresponding SARs for these two thresholds were 20 and 30 W/kg, respectively.

There is general agreement that high-power density exposure can affect the testes and ovaries. There are reports, however, that chronic "low-level" exposure can result in im-pairment of spermatogenesis and reproductive function without measurable temperature increase of the testes.[109,155] These responses can nevertheless be related to the heating of the organs. The sensitivity of the testes to heat is well known.[156] Comparable heating of the scrotum in rats with 2.45-GHz CW MWs or by immersion in water to temperatures of 36, 38, 40, and 42°C resulted in comparable damage at each temperature.[157,158] Gonadal effects of RF exposure are summarized in Table 9.[61]

B. Embryonic Development

There are a few reports that suggest particular combinations of frequency, duration, and power density produce effects on embryonic development and postnatal growth. Alterations in development have been reported in insects,[162] chick embryos,[163,164] and rodents.[154,165-167]

Olsen[168] has provided a thermal basis for explanation of the MW-induced teratogenic effects in the meal worm *Tenebrio molitor* reported by Carpenter and Livstone,[162] Lindauer et al.,[169] and Liu et al.[170] Olsen[171] and Pickard and Olsen[172] noted the threshold of terato-genesis in *Tenebrio* to be associated with a rise of pupal temperature in excess of 10°C.

Van Ummersen[163,164] reported inhibition of growth and development of chick embryos exposed to 2.45-GHz (CW), 200 to 400 W/m², 4.5 min to 5 hr (percent T up to 19°C). Since all embryos in which effects were produced experienced a significant temperature increase, the observed deleterious effects are, no doubt, due to heating.

Quail embryos exposed to 300 W/m², 2.45 GHz (CW) for 4 hr on each of the first 5 days of incubation did not develop gross deformities or changes in hatchability that could be correlated with exposure.[173] MW-induced temperature within the eggs varied from 34 to 37°C, which is the normal incubation temperature.

Table 9
EFFECTS OF RF EXPOSURE ON TESTES

Effects	Test animal	Frequency (GHz)	Exposure conditions Intensity (W/m²)	SAR (W/kg)	Duration (days × min)	Ref.
No change	Mouse	1.7	100	15	1 × <100	57, 87
Abnormal germinal cells, normal interstitial cells			100	15	1 × 100	
All tissue necrotic, altered spermatogenesis			500	75	1 × 30—40	
Scrotal skin burns			2000	300	1 × 20	
"Minimal" injury	Mouse	3.0	500	50	1 × 20	
No change in tissue, sperm	Mouse	2.45	<370	<8	Many × 16 hr	159
Abnormal spermatogenic tissue	Rat	2.45	800	16	1 × 10—73	160
				16	5 × 10—73	
No change	Rat	2.45	50	0.9 — 4.5	Many × 240	74
No change			100	2	5 × 360	
Temporay sterility			280	5.6	20 × 240	
Testicular degeneration	Mouse	9.27 (pulsed)	1000		295 × 4.5	98
Testicular lesions	Mouse	10.0 (pulsed)	4000		5 min	151
No change	Mouse	10 (pulsed)	3.44		20—50 Times prior to mating × 30 min	158
Testicular damage	Rat	24	250		(5—15 Min)	161

Adapted from U.S. Environmental Protection Agency, Biological Effects of Radiofrequency Radiation, Elder, J. A. and Cahill, D. F., Eds., Rep. No. EPA-600/8-83-026F, Research Triangle Park, N.C., 1984.

In another study Japanese quail embryos were exposed during the first 12 days of development to 2.45-GHz MWs at an incident power density of 50 W/m^2 and SA rate of 4.03 W/kg. No gross deformities were observed in the exposed quail when examined and sacrificed at 24 to 36 hr after hatch. No significant changes in the total body weight or weights of the heart, liver, gizzard, adrenals, and pancreas were found in the treated birds. Hematological parameters were also measured in the study. The results showed a statistically significant increase in hemoglobin and statistically significant decrease in monocytes in birds exposed to MWs. No statistically significant changes in hematocrit, red blood cells, total white blood cells, lymphocytes, heterophils, basophils, or eosinophils were detected.[174]

A number of studies, at various MW frequencies, have been concerned with effects of exposure on mammalian embryonic and fetal development. In most of these studies a single acute exposure was administered at a power density that causes an increase in body temperature.[167,175-178] Some investigators, however, have reported studies in which protracted exposures have been given at power density levels which apparently do not cause a significant increase in colonic temperature.[179-182]

Rugh et al.[166,167] reported abnormalities in mouse fetuses exposed in an environmentally controlled waveguide at days 7 to 13 of gestation to 2.45-GHz (CW) MWs in the range of 85 to 112 W/kg (10.3 to 33.5 J/g, 2 to 5 min), equivalent to 1.23 kW/m^2 incident power. Hemorrhages, resorptions, exencephaly, stunting, and fetal death were observed. In a further study, Rugh and McManaway[183,184] demonstrated that lowering the body temperature of the dam with pentobarbital anesthesia could prevent the teratogenic effects of thermal loading with MWs, thus demonstrating the thermal influence in the earlier studies.

Lin et al.[185] studied the effects of repeated exposure of C3H mice to 0.048 GHz. The animals were exposed to 5 W/m^2 (63.25 V/m peak E field) in a TEM exposure chamber for 1 hr/day, 5 days/week, beginning on the 4th to 7th days post partum, for 10 weeks. The formed elements in the blood were not affected by the exposure. The means of body mass of the irradiated and control animals were comparable. No significant differences in lesion onset, incidence, prevalence, extent, or type were observed when repeated RF-exposed animals were compared with sham-control groups. During the period of the study, no cataracts were noted. Fertility differences among irradiated and control animals were not detected. Body growth patterns did not differ among sham and RF-exposed animals.

Employing a multimodal cavity at 2.45 GHz, Chernovetz et al.[176] irradiated pregnant mice with a single "intense" dose of 38 W/kg for 10 min (22.8 J/g) on gestation day 11, 12, 13, or 14. This dose resulted in 10% maternal lethality. The exposure during late organogenesis or early fetal stage caused no change in fetal mortality or morbidity when compared to shams. In another study, Chernovetz et al.[177] noted an increased rate of resorptions in rats after an absorbed dose of 30 W/kg for 20 min once on gestation days 10 to 16.

Berman et al.[179] exposed CD-1 pregnant mice to 2.45 GHz at 34, 136, or 280 W/m^2 100 min daily through gestation days (GD) 6 through 17 for 100 min daily. Estimates of mean dose rate ranged from 2.0 to 22.2 W/kg. Another group was similarly sham-exposed. The mice in half of each group were examined on GD 18. The incidence of pregnancy, the number of live, dead, and resorbed fetuses, and the total number of fetuses were found to be similar for the exposed and sham-exposed mice. The mean body weight of the live fetuses in the RF-exposed group, however, was significantly smaller (by 10%) than those in the sham-exposed group, a finding consonant with their previous results. In addition, ossification of sternal centers was significantly delayed in the RF-exposed mice. The mice in the other half of each group were permitted to come to term. No significant increase in colonic temperature was reported. Significant fetal growth retardation occurred at the highest power density level. A significant increase in central nervous system (CNS) anomalies occurred, but only if the data from three irradiated groups were combined. At 7 days of age, the mean body weight of the suckling mice of the RF-exposed group was also significantly smaller (by 10%) than that for the sham-exposed group. The survival rate was not affected. The authors concluded that MW radiation at a frequency of 2.45 GHz was embryopathic at a power density of 280 W/m² (22.2 W/kg).

Bereznitskaya[154,165] exposed mice to 3 GHz, 100 W/m² which resulted in increased fetal wastage. No definite abnormalities or inborn genetic defects were found. Neonatal mice exposed to 0.105, 0.01927, or 0.0266 GHz pulsed in a magnetic field of 55 A/m and an electric field of 8000 V/m, 40 min/day for 5 days did not show any evidence of alteration in growth and development.[186]

Nelson et al.[187] exposed C3H mice in a rectangular coaxial exposure system at 0.148 GHz at 5 W/m² for 1 hr daily from day 2 through day 19 of gestation, corresponding to an SAR of 0.013 W/kg. The experiment was conducted as three separate replications. In each of the experiments some of the dams were allowed to come to term and the fetuses were assessed for weight at birth and again at 60 days of age, while in others the uterus was extirpated on day 19 of gestation. No differences in percentage of resorbed, stillborn, or abnormal fetuses were observed in any of the three experiments.

The embryofetal toxicity and teratogenicity of 2.45-GHz (CW) MWs at different intensities were investigated in the CD-1 mouse by Nawrot et al.[181] Mice were exposed on GD 1 to 15 at an incident power density of 50 W/m² (SAR of 5 W/kg), and either on days 1 to 6 or 6 to 15 of gestation to 210 W/m² (SAR of 28.14 W/kg) or to 300 W/m² (SAR of 40.2 W/kg) for 8 hr daily. Exposure either on GD 1 to 6 or 6 to 15 to a power density of 210 or 300 W/m² caused an increase in colonic temperature of exposed dams of 1 and 2.3°C, respectively. To distinguish between "thermal" and "nonthermal" effects of 210 or 300 W/m², groups of mice were also exposed to elevated ambient temperature to raise their body temperature to the level of the animals exposed to MWs. Ambient temperatures of 30 and 31°C increased the deep colonic temperature to that obtained with the 210- and 300-W/m² MW exposure, respectively. The mice exposed to higher ambient temperature were handled in exactly the same manner as the MW-exposed mice. A significant reduction in maternal weight gain, either during treatment on days 1 to 6 or 6 to 15 of gestation, was observed in females of all handled groups. Handling plus exposure to elevated ambient temperature (30 or 31°C) during days 6 to 15 of gestation increased this reduction in maternal weight gain. A significant decrease in implantation sites per litter and reduction in fetal weight was noted in the group exposed at 300 W/m² during days 1 to 6 of gestation. Exposure of mice at 300 W/m² (days 6 to 15 of gestation) resulted in a slight but significant increase in the percentage of malformed fetuses, predominantly with cleft palate, when compared to all other groups.

Dietzel et al.[188-190] exposed pregnant rats with a 0.02712-GHz diathermy unit at 55, 70, or 100 W which was sufficient to raise the animals' colonic temperatures to 39, 40, or 42°C. The rats were exposed once between days 1 and 16 of gestation. The fetuses were examined near term. The peak incidence of anomalies was found to occur on gestation days 13 to 14, when 16% of the fetuses were abnormal. This fetal wastage, as in the studies by Rugh et al.,[166,167] are clearly associated with a general body temperature increase in the dam.

In studies of effects of *in utero* exposure of Long-Evans rats to 2.45-GHz CW microwaves, Michaelson et al.[178] found no adverse effects on the dam or offspring when gestation length or litter size were examined. The rats were exposed at 100 or 400 W/m² for 1 hr on day 9 (organogenesis) or day 16 (fetal stage) of gestation. Enhanced maturation, as indicated by adrenocortical response, was suggested.

Laskey et al.[175] observed an increase in resorption rate and a decrease in term fetal weight in offspring of rats exposed to 2.45 GHz for 8 to 13 min on day 2, 2 and 5, 8, or 15 of gestation at 1 kW/m². Maternal body temperature increased 3 to 4°C during the exposure. Smialowicz et al.[191] exposed rats to 2.45-GHz MWs daily for 4 hr/day at a power density of 40 W/m² *in utero* from the 6th day of gestation through 40 days of age. The SAR was determined by twin-well calorimetry for several ages of the animals. Pregnant dams weighing

300 to 350 g had a mean SAR of 0.7 W/kg; animals that were 1 to 5 days of age and weighed 6 to 10 g absorbed approximately 4.7 W/kg. There was no significant difference between the mean body weights of the males (female offspring were not used in this experiment) in the 12 sham-irradiated litters when compared to the mean body weights of the males in the 12 MW-irradiated litters.

Jensh et al.[192] reported on the effects of protracted prenatal exposure of Wistar-derived rats to 2.45-GHz MWs for 8 hr daily throughout gestation. The mean exposure time to incident energy at 200 W/m^2 was 115 hr. No statistically significant differences were observed between control and exposed animals for maternal mass, embryonic and fetal resorption rate, abnormality rate, and the term fetal and placental masses. These authors[193,194] also reported similar experiments in which 0.915-GHz MW were employed. They exposed Wistar rats at 100 W/m^2 for 8 hr daily throughout the period of gestation for an average of 110 hr of exposure. Postexposure colonic temperatures of dams showed no increase over baseline rectal temperatures. No differences were observed in embryonic or fetal death, abnormalities, fetal mass, litter size, placental mass, fetal sex ratio, maternal mass, or maternal gain of mass.

In another study,[195] gravid Wistar rats were exposed 8 hr daily throughout the pregnancy to 0.915-GHz MWs at 100 W/m^2. The average exposure time was 109 hr. In this study, dams were allowed to deliver their offspring. Within 3 days after birth the pups were given three tests for reflexes. After the pups were weaned, a series of performance tests was given. The postweaning tests began on post partum day 60 and were completed by the 90th day. No detrimental effects were observed on performance in a water T-maze, avoidance behavior, open-field behavior, forelimb hanging, or on performance in a 24-hr activity wheel. Jensh et al.[196] also reported no observable teratology in rats exposed throughout gestation to 350-W/m^2 MWs at 6 GHz. The authors of the three studies state that the irradiation was "nonthermal", but no doses or dose rates were given for any of their exposure conditions.

A series of experiments was conducted on rats to assess the teratologic effects of 0.027 GHz because of the prevalence of this frequency in industries using RF heat sealers.[197-201] At various days of gestation the investigators used energy levels sufficient to cause and sustain high body temperatures (up to 43°C, 20 to 40 min). These experiments showed that day 9 was the most sensitive day of pregnancy for the rat. A variety of malformations, including decreased body weight, was found. Thermal (maternal) threshold was 41.5°C. Gross microscopic teratogenic effects were equivalent when maternal temperature was caused by MW or hot-water immersion. The threshold was approximately 12 W/kg for 30 min.

Conover et al.[202] reported on single, 20- and 30-min exposures of fetal rats on the 10th, 12th, or 14th day of gestation to 0.02712 GHz at SARs of 17 to 35 W/kg. Preliminary results suggested that fetal wastage occurred after exposure on the 10th or 14th day of gestation. Grossly observable malformations occurred in animals exposed during the 10th day for 30 min, while visceral abnormalities occurred after exposure on the 10th and 14th days. Fetal mass as well as crown-rump length was smaller numerically in the fetuses exposed to MWs. The same investigators[197] reported on 8 groups of 16 to 28 gravid rats exposed to 0.02712 GHz. The facility for exposure was a RF near-field synthesizer operating in the dominant magnetic field mode at a field strength of 55 A/m. The dams were exposed on gestation days 2, 4, 6, 8, 10, 12, 14, or 16 for 20 to 40 min at an average dose rate of 125 W/kg until their colonic temperature reached 43°C. Eight groups were sham-irradiated for 30 min and one group of 29 rats served as cage controls. Rats irradiated on days 8 to 16 (organogenesis) had a significant increase in gross malformations and a significant decrease in fetal weight and fetal crown-rump length.

Boak et al.[203] administered "shortwave" radiation (0.01 GHz) to rabbits from the 29th day of life through several matings and pregnancies. The total exposure time ranged between 30 and 75 hr, during which the temperatures of the animals were raised to 41 to 42°C. There was no interference with mating, fertilization, or development of the young *in utero*. Litter sizes were not significantly different from those of the control animals.

In a study designed to examine possible effects of chronic RF exposure on mother-offspring behavioral patterns and the EEG, Kaplan et al.[204,205] exposed 33 female squirrels monkeys near the beginning of the second trimester of pregnancy to 2.45-GHz RF in multimode, mode-stirred MW cavities at whole-body SARs of 0.034, 0.34, or 3.4 W/kg (the last value equivalent to about 100 W/m² of plane-wave RF) for 3 hr/day, 5 days/week, until parturition. Eight pregnant monkeys were sham-exposed for the same periods. After parturition, 18 of the RF-exposed dams and their offspring were exposed to RF for an additional 6 months; then the offspring were exposed without the dams for another 6 months. No differences were found between RF- and sham-exposed dams in the numbers of live births or in the growth rates of the offspring. The major difference between RF- and sham-exposed offspring was that four of the five exposed at 3.4 W/kg both prenatally and after birth unexpectedly died before 6 months of age. They apparently had developed a form of pneumonia. Although the numbers of animals used in the behavioral and EEG studies were adequate, the mortality values were too small to place much confidence in statistical inferences. A follow-up study of mortality per se, which involved sufficient numbers of squirrel monkeys for adequate statistical treatment, did not confirm the RF-induced offspring mortality results of the earlier study.

In a recent review, O'Connor[206] notes that with respect to basic design, procedure, and variables assessed, the teratogenic studies reported to date have been more diverse than decisive. Wide variation in exposure parameters makes it difficult to compare the results; additional difficulty is generated, because many of the reports do not contain information on critical variables such as the manner in which the day of gestation was timed. The day on which the animal is sperm- or sperm-plug positive can be timed as day 0, although a more common procedure is to consider this day 1. The manner in which control animals were treated is often not given. Many kinds of controls have been employed including passive cage controls, sham-exposed controls, heat (IR irradiated) controls, and historic controls. In some reports probability statement have been substituted for data from concurrently studied control animals. Multiple control procedures in single experiments have not been used extensively.

Many of the investigators who have reported defects have employed acute, intense irradiation which obviously has placed a thermal burden on the exposed subject. Some studies attempt to control for heat by including controls heated by means of IR. Of importance in this regard are the comparisons with the literature on heat stress.

Many of the teratogenic studies of RF exposure have been performed without a sufficient number of animals, and others lack sophisticated design and thus do not allow for observation of low probability events. One could argue that any increase in the incidence of fetal damage, regardless of how low, should be considered as a possible biologically significant event even if it is not statistically reliable. While few adequate and rigorous designs have been employed, more critical to the assessment of demonstrated teratogenic potential of MWs is the fact that the statistical analyses are usually not given in enough detail to permit evaluation.[206]

If attention is focused, not on procedural questions, but only on similarities in the results of the studies, several trends are apparent. The most common result from fetal exposure to MWs would appear to be a nonspecific, general response, that of reduced or retarded gain of body mass. Without further study it is impossible to know if this decrease is maintained after birth and when, if ever, the exposed stunted fetuses catch up. It is important to note that this general suppression of body mass is the only effect that appears to be common across the range of species that have been studied.

Another more general deleterious response seen in the mammal is increased rate of fetal resorption. The increase may be indicative of malformed fetuses, but the resorbed nature of the fetal material precludes a more fine-grained analysis and thus identification of which, if any, specific structure was damaged. The increased rate of resorption and the range of exposures within which it occurs is remarkably similar to the effects of heat stress. Particularly in the rat, the resorption rate appears to increase within a rather narrow thermal window, the other side of which is febrile death. The majority of defects has been observed following high level, acute exposures with obvious thermal effect.[206]

It does appear that abnormalities in small animal fetuses can be produced with MWs in conjunction, however, with systemic hyperthermia of 2.5 to 5°C above the normal temperature for the species for some time period during specific critical developmental stages. From a survey of the literature, it appears that it is the temperature rise in the fetus, irrespective of the manner in which it is produced, that causes damage. It is important to realize that in all species there is a constantly evolving pattern of maturation during gestation, and in the rat this continues during the first 3 weeks of postnatal life. In interpreting teratological effects of MWs, as with any agent, it is important to realize that many fetal defects such as hemorrhage, resorption, stillbirth, and exencephaly occur spontaneously in mice.[176]

Thermal stress appears to be the primary mechanism by which RF energy absorption exerts a teratogenic action. Chernovetz et al.[176] and others have pointed out evidence that indicates increases in mortality and resorption are probably related to peak body temperature and its duration regardless of the method by which the temperature elevation is elicited. The teratogenic effects of hyperthermia, regardless of the source, have been well documented.[207-212] Rugh and McManaway[183,184] were able to prevent the increase in incidence of teratogenic activity, which they had previously reported, by lowering the maternal body temperature through controlled use of pentobarbital anesthesia.

Concepts of thermoregulation and thermal stress have been discussed by Michaelson[213] and Way et al.[214] The absence of a core temperature change does not necessarily indicate that physiologic adjustments induced by exposure to RF energy are not heat related.[213] The organism has numerous thermoregulatory mechanisms by which it can maintain homeostasis such as modification of skin blood flow, alterations in peripheral conductance, changes in blood temperature, and alterations in evaporative loss mechanisms.[214] This does not imply that the organism is not stressed. The stress induced by maintaining thermal homeostasis may result in abnormal growth and development in the offspring.

In man, infections such as rubella, influenza, and smallpox, occurring during early pregnancy, are known to cause abortions and fetal malformations.[215,216] In general, it appears that any infection giving rise to fever in early stages of pregnancy in man or animals is capable of producing fetal malformation or abortion. It is well known that induction of fever can lead to the early termination of pregnancy.[217] There are numerous reports of abnormalities from the induction of systemic hyperthermia of 2.5 to 5°C above the normal temperature for the species for 1 hr or longer during specific critical developmental stages of the fetus by exposure of the pregnant animals to elevated temperatures and humidity in environmental chambers. Fetal resorption, growth retardation, microphthalmia, and malformations affecting the CNS, musculo-skeletal system, and other organs have been observed in mammalian species, e.g., the guinea pig[211,218] and the rat.[207,210,219] In experiments on rats, hyperthermia of 4 to 4.5°C for 40 to 60 min during specific developmental stages produces increased fetal resorptions, retardation of growth, microphthalmia, anencephaly, and defects of tails, limbs, toes, palate, and body, depending upon the gestational stage at which hyperthermia occurred.[210] These results indicate that the occurrence of fetal malformations in mammals in early pregnancy is probably related, not as much to the viral or bacterial toxemia, but to the fever, hyperthermia occurring at a particular critical stage of organogenesis. The threshold appears to be an elevation of 2.5 to 5.0°C above the normal temperature of the species sustained for an hour or more.

As noted by Marston and Voronina[220] from the standpoint of public health, one must consider the difficulty of extrapolating data from experimental teratology to the human fetus. Such an extrapolation becomes feasible only after detailed analysis of the fine mechanisms of teratogenesis. Also, of great importance is the need for appropriate scaling factors to permit extrapolation of experimental data obtained on small animals, to the human. The concepts of scaling are discussed in Part III, Chapters 1 and 2.

The reports on effects of MW exposure on early development have been reviewed by Baranski and Czerski,[86] who concluded that no serious effects are to be expected at power densities below 100 W/m² under usual exposure conditions. They further note that defects, when observed, are the result of hyperthermia. There are numerous reports of abnormalities from the induction of systemic hyperthermia of 2.5 to 5°C above the normal temperature for the species, by exposure of the pregnant animal to elevated temperatures at specific critical developmental stages of the fetus.[210,211] It would thus appear that in the reports of MW-induced developmental abnormalities, it is the temperature rise in the fetus, irrespective of the manner in which it was produced, that caused the damage.

RF energies must be applied at high SAR (>15 W/kg) to rodents, approaching lethal levels, for teratogenesis to result. High maternal body temperatures are known to be associated with birth defects. There appears to be a threshold for the induction of experimental birth defects when a maternal colonic temperature of 41 to 42°C is reached. Any agent capable of producing elevated internal temperatures in this range is a potential teratogen.

Studies involving prenatal exposures have not shown effects on growth and development. Temperature in the testes of >45°C induced by any modality can cause permanent sterility; from 37 to 42°C mature sperm may be killed with a temporary loss of spermatogenic epithelium. Changes in reproductive efficiency have not been directly associated with RF exposure. Studies concerning the effects of RF exposure on neonatal development have been summarized[61] and presented in Table 10.

C. Developmental Behavior

Few studies have been reported in which postnatal psychophysiologic parameters have been examined in animals exposed prenatally to MWs. Johnson et al.[180] observed behavioral alterations in rats subsequent to protracted (20 hr daily for 19 days during gestation) prenatal exposure to 918 MHz at a power density level of 50 W/m². Such exposure resulted in a significantly lower level of achievement in the Conditioned Avoidance Response test. Jensh et al.[182] were not able to detect significant behavioral changes associated with protracted 0.915-GHz MWs at 100 W/m² in Wistar rats, a power density level which approached that required to increase maternal colonic temperature. Chernovetz et al.[176] subjected mice to a single 10-min exposure to 2.45-GHz radiation at an SAR of 38 W/kg, without observing any postnatal functioning alterations.

VII. EFFECTS ON THE NERVOUS SYSTEM

Transient changes in CNS function have been reported following MW exposure. Although some reports describe the thermal nature of MW energy absorption, others implicate non-thermal or "specific" MW effects at the molecular and cellular level. The first report on the effect of MW energy in the centimeter range on the conditional response activity of experimental animals was made by Gordon et al.[229] In subsequent years, study of "non-thermal" effects of MWs gradually occupied the central role in electrophysiological studies in the Soviet Union.[230]

Yakovleva et al.[231] reported that single and repeated exposures of rats to MWs, 50 to 150 W/m², weakened the "excitation process" and decreased the "functional mobility" of cells in the cerebral cortex. Edematous changes were most often noted throughout the cortex. The greatest number of altered cells was noted with repeated exposures at 150 W/m².

Table 10
EFFECTS OF RF EXPOSURE ON GROWTH AND DEVELOPMENT

Effects	Animal	Exposure conditions				Ref.
		Frequency (GHz)	Intensity (W/cm²)	SAR (W/kg)	Duration (days × min)	
No effect on weight gain	Rat	2.45 (CW)	50	0.7 — 4.7	55 × 240	191
No effect on growth, neurological and immunological development, or mutagenicity	Rat	0.1 (CW)	460	2.8	112 × 240	223
No change in infant mortality	Monkey	2.45 (CW)		3.4	285 × 180	204, 205
No effect on growth	Mouse	2.45 (CW)	100	6—8 (est)	24 × 48	224
No effect on body weight	Infant rat	2.45 (CW)	400	20—60 (est)	6 × 5	225
No effect on growth	Mouse	0.01, 01019 0.026 (CW)	8.9 × 10⁵	0.9, 1.8, 3.6 (est)	1 × 20	186
	Mouse	0.019	1.7 × 10⁵ 11.4 × 10⁵	6.3 (est)	5 × 40	
No effect on weight gain	Mouse	0.148 (CW)	5	0.013	50 × 60	185
Embryonic LD₅₀	Chicken egg	2.45	2000	70	1 × 12	163, 164
Decreased postnatal survival	Mouse	2.45	2800	98	1 × 7	183, 184
			4000	140	1 × 4	
Teratogenesis	Mouse	2.45		104	1 × 4	166, 167
				85—112	1 × 2—5	
No change in teratogenesis	Mouse	2.45		38	1 × 10	176
Increased postnatal survival	Mouse	2.45		38	1 × 10	176
Maternal lethality, resorptions, decreased fetal weight	Rat	2.45		31	1 × 20	177
Decreased fetal weight	Mouse	2.45	280	22	12 × 100	179
No change hatchability, posthatching hemogram, body or organ weights	Japanese quail egg	2.45		14	1 × 1440	221
No change	Rat	2.45		14	1 × 20	222
	Rat	2.45	400	10	1 × 120	178
	Mouse	2.45	34—140	2—8	12 × 100	179

Table 10 (continued)
EFFECTS OF RF EXPOSURE ON GROWTH AND DEVELOPMENT

Effects	Animal	Exposure conditions				Ref.
		Frequency (GHz)	Intensity (W/cm²)	SAR (W/kg)	Duration (days × min)	
Teratogenesis	Japanese quail egg	2.45		4	12	174
No change	Rat	2.45	50	0.7—4.7	Many × 240	191
	Rat	2.45	280	4.2	12 × 100	226
	Rat	0.915	100—350	1—3.5	Many	182
		2.45				195
		7				
Decreased body and brain weight	Rat	0.918	50	2.5	19 × 1200	227
	Rat	2.45	100	2.2	16 × 300	228

Adapted from U.S. Environmental Protection Agency, Biological Effects of Radiofrequency Radiation, Elder, J. A. and Cahill, D. F., Eds., Rep. No. EPA-600/8-83-026F, Research Triangle Park, N.C., 1984.

Seizure response to noise was transiently suppressed in audiogenic-seizure susceptible mice and rats after exposure to 3000-MHz pulsed MWs at an average power density of 100 W/m². [232,233] Changes in olfactory threshold in humans following occupational exposures have been reported. [234-236]

Tolgskaya et al. [237] studied the effects of pulsed and CW 3- and 10-GHz MWs at various intensities on rats. More pronounced morphological changes in the CNS were reported following 3 than 10 GHz at 10 to 100 W/m². Pulsed waves were considered more effective than CW, a phenomenon also reported by Marha. [238] Additional studies comparing pulsed and CW MWs can be found in the monograph by Tolgskaya and Gordon, [239] who noted that the disturbances MWs can cause in conditioned reflex activity and which characterize functional changes in CNS activity are functional in nature, are reversible, and disappear at the same time as the conditional reflex activity of the animals is being restored upon cessation of exposure to MWs. A proliferative reaction of glial cells was also described.

Some investigators suggest that MW energy absorption may affect hypothalamic and midbrain function and also affect cerebral, cortical, and reticular system function. [240,241] According to Gvozdikova et al., [242] the greatest cortical sensitivity occurs in the meter range, less in the decimeter, and least in the centimeter MW band.

A. Electroencephalographic Changes

Several investigators have reported that MW exposure produces alterations in the electroencephalogram (EEG). [243-250] Baldwin et al. [243] exposed the heads of rhesus monkeys to 0.225 to 0.400 GHz CW in a resonant cavity and noted a progressively generalized slowing and some increase in amplitude of EEG patterns accompanied by signs of agitation, drowsiness, akinesia, and nystagmus, as well as autonomic sensory and motor abnormalities. There were signs of diencephalic and mesencephalic disturbances: alternation of arousal and drowsiness, together with confirming EEG signs. The response depended on orientation of the head in the field and reflections from the surrounding enclosure.

EEG tracings in rabbits exposed to 3-GHz (pulsed) 50 W/m² showed slight desynchronization from the motor region; at 200 W/m² variations in the amplitude were observed; 0.300 GHz had a greater effect than 3 GHz. [251,252] It was suggested that pulsed MWs produced a greater effect than CW MWs. Baranski and Edelwejn [248,253] reported that rabbits exposed to 10 GHz (pulsed), 40 W/m² (single exposure) showed no changes in EEG tracings, but exposure to 3 GHz, 70 W/m² 3 hr/day for 60 days produced functional and morphological changes. Changes in conditional reflexes were also reported in rats exposed to 0.07 GHz, 150 V/m, 60 min/day for 4 months. [254]

A review of the literature on EEG effects requires awareness of certain deficiencies in methodology and interpretation. The EEG is difficult to quantitate due to its time-varying waveform. The use of metallic electrodes either implanted in the brain or attached to the scalp also makes many of the reports on EEG or evoked responses (ER) questionable. Johnson and Guy [255] have pointed out that such metallic electrodes grossly perturb the field and produce greatly enhanced absorption of energy in the vicinity of the electrodes. Such enhancement produces artifacts in the biological preparation under investigation. Recording artifacts also result from pickup of fields by the electrodes and leads during the recording of EEGs while the animal is being exposed. For example, Tyazhelov et al. [256] have pointed out that, even for the coaxial electrode described by Frey, [257] diffraction of EM waves is still a major source of error because of the metallic nature and large dimensions of the electrode.

Bawin et al. [258] reported that electromagnetic energy of 0.147 GHz, amplitude modulated at brain wave frequencies, influenced spontaneous and conditioned EEG patterns in the cat at 10 W/m². These amplitude-modulated 0.147-GHz fields induced changes only when the amplitude modulation frequency approached that of physiological bioelectric function rhythms; no effects were seen at modulation frequencies either below 8 Hz or above 16 Hz.

B. Calcium Efflux

Extremely low frequency (ELF) amplitude-modulated RF fields have been used to study calcium efflux changes in chick and rat brains.[258-260] The entire subject of the calcium efflux studies is discussed in Part III, Chapter 5.

C. Histopathology

According to some authors, cellular changes have been found in the nervous system of small animals following MW exposure at 100 W/m².[239,261,262] Degeneration of neurons in the cerebral cortex and tissue changes in the kidney and myocardium of rabbits have been produced by exposure to 0.2 GHz. Head exposure of rabbits to 2.45 GHz resulted in focal lesions in the cerebral cortex; whole-body exposures of rats to 1.43 GHz produced lesions of the brain.[237,263,264]

Exposure of cats for 1 hr to 10 GHz, 4 kW/m² resulted in injury to cerebral and spinal cord nerve cells; changes occurred in tigroid substance (Nissl bodies) and other components of nerve cells.[265] On the other hand, rabbits exposed to 10 GHz (pulsed), 40 W/m² showed no evidence of morphological damage to the brain, but comparable exposure to 3 GHz did produce such changes.[248]

Tolgskaya and Gordon[237] have investigated the influence of pulsed and CW 3 and 10 GHz on the morphology of nervous tissue in rats and rabbits. With exposure to 3 GHz (1.1 kW/m² and 400 W/m²), severe symptoms of overheating were observed, often leading to death. Severe vascular disorders such as edema and hemorrhages in the brain and internal organs were prominent. In repeated but less prolonged exposure, vascular disorders and degenerative changes in internal organs and the nervous system were less severe. With repeated exposures, the animals were better able to withstand successive exposures; they continued to gain weight, body temperature after irradiation quickly recovered, and temperature increase was not evident. Such a response resembles thermal adaptation on a physiological level.

At high field intensities, when death is a result of hyperthermia, the vascular changes are those of hyperemia, hemorrhage, and acute dystrophic manifestations.[266-268] At lower field intensities, the changes are of a more general dystrophic character, and proliferation of the glia and vascular changes are not as prominent.

Albert and DeSantis[269] reported morphological changes in the brains of Chinese hamsters following exposure to 2.45-GHz CW MWs at power densities of 250 and 500 W/m². Exposure durations varied from 30 min to 14 hr/day for 22 days. Both light and electron microscopic examination revealed alterations in the hypothalamus and subthalamic structures of exposed animals, whereas other regions of the brain appeared unaltered. It should be noted that extremely high SAR could be present under these conditions. Peak SAR could reach 40 to 200 W/kg in selected brain regions.

In subsequent studies at 1.76 and 2.45 GHz, 100 and 200 W/m², Albert[270,271] described cytoplasmic vacuolization of neurons, irregular swelling of axons, and decrease in dendritic spines of cortical neurons. The axonal swelling and spine changes were seen only in chronic exposures, whereas neuronal changes were observed in acute exposure. In all studies no signs of permanent degenerative changes were recorded and reversibility was noted 2 hr after exposure.[271,272] The author concluded that while it was possible that higher exposure levels (250 and 500 W/m²) could have resulted in thermal effects, it was unlikely that 100 W/m² would result in significant thermalization of the whole brain. He did not, however, rule out the possibility of hot spots. Exposure of rats to 100 W/m² of 2.45- or 2.8-GHz RF resulted in average hypothalamic temperature increases of 0.4°C or less. This increase is lower than hypothalamic temperature increases observed during the normal activity of an animal.[273,274]

Albert et al.[275] looked at the effects of RF exposure pre- and postnatally on the Purkinje cells of the rat cerebellum. In one experiment, Sprague-Dawley rats were exposed *in utero* to 100-MHz CW RF at 460 W/m^2 (SAR 2.81 W/kg) for 6 hr/day on GD 16 to 21, and then for 4 hr/day for 97 days after birth. Four exposed and four sham-exposed animals were sacrificed 14 months after cessation of irradiation. Quantitative assessment of the cerebella showed that the relative number of Purkinje cells was significantly smaller (12.7%) in experimental animals than in control animals. In another experiment, Sprague-Dawley rats were exposed *in utero* to 2.45-GHz CW RF at 100 W/m^2 (SAR 2 W/kg) for 21 hr/day on GD 17 to 21. Power density measured was variable from 40 to 300 W/m^2 because of group exposure conditions. Half of the litters were used shortly after delivery and the other half 40 days after cessation of irradiation to assess effects of the exposure on cerebellar Purkinje cells. Because of the immaturity of the neonates, the Purkinje cell layer was not clearly displayed and quantitative results could not be obtained. Experimental animals sacrificed 40 days postexposure showed significantly fewer (25.8%) cells than did the controls. In a final experiment, rat pups were exposed to 2.45-GHz RF at 100 W/m^2 for 7 hr/day on postnatal days 6 to 10. Half the pups were sacrificed at the end of exposure and half at 40 days postexposure. Only those sacrificed immediately showed a significant decrease in the relative number of Purkinje cells as compared with sham-exposed controls. Thus, exposure to two frequencies of RF at similar SAR values (2.8 and 2 W/kg) yielded a reduction in the relative number of Purkinje cells for fetuses and newborn rats. Although the change appeared permanent for rats exposed *in utero*, it appeared to be reversible for those exposed postnatally.

In a related study, Albert et al.[276] examined the effects of RF exposure on the Purkinje cells of squirrel monkey cerebella. Pregnant squirrel monkeys were exposed to 2.45-GHz pulsed RF at an equivalent power density of 100 W/m^2 (SAR 3.4 W/kg) 3 hr/day starting in the first trimester of pregnancy. The offspring were exposed similarly for the first 9.5 months after birth. At the end of the irradiation period, seven exposed and seven sham-exposed animals were sacrificed and their cerebella examined. There were no statistically significant differences between control and exposed animals in any of the Purkinje cell parameters examined. Several factors were suggested by Albert et al.[276] to explain the discrepancy between these results and those with the rat. Factors that might have contributed were differences in geometrical configurations of the head, exposure methods, and daily exposure durations, as well as variations in gestational periods and species differences. It is quite possible that high local SAR values in the rat brain, but not in the squirrel monkey brain, at comparable whole-body SARs may be the important factor in these differences.

D. Effects on the Blood-Brain Barrier

The existence of a blood-brain barrier (BBB) in most regions of the brain has been established experimentally, although its specific morphology is still conjectural. This barrier normally provides resistance to movements of large-molecular-weight, fat-insoluble substances (e.g., proteins or polypeptides) from the blood vessels into the surrounding cerebral extracellular fluid, presumably to protect the brain from invasion by various blood-borne pathogens and toxic substances. Several investigators have reported that RF can increase the permeability of the BBB to certain substances of large molecular weight. However, others were unable to confirm such effects, negating the earlier reports.

Sutton et al.[277] used 2.45 GHz to produce selective hyperthermia of the brain in rats. He then studied the integrity of the BBB with horseradish peroxidase (HRP), a protein tracer that can be detected both morphologically and quantitatively. Brains were heated to 40, 42, and 45°C. Barrier integrity was disrupted after heating for more than 45 min at 40°C. Animals with brains heated to 45°C survived for only 8 to 15 min. The most common site of vascular leakage was the white matter adjacent to the granular cell layer of the cerebellum. Sutton concluded that to prevent BBB disruption, brain temperatures must not exceed 40°C in the absence of body-core hypothermia.

Albert et al.[273] also used HRP as a tracer and reported regions of leakage in the microvasculature of the brains of Chinese hamsters exposed to 2.45-GHz MWs at 100 W/m² for 2 to 8 hr. In control animals, extravascular reaction product was found only in brain regions normally lacking a BBB. In a later paper, Albert[272] reported that continuation of these earlier studies indicated that a partial restoration of the BBBs impermeability may have occurred within 1 hr after exposure ceased, and that restoration was virtually complete within 2 hr. Albert believes that the transient changes may be clinically subacute and probably cause no lasting ill effects. It is important to note, however, that such leakage of the microvasculature of the brain occurs irregularly; this was observed in approximately 50% of exposed and 20% of control animals studied by Albert.[270,272]

In another study, Albert[272] exposed 52 animals (34 Chinese hamsters and 18 rats) to 2.8-GHz RF for 2 hr at 100 W/m². Of these, 30 were euthanized immediately, 11 at 1 hr after exposure, and 11 at 2 hr after exposure. Another 20 animals (12 hamsters and 8 rats) were sham-exposed. Leakage of HRP in some brain regions was reported for 17 of the 30 animals euthanized immediately after RF exposure and for 4 of the 20 sham-exposed animals. Fewer areas of increased BBB permeability were evident for animals euthanized 1 hr after RF exposure, and except for one rat, virtually no leakage of HRP was seen for the animals euthanized 2 hr after RF exposure. These results indicate that increased BBB permeability due to RF exposure at levels insufficient to denature brain tissue is a reversible effect. (Albert suggests that such BBB changes may be clinically subacute and would probably cause no lasting ill effects.) However, the increased BBB permeability seen in 4 of the 20 sham-exposed animals may indicate that factors other than RFR in the experimental procedure could alter the BBB. One possible confounding point in the use of injected HRP as a tracer is the existence of endogenous peroxidase, the detection of which could yield false-positive results. Also, no positive (BBB-altering) control agent was used in these studies for comparative purposes.

Comparable results were reported by Albert and Kerns[278] for 51 Chinese hamsters exposed to 2.45-GHz CW RF for 2 hr at 100 W/m². There were 39 original sham-irradiated controls. Of the RF-exposed animals, 12 were allowed to recover for either 1 or 2 hr prior to HRP injection and subsequent fixation. This study appears to be an extension (for Chinese hamsters) of the work reported by Albert[272] for 2.8-GHz RF exposure, with the same conclusions as that paper.

Oscar and Hawkins[279] reported that single exposure of rats to 1.3-GHz pulsed MWs for 20 min (1 to 20 W/m²) induces a temporary change in the permeability of small inert polar molecules across the BBB of rats. Increases in permeability were observed for mannitol and inulin, but not for dextran, both immediately and 4 hr after exposure, but not 24 hr after exposure. They found statistically significant changes in the brain uptake index (BUI) at average power densities less than 30 W/m². They also found that, depending on the specific pulse characteristics used, pulsed RF could be more or less effective in altering BBB permeability than CW RF of the same average power density. For pulses of long duration and high pulse power density, but only a few pulses per second, mannitol permeation could be induced at average power densities as low as 0.3 W/m². The authors consider the possibility of local heating due to hot spots since the greatest BBB alteration occurs in the cerebellum and medulla or close to the neck region of rats.

Because the Oldendorf technique[280] which was used does not permit discriminating between local cerebral blood flow (LCBF) and small BBB permeability changes, Oscar et al.[281] used a different technique utilizing ¹⁴C-iodoantipyrine to measure LCBF in rats. Male rats weighing 250 to 300 g were individually sham-exposed for 5, 30, and 60 min, or individually exposed for 5, 15, 30, 45, or 60 min to pulsed RF at 150 W/m² average power density. Carrier frequency was 2.8 GHz, pulse rate was 500 pps, and pulse width was 2 μsec. Within 5 min after sham or RF exposure, the previously venous-catheterized, but conscious, animals were infused with isotonic saline containing 5 μCi/mℓ of ¹⁴C-iodoantipyrine. After start of infusion (50 sec), the rats were decapitated. Brain regions were dissected out and assayed for radioactivity by routine liquid scintillation counting. Local cerebral blood flow was then calculated by established procedures. The results showed that MW exposure caused a significant increase in LCBF (minimum 39%, some well over 100%) in all 17 brain regions sampled. Because of these findings, the authors indicated that their earlier reported ratio measurements[279] may be overly high. However, the ^{14}C-iodoantipyrine results demonstrated an alteration of brain activity at 150 W/m^2. A direct heating effect on LCBF and not the BBB per se is thus indicated.

Preston et al.,[282] using methods similar to those of Oscar and Hawkins,[279] attempted to determine whether exposure to 2.45-GHz CW RF increased BBB permeability to ^{14}C-labeled D-mannitol. They exposed rats to 1, 5, 10, 50, or 100 W/m^2, with sham-exposed rats for controls, and found no evidence that RFR exposure increased the permeability of the BBB for mannitol. In a second series, rats were exposed to 3, 10, 30, 100, and 300 W/m^2. Again, there were no differences between results from exposed and sham-exposed animals. Like Oscar et al.,[281] Preston et al.[282] believed changes in LCBF confounded the results of earlier studies.

Chang et al.[283] used a technique involving ^{131}I-labeled albumin to investigate alterations of the BBB in dogs. The heads of dogs were exposed to various average power densities between 20 and 2000 W/m^2. In general, no statistically significant differences were found between exposed and sham-exposed animals, but the number of animals used in this study was too small to ascribe a high level of statistical confidence.

Attempts to duplicate the findings of Oscar and Hawkins[279] have yielded equivocal results[283] or failure.[282-289] Merritt et al.[289] exposed rats for 30 min to 1.2-GHz pulsed RF at peak power densities in the range from 20 to 750 W/m^2 and 0.5 duty cycle, corresponding to average power densities of 10 to 380 W/m^2, or for 35 min to 1.3-GHz pulsed or CW RF at average power densities in the range from 1 to 200 W/m^2. They examined brain slices under UV light for transfer of fluorescein and under white light for transfer of Evans blue dye (a visual tracer) across the BBB, and chemically analyzed various brain regions for fluorescein content. They also measured the brain uptake of ^{14}C-labeled D-mannitol and determined the BUI values. To validate these detection methods, they used hypertonic urea, known to alter the BBB, as an alternative agent to RF. Last, sham-exposed rats were heated for 30 min in a 43°C oven to approximate the hyperthermia obtained at 380 W/m^2. In their examination of brain slices, Merritt et al.[289] found no evidence of fluorescein or Evans blue dye transfer across the BBB of RF-exposed rats, whereas penetration of the BBB was apparent for rats treated with urea instead of RF. The analyses of fluroescein content corroborated these findings. However, fluorescein uptake was higher for the sham-exposed rats that were heated in the oven, an indication that hyperthermia of the brain is necessary to alter BBB permeability. In the ^{14}C-mannitol study of the various brain regions, there were no significant differences in BUI between RF- and sham-exposed rats, whereas BUI changes were evident for rats treated with urea. Also, the results showed no evidence of the power density window reported by Oscar and Hawkins.[279] No change in the BBB occurred unless brain temperature was increased by 4°C.

By using a small, dielectrically loaded coaxial applicator, Lin and Lin[290,291] were able to irradiate only the heads of anesthetized adult male Wistar rats with pulsed 2.45-GHz RF (10 μsec, 500 pps) for 20 min at average power densities of 5 W to 30 kW/m^2. The distribution of absorbed MW energy inside the head was determined by thermographic procedures, and average SARs were found to range from 0.04 to 240 W/kg. Evans blue dye was injected into a catheterized femoral vein following sham or RFR exposure, and 5 min later the animal was perfused via the left ventricle with normal saline. The brain was removed, examined, and scored for degree of tissue staining by the tracer. For average power densities up to and including 26 kW/m^2 (200 W/kg), staining was not significantly different between exposed and control animals. For exposures at 30 kW/m² (240 W/kg), extravasation of Evans blue dye could be seen in the cortex, hippocampus, and midbrain. The degree of staining decreased with increasing time to euthanasia postexposure, indicating that the effect was reversible. No alteration in the BBB was evident unless brain temperature was increased by 4°C.

It is important to realize that the methods used to investigate BBB permeability are still controversial. Permeability changes in cerebral blood vessels occur under various conditions, including those that produce heat necrosis.[292] Most techniques used to measure BBB permeability in fact measure the net influence of several variables on brain uptake, and do not differentiate among the effects of changes in the vascular space, alterations of blood flow, and membrane permeability.[279]

The uncertainty in most earlier research on the influence of RF on the BBB is no doubt related to significant artifacts introduced by the kinds of biological techniques used. Several investigators have indicated that exposure to RF may alter the size of vascular and extra-vascular volumes and cerebral blood flow rate, thereby yielding changes in the BUI that are not necessarily related to BBB permeability alterations. Blasberg[293] reviewed many of the methods previously used for investigating BBB changes and the problems associated with these methods. Rapoport et al.[294] developed a method for measuring cerebrovascular permeability to ¹⁴C-labeled sucrose that yields results independent of cerebral blood flow rate. As already noted, Oscar et al.[281] confirmed experimentally that LCBF is increased in the rat brain by exposure to pulsed RFR at 150 W/m² average power density.

Recent findings such as those mentioned above indicate that little quantitative confidence can be placed in the results of early experiments on RF-induced BBB alterations. Qualitatively, hyperthermic levels of RFR clearly can alter the permeability of the BBB. Reviews of this subject have been published by Justesen[295] and Williams and associates.[284-288]

E. Influence of Drugs

A pharmacodynamic approach has been taken by some investigators in the study of MW exposure effects on the CNS and behavior.[296-298] Following exposure to 10-cm pulsed MWs altered sensitivity to neurotropic drugs was noted. Decreased tolerance of rabbits to pentylenetetrazole and increased tolerance to strychnine were observed after a single exposure to 200 W/m². Repeated exposures at 70 W/m² produced a decreased tolerance to pentylenetetrazole, strychnine, and acetophenetidin.[291]

Servantie et al.[299] reported that exposure of rats to 3 GHz at 50 W/m² for several days resulted in an altered reaction to pentylenetetrazole. Using curare-like compounds, a neuromuscular site of action for this MW effect was implicated. Edelwejn[297] observed alterations in the effects of chlorpromazine and/or D-tubocurarine on EEG recordings in rabbits repeatedly exposed to 70 W/m². The author concluded that synaptic structures at the level of the brain stem are affected by MWs. It should be noted that drugs such as pentylenetetrazole and chlorpromazine per se influence physiologic thermoregulation.

A summary of the reports of RF-induced effects on neural tissues and the nervous system in general is presented in Table 11.[61]

VIII. BEHAVIORAL EFFECTS

Studies have been conducted on the effects of RF on performance of trained tasks or operant behavior by rats, rhesus, and squirrel monkeys.[316-328] All of the studies indicated that the exposure would result in suppressed performance of the trained task, and that an energy power density/dose threshold for achieving the suppression existed. Depending on duration and other parameters of exposure, the threshold power density for affecting trained behavior ranged between 50 and 500 W/m².

Justesen and King[322] utilized a 2.45-GHz (CW) multimodal resonating cavity to investigate

Table 11
EFFECTS OF RF ON THE NERVOUS SYSTEM

Effects	Species	Frequency (GHz)	Intensity (W/m²)	SAR (W/kg)	Duration (days × min)	Ref.
Desynchronized EEG	Rabbits	3 (PW)	200 (av)	3.0 (est)	1 × 20	296
Greater effect of CNS stimulating drugs	Rabbits	3 (PW)	70 (av)	1.0 (est)	24—26 × 180	296
Biphasic effect of latency to a convulsive drug effect	Mice	3 (PW)	50	5	8—36 × Unknown	299
Decreased effect of paralyzing drugs	Rats	3 (PW)	50	1	10—15 × Unknown	299
Changes in EEG patterns of unanesthetized animals	Rabbits	9.3 (CW)	7—28	0.1—0.3 (est)	1 × 5	300
Potentiation of drug response	Male rats	2.45 (PW)	10 (av)	0.2 (est)	1 × 30	301, 302
Decreased hypothalamic NE, DA, and hippocampal serotonin in hyperthermic animals	Rats	1.6 (CW)	800	24 (est)	1 × 10	303
Decreased hypothalamic NE, DA	Rats	1.6 (CW)	200, 800	6—24 (est)	1 × 10	304
No effect on neurotransmitter levels	Rats	1.6 (CW)	100	3.0 (est)	1 × 10	304
No effect on GABA content	Rats	2.86 (PW)	800	16.0 (est)	1 × 5	305
			400	8.0 (est)	1 × 20	
			100	2.0 (est)	5 × 480	
			100		40 × 240	
Swollen neurons in hypothalamus and subthalamus	Chinese hamsters	2.45 (CW)	500 / 250	15 (est) / 7.5	1 × 30	269
Swollen neurons in hypothalamus and subthalamus	Chinese hamsters	1.7 (CW)	100	3 (est)	22 × 9840	306
Myelin figures in dendrites 6 weeks postexposure	Female rats	2.45 (CW; multimodal cavity)	100	2.3 (est)	110 × 300	307
Increased permeability of BBB to fluorescein	Rats	1.2 (CW)	24	1.0 (est)	1 × 30	308
		1.2 (PW)	2 (av)	0.08 (est)	1 × 30	

Table 11 (continued)
EFFECTS OF RF ON THE NERVOUS SYSTEM

Effects	Species	Exposure conditions				Ref.
		Frequency (GHz)	Intensity (W/m²)	SAR (W/kg)	Duration (days × min)	
Myelin degeneration and metabolic alterations; glial cell proliferation	Guinea pigs, rabbits	3 (CW; PW) 3 (CW; PW) 3 (CW; PW)	35 250 50	0.5 (est) 3.5 0.4 (est)	90 × 180 1 × 180 90 × 180	309
Focal areas of increased BBB permeability to peroxidase	Chinese hamsters, rats	2.8 (CW) 2.8 (CW)	100	1.9 0.9	1 × 120	272
Increased peroxidase in brain, absent after recovery period	Chinese hamsters	2.45 (CW)	100	2.5	1 × 120	278
Increased peroxidase in brain	Chinese hamsters	2.45 (CW)	100	2.5	1 × 120	270
Brain temperature elevation (40—45°C); increased permeability of BBB	Rats	2.45 (CW)	(80 W)	—	1 × 10—30	310
Increased permeability of BBB (mannitol and inulin)	Rats	1.3 (CW) 1.3 (CW)	10 3 (av)	0.4 0.1	1 × 20 1 × 20	279
EEG effects seen after chronic, but not acute, exposures	Rabbit	0.001—0.01 (AM) 0.001—0.01 (AM)	(60—500) (V_{rms}/m) (90—500) (V_{rms}/m)	10^{-5}—10^{-4} 10^{-4}—10^{-3}	1 × 120—180 20—30 × 120—180	311
Change of predominant EEG frequencies	Cat	0.147 (AM)	10	0.015 (est)	Varying	312
Reversible neuronal morphology alterations	Rat	3	100	2 (est)	35 × 30	313
	Rat	3	100	2 (est)	35 × 30	239
Milk pyknosis of hippocampal neurons, increased brain, and colonic temperature	Rat	2.45	(60—90 W)	Head only	1 × 2.5—7	314
Increased brain serotonin turnover rate	Rat	3	400	8.0 (est)	1 × 60)	315
	Rat	3	100	2.0 (est)	7 × 480	315

No decrease in cerebellar Purkinje cells in offspring	Squirrel monkey	2.45	100	3.4	368 × 180	276
Decreased cerebellar Purkinje cells after perinatal exposure	Rat	2.45 0.100	100 460	2.0 2.7	5 × 1260 110 × 240	275

Note: AM = amplitude modulation, NE = norepinephrine, DA = dopamine, BBB = blood-brain barrier, EEG = electroencelphalogram, CNA = central nervous system, GABA = gamma-aminobutyric acid.

Adapted from U.S. Environmental Protection Agency, Biological Effects of Radiofrequency Radiation, Elder, J. A. and Cahill, D. F., Eds., Rep. No. EPA-600/8-83-026F, Research Triangle Park, N.C., 1984.

conditioned operant behavior in rats. They used a recurrent cycle of exposure, 5 min on and 5 min off, over a 60-min period at average absorbed energy rates of 3.0, 6.2, and 9.2 W/kg. The performance of the animal usually stopped near the end of the 60-min test period during exposure with an energy absorption rate of 6.2 W/kg; at 9.2 W/kg, this effect occurred much earlier in the test period. Hunt et al.,[323] also using a multimodal resonating cavity to expose rats to 2.45 GHz (pulsed), found effects on exploratory activity, swimming, and discrimination performance of vigilance task, after a 30-min exposure at about 6 W/kg.

Lobanova[324] exposed rats to 3-GHz pulsed MWs, after which the rats were tested for swimming time. A decrease in endurance was noted after exposure to power-time combinations ranging from 1 kW/m² for 5 min to 100 W/m² for 90 min.

Lin et al.[329] exposed rats to 0.918 GHz (CW) at levels of 100, 200, or 400 W/m² for 30 min. No effects on response rates were noted at the two lower levels, but at 400 W/m² the performance of the animal decreased after 5 min of exposure and ceased after about 15 min of exposure. The average energy absorption rate measured thermographically was 2.1 W/kg/W/m² incident or 8.4 W/kg absorbed at 400 W/m².

Diachenko and Milroy[330] studied the effects of pulsed and CW MWs on an operant behavior in rats trained to perform a lever-pressing response on a DRL (Differential Reinforcement of Low Rate) schedule and tested immediately after 1 hr daily exposure to 10, 50, 100, and 150 W/m², 2.45-GHz MWs. No behavioral effects were found at these levels. The subjects exposed to 100 W/m², while showing no significant decrement in performance, did show signs of heat stress.

Thomas et al.[331] reported response-rate changes in rats exposed between 50 and 200 W/m² to 2.86-GHz (CW) and 9.6-GHz (pulsed) MWs. Response rates increased in five of ten tests. Decrements in timing of responses were suggested as indicating that "low-level" MWs produce effects on the CNS.

Galloway[321] studied the performance of four monkeys when using a 2.45-GHz (CW) waveguide applicator (total absorbed power of 10, 15, or 25 W) applied to the head. The duration of the irradiation was 2 min or until convulsions began (20 and 25 W produced convulsions). Because of skin burns, only two subjects completed this series of experiments. There were no performance decrements in a discrimination task that the subjects performed immediately after exposure. Acquisition of a new task during the first ten trials of training was impaired at 25 W, which resulted in convulsions.

Roberti et al.[332] measured the running time of rats in an electrifiable runway in which each subject was trained to peak performance. Exposure for 185 hr at 10.7 GHz (CW), 3 GHz (CW), and 3 GHz pulsed, to 10 W/m², caused no performance decrements. No change in baseline performance was noted when rats were irradiated with 3 GHz pulsed for 17 days at a power density of 250 W/m².

Thomas and Maitland[298] investigated the effects of pulsed 2.45-GHz MWs, 1 W/m² in combination with dextroamphetamine on behavior in rats. Both acute and repeated exposures modified the normal dose-effect function. The maximum drug effect was obtained at lower MW exposures as the drug dose was increased.

Galloway and Waxler[333] employed a serotonin-depleting drug (fenfluramine) to investigate the effect of 2.45-GHz CW RF applied to the head of 7- to 9-kg rhesus monkeys at integral dose rates of 1 to 15 W. Combinations of the drug and RF at an integral dose rate of 15 W resulted in behavioral deficits, whereas the drug or RF alone up to 15 W failed to produce this effect. In respect to drugs such as dextroamphetamine and fenfluramine, their influence on thermal regulation may be significant in these results.

In contrast to the previous studies in which postexposure behavior was measured in animals treated with neuroactive drugs, deLorge[334] exposed rats, squirrel monkeys, and rhesus monkeys to 2.45-GHz MWs under far-field conditions while the animals were performing on operant schedules for food reinforcement. Exposure sessions lasted 60 min and were repeated on a daily basis. Stable performance on the operant schedules was disrupted in all three species at power densities positively correlated with the body mass of the animals. The behavior of the squirrel monkeys was disrupted at the middle level of power densities. When the averages of these power densities (280, 450, and 670 W/m²) are plotted as a function of body mass (0.3, 0.7, and 5 kg) a semilog relationship becomes evident. Extrapolation along the resulting curve could permit prediction of the power densities needed to disrupt ongoing operant behavior in larger animals. The power densities associated with behavioral disruption approximated those power densities that produced an increase in colonic temperature of at least 1°C above control levels in the corresponding animals. These data support the need for scaling factors to extrapolate from small animals to larger animals.

Most of the research on the nervous system and behavior has been carried out in rodents and other lower animals. Behavior among animal species reflects adaptive brain-behavior patterns. Behavioral thermal regulation is seen as an attempt to maintain a nearly constant internal thermal environment. Changes in body temperature bring about not only autonomic drives, but also behavioral drives.[335] That MWs can influence behavioral thermoregulation has been shown by Stern et al.[336] and Adair.[337] This subject is treated in greater detail in Part III, Chapter 3.

Behavioral responses are not necessarily manifestations of specific changes in the CNS and may be a function of direct or indirect action of MWs on other body systems. Extrapolation of brain-behavior functions from lower animals to man is thus subject to many difficulties.

In assessing the significance of the reported behavioral changes, it is important to recognize certain fundamental factors. The resting metabolic rate for rats is approximately 7 W/kg. When this level is exceeded, disruption of behavior could be elicited. In most of the reports, alterations in the behavior of rats were observed with exposures at average energy absorption rates of 5 to 8 W/kg or greater, i.e., at similar exposure levels to those that produce increases in circulating corticosterone (CS) concentrations in rats.[338] Behavioral changes may be related to more subtle heat alterations within the body. Heat may produce a general debilitation effect or a decreased motivation for food, since it has been shown that rats maintained in hot environments eat less food[339] and rats show decreased response and food reinforcement frequency on an operant schedule when environmental temperature is 35°C, but not 25°C.[340] Behavioral responses may be influenced by the interaction of the organism with the environment.

A. Behavioral Thermoregulation

The regulation of body temperature can be accomplished by complex patterns of responses of the skeletal musculature to heat and cold which modify the rates of heat production and/or heat loss (e.g., by exercise, change in body conformation, in the thermal insulation of bedding and [in man] of clothing, and by the selection of an environment which reduces thermal stress)[341] (see also Part III, Chapter 3).

Thermoregulation is part of a complex control system involving circulation, metabolism, and respiration, as well as neural structures. Temperature signals from cutaneous thermoreceptors reach the somatosensory region of the cerebral cortex. The main processing of thermal signals and generation of a controlling signal for the effector part of thermoregulation takes place in the hypothalamus. Adair et al.[342] found minimal changes in hypothalamic temperature in conjunction with alteration of thermoregulatory behavior in squirrel monkeys exposed to 2.45-GHz, RF, 40 to 200 W/m² (SAR 0.2 to 1 W/kg).

Successful strategies or mechanisms to maintain body temperatures in a narrow and desirable range in a complex and varying thermal environment are termed thermoregulatory. Although such strategies or mechanisms are found in great variety, they fall into two main categories: voluntary and behavioral adjustments, and involuntary physiological adjustments.

The limits of effectiveness of involuntary physiological thermoregulation are rather narrow and we must rely on behavioral methods of thermoregulation over most of the range of environmental temperatures to which we are often exposed. Changes in body temperatures bring about not only autonomic drives, but also behavioral drives.[335]

Thermal motivation arises in situations of thermal stress. On the warm side, it is the uncomfortable feeling of excessive warmth and the desire for temperature reduction; on the cold side, it involves he unpleasant feeling of being too cold and the desire for temperature increase. The biological significance of thermal motivation is that, by acting in such a way as to minimize thermal discomfort and maximize thermal comfort, the organism tends to escape from situations of thermal stress and to locate itself in a physiologically neutral thermal environment, thereby solving the problem of physiological temperature regulation.[343]

Smialowicz et al.[344] made unrestrained, unanesthetized mice hypothermic by injecting them with 5-hydroxytryptamine (5-HT) in a controlled environment of 22°C and 50% relative humidity. Mice injected with saline were used as controls. Colonic temperatures were measured prior to injection. Following injection, groups of mice were exposed to 2.45-GHz CW RF for 15 min at 100, 50, or 10 W/m² (equivalent SARs 7.2, 3.6, and 0.7 W/kg) or were sham exposed, after which their colonic temperatures were measured again. The experiments were performed with BALB/c and CBA/J mice. For saline-injected mice of either type, there were no significant colonic-temperature differences between mice exposed at 100 W/m² and sham-exposed mice. For BALB/c mice made hypothermic with 5-HT, colonic temperatures were significantly higher for those exposed at all three power densities than those sham exposed, and the differences increased monotonically with power density. The results for the CBA/J mice were similar, but the increases were statistically significant only at 50 and 100 W/m². The investigators conclude that subtle heating by RF can alter the thermoregulatory capacity of hypothermic mice, whereas the colonic temperature of normal (saline-injected) mice was not significantly altered by exposure at 100 W/m².

Adair and Adams[337] exposed squirrel monkeys to 2.45-GHz CW RF for 10 or 90 min in ambient temperatures of 15, 20, or 25°C. The power densities ranged from 25 to 100 W/m² (SARs 0.4 to 1.5 W/kg). Skin and colonic temperatures were monitored continuously during exposure. The metabolic heat production was calculated from the oxygen deficit in the expired air of each monkey. At all three ambient temperatures, 10-min exposure of two monkeys to a threshold power density of 40 W/m² and one monkey to 60 W/m² reliably initiated a reduction of their metabolic heat production, and the magnitudes of the reduction were linear functions of the power density above the threshold values. At exposure termination, the metabolic heat production often rebounded sharply and overshot normal levels. For the 90-min exposures at 20°C, the initially large reduction of metabolic heat production gradually diminished towards normal levels, apparently to ensure precise regulation of internal body temperature at the normal value.

Adair[345] also exposed four squirrel monkeys to 2.45-GHz CW RF in warm ambient temperatures ranging from 32 to 35°C. After an initial 90-min or longer equilibration period, each monkey was exposed for 10-min periods to power densities in an increasing sequence from 25 to 200 W/m², with sufficient time between exposures for reestablishment of equilibrium. The colonic temperature and the skin temperature at the abdomen, tail, leg, and foot were monitored continuously. As in the previous investigation, the metabolic heat production was determined from the oxygen deficit in the expired air. In addition, thermoregulatory sweating from the foot was determined by sensing the dewpoint of the air in a special boot over the foot. The results indicate that at ambient temperatures below about 36°C, at which sweating in a sedentary monkey may occur spontaneously, the threshold power density (or SAR) for initiating thermoregulatory sweating from the foot decreased with decreasing ambient temperature.

Adair and Adams[346] equilibrated squirrel monkeys for a minimum of 2 hr to constant environmental temperatures (22 to 26.5°C), cool enough to ensure that the cutaneous blood vessels in the tail and extremities were fully vasoconstricted (an effect produced by the thermoregulatory system to minimize heat loss). The monkeys then underwent 5-min exposures to 2.45-GHz RF at successively higher power densities, starting at 25 to 40 W/m², until vasodilation in the tail occurred, as evidenced by an abrupt and rapid temperature increase in the tail skin. For example, a monkey equilibrated to 25°C exhibited tail vasodilation when exposed to RF at 100 W/m² (whole-body SAR 1.5 W/kg), whereas it did not when exposed to IR radiation at the equivalent power density, an indication that the effect resulted from stimulation of thermosensitive elements of the thermoregulatory system by the RF rather than from heating of the tail skin. RF exposure at higher power densities was required to cause tail vasodilation in monkeys equilibrated to lower environmental temperatures. Specifically, an increase of 30 to 40 W/m² was found necessary for every 1°C reduction in environmental temperature.

Behavioral reactions to RF exposure are summarized in Table 12.

IX. NEUROENDOCRINE EFFECTS

A. Mechanisms of Interaction

Response of the endocrine system of rats to whole-body exposure to MWs has been studied in recent years in great detail, most of the reports are based on relatively short exposures at modest- to high-power densities.[225,372-377] A comprehensive review of neuroendocrine responses to MW exposure has been presented by Lu et al.[378]

Some investigators believe that endocrine changes result from stimulation of the hypothalamic-hypophysial system due to thermal interactions at the hypothalamic or immediately adjacent levels of organization, the hypophysis itself (pituitary), or the particular endocrine gland or end-organ under study. According to other investigators, the observed changes have been interpreted as being the result of direct MW interactions with the CNS. In any case, one cannot consider neuroendocrine perturbations as necessarily pathologic, because the function of the neuroendocrine system is to maintain homeostasis and hormone levels will fluctuate to maintain such organismic stability.

Effects of RF exposure on endocrine function are generally consistent with both immediate and long-term responses to thermal input and to nonspecific stress, which can also arise from thermal loading. Changes found in plasma levels of CS and growth hormone are typical reactions of animals to nonspecific stress; indeed, great care is required in performing experiments to ensure that the changes in hormone level do not result from stress caused by handling of the animals or the novelty of the experimental situation.

B. Hypothalamic-Hypophysial-Adrenal Response

Several investigators have reported biochemical and physiological changes as a result of MW exposure which suggest an adrenal effect. According to Petrov and Syngayevskaya,[379] 3 and 24 hr after dogs were exposed to 3 GHz, 100 W/m², the serum corticosteroid content had increased by 100 to 150% above the original level. Serum potassium was decreased by 5 to 10% and sodium was increased by the same amount. They also noted that the susceptibility of rats to MW exposure was sharply increased 1 week after bilateral adrenalectomy. Chronic exposure of animals to MWs (CW or pulsed) was accompanied by reduced cholinesterase activity and an increased amount of 17-ketosteroids in the urine, reduced ascorbic acid, and reduced weight of the adrenal glands.

Demokidova[380] reported increased adrenal and pituitary gland weight in rats exposed 1.5 months to 0.0697 GHz, 12 V/m, 1 hr/day and increased adrenal weight in infant rats exposed to 48 V/m, 4 hr/day. The same author, however, reported decreased weight of the adrenal glands in infant rats exposed to 0.01488 GHz at 70 V/m.

Table 12
BEHAVIORAL REACTIONS TO RF EXPOSURE

Effects	Species	Exposure conditions				Ref.
		Frequency (GHz)	Intensity (W/m²)	SAR (W/kg)	Duration (days × min)	
Decreased exploratory activity and swimming speed, ΔT increase 2.5°C	Rat	2.45 (PW, multi-)modal cavity 120 Hz, AM	?	6.3	1 × 30	323
No effect on spontaneous activity nor forced running	Rat	10.7 (CW) / 3 (CW) / 3 (PW) / 3 (PW)	6—9 / 5—10 / 15—20 / 240—260 (av)	0.2 / 0.3 / 0.6 / 8.3	7.7 × 1440 / 7.7 × 1440 / 7.7 × 1440 / 17 × 1440	332
Increased locomotor activity	Rat	2.45 (CW, multimodal cavity)	100	2.3	110 × 300	347
Decreased spontaneous activity and food intake	Rat	0.918 (CW)	100	3.6	21 × 600	348
No effect on spontaneous activity nor food intake	Rat	0.918 (CW)	25	1.0	91 × 600	125
Decreased activity on stabilimetric platform, no significant increase in wheel running	Rat	2.45 (CW)	50	1.2	80 × 480	349
Increased activity on stabilimetric platform and in wheel running	Rat	0.915 (CW)	50	2.5	80 × 480	350
Decreased time on treadmill and inclined rod, decreased exploratory activity, increased then decreased shock sensitivity; decreased activity and shock sensitivity persisted 90 days after exposure	Rat	2.375 (CW)	5	0.1	30 × 420	351
Colonic temperature rise = 0.37°C before start of test; ΔT = 1.5°C with microwaves	Rat	2.45 (PW, multi-)modal cavity, 60 and 12 Hz AM)		420 / 220	10 × 0.17 / 10 × 0.5	352
Response decreased during exposure on random interval schedule (lowest intensity for effect, ΔT = 1.8°C	Rat	0.500 (CW)	250	10	1 × 11	353

Effect	Species	Frequency GHz (waveform)	Power density	SAR	Exposure	Ref.
Response decreased during exposure (maximum effect) on random interval schedule, ΔT = 1.8°C	Rat	0.600 (CW)	100	7.5	1 × 55	318
Decreased observing responses on vigilance task, ΔT = 2°C	Rhesus monkey	2.45 (120 Hz, AM)	720	5.0	1 × 60	354
No effect on observing responses	Squirrel monkey	2.45 (120 Hz, AM)	160	1.1	1 × 20	355
Decreased observing responses on vigilance task			500	2.8	1 × 30	356
No effect on observing responses	Rat	1.28 (PW)	100	0.6 — 1.7	1 × 60	357
Decreased observing responses on vigilance task	Rat	5.62 (PW)	260	2.5 / 4.9	1 × 60 / 1 × 40 / 1 × 4	
Response rate decreased on fixed interval schedule in rats with high base-line rates, spending time away from lever	Rat	2.45 (120 Hz, AM)	375	7.5	1 × 60	316
No effect on response rate	Rhesus monkey	1.2 (CW)	88—184	1.8—3.7	1 × 60	358
No effect on visual tracking task			200	1.6	1 × 120	358
Response rate decreased on FR and increased on DRL schedules	Rat (120 days)	2.45 (CW) / 2.86 (PW) / 9.60 (PW)	50 / 50 / 50	1.4 / 1.4 / 1.5	1 × 30 / 1 × 30 / 1 × 30	331
Decreased length of runs and fewer reinforcers on FCN schedule	Rat	2.45 (CW)	50	?	1 × 30	359
Decreased response rate on FR operant schedule	Rat	2.45 (CW)	100	2.7	1 × 900	320
Increased rate of missing observing responses on vigilance task	Young Rat	2.45 (PW, AM, multimodal cavity)		6.5	1 × 30	323
Decreased rate of responding on repeated acquisition task	Rat	2.80 (PW)	50	0.7	1 × 30	360
Increased response rates in extinction, decreased stimulus control, no effect on Sidman avoidance	Rat	2.45 (CW, multimodal cavity)	100	2.3	110 × 300	347
No effect on flavor aversion test	Rat	0.918 (CW)	100	3.9	21 × 600	348
	Rat	0.918 (CW)	25	1.0	91 × 600	125
Microwaves detected as stimulus	Rat	2.45 (PW 120 Hz, AM, multimodal cavity)		0.6—2.4	1 × 1	361
	Rat	0.918 (PW)	150	7.5	1 × 0.5	362

Table 12 (continued)
BEHAVIORAL REACTIONS TO RF EXPOSURE

Effects	Species	Exposure conditions				Ref.
		Frequency (GHz)	Intensity (W/m^2)	SAR (W/kg)	Duration (days × min)	
Spending more time in shielded vs. unshielded compartment	Rat	1.20 (CW)	2	0.2	4 × 30	363
Spending equal time in shielded vs. unshielded compartment	Rat	1.20 (CW)	24	2.2	3 × 30	363
Spending more time in shielded vs. unshielded compartment (occurred in first of seven sessions)	Rat	1.20 (PW)	4	0.4	1 × 90	308
Spending more time in unirradiated compartment	Rat	2.80 (PW)	95	2.1	9 × 60	364
Decrease in SAR at 24°C	Mouse	2.45 (CW)		28	1 × 15	365
Decrease in SAR when ambient temperature increased from 20 to 35°C	Mouse	2.45 (CW)		43.6—0.6	1 × 20	366
No preferential orientation of rats or mice in far field of plane wave	Rat (200—360) Mouse	2.45 (CW) 2.45 (CW)	150 150	3.3 6.2—12.3 (Depending on orientation)	1 × 50 1 × 60	367
Cannot take specific action to reduce intensity of irradiation	Rat	0.918 (PW, 60 Hz, AM, multimodal cavity)		60	5 × 2	368
Augmentation of increased response rates produced by chlordiazepoxide	Rat	2.45 (PW)	10	0.2	1 × 30	301
Shift to left of dose-response curve for d-amphetamine in DRL schedule	Rat	2.45 (PW)	10 10	0.2 0.2	1 × 30 4 × 30	298
No effect on dose response curve for chlorpromazine or diazepam	Rat	3.8 (PW)	10	0.2	1 × 30	302
Chlordiazepoxide reduced responses, decreased avoidance responses, and increased escape responses to microwaves	Mouse	2.45 (CW)		46	1 × 30	369

Response rate decreases were augmented after exposures at higher ambient temperatures	Rat	2.45 (CW)	100	2.0	1 × 930	370
Reduced responding for heat lamp in a cold room	Rat	2.45 (CW)	50	1.0	1 × 15	336
Selection of a lower ambient air temperature	Squirrel monkey	2.45 (CW)	60	1.0	1 × 10	371

Note: If measured SAR was not reported, SAR was estimated when possible.

Adapted from U.S. Environmental Protection Agency, Biological Effects of Radiofrequency Radiation, Elder, J. A. and Cahill, D. F., Eds., Rep. No. EPA-600/8-83-026F, Research Triangle Park, N.C., 1984.

The pituitary gland in female mice retained its gonadotropic function when exposed to 3 GHz (100 W/m²) twice daily for 5 months, although its activity was reduced in comparison with that of nonexposed animals.[154] Tolgskaya and Gordon[239] noted the reversibility of changes in the neurosecretory function of the hypothalamus when exposure was terminated. Rabbits exposed to 3 GHz (500 to 600 W/m²) 4 hr daily for 20 days tended to show a decline in the amount of urinary 17-hydroxycorticosteroids (17-OHCS) at the beginning of exposure, followed by a gradual return to normal.[381] No change was evident in the excretion of 17-ketosteroids in the urine.

In rats exposed to MWs of varying intensity, no quantitative changes in CS were found in the adrenals and blood plasma.[382] Prepubescent hypophysectomized rats displayed no differences in adrenal growth rate when treated with pituitary homogenates collected either from rats exposed to MWs or from control rats.

Rats exposed to 2.45 GHz (CW), 100 W/m² for 4 hr showed no change in adrenal weight, phenylethanolamine-*N*-methyl transferase (PNMT) activity, or epinephrine levels.[383] After 16 hr of exposure, however, decrease in adrenal epinephrine (32%) was significant and PNMT activity was elevated (25%). There were no statistically significant differences ($p > 0.1$) in adrenal or plasma CS levels between exposed and sham-exposed animals. It should be noted, however, that similar alterations in epinephrine levels can occur in rats subjected to a stressor such as immobilization or acute exposure to cold.

Increased adrenal function due to MW exposure has been correlated with colonic temperature increase in rats.[338] Plasma CS levels in hypophysectomized rats exposed at 600 W/cm² for 60 min were below control levels. When rats were pretreated with dexamethasone before being exposed at 500 W/m² for 60 min the CS response was suppressed. These results suggest that the MW-induced CS response observed in intact rats is dependent on adrenocorticotropic hormone secretion by the pituitary, i.e., the adrenal gland is not the primary endocrine gland stimulated by MWs. The evidence obtained in these experiments is consistent with the hypothesis that the stimulation of the adrenal axis in MW-exposed rats is a systemic, integrative process due to a general hyperthermia.[338]

C. Hypothalamic-Hypophysial-Thyroid Response

The literature offers comparatively few experimental studies of the effect of RF or MWs on the thyroid. No alterations in thyroid structure or function attributable to MW exposure were noted in rats subjected to 2.45 GHz, CW, 10 W/m² continuously for 8 weeks or 100 W/m², 8 hr/day for 8 weeks.[384] On the other hand, a stimulatory influence of 50 W/m² on the trapping and secretory function of the thyroid gland of rabbits has been reported.[385] These functional changes were in agreement with altered histology of the thyroids.

In rats exposed for 16 hr to 2.45 GHz (CW) at 100 to 250 W/m², tests of thyroid function in general showed no statistically significant deviations from the norm, except that in animals with a 1.0 to 1.7°C increase in colonic temperature there was a reduction in the ability of the thyroid to concentrate iodide.[383] Decreased thyroid gland weight was noted in infant rats exposed to 0.0697 GHz, 48 V/m, 4 hr/day, 1.5 months.[380]

Increased thyroid hormone secretion has been correlated with MW-induced thyroid temperature increase in dogs.[386,387] Vetter[388] found that serum protein levels increased as a function of power density, indicating an alteration of protein synthesis or catabolism; levels of thyroid hormone decreased as power density of 2.45-GHz CW was increased from 50 to 250 W/m². In agreement is the finding of Lu et al.,[389] who reported that serum thyroxine levels were transiently elevated after exposure of rats at 10 W/m² (2.45 GHz) and were depressed after exposure at 200 W/m². None of the reported alterations was irreversible or resulted in morbidity.

Perturbation of the thyroid gland may be the result of an indirect effect, the thermal stress on the body producing a hypothalamic-hypophysial response. This is consistent with MW-induced thermal stimulation of hypothalamic-hypophysial-thyroid (HHT) activity.[390] McLees and Finch[391] point out that temperature elevation and heat stress have been associated with alterations in radioactive iodine (RAI) turnover rate. The HHT axis has been shown to be sensitive to environmental temperature.[392] Differences in rate of temperature change or alteration in thermal gradients could also result in qualitative differences in endocrine response.

In a study described as a delineation of acute neuroendocrine responses to MW exposure, Lu et al.[393] exposed "gentled" rats to the same frequency at power densities ranging from 10 to 700 W/m², 8 hr, at an environmental temperature maintained at 24°C. Sham-exposed rats were used as controls. After treatment, the rats were decapitated, colonic temperatures were taken, and blood was collected for assays of thyroxine (T-4), thyrotropin (TSH), growth hormone (GH), and CS. For exposure of 1 hr, colonic temperature increased with power density at 200 W/m² and higher, but consistent elevation of serum CS did not occur below 500 W/m². Lower serum TSH and GH levels also occurred at this and higher power densities. Significant serum T-4 elevations were noted at 400 and 700 W/m², but they were not consistently related to power density. For sham exposures and exposures at 10 to 200 W/m² for longer durations (2 to 8 hr), the results were rather equivocal, presumably, because such exposures encompassed significant portions of the circadian cycle. Specifically, in the sham-exposed rats, the level of T-4 did not change significantly with exposure duration, and significant increases of CS and decreases of TSH and GH were seen, so it was difficult to discern consistent differences in these hormones ascribable to RF exposure.

The most sensitive parameter measured proved to be the colonic temperature. For example, for rats exposed at 200 W/m², colonic temperature increases were consistent for any exposure duration, and smaller increases were noted for exposures at 1.2 kW/m² for 2 hr and 10 W/m² for 4 hr. The investigators suggest that the divergent responses may be due to two different mechanisms that are dependent on RF intensity and the time of the exposures relative to circadian rhythms.

In a review article, Lu et al.[378] discuss evidence for the existence of threshold intensities for various RF-induced neuroendocrine effects. They indicate that such thresholds are dependent on the intensity and duration of exposure and can be different for each endocrine parameter.

Abhold et al.[394] exposed rats to 2.45-GHz CW RF for 8 hr (continuously) at 20 or 100 W/m² (SAR 0.44 or 2.2 W/kg with the long body axis of the rat parallel to the electric component of the RF). Other rats were sham exposed for the same period, and still others served as untreated controls. Within 15 min after treatment, the rats were euthanized and their blood was assayed for serum T-4, triiodothyronine (T-3), and T-3 uptake. In addition, serum CS concentrations were measured. The results indicated that concentrations of T-4, T-3, and T-3 uptake were not altered by the treatments. However, the rats that were sham-exposed or exposed to RF at 20 W/m² had higher levels of CS than the untreated rats, whereas the rats exposed to 100 W/m² had values similar to those of the untreated rats. The changes observed were more likely ascribable to stress and other factors in the experimental protocols than the RF exposure.

In summary, although some effects of RF exposure on the endocrine system appear to be relatively straightforward and predictable from physiological considerations, other more subtle effects require further study, notably those related to the interactions among the pituitary, adrenal, thyroid, and hypothalamus and/or their secretions. Part of the problem in interpreting results appears to arise from uncertainties regarding stressors inadvertently introduced into the experimental design and the response of the animal to such stressors. Animals placed in novel situations are much more prone to exhibit stress responses than animals that have been adapted to the situation. However, there may be large variations in adaptation among animals in a given situation or among experimental situations in different laboratories. Moreover, the use of sham-treated controls may not always reduce the problem.

A review of the effects of RF on physiologic regulation has been published by Michaelson.[395] Neuroendocrine effects of RF exposure are summarized in Table 13.

X. CARDIOVASCULAR EFFECTS

Several investigators report that exposure to MWs may result in direct or indirect effects on the cardiovascular system. Some authors suggest that exposure to MWs at intensities which do not produce appreciable hyperthermal effects may lead to functional changes with acute as well as chronic exposures.

Cooper et al.[398,399] and Pinakatt et al.[400,401] studied the influence of various drugs in pentobarbital-anesthetized rats exposed to 2.45 GHz, 800 W/m² for 10 min which resulted in a 40.5°C colonic temperature. Pyridoxine and digitoxin did not alter blood pressure or heart rate. Ouabain abolished the circulatory reaction in exposed animals and increased the stroke volume in control animals, while blood pressure and heart rate remained unchanged. Reserpine, vagotomy, and pharmacological gang3ioplegia diminished the reaction to MW-induced hyperthermia. A comprehensive review of the literature[61] (Table 14) shows very clearly the high SAR required to elicit cardiovascular responses.

XI. EFFECTS ON HEMATOPOIESIS AND HEMATOLOGY

A number of investigators state that the blood and blood-forming system are not affected by acute or chronic MW exposure.[149,412-417] Effects on hematopoiesis have nevertheless been reported.[236,379,390,418-430] A summary of studies on hematologic effects of RF exposure is presented in Table 15.[61]

Hyde and Friedman[415] exposed anesthetized female CF-1 mice to 3 GHz, 200 W/m² and 10 GHz, 170, 400, or 600 W/m² up to 15 min. No significant effect on total or differential leukocyte count or hemoglobin concentration was noted immediately, 3, 7, or 20 days after exposure. There were no changes in femoral bone marrow other than a variable, but slight, increase in the eosinophil series of the exposed animals which was not reflected in peripheral blood counts.

Miro et al.[431] exposed Swiss albino mice to pulsed 3.105-GHz at 200 W/m² average power density (4 kW/m² peak) for 145 hr. Stimulation of splenic lymphopoiesis and increased ³⁵S methionine incorporation in spleen, thymus, and liver were found. The authors interpret these results as a sign of stimulation of cells belonging to the reticuloendothelial system.

Kitsovskaya[422] subjected rats to 3 GHz, 100 W/m², 60 min/day for 216 days; 400 W/m², 15 min/day for 20 days; 1 kW/m², 5 min/day for 6 days. At 400 W/m² and 1 kW/min², total RBC, WBC, and absolute lymphocytes were decreased; granulocytes and reticulocytes were elevated. At 100 W/m², total WBC and absolute lymphocytes decreased, and granulocytes increased. Bone marrow examination revealed erythroid hyperplasia at the higher power densities. Decreased leukocyte count and phagocytic activity had been reported in rats and rabbits exposed to 0.05 GHz, 0.5 to 6 V/m, 10 to 12 hr/day, 180 days.[432]

Baranski[424,425] exposed guinea pigs and rabbits to 3 GHz (pulsed or CW), 35 W/m² for 3 months, 3 hr daily. Increases in absolute lymphocyte counts, abnormalities in nuclear structure, and mitosis in the erythroblastic cell series in the bone marrow and in lymphoid cells in lymph nodes and spleen were observed. No alteration in the granulocyte series was noted. Baranski suggests that extrathermal complex interactions seem to be the underlying mechanism for the changes.

Spalding et al.[417] exposed mice to 0.8 GHz 2 hr daily for 120 days in a waveguide at an incident power density of 430 W/m². Red and white blood cell count, hematocrit, hemoglobin, growth, voluntary activity, and life span remained normal.

In dogs exposed whole-body to 2.8 GHz pulsed, 1 kW/m² for 6 hr, there was a marked

Table 13
NEUROENDOCRINE RESPONSE TO RF EXPOSURE

Effects	Species	Frequency (GHz)	Intensity (W/m²)	SAR (W/kg)	Duration (days × min)	Δ (°C)	Ref.
			Exposure conditions				
Increased thyroxine and triiodothyronine	Dog	2.45 (CW)	720—2360	58—190	1 × 120	2—8 (Thyroid temp)	386, 387
No effect on thyroid gland or thyroid hormone	Rat	2.45 (CW)	10, 100, 1000	0.25—25	1 × 10—45	≤100 W/m² = 0	384
				0.25—2.5	56 × 480	1000 W./m² ≥5	
No effect on thyroid function	Rat	2.45 (CW)	100, 200, 250	2, 5, 5, 6.5 (est)	1 × 240—960	≤200 W/m² = 0	383
						250 W/m² = 1.7	
Decrease in serum-protein-bound iodide, thyroxine, and thyroxine/serum ratio	Rat	2.45 (CW)	150	3.8 (est)	2.5 × 1440	0	383
Increase in thyroid hormone	Rabbit	3 (PW)	50	0.25—0.75 (est)	48 × 180	Not reported	385
Decrease in serum thyroxine levels	Rat	2.45 (CW)	200	5	1 × 240—480	0—0.6	389
			(Neg 10—100)	0.25—25	1 × 60—480	0.6 — 1.4	
Increase in CS levels	Rat	2.45 (120 Hz AM)	400—700	8.4—14	1 × 60	1.3—3.0	396
			(Neg 10—200)	(Neg 0.21—4.2)		0—0.6	
Decrease in TSH levels			100, 400—700	2.1—14.7		0—3.0	
			(Neg 10—50, 200)	(Neg 0.21—4.2)		0—0.6	
Increase in CS levels			100—400	2.1—8.4	1 × 240	0.3 — 2.1	
			(Neg 10—50)	(Neg 0.2—1.0)			
Decrease in TSH levels			250, 400	0.6—2.1			
			(Neg 10—200)	0—1			
No effect on thyroid, pituitary, or adrenal gland weights or growth hormone levels	Rat	2.45 (CW)	10—200	0.25—25	1 × 60—480	0—1.4	389
No effect on thyroid, anterior pituitary gland, adrenal, prostate, or testes weights; no change in follicle-stimulating hormone or gonadotropic hormone levels	Rat	2.86—2.88 (CW)	100	1—2 (est)	36 × 360	Not reported	377

Table 13 (continued)
NEUROENDOCRINE RESPONSE TO RF EXPOSURE

Effects	Species	Exposure conditions Frequency (GHz)	Intensity (W/m²)	SAR (W/kg)	Duration (days × min)	Δ (°C)	Ref.
Increase in leutinizing hormone	Rat	2.86—2.88 (CW)	100	1—2 (est)	36 × 360	Not reported	377
Increased adrenal weights and significant adrenal response	Infant rat	2.45 (CW)	400	20 — 60 (est)	6 × 5	1.5 — 2.5	225
Increased plasma CS levels	Rat	2.45 (CW)	500, 600 (Neg 130—400)	11.5, 13.8 (est) (Neg 3.0—9.2)	1 × 30—60	130 W/m² = 0.5 200 W/m² = 0.7 300 W/m² = 0.9—1.4 400 W/m² = 1.3—1.4 500 W/m² = 1.6—2.4 600 W/m² = 2.5—2.9	338
			200—400 (Neg 130)	4.6—9.2 (est) (Neg 3) (est)	1 × 120		
Increase in CS	Rat	2.45 (AM, 120 Hz)	500, 600	8.3, 9.6	1 × 60	Not reported	397
No effect on serum CS levels	Rat	0.918 (CW)	25	1.0	91 × 600		125

Note: Δ (°C) = Colonic temperature increase; and Neg = effect not found at value indicated.

Adapted from U.S. Environmental Protection Agency, Biological Effects of Radiofrequency Radiation, Elder, J. A. and Cahill, D. F., Eds., Rep. No. EPA-600/8-83-026F, Research Triangle Park, N.C., 1984.

Table 14
EFFECTS OF RF EXPOSURE ON CARDIAC FUNCTION

Effects	Species	Exposure conditions					Ref.
		Frequency (GHz)	Intensity (W/m²)	SAR (W/kg)	Duration (days × min)		
Bradycardia develops after whole-body exposure, along with hyperthermia	Rat	2.45 (PW)	280, 480	6.5, 11.1	1 × 30		129
Exposure to head promotes tachycardia; exposure to back raises respiratory rate, but not heart rate	Rabbit	2.4 (CW and PW)	200	3	1 × 60		402
Increased respiration	Rabbit	2.4	400—1000	8—20 (est)	1 × 20		403
Increased heart rate from dorsal exposure of the head	Rabbit	2.4	1000	20 (est)	1 × 20		403
Alterations in ECG (shortening of QT interval increased height of T-wave, appearance of U-wave)	72-Hr chick heart	24 (PW)	740		1 × 3		404
No effect on heart rate that cannot be attributed to MW heating	Quail embryo	2.45 (CW and PW)	0.32	0.3—30	1 × 5—10		405
Pulses synchronized with each R-wave do not affect heart rate	Frog	1.42—10 (PW)			(100-µsec pulses)		406
Synchronized pulses with QRS complex cause increase in heart rate with some arrhythmias	Frog	1.425 (PW)	0.006		(10-µsec pulses)		407
Increased heart rate No effect on heart rate	Rabbit	2.45 (CW)	800 50	12 0.3 and 0.093	10 × 20 10 × 20		408
Low power levels cause bradycardia in the isolated heart	Turtle	0.960 (CW)		2—10	1 × 60		409
Causes slight decrease in the isolated heart rate	Rat	0.960 (CW)		1.3 and 2.1	1 × 5—10		410
Synchronized exposures with ECG have no effect on heart rate	Frog	1.42—3 (PW)	0.006		(2-, 10-, 150-µsec pulses)		411

Table 15
HEMATOLOGIC AND HEMATOPOIETIC EFFECTS OF RF

Effects	Species	Frequency (GHz)	Intensity (W/m²)	SAR (W/kg)	Duration (days × min)	Ref.
Increased: WBC, lymphs, PMN, RBC, Hct, and Hgb	Rat	24 (PW)	100 / 200	3.0[a]	1 × 180 / 1 × 420	434
No change	Dog	24 (PW)	240	1[a]	400 × 400—900	435
No change	Rat	3 (PW)	100	2[a]	216 × 60	422
Decreased: RBC, WBC, and lymphs			400	8[a]	20 × 15	
Increased: PMN			1000	20[a]	6 × 5	
Decreased: lymphs and eos	Dog	2.8 (PW)	1000	4[a]	1 × 360	419
Decreased: WBC, PMN, and eos		2.8 (PW)	1650	6[a]	1 × 120	
Decreased: WBC, lymphs, and eos		1.28 (PW)	1000	4.5[a]	1 × 360	
Increased: PMN						
Decreased: lymphs	Mouse	0.200 (CW)	1650	25[a]	1 × 360	417
No change		0.800	430	12.9[a]	175 × 120	424
Increased: lymphs and mitotic index of lymphoid cells	Guinea pig	3 (CW or PW)	35	0.5[a]	120 × 180	425
Increased: RBC, Hct, and Hgb	Rat	2.4 (CW)	100	2[a]	30 × 120	436
No change	Rat	2.4 (CW)	50	1[a]	90 × 60	437
Increased: eosinophils	Rabbit	2.45 (CW)	100	1.5	180 × 1380	438
Increased: WBC, CFU	Mouse	2.45 (CW)	1000	70[a]	1 × 5	439
Decreased: ^{59}Fe uptake						
Accelerated recovery following X-irradiation; increased erythropoiesis and myelopoiesis	Mouse	2.45 (CW)	1000	70[a]	1 × 5	440
Accelerated recovery from X-irradiation	Dog	2.8 (CW)	1000	4[a]	1 × 3600	441 442
Increased: PMN and RBC	Chinese hamster	2.45 (CW)	600	28[a]	1 × 30	443
Decreased: lymphs						
Accelerated recovery from X-irradiation	Rat	0.425 (CW)	100	3—7	47 × 240	444
Increased: lymphs						

391

Effect	Animal	Frequency (GHz) (wave)	Power density	SAR	Duration	Ref.
Decreased: PMN (not reproduced consistently)						
No change	Rat (perinatal exposure)	2.45 (CW)	50	1—5	57 × 240	191
	Rat (perinatal exposure)	0.1 (CW)	460	2—3	57 × 240	223
Decreased Hct, WBC, and lymphs	Quail egg	2.45 (CW)	300	14	1 × 1440	221
	Rat (young)	2.736 (PW)	244	5— / 25[a]	35 × 240	445
No change	Mouse	2.45 (CW)	300	22	22 × 30	446
Decreased: lymphs	Mouse	0.026 (CW)	86, 100	13[a]		447
Increased: PMN						

Note: WBC = white blood cell, PMN = polymorphonuclear leukocytes, RBC = red blood cell, Hct = hematocrit, Hgb = hemoglobin, CFU = colony-forming unit, lymphs = lymphocytes, and eso = eosinophils.

[a] SAR estimated

Adapted from U.S. Environmental Protection Agency, Biological Effects of Radiofrequency Radiation, Elder, J. A., and Cahill, D. F., Eds., Rep. No. EPA-600/8-83-026F, Research Triangle Park, N.C., 1984.

decrease in lymphocytes and eosinophils.[421] The neutrophils remained slightly increased at 24 hr postexposure, while eosinophil and lymphocyte values returned to normal levels. Following 2 hr of exposure at 1.65 kW/m², there was a slight leukopenia and decrease in neutrophils. When the exposure was of 3-hr duration, leukocytosis was evident immediately after exposure and was more marked at 24 hr, reflecting the neutrophil response. After exposure to 1.285 GHz pulsed 1 kW/m² for 6 hr, there was an increase in leukocytes and neutrophils. At 24 hr, the neutrophil level was still noticeably increased. Lymphocyte and eosinophil values were moderately depressed initially, but at 24 hr slightly exceeded their initial value. An exposure of 6 hr to 0.2 GHz (CW), 1.65 kW/m², resulted in a marked increase in neutrophils and a mild decrease in lymphocytes. On the following day, the leukocyte count was further increased, and the lymphocytes markedly increased. Such shifts in the white blood cell picture are consistent with focal thermal lesions to be expected under these exposure conditions.

Exposure of mice to 2.45 GHz, 1 kW/m² for 5 min resulted in a decrease followed by an increase in ^{59}Fe uptake in the spleen and bone marrow.[427] Alteration in ferrokinetics was also found in rabbits and guinea pigs exposed to 3 GHz at 10 or 30 W/m², 2 to 4 hr daily, 14 to 79 days.[428,429]

Possible alteration in the circadian rhythm of bone-marrow mitoses in guinea pigs and mice was noted. No effects were seen on precursors of granulocytes and only minimal effects were found on the erythroid series, but pronounced phase shifts were noted in the pool of stem cells. In inbred Swiss albino mice exposed once for 4 hr at 5 W/m² to pulsed 3-GHz MWs, the diurnal rate of proliferation of the stem-cell population was amplified and the phase shifted from that of controls.[429] A comparable phase shift in the circadian rhythm of body temperature was observed in rats exposed to 2.45 GHz, CW, 10 W/m², 1 to 8 hr, suggesting an interrelated example of physiologic regulation.[389]

In evaluating reports of hematological changes (Table 15) one must be cognizant of the relative distributions of blood cells in a population of animals or humans and the thermal influence on these alterations. Early and sustained leukocytosis in animals to thermogenic levels of MWs may be related to stimulation of the hematopoietic system, leukocytic mobilization, or recirculation of sequestered cells. Eosinopenia and transient lymphocytopenia with rebound or overcompensation, when accompanied by neutrophilia, may be indicative of increased hypothalamic-hypophysial adrenal function as a result of thermal stress.[433]

XII. EFFECTS ON THE IMMUNE RESPONSE

In recent years, considerable interest has developed in the relationship of MW exposure and alteration of the immune response.[446-455,462,470-475] Because of the emphasis placed on the "hazard" aspects of the reports of lymphoblastoid transformations and immunological consequences of MW exposure, it is important to place this in perspective. In essence, the reports do not negate the thermal influences of MW energy absorption. If the reports are confirmed, they may not be indicative of a hazard, but actually portend an exciting and important possibility for therapy of infectious diseases or cancer which have been shown to be influenced by hyperthermia.

A. Lymphoblastoid Transformation

Lymphoblastoid transformation, in vitro, after free field exposure to 3-GHz pulsed MWs, 70 W/m², for 4 hr daily and 200 W/m² for 15 min daily, 3 to 5 days, has been reported.[451] At this power density, the temperature of the media increased by 0.5°C after 15 min, and after 20 min the increase was 1°C. Changes in the mitotic index was dependent on the exposure time. Although a 5-min exposure did not influence the proportion of dividing cells, slight differences compared with controls were observed after 10- and 15-min exposures and significant differences were seen following 3- or 4-hr exposures at 70 W/m².

Czerski[461] exposed human lymphocyte suspensions to 3 GHz for various periods of time at several power densities. Lymphoblastoid transformations were observed, but no correlation between the cellular changes and either duration of exposure or incident power were noted.

Huang et al.[42] reported that lymphocytes from Chinese hamsters exposed from 50 to 450 W/m², 2.45-GHz CW, 15 min/day for 5 days showed changes in blast transformation and mitosis. No chromosomal aberrations were evident. These studies noted increased but reversible transformation of lymphocytes (without mitogenic stimulation), related to the power density, but a decreased proportion of mitogen-stimulated cells in mitosis. A thermal effect within a range that might normally be managed and dissipated by the animals could not be excluded. These investigators called attention to the changes over a range of body temperatures of less than 2°C and suggested that such limited hyperthermia might be of considerable interest to investigators of immunological effects of MW radiation.

MWs have been reported to induce an increase in the frequency of complement receptor-bearing lymphoid spleen cells in mice.[448] Although the significance of this is not fully assessed, it may represent a maturation of B lymphocytes.

Wiktor-Jedrzejczak et al.[448,449] exposed adult male mice to 2.45-GHz CW at an absorbed dose of 12 to 15 W/kg for 30 min in an environmentally controlled waveguide facility and then measured the function of different classes of lymphocytes in vitro. Such exposure failed to produce any detectable changes in function of T lymphocytes or increase in DNA, RNA, or protein synthesis, as measured by incorporation of tritiated-thymidine, -uridine, and -leucine by spleen, bone marrow, and peripheral blood lymphocytes in vitro. However, the maturation of B lymphocytes from the spleen of exposed mice was stimulated. Consistent with this effect on B lymphocytes are the results reported by Czerski and associates.[428,429,461]

Smialowicz[450] examined the proliferative capacity of lymphocytes that are responsible for cellular immune responses (T cells) and humoral immune responses (B cells) following 2.45-GHz exposure in vitro. The ability of mouse spleen lymphocytes to undergo blast transformation in response to mitogens that selectively stimulate either T or B cells was measured by the incorporation of ³H-thymidine into DNA. No consistent difference was found between the blastogenic response of exposed (100 W/m², 19 W/Kg, 1 to 4 hr) and control cells.

Mice exposed to 2.45-GHz (CW), 50 to 350 W/m² (SAR = 4 to 25 Wk/g) for 1 to 22 consecutive days (15 to 30 min/day) showed no consistent significant alterations in several parameters such as mitogen-stimulated response of T- and B-splenic lymphocytes, enumeration of the frequencies of T- and B-splenic lymphocytes, and the primary antibody response of mice to sheep erythrocytes.[446] Subsequently, Schlagel et al.[452] reported that RF-induced complement-receptor-positive (CR⁺) lymphocyte increase was under genetic control. Smialowicz et al.[453] showed, in addition, the age and strain of the mouse, the RF exposure characteristics (waveguide vs. far field), and the environmental conditions are all sources of variation that affect the CR⁺ cell appearance.

B. Lymphocyte Mitotic Stimulation

Smialowicz et al.[191] exposed rats *in utero* and neonatally through 40 days of age (4 hr/day, 7 days/week) in a controlled environment to either 2.45 GHz (CW, 50 W/m², SAR = 1 to 5 W/kg) or to 0.425 GHz (CW, 100 W/m², SAR = 3 to 7 W/kg). At 40 days of age significant increases in the response of lymphocytes from exposed rats to in vitro stimulation with several mitogens was observed in several experiments. While these results have not been consistently reproduced, the trend in the results suggests that chronic exposure during fetal and neonatal development may change either the frequency or responsiveness of lymphocyte subpopulations. The biological significance of these observed changes is unknown. The mechanism by which these changes are initiated may be related to a thermally induced stress response. Similar responses have been observed in animals following prolonged exposure to nonspecific stressors.

Hamrick[114] examined the response of mammalian lymphocytes exposed to 2.45-GHz (CW) in cultures at 200 W/m^2 (7 W/kg) for 48 hr. Changes in the stimulation caused by phytohemagglutinin under control and exposed conditions were tested. No effects of exposure were detected. Also, no effect on DNA was found at power densities as high as 67 to 160 W/kg. It was concluded that 2.45-GHz CW MW exposure has very little, if any, effect other than that of heating on the secondary structure of DNA as determined by comparison of thermal denaturation curves.

Decrease in natural killer (NK) activity in lymphocytes of hamsters after exposure to 13 W/kg, 2.45-GHz CW for 1 hr was reported by Yang et al.[454] Roberts et al.[455] examined the effects of exposure to 2.45-GHz continuous wave RF on human mononuclear synthesis after exposure to RF at SARs up to 4 W/kg. They also measured effects on synthesis of specific host defense proteins, namely, interferons, and examined for morphological lymphoblastoid transformation as well as changes indicated by incorporation of radiolabeled precursors. Leukocytes were exposed in 37°C chambers without attempts to counteract RF-induced heating, and final culture temperatures were approximately 0.9°C higher than those of sham-exposed cultures, similar to changes induced by Stodolnik-Baranska.[451] Such exposures resulted in no detectable effects on viability or on unstimulated or mitogen-stimulated DNA, RNA, total protein, or interferon synthesis by the human mononuclear leukocytes.[455]

In many of the studies on MW effects, especially in vivo, varying conditions of exposure have been used, often with sufficient power density that thermal effects may be the predominant, if not the only, factor. These are summarized in Tables 16 and 17. This variation in application of MWs may account for much of the diversity in data regarding the effects of such treatments on components of the hematopoietic system.

C. Adaptation

Czerski and associates[428,429] have reported that inbred Swiss mice, immunized with sheep red blood cells and exposed 2 hr/day for 6 or 12 weeks to 5 W/m^2 of 2.95 GHz, showed increased serum hemagglutinin titers and antibody-producing cells in lymph node homogenates. The increase was greater in mice exposed 6 weeks than in those exposed 12 weeks. Similar results were obtained in rabbits exposed 2 hr/day to 30 W/m^2 for 6 months with the maximum increase occurring after exposure of 1 or 2 months, and then returning to control values. According to Czerski,[456] this may indicate that after a period of response, the animals become adapted to the MWs. The phenomenon of physiologic adaptation or decreased reaction as a result of repeated exposure to microwaves has also been reported by others.[86,423,433,457,458,459]

Baranski[425] exposed adult guinea pigs 3 hr/day for 3 months to 35 W/m^2 of 3 GHz and found an increase in lymphopoiesis over controls, as indicated by increased incorporation of ^{3}H-thymidine and increased mitotic indices. No differences could be detected between pulsed or CW MWs of the same average power level. Twofold increases in lymphocyte numbers were found in the spleen and lymph nodes.

D. Influence of Hyperthermia

RF-induced hyperthermia in mice has been associated with transient lymphopenia and neutrophilia, with a relative increase in splenic T and B lymphocytes, and with decreased in vivo local delayed hypersensitivity.[447,460,462] The latter was not affected by a comparable increase in core temperature produced by warm air. Reduced thymic mass and cell density,[447,462] suppressed inflammatory response,[447,460] and suppressed allograft transplant rejection[463] have been reported. Such alterations in lymphocyte distribution and function are concomitant with a state of immunosuppression. Qualitatively, similar changes can be induced by administration of synthetic glucocorticoids or CS.[447,462] Liburdy[462] reported elevated plasma CS levels in mice following exposure to RF energy sufficient to cause hyperthermia.

Table 16

IMMUNOLOGIC EFFECTS OF IN VIVO RF EXPOSURE

Effects	Species	Frequency (GHz)	Intensity (W/m²)	SAR (W/kg)	Duration (days × min)	Ref.
Increase in lymphoblasts in lymph nodes and increased response to SRBC	Mouse	2.95 (PW)	5	0.5^a	42 × 120	456
Increase in "spontaneous" lymphoblast transformation of cultured lymphocytes	Rabbit	2.95 (PW)	50	0.8^a	24—48 × 120	456
Increase in lymphoblasts in spleen and lymphoid tissue	Mouse	3.105 (PW)	20	2^a	6—8700	431
Increased transformation of unstimulated cultured lymphocyte and decreased mitosis in PHA-stimulated lymphocyte cultures	Chinese hamster	2.45 (CW)	50, 150, 300, or 450	2.3, 6.9, 13.8, or 20.7	5 × 15	42
Transient decrease and increased response of cultured lymphocytes to PHA, Con A, and LPS	Mouse	2.45 (CW)	50 or 150	3.6 or 10	1—17 × 30	470
Increase in CR^+, Fc^+, and Ig^+ spleen cells; increased response to B-cell mitogens; decrease in primary response to SRBC	Mouse	2.45 (CW)	—	14	1 or 3 × 30	448, 449, 471
Increase in Cr^+ and Fc^+ spleen cells	Mouse	2.45 (CW)	—	11.8 5	1 × 15 1 × 30	472
Increase in response of cultured lymphocytes to T- and B-cell mitogens	Rat	2.45 (CW) 0.425 (CW)	50 100	1—5 3—7	57 × 240 47 × 240 (perinatal exp)	191 444
No change	Mouse Rat	2.45 (CW) 0.1 (CW)	50—350 460	4—25 2—3	1—22 × 15 or 30 57 × 240	446 223
Increase in T and B lymphocytes in spleen; decrease in DTH	Mouse	0.026 (CW)	8,000	5.6^a	1 × 15 10 × 15	462
Reduction of lymphocyte traffic from lung to spleen	Mouse	2.6 (CW)	50 or 250	3.8 or 19	1 × 60	473

Table 16 (continued)
IMMUNOLOGIC EFFECTS OF IN VIVO RF EXPOSURE

Effects	Species	Exposure condition				Ref.
		Frequency (GHz)	Intensity (W/m²)	SAR (W/kg)	Duration (days × min)	
Decreased response to PWM	Rabbit	2.45 (CW)	100	1.5	180 × 1,380	438
No change	Quail	2.45 (CW)	50	4.03	12 × 1,440	474
Decrease in tumor development	Mouse	2.45 (CW)	— (Near-field application)	35	11—14 Day of gestation or 11—14 and 19—45 × 20	97
Decreased granulocytic response	Rabbit	3 (CW)	30	0.5	42—84 × 360	475
Tumor regression and increase in antitumor antibodies and anti-BSA	Rabbit	1.356	(Near-field application)	(Local hyperthermia)	1 × 10 — 15	476
Tumor inhibition and immune stimulation	Rat	2.45 (CW)	(200 W)	(Local hyperthermia)	3 or 6 × 45	477
Increased tumoricidal activity in lymphocytes and macrophages	Mouse	1.356	6,000—9,000	(Local hyperthermia)	1 × 5	478
Tumor regression	Mouse	3 (CW)	400	28[a]	1—14 × 120	479
Increase in lung cancer colonies and inhibition of contact sensitivity to oxazolone	Mouse	2.45 (CW)	500	36[a]	4, 7, 10, or 14 × 120	480
Decrease in response to BSA	Rabbit	1.356	(Near-field application)	(Local hyperthermia)	3 × 60	481
Decrease in CFU for erythroid and granulocyte-macrophage series	Mouse	2.45 (CW)	150	10	9 × 30	470
Reduction in CFU granulocyte-macrophage precursors exposed in vitro	Mouse	2.45 (CW)	600—10,000	120—2,000	1 × 15	482

Note: SRBC = sheep red blood cells, PHA = phytohemagglutinin, Con A = concanavalin A, LPS = lipopolysaccharide, CR$^+$ = complement-receptor positive, Ig$^+$ = immunoglobulin positive, Fc$^+$ = Fc portion of immunoglobulin, DTH = delayed-type hypersensitivity, PWM = pokeweed mitogen, BSA = bovine, and CFU = colony-forming unit.

[a]   SAR estimated.

Adapted from U.S. Environmental Protection Agency, Biological Effects of Radiofrequency Radiation, Elder, J. A. and Cahill, D. F., Eds., Rep No. EPA-600/8-83-026F, Research Triangle Park, N.C., 1984.

Table 17
IMMUNOLOGICAL EFFECTS OF IN VITRO RF EXPOSURE

Effects or end point	Experimental system	Exposure conditions				Ref.
		Frequency (GHz)	Intensity (W/m²)	SAR (W/kg)	Duration (days × min)	
Increased blastogenesis of exposed lymphocytes in vitro	Human lymphocytes	3 (PW)	70, 140	a	3—5 × 240, 3—5 × 15	483
Increased blastogenesis	Human lymphocytes	10	50—150	a	Observed effect only when culture temperature approached 38°C	86
No effects on viability, DNA, RNA, total protein, interferon synthesis	Human lymphocytes	2.45 (CW)	—	4	Variable	455
No change in mitogen response to PHA, Con A, or LPS	Mouse spleen cells	2.45 (CW)	100	19	1 × 60, 120, or 240	450
No change in mitogen response to PHA	Rat blood lymphocytes	2.45 (CW)	50, 100, or 200	0.7, 1.4, or 2.8	1 × 240, 1440, or 2640	123
No change in viability or growth	Human lympho-blast cell lines (Daudi and HSB₂)	2.45 (CW)	100—5000	25—1200	1 × 15	484
Decreased macrophage phagocytosis	Mouse macrophage	2.45 (CW)	500	15 J/min	1 × 30	485
Liberation of intracellular hydrolytic enzymes and increased death	Rabbit granulocytes	3 (CW)	10 or 50	a	1 × 15, 30, or 60	486

Table 17a

Effects or end point measured	Experiment system	Frequency (GHz)	Exposure conditions		Duration	Ref.
				SAR (W/kg)		
No change in growth, CFU, of various strains of exposed cultures under several growth conditions	*E. coli*	2.45 (CW)	(Anechoic chamber — far field)	0.0075—75	240 min	71
No change in survival curves (measuring CFU) of exposed cultures	*E. coli* *B. subtilis* spores	2.45	(Microwave oven)	~400	1 min	487
No effect on colony-forming ability	*E. coli*	2.6—4.0 (CW)	(Waveguide)	29	8 hr	70
Temporary decrease in virulence (>6 hr) of bacteria for its host cells; recovery within 24 hr at 37°C	*A. tumefaciens*	10 (CW)	(Cavity)	~1	30, 60, 230 min	488

Note: PHA = phytohemagglutinin, Con A = concanavalin A, LPS = lipopolysaccharide, and CFU = colony-forming unit.

a   Unable to calculate SAR.

Adapted from U.S. Environmental Protection Agency, Biological Effects of Radiofrequency Radiation, Elder, J. A. and Cahill, D. F., Eds., Rep. No. EPA-600/8-83-026F, Research Triangle Park, N.C., 1984.

A similar response has been reported by Lotz and Michaelson,[338] who showed a correlation between MW-induced body heating and CS levels in the blood of rats. These studies suggest that exposure to MWs of sufficient intensity results in stimulation of the hypophysial-adrenal axis which could affect the immune system.[462]

It would thus appear that RF exposure initially causes a general stimulation of the immune system, but if the exposure continues, the stimulatory effect disappears, suggesting a phase of adaptation to continued RF exposure. There is a body of literature on the influence of heat per se on immunity. Significant influences of microwaves on immune responsiveness would be expected on the basis of the known effects of hyperthermia. Although there has been some uncertainty whether fever enhances host resistance to infection,[464,465] recent evidence suggests that fever may enhance survival after infection in an animal model.[466] Cell-mediated immunity plays a role in defense against facilitative intracellular bacteria, viruses, and certain other infectious agents.[467,468] Roberts and Steigbigel[469] have shown that increased temperature (38.5°C) enhances human lymphocyte response to mitogen (PHA) and antigen (streptokinase-streptodornase) and enhances, but does not accelerate, certain bactericidal functions of human phagocytic leukocytes. One, therefore, has to be circumspect in assessing the mechanisms of MW exposure related to alterations in immune processes. It may be that if MWs do, in fact, increase the proportion of lymphocytes undergoing transformation, it may not in itself be harmful, but actually beneficial. The immune system has a considerable redundancy and adaptability. Perturbations of the immune system may not have clinical significance.

XIII. AUDITORY RESPONSE

''Microwave hearing'' is sensed in the head as a clicking or buzzing due to pulsed MW at rather low power densities.[489,490] It appears that this phenomenon is one of thermoelastic expansion.[319,491-495] This is discussed in detail in Part III, Chapter 5.

XIV. OCULAR EFFECTS

During the past 25 years, numerous investigations in animals and several surveys among human populations have been devoted to assessing the relationship of MW exposure to the subsequent development of cataracts. It is significant that of the many experiments on rabbits by several investigators using various techniques, a power density above 1 kW/m² for 1 hr or longer appears to be the lowest time-power threshold in the tested frequency range of 0.2 to 10 GHz. In other species of animals such as dogs and nonhuman primates, the threshold for experimental MW-induced cataractogenesis appears to be even higher. This threshold is a time-power threshold, that is, the higher the power density, the shorter is the time threshold and vice versa, down to a certain minimum power density. All of the reported effects of MW radiation on the lens can be explained on the basis of thermal injury.

A. Threshold for Opacity in Rabbits

The most extensive investigations in this area were performed by Carpenter and van Ummersen[496] and Carpenter et al.[497] Later work by Guy et al.[498-500] and Kramar et al.[501,502] has shown that in rabbits exposed to 2.45 GHz, the threshold for cataract production is 1.5 kW/m² for 100 min. The data suggest that an intraocular temperature of at least 43°C must be obtained to induce cataracts, although exceptions may be found in the literature.[501] These investigators also found that single potentially cataractogenic exposures will not injure the eye under conditions of controlled general hypothermia, and exposure to 1 kW/m², 2 hr/day, for 4 to 9 days produced no cataracts as evidenced by periodic examinations for 6 months after exposure. Guy et al.[499] exposed rabbits to 0.918 GHz, 4.66 kW/m² for 15 min and 1.17 W/m² for 100 min with no evidence of cataract, and concluded the threshold for cataractogenesis is higher for this frequency in comparison to 2.45 GHz.

Appleton et al.[503] exposed rabbits to 3 GHz, 1 or 2 kW/m² for 15 to 30 min. Examination daily for 14 days, weekly for 1 month, and monthly for a year revealed no ocular changes. At power densities of 3, 4, or 5 kW/m² for 15 min, acute ocular changes involving, especially, the conjunctiva and iris occurred during exposure. In a comparable study by Hirsch et al.,[504] rabbits were repeatedly exposed to 3 GHz once daily for a month. Clinical examinations were carried out for 1 year afterward. No changes occurred at power densities under 3 kW/m².

B. Biochemical Changes

Reports of decreased enzymatic activity[496,505] may quite likely be due to thermal inactivation with resultant alterations in metabolism. Decreases in ascorbic acid concentration in the lens have been cited as being the first biochemical indication of opacity formation:[506,507] progressive clouding of the lens is associated with decreases in ascorbic acid below 60 μg/g of lens tissue. All of these effects could be fully reparable until the altered metabolism has produced a permanent opacification of the lens. Latent periods and time-power thresholds would be in agreement with a mechanism of this nature.

C. Thermal Aspects of Microwave Cataractogenesis

Guy et al.,[498-500] Kramar et al.,[501,502] and Taflove and Brodwin[508] have computed the MW energy deposition and induced temperatures in the eyes of rabbits and a model of the human eye. They have indicated that a distinct hot spot exceeding 40.4°C probably occurs deep within the eye at a frequency of 1.5 GHz, when the power density would be cataractogenic (i.e., greater than 1.99 kW/m²).

Paulsson[509] measured the absorption of 0.915-, 2.45-, and 9.0-GHz MW energy in a model of the human head. The absorbed power showed an essentially exponential decrease with the distance from the cornea at 9.0 GHz. At 2.45 and 0.915 GHz maximum absorbed power occurred within the eyeball.

D. Cumulative Effect

The possibility of a cumulative effect on the lens from repeated "subthreshold" exposures of the rabbit eyes to MW has been suggested by Carpenter and associates.[496,497] No cumulative rise in temperature can occur if the intervals between exposure exceed the time required for the tissue to return to normal temperature. The cumulative effect to be anticipated, therefore, is the accumulation of damage resulting from repeated exposures, each of which is individually capable of producing some degree of damage.[510]

McAfee et al.[511] trained monkeys to face an RF source and then exposed them to 9.3-GHz pulsed RF at average and peak power densities of 1.5 and 2.86 kW/m², respectively. Each of 12 monkeys was exposed for up to 20 min/day for 30 to 40 sessions over several months. A total of 75 monkeys, neither exposed nor sham-exposed, served as controls. No cataracts or corneal lesions were seen in any of the 12 exposed animals up to 12 months after exposure.

The existence of a cataractogenesis threshold of about 1.5 kW/m² is regarded as evidence that single or multiple exposure for indefinitely long durations at average power densities well below the threshold would not cause eye damage to humans or any other species. This conclusion is supported by results of an investigation by Chou et al.[512] They exposed one group of six rabbits to 2.45-GHz CW at 150 W/m² for 2 hr/day over several months, and another group to pulsed RF at the same frequency and average power density (10-μsec pulses at a repetition rate of 100 pulses per second, comprising a duty cycle of 0.001 and a pulse power density of 1.5 kW/m²); a third group was sham irradiated. Periodic eye examinations for cataract formation yielded no statistically significant differences among the three groups.

Guy et al.[513] exposed four rabbits to 2.45-GHz CW RF at 100 W/m² (maximum whole-body SAR of 17 W/kg) 7 days/week, 23 hr/day, for 6 months. For controls, four other rabbits were sham exposed for the same durations. Periodic eye examinations with a slit-lamp microscope showed normal aging changes in the lenses and no significant differences between the exposed and control groups.

In general, the results from animal studies indicate that RF cataractogenesis is essentially a gross thermal effect that has a threshold power density at which the difference between the rates of heat generation by RF and heat removal is large enough to result in damage to the lens of the eye. The mean threshold values probably vary to some extent from species to species, but are of the order of 1 kW/m².

Analysis of available data indicates that when repeated exposures are near threshold and within the time frame of the latency, cataract may appear. When the time interval between exposures is longer than the latency period, the lesion does not appear unless the power density/time relationship is well above threshold. The so-called "cumulative effect" is only a phenomenon confined to near-threshold exposures. No one has yet been able to produce cataracts, even by repetitive exposures, when the power density is below 800 to 900 W/m².

E. Extrapolation to the Human

After 25 years of studies of the effects of MWs on the ocular lens, primarily in the rabbit, the principal conclusions are

1. The acute thermal insult from high-intensity MW fields is cataractogenic in the rabbit if intraocular temperature is greater than 43°C.
2. The MW exposure threshold is between 1 and 1.5 kW/m² applied for about 60 to 100 min.
3. There does not appear to be a cumulative effect from MW exposure unless each single exposure is sufficient to produce some irreparable degree of injury to the lens.

That opacity of the ocular lens can be produced in rabbits by exposure to MW is well established. Extrapolating results from animal studies to humans is difficult, because the conditions, durations, and intensities of exposure are usually quite different. Several cases of alleged cataract formation in humans exposed to MW have been reported in the literature, but precise exposure parameters are generally impossible to determine. It is also difficult to relate cause and effect, because lens imperfections do occur in otherwise healthy individuals, especially with increasing age. Numerous drugs, industrial chemicals, and certain metabolic diseases are associated with cataracts.

XV. CRITIQUE

Elucidation of the biological effects of MW exposure requires a careful review and critical analysis of the available literature. This requires differentiating established effects and mechanisms from speculative and unsubstantiated reports. Most of the experimental data support the concept that the effects of MW exposure are primarily, if not only, a response to hyperthermia or altered thermal gradients in the body. There are, nevertheless, large areas of confusion, uncertainty, and actual misinformation.

Although there is considerable agreement among scientists concerning the biological effects and potential hazards of MWs, there are areas of disagreement. There also is a serious philosophical question about the definition of hazard. One objective definition of injury is an irreversible change in biological function as observed at the organ or system level. With this definition, it is possible to define a hazard as a probability of injury on a statistical basis. It is important to distinguish between the intensities of RF energy at which injury may be sustained and those that produce nonharmful effects or merely perception with no further pathophysiologic consequences. All effects are not necessarily hazards. In fact, some effects may have beneficial applications under appropriately controlled conditions. RF-induced changes must be understood sufficiently so that their clinical significance can be determined, their hazard potential assessed, and the appropriate benefit/risk analyses applied. It is important to determine whether an observed effect is irreparable, transient, or reversible, disappearing when the electromagnetic field is removed either immediately or after some interval of time. Of course, even reversible effects are unacceptable if they transiently impair the ability of the individual to function properly or to perform a required task.

A critical review of studies into the biological effects of RF shows that many past investigations suffered from inadequate experimental facilities and energy measurement skills, or insufficient control of the biological specimens and the criteria for biological change. More sophisticated conceptual approaches and more rigorous experimental designs must be developed. There is a great need for systematic and quantitative comparative investigation of the biological effects using well-controlled experiments.

For RF bioeffects, body size of the experimental animal, as well as its orientation relative to the incident field vectors, must be taken into account. Since body-absorption cross-sections and internal heating patterns can differ widely, as shown in Part III, Chapters 1 and 2, an investigator may think he is observing a low-level or a nonthermal effect in one animal because the incident power is low, while in actuality the animal may be exposed to as much absorbed power in a specific region of the body due to resonance or focusing effects as another larger animal is with much higher incident power at certain frequencies. In performing experimental studies on animals for extrapolation to man, interspecies scaling factors must be considered.

The question of whether reported CNS changes in man (if they are validated) would be important enough to affect his performance should be resolved. Better understanding of local, regional, and whole-body thermal regulation is required. More precise and better controlled long-term, low-level laboratory studies have been suggested. Such studies have to be rigidly controlled to obviate circadian rhythm and biological drift over time, which will influence responses.

Particular attention must be paid to instrumentation problems, such as developing more adequate probes for making measurements in the presence of electromagnetic fields, as discussed in Part III, Chapter 1. Field strength and electrophysiological and thermal probes are essential which will give artifact-free readings, which will not distort the field in any way, and which will not give rise to inadvertent stimulation of the tissue due to induced currents.

ACKNOWLEDGMENT

This paper is based on work performed under Contract No. DE-AC02-76EV03490 with the U.S. Department of Energy and PHS Grant No. ES03239 at the University of Rochester Department of Radiation Biology and Biophysics and has been assigned Report No. UR-3490-2252.

REFERENCES

1. **Durney, C. H., Johnson, C. C., Barber, P. W., Massoudi, H., Iskander, M. F., Lords, J. L., Ryser, D. K., Allen, S. J., and Mitchell, J. C.,** *Radiofrequency Radiation Dosimetry Handbook,* 2nd ed., USAF Report SAM-TR-78-22, Brooks Air Force Base, Tex., 1978.
2. **Gandhi, O. P.,** Conditions of strongest electromagnetic power deposition in man and animals, *IEEE Trans. Microwave Theory Tech.,* 23, 1021, 1975.
3. **Gandhi, O. P.,** State of knowledge for electromagnetic absorbed dose in man and animals, *Proc. IEEE,* 68, 24, 1980.
4. **Gandhi, O. P., Hunt, E. L., and D'Andrea, J. A.,** Deposition of electromagnetic energy in animals and in models of man with and without grounding and reflector effects, *Radio Sci.,* 12(S), 39, 1977.
5. **Guy, A. W.,** Quantitation of induced electromagnetic field patterns in tissue and associated biologic effects, in *Biologic Effects and Health Hazards of Microwave Radiation,* Czerzki, P. et al., Eds., Polish Medical Publishers, Warsaw, 1974, 203.
6. **Guy, A. W., Webb, M. D., and Sorensen, C. C.,** Determination of power absorption in man exposed to high frequency electromagnetic fields by thermographic measurements on scale models, *IEEE Trans. Biol. Med. Eng.,* 23, 361, 1976.
7. **Massoudi, H.,** Long Wavelength Analysis of Electromagnetic Power Absorption by Prolate Spheroidal and Ellipsoidal Models of Man, Ph.D. thesis, University of Utah, Salt Lake City, 1976.
8. **Massoudi, H., Durney, C. H., and Johnson, C. C.,** Long-wavelength electromagnetic power absorption in ellipsoidal models of man and animals, *IEEE Trans. Microwave Theory Tech.,* 25, 47, 1977.
9. **Anne, A.,** Scattering and Absorption of Microwaves by Dissipative Dielectric Objects: The Biological Significance and Hazard to Mankind, Ph.D. thesis, University of Pennsylvania, Philadelphia, 1963.
10. **Anne, A., Saito, M., Salati, O. M., and Schwan, H. P.,** Relative microwave absorption cross sections of biological significance, in *Proc. 4th Annu. Tri-Service Conf. Biol. Effects of Microwave Radiating Equipment; Biological Effects of Microwave Radiation,* Peyton, M. F., Ed., Plenum Press, New York, 1962, 153.
11. **Anne, A., Saito, M., Salati, O. M., and Schwan, H. P.,** Penetration and Thermal Dissipation of Microwaves in Tissues, RADC-TDR-62-244, Cont. AF 30 (602)-2344 (ASTIA 284 981), University of Pennsylvania, Philadelphia, 1962.
12. **Schwan, H. P.,** Characteristics of absorption and energy transfer of microwaves and ultrasound in tissues, in *Medical Physics,* Vol. 3, Glasser, O., Ed., Year Book Medical, Chicago, 1960, 1.
13. **Schwan, H. P.,** Radiation biology, medical applications and radiation hazards, in *Microwave Power Engineering,* Vol. 2, Okress, E. C., Ed., Academic Press, New York, 1968, 213.
14. **Schwan, H. P.,** Interaction of microwave and radio frequency radiation with biological systems, *IEEE Trans. Microwave Theory Tech.,* 19, 146, 1971.
15. **Schwan, H. P.,** Principles of interaction of microwave fields at the cellular and molecular level, in *Biological Effects and Health Hazards of Microwave Radiation,* Czerski, P. et al., Eds., Polish Medical Publishers, Warsaw, 1974, 152.
16. **Schwan, H. P.,** Dielectric properties of biological materials and interaction of microwave fields at the cellular and molecular level, in *Fundamental and Applied Aspects of Non-Ionizing Radiation,* Michaelson, S. M. et al., Eds., Plenum Press, New York, 1975, 3.
17. **Schwan, H. P. and Piersol, G. M.,** The absorption of electromagnetic energy in body tissues, a review and critical analysis. I. Biophysical aspects, *Am. J. Phys. Med.,* 33, 371, 1954.
18. **Schwan, H. P. and Piersol, G. M.,** The absorption of electromagnetic energy in body tissues, a review and critical analysis. II. Physiological and clinical aspects, *Am. J. Phys. Med.,* 34, 425, 1955.
19. **Guy, A. W.,** Analyses of electromagnetic fields induced in biological tissues by thermographic studies on equivalent phantom models, *IEEE Trans. Microwave Theory Tech.,* 19, 205, 1971.
20. **Guy, A. W.,** Electromagnetic fields and relative heating patterns due to a rectangular aperture source in direct contact with bilayered biological tissue, *IEEE Trans. Biomed. Eng.,* 13, 16, 1971.
21. **Guy, A. W. and Lehmann, J. F.,** On the determination of an optimum microwave diathermy frequency for a direct contact applicator, *IEEE Trans. Biomed. Eng.,* 13, 76, 1966.
22. **Guy, A. W., Lehmann, J. F., McDougall, J. A., and Sorensen, C. C.,** Studies on therapeutic heating by electromagnetic energy, in *Thermal Problems in Biotechnology,* American Society of Mechanical Engineers, United Engineering Center, New York, 1968, 26.
23. **Lehmann, J. F.,** Diathermy, in *Handbook of Physical Medicine and Rehabilitation,* Krusen, F. H., Kottke, F. J., and Ellwood, P. M., Eds., W. B. Saunders, Philadelphia, 1971, 273.
24. **Lehmann, J. F., Guy, A. W., Johnston, V. C., Brunner, G. D., and Bell, J. W.,** Comparison of relative heating patterns produced in tissues by exposure to microwave energy at frequencies of 2450 and 900 megacycles, *Arch. Phys. Med.,* 43, 69, 1962.
25. **Lehmann, J. F., McMillan, J. A., Brunner, G. D., and Johnston, V. C.,** Heating patterns produced in specimens by microwaves of the frequency of 2456 megacycles when applied with the ''A'', ''B'', and ''C'' directors, *Arch. Phys. Med.,* 43, 538, 1962.

26. **Lehmann, J. F., Brunner, G. D., McMillan, J. A., Silverman, D. R., and Johnston, V. C.,** Modification of heating patterns produced by microwaves at the frequencies of 2456 and 900 Mc by physiologic factors in the human, *Arch. Phys. Med. Rehabil.,* 45, 555, 1964.

27. **Lehmann, J. F., Johnston, V. C., McMillan, J. A., Silverman, D. R., Brunner, G. D., and Rathbun, L. A.,** Comparison of deep heating by microwaves at frequencies of 2456 and 900 megacycles, *Arch. Phys. Med.,* 46, 307, 1965.

28. **Lehmann, J. F., Silverman, D. R., Baum, B. A., Kirk, N. L., and Johnston, V. C.,** Temperature distributions in the human thigh, produced by infrared, hot pack and microwave application, *Arch. Phys. Med. Rehabil.,* 47, 291, 1966.

29. **Schaffer, M. B.,** The Thermal Response of Small Animals to Microwave Radiation, The Rand Corporation (Rep. Rand-P-2558-1), Santa Monica, Calif., 1962.

30. **Hoeft, L. O.,** Microwave heating: a study of the critical exposure variables for man and experimental animals, *Aerosp. Med.,* 37, 621, 1965.

31. **McMahon, T.,** Size and shape in biology, *Science,* 179, 1201, 1973.

32. **Adolph, E. F.,** Quantitative relations in the physiological constitutions of mammals, *Science,* 109, 579, 1949.

33. **Johnson, C. C.,** Recommendations for specifying EM wave irradiation conditions in bioeffects research, *J. Microwave Power,* 10, 249, 1975.

34. NCRP, Radiofrequency Electromagnetic Fields: Properties, Quantities and Units, Biophysical Interactions and Measurements, Rep. No. 67, National Council on Radiation Protection and Measurements, Washington, D.C., 1981.

35. **Stuchly, M. A.,** Health Aspects of Radiofrequency and Microwave Radiation Exposure, Part 2, Department of National Health and Welfare, Ottawa, 1978.

36. **Johnson, C. C., Durney, C. H., Barber, P. W., Massoudi, H., Allen, S. J., and Mitchell, J. C.,** *Radiofrequency Radiation Dosimetry Handbook,* SAM-TR-76-35, Brooks Air Force Base, Tex., AFSC, AMD, SAM, 1976.

37. **Heller, J. H.,** Cellular effects of microwave radiation, in Biological Effects and Health Implications of Microwave Radiation, Symp. Proc., Cleary, S. F., Ed., Public Health Service BRH/DBE 70-21, U.S. Department of Health, Education, and Welfare, 1970, 116.

38. **Janes, D. E., Leach, W. M., Mills, W. A., Moore, R. T., and Shore, M. L.,** Effects of 2450 MHz microwaves on protein synthesis and on chromosomes in Chinese hamsters, *Nonioniz. Radiat.,* 1, 125, 1969.

39. **Yao, K. T. S.,** Microwave radiation-induced chromosomal aberrations in corneal epithelium of Chinese hamsters, *J. Heredity,* 69, 409, 1978.

40. **Chen, K. M., Samuel, A., and Hoopingavner, R.,** Chromosomal abberrations of living cells induced by microwave radiation, *Environ. Lett.,* 6, 37, 1974.

41. **Yao, K. T. S.,** Cytogenetic consequences of microwave irradiation on mammalian cells incubated *in vitro,* *J. Heredity,* 73, 133, 1982.

42. **Huang, A. T., Engle, M. E., Elder, J. A., Kinn, J. B., and Ward, T. R.,** The effect of microwave radiation (2450 MHz) on the morphology and chromosomes of lymphocytes, *Radio Sci.,* 12(S), 173, 1977.

43. **Yao, K. T. S.,** Cytogenetic consequences of microwave incubation of mammalian cells in culture, *Genetics,* 83 (Suppl.), 584 (Abstr.), 1976.

44. **Manikowska, E., Luciani, J. M., Servantie, B., Czerski, P., Obrenovitch, J., and Stahl, A.,** Effects of 9.4 GHz microwave exposure on meiosis in mice, *Experientia,* 35, 388, 1979.

45. **Savage, J. R. K.,** Use and abuse of chromosomal aberrations as an indicator of genetic damage, *Int. J. Environ. Stud.,* 1, 233, 1977.

46. **Webb, S. J. and Booth, A. D.,** Absorption of microwaves by microorganisms, *Nature,* 222, 1199, 1969.

47. **Webb, S. J. and Booth, A. D.,** Microwave absorption by normal and tumor cells, *Science,* 174, 72, 1971.

48. **Stamm, M. E., Winters, W. D., Morton, D. L., and Warren, S. L.,** Microwave characteristics of human tumor cells, *Oncology,* 29, 294, 1974.

49. **Dardanoni, L., Torregrossa, V., Tamburello, C., Zanforlin, L., and Spalla, M.,** Biological effects of millimeter waves at spectral singularities, in *Electromagnetic Compatibility, 3rd Wroclaw Symp. on Electromagnetic Compatibility,* Wydawnictwo Politechniki, Wroclawskiej, Breslau, Poland, 1976, 308.

50. **Lee, R. A. and Webb, S. J.,** Possible detection of *in vivo* viruses by fine-structure millimeter microwave spectroscopy between 68 and 76 GHz, *IRCS Med. Sci.,* 5, 222, 1977.

51. **Devyatkov, N. D., et al.,** Highlights of the papers presented on millimeter wave biological effects, *Sov. Phys. — USPEKHI,* 4, 568, 1974.

52. **Stamm, M. E., Warren, S. L., Rand, R. W., et al.,** Microwave therapy experiments with B-16 murine melanoma, *IRCS Med. Sci.,* 3, 392, 1975.

53. **Berteaud, A. J., Dardalhon, M., Rebeyrotte, N., and Averbeck, D.,** Action d'un rayonnement electromagnetique a longueur d'onde millimetrique sur la croissance bacterienne, *C.R. Acad. Sci. Ser. D,* 281, 843, 1975.

54. **Grundler, W. and Keilmann, F.,** Sharp resonances in yeast prove nonthermal sensitivity to microwaves, *Phys. Rev. Lett.,* 51, 214, 1983.

55. **Grundler, W. and Keilmann, F.,** Nonthermal effects of millimeter microwaves on yeast growth, *Z. Naturforsch.,* 33C(1/2), 15, 1978.

56. **Blackman, C. F., Surles, M. C., and Benane, S. G.,** The effects of microwave exposure on bacteria mutation reduction, Vol. 1, Publ. (FDA) 77-8010, Symp. Biol. Eff. of E.M. Waves, U.S. Department of Health, Education and Welfare, Rockville, Md., 1976, 406.

57. **Varma, M. M. and Traboulay, E. A.,** Evaluation of dominant lethal test and DNA studies in measuring mutagenicity caused by nonionizing radiation, in Biological Effects of Electromagnetic Waves, Vol. 1, Publ. (FDA) 77-8010, Johnson, C. C. and Shore, M. L., Eds., U.S. Department of Health, Education, and Welfare, Rockville, Md., 1976, 386.

58. **Anderstam, B., Hamnerius, Y., Hussain, S., and Ehrenberg, L.,** Studies of possible genetic effects in bacteria of high frequency electromagnetic fields, *Hereditas,* 98, 11, 1983.

59. **Mittler, S.,** Failure of 2 and 10 meter ratio waves to induce genetic damage in *Drosophila melanogaster, Environ. Res.,* 11, 326, 1976.

60. **Pay, T. L., Beyer, E. C., and Reichelderfer, C. F.,** Microwave effects on reproductive capacity and genetic transmission in *Drosophila melanogaster, J. Microwave Power,* 7, 75, 1972.

61. U.S. Environmental Protection Agency, Biological Effects of Radiofrequency Radiation, Elder, J. A. and Cahill, D. F., Eds., Rep. No. EPA-600/8-83-026F, Research Triangle Park, N.C., 1984.

62. **Varma, M. M. and Traboulay, A. E.,** Comparison of native and microwave irradiated DNA, *Experientia,* 33, 1649, 1977.

63. **McRee, D. I., MacNichols, G., and Livingston, G. K.,** Incidence of sister chromatid exchange in bone marrow cells of the mouse following microwave exposure, *Radiat. Res.,* 85, 304, 1981.

64. **Alam, M. T., Barthakur, N., Lambert, N. G., and Kasatiya, S. S.,** Cytological effects of microwave radiation in Chinese hamster cells *in vitro, Can. J. Genet. Cytol.,* 20, 23, 1978.

65. **McLees, B. D., Finch, E. D., and Albright, M. L.,** An examination of regenerating hepatic tissue subjected to radiofrequency irradiation, *J. Appl. Physiol.,* 32, 78, 1972.

66. **Dutta, S. K., Hossain, M. A., Ho, H. S., and Blackman, C. F.,** Effects of 8.6-GHz pulsed electromagnetic radiation on an *Escherichia coli* repair-deficient mutant, in *Electromagnetic Fields in Biological Systems,* Stuchly, S. S., Ed., IMPI, Edmonton, Alberta, Canada, 1979, 76.

67. **Dutta, S. K., Nelson, W. H., Blackman, C. F., and Brusick, D. J.,** Cellular effects in microbial tester strains caused by exposure to microwaves or elevated temperatures, *J. Environ. Pathol. Toxicol.,* 3, 195, 1980.

68. **Dardalhon, M., Averbeck, D., and Berteaud, A. J.,** Determination of thermal equivalent of millimeter microwaves in living cells, *J. Microwave Power,* 14, 307, 1979.

69. **Dardalhon, M., Averbeck, D., and Berteaud, A. J.,** Action des Ondes Centimetriques Seules ou Combinees avec les Rayons Ultra Violets sur les Cellules Eucaryotiques, in *URSI Int. Symp. Proc., Ondes Electromagnetiques et Biologie,* Berteaud, A. J. and Servantie, B., Eds., Paris, 1980, 17.

70. **Correlli, J. C., Gutmann, R. J., Kohazi, S., and Levy, J.,** Effects of 2.6-4.0 GHz microwave radiation on *E. coli* B, *J. Microwave Power,* 12, 141, 1977.

71. **Blackman, C. F., Benane, S. G., Weil, C. M., and Ali, J. S.,** Effects of nonionizing electromagnetic radiation on single-cell biologic systems, *Ann. N.Y. Acad. Sci.,* 247, 352, 1975.

72. **Hamnerius, Y., Olofsson, H., Rasmuson, A., and Rasmuson, B.,** A negative test for mutagenic action of microwave radiation in *Drosophila melanogaster, Mutation Res.,* 68, 217, 1979.

73. **Mittler, S.,** Failure of chronic exposure to nonthermal FM radio waves to mutate *Drosophila, J. Heredity,* 68, 257, 1977.

74. **Berman, E., Carter, H. B., and House, D.,** Tests of mutagenesis and reproduction in male rats exposed to 2450-MHz (CW) microwaves, *Bioelectromagnetics,* 1, 65, 1980.

75. **Smolyanskaya, A. Z. and Vilenskaya, R. L.,** Effects of millimeter-band electromagnetic radiation on the functional activity of certain genetic elements of bacterial cells, *Trans. Usp Fiz. Nauk,* 110, 571, 1973.

76. **Mickey, G. H. and Koerting, L.,** Chromosome breakage in cultured Chinese hamster cells induced by radiofrequency treatment, *Environ. Mutagen Soc.,* 3, 25, 1970.

77. **Blevins, R. D., Crenshaw, R. C., Houghland, A. E., and Clark, C. E.,** The effects of microwave radiation and heat on specific mutants of *Salmonella typhimurium* LT2, *Radiat. Res.,* 82, 511, 1980.

78. **Baranski, S., Debiec, H., Kwarecki, K., and Mezykowski, T.,** Influence of microwaves on genetical processes of *Aspergillus nidulans, J. Microwave Power,* 11, 146, 1976.

79. **Saunders, R. D., Darby, S. C., and Kowalczuk, C. I.,** Dominant lethal studies in male mice after exposure to 2.45 GHz microwave radiation, *Mutation Res.,* 117, 345, 1983.

80. **Dutta, S. K., Nelson, W. H., Blackman, C. F., and Brusick, D. J.,** Lack of microbiol genetic response to 2.45 GHz CW and 8.5 to 9.6 GHz pulsed microwaves, *J. Microwave Power,* 14, 275, 1979.

81. **Livingston, G. K., Johnson, C. C., and Dethlefsen, L. A.,** Comparative effects of water bath and microwave-induced hyperthermia on cell survival and sister chromatid exchange in Chinese hamster ovary cells, in Abstr. 1977 Int. Symp. on the Biological Effects of Electromagnetic Waves, Airlie, Va., 1977, 106.

82. **McRee, D. I., MacNichols, G., and Livingston, G. K.,** Incidence of sister chromatid exchange in bone marrow cells of the mouse following microwave exposure, *Radiat. Res.,* 85, 340, 1981.

83. **Meltz, M. L. and Walker, K. A.,** Genetic effects of microwave exposure on mammalian cells *in vitro,* in USAF Radiofrequency Radiation Bioeffects Research Program — A Review, Mitchell, J. C., Ed., Aeromed. Rev. 4-18, Rep. No. SAM-TR-81-30, 1981, 201.

84. **Brown, R. F., Marshall, S. V., and Hughes, C. W.,** Effects of RFR on excision-type DNA repair *in vivo,* in USAF Radiofrequency Radiation Bioeffects Research Program — A Review, Mitchell, J. C., Ed., Aeromed. Rev. 4-18, Rep. No. SAM-TR-81-30, 1981, 184.

85. **Mickey, G. H.,** Electromagnetism and its effect on the organism, *N.Y. State J. Med.,* 63, 1935, 1963.

86. **Baranski, S. and Czerski, P.,** *Biological Effects of Microwaves,* Dowden, Hutchinson & Ross, Stroudsburg, Pa., 1976.

87. **Varma, M. M., Dage, E. L., and Joshi, S. R.,** Mutagenicity induced by nonionizing radiation in Swiss male mice, in Biological Effects of Electromagnetic Waves, Vol. 1, Johnson, C. C. and Shore, M. L., Eds., Publ. (FDA) 77-8010, U.S. Department of Health, Education, and Welfare, Rockville, Md., 1976, 397.

88. **Leach, W. M.,** On the induction of chromosomal aberrations by 2450 MHz microwave radiation, *J. Cell Biol.,* 70(S), 387a (Abstr.), 1976.

89. **Liu, L. M., Nickless, F. G., and Cleary, S. F.,** Effects of microwave radiation on erythrocyte membranes, *Radio Sci.,* 14(6S), 109, 1979.

90. **Janiak, M. and Szmigielski, S.,** Injury of cell membranes in normal and SV40-virus transformed fibroblasts exposed *in vitro* to microwave (2450 MHz) or water-bath hyperthermia (43°C), in Abstr. of 1977 Int. Symp. on the Biological Effects of Electromagnetic Waves, Airlie, Va., 1977.

91. **Guy, A. W.,** A method for exposing cell cultures to electromagnetic fields under controlled conditions of temperature and field strength, *Radio Sci.,* 12(6S), 87, 1977.

92. **Michaelson, S. M.,** Biological effects of microwave exposure, in Biological Effects and Health Implications of Microwave Radiation, Symp. Proc., Cleary, S. F., Ed., Public Health Service BRH/DBE 70-2, U.S. Department of Health, Education and Welfare, 1970, 35.

93. **Michaelson, S. M.,** Biologic and pathophysiologic effects of exposure to microwaves, in *Microwave Bioeffects and Radiation Safety,* Stuchly, M. A., Ed., Trans. IMPI, Vol. 8, The International Microwave Power Institute, Edmonton, Alberta, Canada, 1978, 55.

94. **Stodolnik-Baranska, W.,** The effects of microwaves on human lymphocyte cultures, in *Biologic Effects and Health Hazards of Microwave Radiation,* Czerski, P. et al., Eds., Polish Medical Publishers, Warsaw, 1974, 189.

95. **Szmigielski, S., Szydinski, A., Pietraszek, A., and Bielec, M.,** Acceleration of cancer development in mice by long-term exposition to 2450 MHz microwave fields, in *URSI Int. Symp. Proc., Ondes Electromagnetiques et Biologie,* Berteaud, A. J. and Servantie, B., Eds., Paris, 1980.

96. **Szmigielski, S., Szudzinski, A., Pietraszek, A., Bielec, M., Jahiak, M., and Wrembel, J. K.,** Accelerated development of spontaneous and benzopyrene-induced skin cancer in mice exposed to 2450 MHz microwave radiation, *Bioelectromagnetics,* 3, 179, 1982.

97. **Preskorn, S. H., Edwards, W. D., and Justesen, D. R.,** Retarded tumor growth and augmented longevity in mice after fetal irradiation by 2450 MHz microwaves, *J. Surg. Oncol.,* 10, 483, 1978.

98. **Prausnitz, S. and Susskind, C.,** Effects of chronic microwave irradiation in mice, *IRE Trans. Biomed. Electron.,* 9, 104, 1962.

99. **Roberts, N. J., Jr. and Michaelson, S. M.,** Microwave and neoplasia in mice: analysis of a reported risk, *Health Phys.,* 44, 430, 1983.

100. **Dumansky, Y. D. and Rudichenko, V. F.,** Dependence of the functional activity of liver mitochondria with super-high frequency radiation, *Hyg. Sanit.,* 16, 1976.

101. **Elder, J. A., Ali, J. S., Long, M. D., and Anderson, G. E.,** A coaxial air line microwave exposure system: respiratory activity of mitochondria irradiated at 2-4 GHz, in Biological Effects of Electromagnetic Waves, Publ. (FDA) 77-8010, Vol. 1, U.S. Department of Health, Education and Welfare, Rockville, Md., 1976, 352.

102. **Straub, K. D. and Carver, P.,** Effects of electromagnetic fields on microsomal ATPase and mitochondrial oxidative phosphorylation, *Ann. N.Y. Acad. Sci.,* 247, 292, 1975.

103. **Ward, T. R., Allis, J. W., and Elder, J. A.,** Measure of enzymatic activity coincident with 2450 MHz microwave exposure, *J. Microwave Power,* 10, 315, 1975.

104. **Allis, J. W.,** Irradiation of bovine serum albumin with a crossed-beam exposure-detection system, *Ann. N.Y. Acad. Sci.,* 247, 312, 1975.

105. **Belkhode, M. L., Johnson, D. L., and Muc, A. M.,** Thermal and athermal effects of microwave radiation on the activity of glucose-6-phosphate dehydrogenese in human blood, *Health Phys.,* 26, 45, 1974.

106. **Bini, M., Checcucci, A., Ignesti, A., Millanta, L., Rubino, N., Camici, S., Manao, G., and Ramponi, G.,** Analysis of the effects of microwave energy on enzymatic activity of lactate dehydrogenase (LDH), *J. Microwave Power,* 13, 96, 1978.

107. **Albert, E. N., McCullars, G., and Short, M.,** The effect of 2450 MHz microwave radiation on liver adenosine triphosphate (ATP), *J. Microwave Power,* 9, 205, 1974.

108. **Allis, J. W. and Fromme, M. L.,** Activity of membrane-bound enzymes exposed to sinusoidally modulated 2450 MHz microwave radiation, *Radio Sci.,* 14, 85, 1979.

109. **Dumansky, Y. D., Serdyuk, A. M., Litvinova, C. I., Tomashevskaya, L. A., and Popovich, V. M.,** Experimental research on the biological effects of 12-centimeter low-intensity waves, in *Health in Inhabited Localities,* 2nd ed., Kiev, 1972, 29.

110. **Malyshev, V. T. and Tkachenko, M. I.,** Activity of ferments on the mucous membrane of the small intestine under the influence of an SHF field, in *Physiology and Pathology of Digestion,* Kishenev, 1972, 186.

111. **Wangemann, R. T. and Cleary, S. F.,** The *in vivo* effects of 2.45 GHz microwave radiation on rabbit serum components, *Radiat. Environ. Biophys.,* 13, 89, 1976.

112. **Fulk, D. W. and Finch, E. D.,** Effects of Microwave Irradiation *In Vivo* on Rabbit Blood Serum, Rep. No. 5, Project MF 51.524.015-0001BD7X, Naval Medical Research Institute, Bethesda, Md., 1972.

113. **Swiecicki, W. and Edelwejn, Z.,** The influence of 3 cm and 10 cm microwave irradiation in blood proteins in rabbits, *Med. Lotnicza,* 11, 54, 1963.

114. **Hamrick, P. E.,** Thermal denaturation of DNA exposed to 2450 MHz CW microwave radiation, *Radiat. Res.,* 56, 400, 1973.

115. **Baranski, S.,** Histological and histochemical effects of microwave irradiation on the central nervous system of rabbits and guinea pigs, *Am. J. Phys. Med.,* 51, 182, 1972.

116. **Nikogosyan, S. V.,** Influence of UHF on the cholinesterase activity in the blood serum and organs in animals, in *The Biological Action of Ultrahigh Frequencies,* Letavet, A. A. and Gordon, Z. V., Eds., JPRS 12471, 1962.

117. **Revutsky, E. L. and Edelman, F. M.,** Effects of centimeter and meter electromagnetic waves in the content of biologically active substances in human blood, *Philos. J. Ukr. Acad. Sci.,* 10, 379, 1964.

118. **Olcerst, R. B. and Rabinowitz, J. R.,** Studies on the interaction of microwave radiation with cholinesterase, *Radiat. Environ. Biophys.,* 15, 289, 1978.

119. **Ho, H. S. and Edwards, W. P.,** Oxygen-consumption rate of mice under differing dose rates of microwave radiation, *Radio Sci.,* 12(6S), 131, 1977.

120. **Elder, J. A. and Ali, J. S.,** The effect of microwaves (2450 MHz) on isolated rat liver mitochondria, *Ann. N.Y. Acad. Sci.,* 247, 251, 1975.

121. **Paulsson, L.-E., Hamnerius, Y., and McLean, W. G.,** The effects of microwave radiation on microtubules and axonal transport, *Radiat. Res.,* 70, 212, 1977.

122. **Allis, J. W., Fromme, M. L., and Janes, D. E.,** Pseudosubstrate binding to ribonuclease during exposure to microwave radiation at 1.70 and 2.45 GHz, in Biological Effects of Electromagnetic Waves, Vol. 1, Johnson, C. C. and Shore, M. L., Eds., HEW Publ. (FDA) 77-8010, Department of Health, Education and Welfare, Rockville, Md., 1976, 366.

123. **Hamrick, P. E. and Fox, S. S.,** Rat lymphocytes in cell culture exposed to 2450 MHz (CW) microwave radiation, *J. Microwave Power,* 12, 125, 1977.

124. **Henderson, H. M., Hergenroeder, K., and Stuchly, S. S.,** Effect of 2450 MHz microwave radiation on horseradish peroxidase, *J. Microwave Power,* 10, 27, 1975.

125. **Lovely, R. H., Myers, D. E., and Guy, A. W.,** Irradiation of rats by 918-MHz microwaves at 2.5 mW/cm²: delineating the dose-response relationship, *Radio Sci.,* 12(6S), 139, 1977.

126. **Chamness, A. F., Scholes, H. R., Sexauer, S. W., and Frazer, J. W.,** Metal ion content of specific areas of the rat brain after 1600 MHz radio frequency irradiation, *J. Microwave Power,* 11, 333, 1976.

127. **Ho, H. S. and Edwards, W. P.,** The effect of environmental temperature and average dose rate of microwave radiation on the oxygen-consumption rate of mice, *Radiat. Environ. Biophys.,* 16, 325, 1979.

128. **Sanders, A. P., Schaefer, D. J., and Joines, W. T.,** Microwave effects on energy metabolism of rat brain, *Bioelectromagnetics,* 1, 171, 1980.

129. **Phillips, R. D., Hunt, E. L., Castro, R. D., and King, N. W.,** Thermoregulatory metabolic and cardiovascular response of rats to microwaves, *J. Appl. Physiol.,* 38, 630, 1975.

130. **Adair, E. R. and Adams, B. W.,** Adjustments in metabolic heat production by squirrel monkeys exposed to microwaves, *J. Appl. Physiol.,* 52(4), 1049, 1982.

131. **Boggs, R. F., Sheppard, A. P., and Clark, A. J.,** Effects of 2450 MHz microwave radiation on human blood coagulation processes, *Health Phys.,* 22, 217, 1972.

132. **Ismailov, E. Sh.,** Effect of ultrahigh frequency electromagnetic radiation on electrophoretic mobility of erythrocytes (transl.), *Biophysics,* 22, 510, 1978; *Biofizika,* 22, 493, 1977.

133. **Ismailov, E. Sh.,** Mechanism of effects of microwaves on erythrocyte permeability for potassium and sodium ions, *Biol. Nauki,* 3, 58, 1971; (English trans.: JPRS 72606, p. 38, January 12, 1979.

134. **Hamrick, P. E. and Zinkl, J. G.,** Exposure of rabbit erythrocytes to microwave radiation, *Radiat. Res.,* 62, 164, 1975.

135. **Peterson, D. J., Partlow, L. M., and Gandhi, O. P.,** An investigation of the thermal and athermal effects of microwave irradiation on erythrocytes, *IEEE Trans. Biomed. Eng.,* 26, 428, 1979.

136. **Olcerst, R. B., Belman, S., Eisenbud, M., Mumford, W. W., and Rabinowitz, J. R.,** The increased passive efflux of sodium and rubidium from rabbit erythrocytes by microwave radiation, *Radiat. Res.,* 82, 244, 1980.

137. **Mirutenko, V. I.,** Investigating local thermal effect of electromagnetic (3 cm) waves on animals, *Fiziol. Zh. Akad. Nauk. UKR SSR,* 8, 382, 1962.

138. **Mirutenko, V. I.,** Effect of blood circulation on the distribution of heat, and the magnitude of the thermal effect during action of a SHF-UHF electromagnetic field on animals, *Fiziol. Zh. Akad. Nauk. UKR SSR,* 10, 641, 1964.

139. **Cook, H. F.,** The pain threshold for microwave and infra-red radiation, *J. Physiol.,* 118, 1, 1952.

140. **Herrick, J. F. and Krusen, F. H.,** Certain physiologic and pathologic effects of microwaves, *Electron. Eng.,* 72, 239, 1953.

141. **Krusen, F. H., Herrick, J. F., Leden, U., and Wakim, K. G.,** Microkymatotherapy: preliminary report of experimental studies of the heating effect of microwaves (radar) in living tissues, *Proc. Mayo Clin.,* 22, 209, 1947.

142. **Rae, J. W., Jr., Herrick, J. F., Wakim, K. G., and Krusen, F. H.,** A comparative study of temperatures produced by microwave and short wave diathermy, *Arch. Phys. Med.,* 30, 199, 1949.

143. **Semenov, A. I.,** The effect of UHF on the temperature of rabbit femoral tissues, *Byull. Eksper. Biol. Med.,* 60, 64, 1965.

144. **Engle, J. P., Herrick, J. F., Wakim, K. G., Grindlay, J. H., and Krusen, F. H.,** The effects of microwaves on bone and bone marrow and on adjacent tissues, *Arch. Phys. Med.,* 31, 453, 1950.

145. **Gersten, J. W., Wakim, K. G., Herrick, J. F., and Krusen, F. H.,** Effect of microwave diathermy on the peripheral circulation and on tissue temperature in man, *Arch. Phys. Med.,* 30, 7, 1949.

146. **Gordon, Z. V.,** The problem of the biological action of UHF, *Tr. Gig. Tr. Prof. AMN SSR,* 1, 5, 1960.

147. **Gordon, Z. V. and Lobanova, Y. E.,** The temperature reaction of animals to exposure to SHF, in *The Biological Action of Super High-Frequency Fields,* Letavet, A. A. and Gordon, Z. V., Eds., Academy of Medical Science, Moscow, 1960.

148. **Tyagin, N. W.,** Study of the thermal effect of SHF-UHF electromagnetic fields on various animals using the thermometric method, *Tr. Voyenno Med. Akad. Kirov. (U.S.S.R.),* 73, 9, 1957.

149. **Tyagin, N. V.,** Change in the blood of animals subjected to a SHF-UHF field, *Voyenno-Medit. Akad. Kirov Leningrad,* 73, 116, 1957.

150. **Ely, T. S., Goldman, D., Hearon, J. Z., Williams, R. B., and Carpenter, H. M.,** Heating characteristics of laboratory animals exposed to ten centimeter microwaves, U.S. Naval Medical Research Institute (Res. Rep. Proj. NM 001-056.13.02), Bethesda, Md., *IEEE Trans. Biomed. Eng.,* 11, 123, 1964.

151. **Gorodetskaya, S. F.,** The effect of centimeter radio waves on mouse fertility, *Fiziol. Zh.,* 9, 394, 1963.

152. **Imig, C. J., Thomson, J. D., and Hines, H. M.,** Testicular degeneration as a result of microwave irradiation, *Proc. Soc. Exp. Biol.,* 69, 382, 1948.

153. **Miro, L., Loubiere, R., and Pfister, A.,** Studies of visceral lesions observed in mice and rats exposed to UHF waves; a particular study of the effects of these waves on the reproduction of these animals, *Rev. Med. Aeronaut. (Paris),* 4, 37, 1965.

154. **Bereznitskaya, A. N. and Kazbekov, I. M.,** Studies on the reproduction and testicular microstructure of mice exposed to microwaves, in *Biological Effects of Radiofrequency Electromagnetic Fields,* Gordon, Z. V., Ed., No. 4, Moscow 221-229 (JPRS63321), 1974.

155. **Saunders, R. D. and Kowalczuk, C. I.,** Effects of 2.45 GHz microwave radiation and heat on mouse spermatogenic epithelium, *Int. J. Radiat. Biol.,* 40(6), 623, 1981.

156. **vanDemark, W. R. and Free, J. R.,** Temperature effects, in *The Testis,* Vol. 3, Johnson, A. D., Gomes, W. R., and vanDeman, M. L., Eds., Academic Press, New York, 1973, 233.

157. **Muraca, G. J., Jr., Ferri, E. S., and Buchta, F. L.,** A study of the effects of microwave irradiation of the rat testes, in Biological Effects of Electromagnetic Waves, Vol. 1, Johnson, C. C. and Shore, M. L., Eds., Publ. (FDA) 77-8010, Department of Health and Welfare, Rockville, Md., 1977, 484.

158. **Haidt, S. J. and McTighe, A. H.,** The effect of chronic, low-level microwave radiation on the testicles of mice, in 1973 IEEE-G-MIT Int. Microwave Symp., Maley, S. W., Ed., 1973, 324.

159. **Cairnie, A. B., Hill, D. A., and Assenheim, H. M.,** Dosimetry for a study of effects of 2.45-GHz microwaves on mouse testes, *Bioelectromagnetics,* 1, 325, 1980.

160. **Muraca, G. J., Ferri, E. S., and Buchta, F. L.,** A study of the effects of microwave irradiation of the rat testes, in Biological Effects of Electromagnetic Waves, Vol. 1, Johnson, C. C. and Shore, M. L., Eds., DHEW Publ. (FDA) 77-8010, Department of Health, Education and Welfare, Rockville, Md., 1976, 484.

161. **Gunn, S. A., Gould, T. C., and Anderson, W. A. D.,** The effect of microwave radiation on morphology and function of rat testis, *Lab. Invest.,* 10(2), 301, 1961.

162. **Carpenter, R. L. and Livstone, E. M.,** Evidence for nonthermal effects of microwave radiation: abnormal development of irradiated insect pupae, *IEEE Trans. Microwave Theory Tech.,* 19, 173, 1971.

163. **Van Ummersen, C. A.,** The effect of 2450 mc radiation on the development of the chick embryo, in *Biological Effects of Microwave Radiation,* Vol. 1, Peyton, M. F., Ed., Plenum Press, New York, 1961, 201.

164. **Van Ummersen, C. A.,** An Experimental Study of Development Abnormalities Induced in the Chick Embryo by Exposure to Radio Frequency Waves, Ph.D. thesis, Tufts University, Medford, Mass., 1963.

165. **Bereznitskaya, A. N.,** Research on the reproductive function in female mice under the impact of low-intensity radio waves of different ranges, in Industrial Health and Biological Effects of Radiofrequency Electromagnetic Waves, Material of the 4th All-Union Symp., October 17 to 19, 1972, Moscow, 1972.

166. **Rugh, R., Ginns, E. I., Ho, H. S., and Leach, W. M.,** Are microwaves teratogenic? in *Biologic Effects and Health Hazards of Microwave Radiation,* Czerski, P. et al., Eds., Polish Medical Publishers, Warsaw, 1974, 98.

167. **Rugh, R., Ginns, E. I., Ho, H. S., and Leach, W. M.,** Responses of the mouse to microwave radiation during estrous cycle and pregnancy, *Radiat. Res.,* 62, 225, 1975.

168. **Olsen, R. G.,** Constant dose microwave irradiation of insect pupae, *Radio Sci.,* 17(5S), 145, 1982.

169. **Lindauer, G. A., Liu, L. M., Skewes, G. W., and Rosenbaum, F. J.,** Further experiments seeking evidence of nonthermal biological effects of microwave radiotion, *IEEE Trans. Microwave Theory Tech.,* 22, 790, 1974.

170. **Liu, L. M., Rosenbaum, F. J., and Pickard, W. F.,** The relation of teratogenesis in *Tenebrio molitor* to the incidence of low level microwaves, *IEEE Trans. Microwave Theory Tech.,* 23, 929, 1975.

171. **Olsen, R. G.,** Insect teratogenesis in a standing-wave irradiation system, *Radio Sci.,* 12(6S), 199, 1977.

172. **Pickard, W. F. and Olsen, R. G.,** Developmental effects of microwaves in *Tenebrio molitor:* experiments to detect possible influences of radiation frequency and of culturing protocols, in *Abstr. of the 1977 USNC/URSI Symp. on Biological Effects of Electromagnetic Waves,* National Academy of Sciences, Washington, D.C., 1977, 66.

173. **McRee, D. I., Hamrick, P. E., Zinkl, J. E., Thaxton, P., and Parkhurst, C. R.,** Some effects of exposure of the Japanese quail embryo to 2.45 GHz microwave radiation, *Ann. N.Y. Acad. Sci.,* 247, 377, 1975.

174. **McRee, D. I. and Hamrick, P. E.,** Exposure of Japanese quail embryos to 2.45 GHz microwave radiation during development, *Radiat. Res.,* 71, 355, 1977.

175. **Laskey, J., Dawes, D., and Howes, M.,** Progress report on 2450 MHz irradiation of pregnant rats and the effect on the fetus, in Radiation Bioeffects Summary Rep. PHS, Publ. BRH/DBE-70, U.S. Department of Health, Education and Welfare, Rockville, Md., 1970, 167.

176. **Chernovetz, M. E., Justesen, D. R., King, N. W., and Wagner, J. E.,** Teratology, survival, and reversal learning after fetal irradiation of mice by 2450 MHz microwave energy, *J. Microwave Power,* 10, 391, 1975.

177. **Chernovetz, M. E., Justesen, D. R., and Oke, A. F.,** A teratologic study of the rat: microwave and infrared radiations compared, *Radio Sci.,* 12(6S), 191, 1977.

178. **Michaelson, S. M., Guillet, R., Catallo, M. A., Small, J., Inamine, G., and Heggeness, F. W.,** Influence of 2450 MHz microwaves on rats exposed *in utero, J. Microwave Power,* 11, 165, 1976.

179. **Berman, E., Kinn, J. B., and Carter, H. B.,** Observations of mouse fetuses after irradiation with 2.45 GHz microwaves, *Health Phys.,* 35, 791, 1978.

180. **Johnson, R. B., Mizumori, S., and Lovely, R. H.,** Adult behavioral deficit in rats exposed prenatally to 918-MHz microwaves, in Developmental Toxicology of Energy-Related Pollutants, DOE Symp., Ser. 47, Department of Energy, Washington, D.C., 1977, 281.

181. **Nawrot, P. S., McRee, D. I., and Staples, R. E.,** Effects of 2.45 GHz CW microwave radiation on embryofetal development in mice, *Teratology,* 24, 303, 1981.

182. **Jensh, R. P., Weinburg, L., and Brent, R. L.,** Teratologic studies of prenatal exposure of rats to 915 MHz microwave radiation, *Radiat. Res.,* 92, 160, 1982.

183. **Rugh, R. and McManaway, M.,** Can electromagnetic waves cause congenital anomalies?, in Int. IEEE/AP-S USNC/URSI Symp., Amherst, Mass., 1976, 143.

184. **Rugh, R. and McManaway, M.,** Anesthesia as an effective agent against the production of congenital anomalies in mouse fetuses exposed to electromagnetic radiation, *J. Exp. Zool.,* 197, 363, 1976.

185. **Lin, J. C., Nelson, J. C., and Ekstrom, M. E.,** Effects of repeated exposure to 148 MHz radiowaves on growth and hematology of mice, *Radio Sci.,* 14, 173, 1979.

186. **Stavinoha, W. B., Medina, M. A., Frazer, J., Weintraub, S. T., Ross, D. H., Modak, A. L., and Jones, D. J.,** The effects of 19 megacycle irradiation on mice and rat, in Biological Effects of Electromagnetic Waves, Vol 1, Johnson, C. C. and Shore, M. L., Eds., Publ. (FDA) 77-8010, Department of Health, Education and Welfare, Rockville, Md., 1975, 431.

187. **Nelson, J. C., Lin, J. C., and Ekstrom, M. E.,** Teratogenic effects of RF radiation on mice, presented at Bioelectromagnetics Symp., Seattle, Wash., June 18 to 22, 1979.

188. **Dietzel, F. and Kern, W.,** Abortion following ultra-shortwave hyperthermia animal experiments, *Arch. Gynakol.,* 209, 445, 1970.

189. **Dietzel, F., Kern, W., and Steckenmesser, R.,** Deformity and intrauterine death after short-wave therapy in early pregnancy in experimentsl animals, *Muench. Med. Wochenschr.,* 114, 228, 1972.

190. **Dietzel, F.,** Effects of nonionizing electro-magnetic radiation on the development and intrauterine implantation of the rat, *Ann. N.Y. Acad. Sci.,* 247, 367, 1975.

191. **Smialowicz, R. J., Kinn, J. B., and Elder, J. A.,** Perinatal exposure of rats to 2450 MHz (CW) microwave radiation: effects on lymphocytes, *Radio Sci.,* 14, 147, 1979.

192. **Jensh, R. P., Vogel, W. H., and Brent, R. L.,** Postnatal functional analysis of prenatal exposure of rats to 915 MHz microwave radiation, *J. Am. Coll. Toxicol.,* 1, 73, 1982.

193. **Jensh, R. P., Weinberg, I., and Brent, R. L.,** An evaluation of the teratogenic potential or protracted exposure of pregnant rats to 2450 MHz microwave radiation. I. Morphologic analysis at term, *J. Toxicol. Environ. Health,* 11, 23, 1983.

194. **Jensh, R. P., Vogel, W. H., and Brent, R. L.,** An analysis of the teratogenic potential or protracted exposure of pregnant rats to 2450 MHz microwave radiation. II. Postnatal psychophysiologic analysis, *J. Toxicol. Environ. Health,* 11, 37, 1983.

195. **Jensh, R. P. and Ludlow, J.,** Behavioral teratology: application in low dose chronic microwave irradiation studies, in *Advances in the Study of Birth Defects,* Vol. 4, Persand, T. V. N., Ed., MTP Press, Lancaster, England, 1980, chap. 8.

196. **Jensh, R. P., Vogel, W. H., Ludlow, J., and McHugh, T.,** Studies concerning the effects of low dosage prenatal 6000 MHz microwave radiation on growth and development in the rat, *Teratology,* 19(2), 32A (Abstr.), 1979.

197. **Lary, J. M., Conover, D. L., Foley, E. D., and Hanser, P. L.,** Teratgoenicity of 27.12 MHz radiofrequency radiation in rats, *Teratology,* 19, 36A, 1979.

198. **Lary, J. M., Conover, D. L., Foley, E. D., and Hanser, P. L.,** Teratogenicity of 27.12 MHz radiofrequency radiation in rats, *Bioelectromagnetics,* 2, 402, 1980.

199. **Conover, D. L., Lary, J. M., and Hanser, P. L.,** Thermal threshold for teratogenic response in rats irradiated at 27.12 MHz, *Bioelectromagnetics,* 1, 204, 1979.

200. **Lacy, K. K., Desesso, J. M., and Lary, J. M.,** A comparison of the teratogenic effects of radiofrequency radiation and hyperthermia: gross evaluation, *Teratology,* 21, 51A, 1980.

201. **Lacy, K. K., Desesso, J. M., Sadler, T. W., and Lary, J. M.,** A comparison of the teratogenic effects of radiofrequency radiation and hyperthermia: light microscopic evaluation, *Teratology,* 21, 52A, 1980.

202. **Conover, D., Lary, J. M., and Foley, E.,** Induction of teratogenic effects in rats by 27.12 MHz RF radiation, presented at the 1978 Symp. on Electromagnetic Fields in Biological Systems, Ottawa, June 27 to 30, 1978.

203. **Boak, R. A., Carpenter, C. M., and Warren, S. L.,** Studies on the physiological effects of fever temperatures. II. The effect of repeated short wave (30 meter) fevers on growth and fertility of rabbits, *J. Exp. Med.,* 56, 725, 1932.

204. **Kaplan, J. N.,** Study of the lethal effects of microwaves in the developing squirrel monkey, Final Report for Contract No. 68-02-3210, U.S. Environmental Protection Agency, 1981.

205. **Kaplan, J., Polson, P., Rebert, C., Lunan, K., and Gage, M.,** Biological and behavioral effects of prenatal and postnatal exposure to 2450-MHz electromagnetic radiation in the squirrel monkey, *Radio Sci.,* 17(5S), 135, 1982.

206. **O'Connor, M. E.,** Mammalian teratogenesis and radiofrequency fields, *Proc. IEEE,* 68, 56, 1980.

207. **Hsu, C. Y.,** Influence of temperature on development of rat embryos, *Anat. Rec.,* 100, 79, 1948.

208. **Fernandez-Cano, L.,** Effect of increase or decrease of body temperature and hypoxia on pregnancy in the rat, *Fertil. Steril.,* 9, 455, 1958.

209. **Pleet, H., Graham, J. M., and Smith, D. W.,** Central nervous system and facial defects associated with maternal hyperthermia at four to 14 weeks gestation, *Pediatrics,* 67, 785, 1981.

210. **Edwards, M. J.,** Congenital malformations in the rat following induced hyperthermia during gestation, *Teratology,* 1, 173, 1968.

211. **Edwards, M. J.,** Congenital defects in guinea pigs: fetal resorptions, abortions, and malformations following induced hyperthermia during early gestation, *Teratology,* 2, 313, 1969.

212. **Howarth, B., Jr.,** Embryonic survival in adrenalectomized rabbits following exposure to elevated ambient temperature and constant humidity, *J. Anim. Sci.,* 28, 80, 1969.

213. **Michaelson, S. M.,** Physiologic regulation in electromagnetic fields, *Bioelectromagnetics,* 3, 91, 1982.

214. **Way, W. I., Kritikos, H., and Schwan, H.,** Thermoregulatory physiologic responses in the human body exposed to microwave radiation, *Bioelectromagnetics,* 2, 341, 1981.

215. **Wilson, G. J.,** Experimental studies on congenital malformations, *J. Chron. Dis.,* 10, 111, 1959.

216. **Gruenwald, P.,** Mechanisms of abnormal development, *Arch. Pathol.,* 44, 398, 1947.

217. **Cameron, J. A.,** Termination of early pregnancy by artificial fever, *Proc. Soc. Exp. Biol. Med.,* 52, 76, 1943.

218. **Edwards, M. J.,** Congenital defects in guinea pigs following induced hyperthermia during gestation, *Arch. Pathol.,* 84, 42, 1967.

219. **Garrison, L. H.,** The effect of fever on the development of the rat incisor, *J. Dent. Res.,* 19, 215, 1940.

220. **Martson, L. V. and Voronina, V. M.,** Experimental study of the effect of a series of phosphororganic pesticides (Dipterex and Imidan) on embryogenesis, *Environ. Health Perspect.,* 13, 121, 1976.

221. **Hamrick, P. E. and McRee, D. I.,** Exposure of the Japanese quail embryo to 2.45-GHz microwave radiation during the second day of development, *J. Microwave Power,* 10, 211, 1975.

222. **Chernovetz, M. E., Justesen, D. R., and Levinson, D. M.,** Acceleration and deceleration of fetal growth of rats by 2450-MHz microwave radiation, in *Electromagnetic Fields in Biological Systems,* Stuchly, S. S., Ed., IMPI, Edmonton, Alberta, Canada, 1979, 175.

223. **Smialowicz, R. J., Ali, J. S., Berman, E., Bursian, S. J., Kinn, J. B., Liddle, C. G., Reiter, L. W., and Weil, C. M.,** Chronic exposure of rats to 100-MHz (CW) radiofrequency radiation: assessment of biological effects, *Radiat. Res.,* 86, 488, 1981.

224. **McAfee, R. D., Braus, R., Jr., and Fleming, J., Jr.,** The effect of 2450 MHz microwave irradiation on the growth of mice, *J. Microwave Power,* 8, 111, 1973.

225. **Guillet, R. and Michaelson, S. M.,** The effect of repeated microwave exposure on neonatal rats, *Radio Sci.,* 12(6S), 125, 1977.

226. **Berman, E., Carter, H. B., and House, D.,** Observations of rat fetuses after irradiation with 2450-MHz (CW) microwaves, *J. Microwave Power,* 16, 9, 1981.

227. **Johnson, R. B., Mizumori, S., and Lovely, R. H.,** Adult behavior deficit in rats exposed prenatally to 918-MHz microwaves, in Developmental Toxicology of Energy-Related Pollutants, Mahlum, D.C., Sikov, M. R., Hackett, P. L., and Andrew, F. D., Eds., DOE Symp. Ser. 47, Department of Energy, Washington, D. C., 1978, 281.

228. **Shore, M. L., Felten, R. P., and Lamanna, A.,** The effect of repetitive prenatal low-level microwave exposure on development in the rat, in Symp. on Biological Effects and Measurement of Radio Frequency/ Microwaves, Hazzard, D. G., Ed., HEW Publication (FDA) 77-8026, Department of Health, Education and Welfare, Rockville, Md., 1977, 280.

229. **Gordon, Z. V., Lobanova, Y. A., and Tolgskaya, M. S.,** Some data on the effect of centimeter waves (experimental studies), *Gig. Sanit. (U.S.S.R.),* 12, 16, 1955.

230. **Novitskiy, Y. I., Gordon, Z. V., Presman, A. S., and Kholodov, Y. A.,** Radio Frequencies and Microwaves, Magnetic and Electrical Fields, National Aeronautics and Space Administration (NASA TT F-14.021), Washington, D.C., 1971.

231. **Yakovleva, M. I., Shlyafer, I. P., and Tsvetkova, I. P.,** On the question of conditioned cardiac reflexes, the functinal and morphological state of cortical neurons under the effect of superhigh-frequency electromagnetic fields, *Zh. Vyssh. Nervn. Deyat. (U.S.S.R.),* 18, 973, 1968.

232. **Kitsovskaya, I. A.** An investigation of the interrelationships between the main nervous processes in rats on exposure to SHF fields of various intensities, *Tr. Gig. Tr. Prof. AMN SSSR,* 1, 75, 1960.

233. **Kitsovskaya, I. A.,** The effect of radiowaves of various ranges on the nervous system (sound stimulation method), in *On the Biological Effect of Radio-Frequency Electromagnetic Fields,* Moscow, 1968, 81.

234. **Lobanova, Y. A. and Gordon, Z. V.,** The study of olfactory sensitivity in persons exposed to SHF, *TR. Gig. Prof. AMN SSR,* 1, 52, 1960.

235. **Fukalova, P. P.,** The sensitivity of olfactory and optic analyzers in persons exposed to the effect of constantly-generated SW and USW, *Tr. Gig. Prog. AMN SSR (Moscow),* 2, 144, 1964.

236. **Goncharova, N. N., Karamyshev, V. B., and Maksimenko, N. V.,** Occupational hygiene problems in working wth ultrashort-wave transmitters used in TV and radio broadcasting, *Gig. Tr. Prof. Zabol.,* 10, 10, 1966.

237. **Tolgskaya, M. S. and Gordon, Z. V.,** Changes in the receptor and interoreceptor apparatuses under the influence of UHF, in *The Biological Action of Ultrahigh Frequencies,* Letavet, A. A. and Gordon, Z. V., Eds., Academy of Medical Science, Moscow, 1960, 104.

238. **Marha, K.,** Biological effects of rf electromagnetic waves, *Prac. Lek. (Prague),* 15, 387, 1963.

239. **Tolgskaya, M. S. and Gordon, Z. V.,** *Pathological Effects of Radio Waves,* Meditsina Press, Moscow, 1971; transl. Consultants Bureau, New York, 1973.

240. **Thompson, W. D. and Bourgeois, A. E.,** Effects of Microwave Exposure on Behavior and Related Phenomena, Primate Behavior Lab., Aeromedical Research Lab. Report, (ARL-TR-65-20; AD 489245), Wright-Patterson AFB, Ohio, 1975.

241. **Yermakov, Y. V.,** On the mechanism of developing astheno-vegetative disturbance under the chronic effect of a SHF-field, *Voyenno-Medit. Zh. (U.S.S.R.),* 3, 42, 1969.

242. **Gvozdikova, Z. M., Anan'yev, V. M., Zenina, I. N., and Zak, V. I.,** Sensitivity of the rabbit central nervous system to a continuous (nonpulsed) ultrahigh frequency electromagnetic field, *Byull. Eksp. Biol. Med. (Moscow),* 58, 63, 1964.

243. **Baldwin, M. S., Bach, S. A., and Lewis, S. A.,** Effects of radiofrequency energy on primate cerebral activity, *Neurology,* 10, 178, 1960.
244. **Kholodov, Y. A.,** Changes in the electrical activity of the rabbit cerebral cortex during exposure to a UHF-HF electromagnetic field. II. The direct action of the UHF-HF field on the central nervous system, *Byull. Eksp. Biol. Med. (Moscow),* 56, 42, 1963.
245. **Kholodov, Y. A.,** The influence of a VHF-HF electromagnetic field on the electrical activity of an isolated strip of cerebral cortex, *Byull. Eksp. Biol. Med. (Moscow),* 57, 98, 1964.
246. **Kholodov, Y. A.,** *The Effect of Electromagnetic and Magnetic Fields on the Central Nervous System,* NASA TT-F-465, Nauka Press, Moscow, 1966.
247. **Livanov, M. N., Tsypin, A. B., Grigoriev, Y. G., Kruschev, U. G., Stepanov, S. M., and Anen'yev, A. M.,** The effect of electromagnetic fields on the bioelectric activity of cerebral cortex in rabbits, *Byull. Eksp. Biol. Med.,* 49, 63, 1960.
248. **Baranski, S. and Edelwejn, Z.,** Electroencephalographic and morphological investigations on the influence of microwaves on the central nervous system, *Acta Physiol. Pol.,* 18, 423, 1967.
249. **Bawin, S. M., Gavalas-Medici, R. J., and Adey, W. R.,** Effects of modulated very high frequency fields on specific brain rhythms in cats, *Brain Res.,* 58, 365, 1973.
250. **Servantie, B., Servantie, A. M., and Etienne, J.,** Synchronization of cortical neurons by a pulsed microwave field as evidenced by spectral analysis of EEG from the white rat, *Ann. N.Y. Acad. Sci.,* 247, 82, 1975.
251. **Chizenkova, R. A.,** Brain biopotentials in the rabbit during exposure to electromagnetic fields, *Fiziol. Zh. SSR (Moscow),* 53, 514, 1967.
252. **Chizenkova, R. A.,** Background and induced activity of neurons of the optical cortex of a rabbit after the action of a SHF field, *Zh. Vyssh. Nervn. Deyat.,* 19, 495, 1969.
253. **Baranski, S. and Edelwejn, Z.,** Experimental morphologic and electroencephalographic studies of microwave effects on the nervous system, *Ann. N.Y. Acad. Sci.,* 247, 109, 1975.
254. **Lobanova, E. A. and Goncharova, A. V.,** Investigation of conditioned-reflex activity in animals (albino rats) subjected to the effect of ultrashort and short radio-waves, *Gig. Tr. Prof. Zabol.,* 15(1), 29, 1971.
255. **Johnson, C. C. and Guy, A. W.,** Non-ionizing electromagnetic wave effects in biological materials and systems, *Proc. IEEE,* 60, 692, 1972.
256. **Tyazhelov, V. V., Tigranian, R. E., and Khizhniak, E. P.,** New artifact-free electrodes for recording of biological potentials in strong electromagnetic fields, *Radio Sci.,* 12(6S), 121, 1977.
257. **Frey, A. H.,** A coaxial pathway for recording from the cat brain during illumination with UHF energy, *Physiol. Behav.,* 3, 363, 1968.
258. **Bawin, S. M. and Adey, W. R.,** Calcium binding in cerebral tissue, in Symp. on Biological Effects and Measurement of Radio Frequency/Microwaves, Hazzard, D. G., Ed., Publ. (FDA) 77-8026, U.S. Department of Health, Education and Welfare, Rockville, Md., 1977.
259. **Blackman, C. F., Elder, J. A., Weil, C. M., Benane, S. G., Eichinger, D. C., and House, D. E.,** Induction of calcium ion efflux from brain tissue by radio-frequency radiation: effects of modulation frequency and field strength, *Radio Sci.,* 14(6S), 93, 1979.
260. **Blackman, C. F., Benane, S. G., Elder, J. A., House, D. E., Lampe, J. A., and Faulk, J. M.,** Induction of calcium-ion efflux from brain tissue by radiofrequency radiation: effect of sample number and modulation frequency on the power-density window, *Bioelectromagnetics,* 1, 35, 1980.
261. **Tolgskaya, M. S.,** Morphological changes in animals exposed to 10 cm microwaves, *Vopr. Kurortol. Fizioter. Lech. Fiz. Kult.,* 1, 21, 1959.
262. **Tolgskaya, M. S. and Gordon, Z. V.,** Comparative morphological characterization of action of microwaves of various ranges, *Tr. Gig. Tr. Prof. AMN SSSR,* 2, 80, 1964.
263. **Oldendorf, W. H.,** Focal neurological lesions produced by microwave irradiation, *Proc. Soc. Exp. Biol. Med.,* 72, 432, 1949.
264. **de Seguin, L. and Castelain, G.,** Action of ultrahigh frequency radiation (wavelength 21 cm) on temperature of small laboratory animals, *C. R. Acad. Sci. (Paris),* 224, 1662, 1947.
265. **Bilokrynytsk'ky, V. S.,** Changes in the tigroid substance of neurons under the effect of radio waves, *Fiziol. Zh.,* 12, 70, 1966.
266. **Baranski, S., Czekalinski, L., Czerski, P., and Haduch, S.,** Experimental research on fatal effect of micrometric wave electromagnetic radiation, *Rev. Med. Aeronaut. (Paris),* 2, 108, 1966.
267. **Dolina, L. A.,** Morphological changes in the central nervous system due to the action of centimeter waves on the organism, *Arkh. Patol.,* 23, 51, 1961.
268. **Minecki, L. and Bilski, R.,** Histopathological changes in internal organs of mice exposed to the action of microwaves, *Med. Pr. (Poland),* 12, 337, 1961.
269. **Albert, E. N. and DeSantis, M.,** Do microwaves alter nervous system-structure? *Ann. N.Y. Acad. Sci.,* 247, 87, 1975.

270. **Albert, E. N.,** Light and electron microscopic observations on the blood brain barrier after microwave irradiation, in Symp. on Biological Effects and Measurements of Radiofrequency/Microwaves, Hazzard, D. S., Ed., Publ. (FDA) 77-8026, U.S. Department of Health, Education and Welfare, Rockville, Md., 1977, 294.

271. **Albert, E. N.,** Ultrastructural pathology associated with microwave induced alterations in blood-brain barrier permeability, Proc. Biol. Eff. E. M. Waves, 19th Gen. Assembly, Int. Union Radio Sci., Helsinki, August 1978.

272. **Albert, E. N.,** Reversibility of the blood brain barrier, *Radio Sci.,* 14(6S), 323, 1979.

273. **Albert, E. N., Grau, L., and Kerns, J.,** Morphologic alterations in hamster blood-brain barrier after microwave irradiation, *J. Microwave Power,* 12, 43, 1977.

274. **Albert, E. N., Brainard, D. L., Randal, J. D., and Jannatta, F. S.,** Neuropathological observations on microwave-irradiated hamsters, Proc. Biol. Eff. E. M. Waves, 19th Gen. Assembly, Int. Union Radio Sci., Helsinki, August 1978.

275. **Albert, E. N., Sherif, M. F., Papadopoulos, N. J., Slaby, F. J., and Monahan, J.,** Effect of nonionizing radiation on the purkinje cells of the rat cerebellum, *Bioelectromagnetics,* 2, 247, 1981.

276. **Albert, E. N., Sherif, M. F., and Papadopoulos, N. J.,** Effect of nonionizing radiation on the purkinje cells of the uvula in squirrel monkey cerebellum, *Bioelectromagnetics,* 2, 241, 1981.

277. **Sutton, C. H., Nunnally, R. L., and Carroll, F. B.,** Protection of the microwave-irradiated brain with body-core hypothermia, *Cryobiology,* 10, 513, 1973.

278. **Albert, E. N. and Kerns, J. M.,** Reversible microwave effects on the blood-brain barrier, *Brain Res.,* 230, 153, 1981.

279. **Oscar, K. J. and Hawkins, T. D.,** Microwave alteration of the blood-brain barrier system of rats, *Brain Res.,* 126, 281, 1977.

280. **Oldendorf, W. H.,** Measurement of brain uptake of radiolabeled substances using a tritiated water internal standard, *Brain Res.,* 24, 372, 1970.

281. **Oscar, K. J., Gruenau, S. P., Folker, M. T., and Rapoport, S. I.,** Local cerebral blood flow after microwave exposure, *Brain Res.,* 204, 220, 1981.

282. **Preston, E., Vavasour, E. J., and Assenheim, H. M.,** Permeability of the blood brain barrier to mannitol in the rat following 2450-MHz microwave irradiation, *Brain Res.,* 174, 109, 1979.

283. **Chang, B. K., Huang, A. T., Joines, W. T., and Kramer, R. S.,** The effect of microwave radiation (1.0 GHz) on the blood-brain barrier in dogs, *Radio Sci.,* 17(5S), 165, 1982.

284. **Williams, W. M., Hoss, W., Formanick, M., and Michaelson, S. M.,** Effects of 2450 MHz microwave energy on the blood-brain barrier to hydrophobic molecules. A. Effect on the permeability to sodium fluoride, *Brain Res. Rev.,* 7, 165, 1984.

285. **Williams, W. M., Del Cerro, M., and Michaelson, S. M.,** Effects of 2450 MHz microwave energy on the blood-brain barrier to hydrophobic molecules. B. Effect on the permeability to HRP, *Brain Res. Rev.,* 7, 171, 1984.

286. **Williams, W. M., Platner, J., and Michaelson, S. M.,** Effects of 2450 MHz microwave energy on the blood-brain barrier to hydrophobic molecules. C. Effect on the permeability to [^{14}C], *Brain Res. Rev.,* 7, 183, 1984.

287. **William, W. M., Lu, S.-T., Del Cerro, M., and Michaelson, S.,** Effects of 2450 MHz microwave energy on the blood-brain barrier to hydrophobic molecules. D. Brain temperature and blood-brain barrier permeability to hydrophilic tracers, *Brain Res. Rev.,* 7, 192, 1984.

288. **Williams, W. M., Lu, S.-T., Del Cerro, M., Hoss, W., and Michaelson, S. M.,** Effects of 2450 MHz microwave energy on the blood-brain barrier: an overview and critique of past and present research, *IEEE Trans. Microwave Theory Tech.,* 32, 808, 1984.

289. **Merritt, J. H., Chamness, A. F., and Allen, S. J.,** Studies on blood-brain barrier permeability after microwave-radiation, *Radiat. Environ. Biophys.,* 15, 367, 1978.

290. **Lin, J. C. and Lin, M. F.,** Studies on microwave and blood-brain barrier interaction, *Bioelectromagnetics,* 1, 313, 1980.

291. **Lin, J. C. and Lin, M. F.,** Microwave hyperthermia-induced blood-brain barrier alterations, *Radiat. Res.,* 89, 77, 1982.

292. **Rozdilsky, B. and Olszewski, J.,** Permeability of cerebral blood vessels studies by radioactive iodinated bovine albumin, *Neurology,* 7, 270, 1957.

293. **Blasberg, R. G.,** Problems of quantifying effects of microwave irradiation on the blood-brain barrier, *Radio Sci.,* 14(6S), 335, 1979.

294. **Rapoport, S. I., Ohno, K., Fredricks, W. R., and Pettigrew, K. D.,** A quantitative method for measuring altered cerebrovascular permeability, *Radio Sci.,* 14(6S), 345, 1979.

295. **Justesen, D. R.,** Microwave irradiation and blood-brain barrier, *Proc. IEEE,* 68, 60, 1980.

296. **Baranski, S. and Edelwejn, Z.,** Studies on the combined effect of microwaves and some drugs on bioelectric activity of the rabbit CNS, *Acta Physiol. Pol.,* 19, 37, 1968.

297. **Edelwejn, Z.,** An attempt to assess the functional state of the cerebral synapses in rabbits exposed to chronic irradiation with microwaves, *Acta Physiol. Pol.*, 19, 897, 1968.

298. **Thomas, J. R. and Maitland, G.,** Microwave radiation and dextroamphetamine: evidence of combined effects on behavior of rats, *Radio Sci.*, 14(6S), 253, 1979.

299. **Servantie, B., Bertharion, G., July, R., Servantie, A. M., Etienne, J., Dreyfus, P., and Escoubet, P.,** Pharmacologic effects of a pulsed microwave field, in *Biologic Effects and Health Hazards of Microwave Radiation*, Czerski, P. et al., Eds., Polish Medical Publishers, Warsaw, 1974, 36.

300. **Goldstein, L. and Sisko, Z.,** A quantitative electroencephalographic study of the acute effects of X-band microwaves in rabbits, in *Biologic Effects and Health Hazards of Microwave Radiation*, Czerski, P. et al., Eds., Polish Medical Publishers, Warsaw, 1974, 128.

301. **Thomas, J. R., Burch, L. S., and Yeandle, S. S.,** Microwave radiation and chlordiazepoxide: synergistic effects on fixed-interval behavior, *Science*, 203, 1357, 1979.

302. **Thomas, J. R., Schrot, J., and Banvard, R. A.,** Behavioral effects of chlorpormazine and diazepam combined with low-level microwaves, *Neurobehav. Toxicol.*, 2, 131, 1980.

303. **Merrit, J. H., Hartzell, R. H., and Frazer, J. W.,** The effect of 1.6 GHz radiation on neurotransmitters in discrete areas of the rat brain, in Biological Effects of Electromagnetic Waves, Vol., 1, Johnson, C. C. and Shore, M. L., Eds., HEW Publ. (FDA) 77-8010, Department of Health, Education and Welfare, Rockville, Md., 1976, 290.

304. **Merritt, J. H., Chamness, A. F., Hartzell, R. H., and Allen, S. J.,** Orientation effects on microwave-induced hyperthermia and neurochemical correlates, *J. Microwave Power*, 12, 167, 1977.

305. **Zeman, G. H., Chaput, R. L., Glazer, Z. R., and Gershman, L. C.,** Gamma-aminobutyric acid metabolism in rats following microwave exposure, *J. Microwave Power*, 8, 213, 1973.

306. **Albert, E. N. and DeSantis, M.,** Histological observations on central nervous system, in Biological Effects of Electromagnetic Waves, Vol. 1, Johnson, C. C. and Shore, M. L., Eds., HEW Publ. (FDA) 77-8010, Department of Health, Education and Welfare, Rockville, Md., 1976, 299.

307. **Switzer, W. G. and Mitchell, D. S.,** Long-term effects of 2.45 GHz radiation on the ultrastructure of the cerebral cortex and on hematologic profiles of rats, *Radio Sci.*, 12, 287, 1977.

308. **Frey, A. H. and Feld, S. R.,** Avoidance by rats of illumination with low power nonionizing electromagnetic energy, *J. Comp. Physiol. Psychol.*, 89, 183, 1975.

309. **Baranski, S.,** Histological and histochemical effect of microwave irradiation on the central nervous system of rabbits and guinea pigs, *Am. J. Phys. Med.*, 51, 182, 1972.

310. **Sutton, C. H. and Carroll, F. B.,** Effects of microwave-induced hyperthermia on the blood-brain barrier of the rat, *Radio Sci.*, 14(6S), 329, 1979.

311. **Takashima, S., Onaral, B., and Schwan, H. P.,** Effects of modulated RF energy on the EEG of mammalian brains, *Radiat. Environ. Biophys.*, 16, 15, 1979.

312. **Bawin, S. M., Galavas-Medici, R. J., and Adey, W. R.,** Effects of modulated very high frequency fields on specific brain rhythms in cats, *Brain Res.*, 58, 365, 1973.

313. **Gordon, Z. V.,** Biological effect of microwaves in occupational hygiene, Israel Program for Scientific Translations, NASA TT F-633, TT 70-50087; NTIS N71-14632, 1970, Jerusalem, Israel.

314. **Austin, G. N. and Horvath, S. M.,** Production of convulsions in rats by high frequency electrical currents, *Am. J. Phys. Med.*, 33, 141, 1954.

315. **Snyder, S. H.,** The effect of microwave irradiation on the turnover rate of serotonin and norepinephrine and the effect of monoamine metabolizing enzymes, Final Report, Contract No. DADA 17-69-C-9144, (NTIS AD-729 161), U.S. Army Medical Research and Development Command, Washington, D.C., 1971.

316. **Sanza, J. N. and de Lorge, J.,** Fixed interval behavior of rats exposed to microwaves at low power densities, *Radio Sci.*, 12(6S), 273, 1977.

317. **deLorge, J.,** Operant Behavior and Colonic Temperature of Squirrel Monkeys (Saimiri sciureus) during Microwave Irradiation, NAMRL-1236 Naval Aerospace Medical Res. Lab., Pensacola, Fla., 1977.

318. **D'Andrea, J. A., Gandhi, O. P., and Lords, J. L.,** Behavioral and thermal effects of microwave radiation at resonant and nonresonant wave lengths, *Radio Sci.*, 12(6S), 251, 1977.

319. **Lin, J. C.,** On microwave-induced hearing sensation, *IEEE Trans. Microwave Theory Tech.*, 25, 605, 1977.

320. **Gage, M.,** Behavior in rats after exposure to various power densities of 2450 MHz microwaves *Neurobehav. Toxicol.*, 1, 137, 1979.

321. **Galloway, W. D.,** Microwave dose-response relationships on two behavioral tasks, *Ann. N.Y. Acad. Sci.*, 247, 410, 1975.

322. **Justesen, D. R. and King, N. W.,** Behavioral effects of low level microwave irradiation in the closed space situation, in Biological Effects and Health Implications of Microwave Radiation, Cleary, S. F., Ed., Symp. Proc., Public Health Service BRH/DBE 70-2, U.S. Department of Health, Education and Welfare, 1970, 154.

323. **Hunt, E. L., King, N. W., and Phillips, R. D.,** Behavioral effects of pulsed microwave radiation, *Ann. N.Y. Acad. Sci.*, 247, 440, 1975.

324. **Lobanova, Y. A.,** Survival and development of animals at various intensities and duration of SHF action, *Tr. Gig. Prof. AMN SSSR,* 1, 61, 1960.

325. **Lebovitz, R. M.,** Pulse modulated and continuous wave microwave radiation yield equivalent changes in operant behavior of rodents, *Physiol. Behav.,* 30, 891, 1983.

326. **Gage, M. I. and Guyer, W. M.,** Interaction of ambient temperature and controlled behavior in the rat, *Radio Sci.,* 17(5S), 179, 1982.

327. **deLorge, J. O.,** Effects of Microwaves on Animal Operant Behavior, Rep. No. NAMRL-1285, Naval Aerospace Medical Research Lab, Pensacola, Fla., 1982.

328. **deLorge, J. O.,** Operant behavior and colonic temperature of Rhesus monkeys, Macaca mulatta, exposed to microwaves at frequencies above and near whole-body resonance, Rep. No. NAMRL-1289, Naval Aerospace Medical Research Lab., Pensacola, Fla., 1983.

329. **Lin, J. C., Guy, A. W., and Caldwell, L. T.,** Thermographic and behavioral studies of rats in the near field of 918-MHz radiations, *IEEE Trans. Microwave Theory Tech.,* 25, 833, 1977.

330. **Diachenko, J. A. and Milroy, W. C.,** The Effects of High Power Pulsed and Low Level CW Microwave Radiation on an Operant Behavior in Rats, Naval Surface Weapons Center, Dahlgren Laboratory, Dahlgren, Va., 1975.

331. **Thomas, J. R., Finch, E. D., Fulk, D. W., and Burch, L. S.,** Effects of low level microwave radiation on behavioral baselines, *Ann. N.Y. Acad. Sci.,* 247, 425, 1975.

332. **Roberti, B., Heebels, G. H., Hendricx, J. C. M., De Greef, A. H. A. M., and Wolthuis, O. L.,** Preliminary investigations of the effects of low-level microwave radiation on spontaneous motor activity in rats, *Ann. N.Y. Acad. Sci.,* 247, 417, 1975.

333. **Galloway, W. D. and Waxler, M.,** Interaction between microwave and neuroactive compounds, in Symp. on Biological Effects and Measurement of Radio Frequency/Microwaves, Hazzard, D., Ed., Publ. (FDA) 77-8026, U.S. Department of Health, Education and Welfare, Rockville, Md., 1977, 62.

334. **deLorge, J.,** Disruption of behavior in mammals of three different sizes exposed to microwaves: extrapolation to larger mammals, in 1978 Symp. on Electromagnetic Fields in Biological Systems, Stuchly, S. S., Ed., Ottawa 1978, 215.

335. **Stolwijk, J. A. J.,** Responses to the thermal environment, *Fed. Proc., Fed. Am. Soc. Exp. Biol.,* 36, 1655, 1977.

336. **Stern, S., Margolin, L., Weiss, B., Lu, S. T., and Michaelson, S. M.,** Microwaves: effect on thermoregulatory behavior in rats, *Science,* 206, 1198, 1979.

337. **Adair, E. R. and Adams, B. A.,** Adjustments in metabolic heat production by squirrel monkeys exposed to microwaves, *J. Appl. Physiol.,* 52, 1049, 1982.

338. **Lotz, W. G. and Michaelson, S. M.,** Temperature and corticosterone relationship in microwave exposed rats, *J. Appl. Physiol.,* 44, 438, 1978.

339. **Hamilton, C. L.,** Interactions of food intake and temperature regulation in the rat, *J. Comp. Physiol. Psychol.,* 56, 476, 1963.

340. **Barofsky, I.,** The effect of high ambient temperature on timing behavior of rats, *J. Exp. Anal. Behav.,* 12, 59, 1969.

341. **Bligh, J.,** Physiologic responses to heat, in *Fundamental and Applied Aspects of Nonionizing Radiation,* Michaelson, S. et al., Eds., Plenum Press, New York, 1975, 143.

342. **Adair, E. R., Adams, B. W., and Akel, G. M.,** Minimal changes in hypothalamic temperature accompany microwave-induced alteration of thermoregulatory behavior, *Bioelectromagnetics,* 5, 13, 1984.

343. **Corbit, J. D.,** Thermal motivation, In "Neural Control of Motivatd Behavior." A report based on a NRP work session, *Neurosci. Res. Prog. Bull.,* 11, 4, 1973.

344. **Smialowicz, R. J., Riddle, M. M., Brugnolotti, P. L., Rogers, R. R., and Comptom, K. L.,** Detection of microwave heating in 5-hydroxytryptamine-induced hypothermic mice, *Radiat. Res.,* 88, 108, 1981.

345. **Adair, E. R.,** Microwaves and thermoregulation, in USAR Radiofrequency Radiation Bioeffects Research Program — A Review, Mitchell, J. C., Ed., Review 4-81, USAF School of Aerospace Medicine, San Antonio, Tex., 1981.

346. **Adair, E. R. and Adams, B. W.,** Microwaves induce peripheral vasodilation in squirrel monkey, *Science,* 207, 1381, 1980.

347. **Mitchell, D. S., Switzer, W. G., and Bronaugh, E. L.,** Hyperactivity and disruption of operant behavior in rats after multiple exposure to microwave radiation, *Radio Sci.,* 12(6S), 263, 1977.

348. **Moe, K. E., Lovely, R. H., Meyers, D. E., and Guy, A. W.,** Physiological and behavioral effects of chronic low level microwave radiation in rats, in Biological Effects of Electromagnetic Waves, Vol. 1, Johnson, C. C. and Shore, M. L., Eds., HEW Publ. (FDA) 77-8010, Department of Health, Education and Welfare, Rockville, Md., 1976, 248.

349. **D'Andrea, J. A., Gandhi, O. P., Lords, J. L., Durney, C. H., Johnson, C. C., and Astle, L.,** Physiological and behavioral effects of chronic exposure to 2450-MHz microwaves, *J. Microwave Power,* 14, 351, 1979.

350. **D'Andrea, J. A., Gandhi, O. P., Lords, J. L., Durney, C. H., Astle, L., Stensaas, L. J., and Schoenberg, A. A.,** Physiological and behavioral effects of prolonged exposure to 915 MHz microwaves, *J. Microwave Power,* 15(2), 123, 1980.

351. **Rudnev, M., Bokina, A., Eksler, N., and Navakatikyan, M.,** The use of evoked potential and behavioral measures in the assessment of enviornmental insult, in Multidisciplinary Perspectives, in Event-Related Brain Potential Research, Otto, D. A., Ed., EPA-600/9-77-043, U.S. Environmental Protection Agency, Research Triangle Park, N.C., 1978, 444.

352. **Bermant, R. I., Reeves, D. L., Levinson, D. M., and Justesen, D. R.,** Classical conditioning of microwave-induced hyperthermia in rats, *Radio Sci.,* 14(6S), 201, 1979.

353. **D'Andrea, J. A., Gandhi, O. P., and Kesner, R. P.,** Behavioral effects of resonant electromagnetic power absorption in rats, in Biological Effects of Electromagnetic Waves, Vol. 1, Johnson, C. C. and Shore, M. L., Eds., HEW Publ. (FDA) 77-8010, Department of Health, Education and Welfare, Rockville, Md., 1976, 257.

354. **deLorge, J. O.,** The effects of microwave radiation on behavior and temperature in rhesus monkeys, in Biological Effects of Electromagnetic Waves, Vol. 1, Johnson, C. C. and Shore, M. L., Eds., HEW Publ. (FDA) 77-8010, Department of Health, Education and Welfare, Rockville, Md., 1976, 158.

355. **deLorge, J.,** Disruption of behavior in mammals of three different sizes exposed to microwaves: extrapolation to larger mammals, in *Electromagnetic Fields in Biological Systems,* Stuchly, S. S., Ed., IMPI, Edmonton, Alberta, Canada, 1979, 215.

356. **deLorge, J.,** Operant behavior and rectal temperature of squirrel monkeys during 2.45-GHz microwave irradiation, *Radio Sci.,* 14(6S), 217, 1979.

357. **deLorge, J. and Ezell, C. S.,** Observing-responses of rats exposed to 1.28- and 5.62-GHZ microwaves, *Bioelectromagnetics,* 1, 183, 1980.

358. **Scholl, D. M. and Allen, S. J.,** Skilled visual-motor performance by monkeys in a 1.2-GHz microwave field, *Radio Sci.,* 14(6S), 247, 1979.

359. **Thomas, J. R., Yeandle, S. S., and Burch, L. S.,** Modification of internal discriminative stimulus control of behavior by low levels of pulsed microwave radiation, in Biological Effects of Electromagnetic Waves, Vol. 1, Johnson, C. C. and Shore, M. L., Eds., HEW Publ. (FDA) 77-8010, Department of Health, Education and Welfare, Rockville, Md., 1976, 201.

360. **Schrot, J., Thomas, J. R., and Banvard, R. A.,** Modification of the repeated acquisition of response sequences in rats by low-level microwave exposure, *Bioelectromagnetics,* 1, 89, 1980.

361. **King, N. W., Justesen, D. R., and Clarke, R. L.,** Behavioral sensitivity to microwave irradiation, *Science,* 172, 398, 1971.

362. **Johnson, R. B., Meyers, D. E., Guy, A. W., Lovely, R. H., and Galambos, R.,** Discriminative control of appetitive behavior by pulsed microwave radiation in rats, in Biological Effects of Electromagnetic Waves, Vol. 1, Johnson, C. C. and Shore, M. L., Eds., HEW Publ. (FDA) 77-8010, Department of Health, Education and Welfare, Rockville, Md., 1976, 238.

363. **Frey, A. H., Feld, S. R., and Frey, B.,** Neural function and behavior: defining the relationship, *Ann. N.Y. Acad. Sci.,* 247, 433, 1975.

364. **Hjeresen, D. L., Doctor, S. R., and Sheldon, R. L.,** Shuttlebox side preference as mediated by pulsed microwave and conventional auditory cues, in *Electromagnetic Fields in Biological Systems,* Stuchly, S. S., Ed., IMPI, Edmonton, Alberta, Canada, 1979, 194.

365. **Monahan, J. C. and Ho, H. S.,** Microwave induced avoidance behavior in the mouse, in Biological Effects of Electromagnetic Waves, Vol. 1, Johnson, C. C. and Shore, M. L., Eds., HEW Publ. (FDA) 77-8010, Department of Health, Education and Welfare, Rockville, Md., 1976, 274.

366. **Monahan, J. C. and Ho, H. S.,** The effect of ambient temperature on the reduction of microwave energy absorption by mice, *Radio Sci.,* 12(6S), 257, 1977.

367. **Gage, M. I., Berman, E., and Kinn, J. B.,** Videotape observation of rats and mice during an exposure to 2450 MHz microwave radiation, *Radio Sci.,* 14(6S), 227, 1979.

368. **Carroll, D. R., Levinson, D. M., Justesen, D. R., and Clarke, R. L.,** Failure of rats to escape from a potentially lethal microwave field, *Bioelectromagnetics,* 1, 101, 1980.

369. **Monahan, J. E. and Henton, W. W.,** The effect of psychoactive drugs on operant behavior induced by microwave radiation, *Radio Sci.,* 14(6S), 233, 1979.

370. **Gage, M. I.,** Microwave irradiation and ambient temperature interact to alter rat behavior following overnight exposure, *J. Microwave Power,* 14, 389, 1979.

371. **Adair, E. R. and Adams, B. W.,** Microwaves modify thermoregulatory behavior in squirrel monkey, *Bioelectromagnetics,* 1, 1, 1980.

372. **Lotz, W. G. and Podgorski, R. P.,** Temperature and adrenocortical responses in Rhesus monkeys exposed to microwaves, *J. Appl. Physiol. Respir. Environ. Exercise Physiol.,* 53, 1565, 1982.

373. **Guillet, R., Lotz, W. G., and Michaelson, S. M.,** Time-course of adrenal response in microwave-exposed rats, in *Proc. 1975 Annu. Meet. USNC/URSI, University of Colorado, Boulder,* National Academy of Sciences, Washington, D.C., 1975, 316.

374. **Houk, W. M., Michaelson, S. M., and Beischer, D. E.,** The effects of environmental temperature on thermoregulatory, serum lipid, carbohydrate, and growth hormone responses of rats exposed to microwaves, in *Proc. 1975 Annu. Meet. USNC/URSI, University of Colorado, Boulder,* National Academy of Sciences, Washington, D. C., 1975.

375. **Travers, W. D. and Vetter, R. J.,** Low intensity microwave effects on the synthesis of thyroid hormones and serum proteins, in *Proc. 1976 Annu. Meet. USNC/URSI, University of Massachusetts, Amherst,* National Academy of Sciences, Washington, D.C., 1976, 91.

376. **Mikolajczyk, H.,** Microwave irradiation and endocrine functions, in *Biologic Effects and Health Hazards of Microwave Radiation,* Czerski, P. et al., Eds., Polish Medical Publishers, Warsaw, 1974, 46.

377. **Mikolajczyk, H.,** Microwave-induced shifts of gonadotrophic activity in anterior pituitary gland of rats, in Biological Effects of Electromagnetic Waves, Vol. 1, Johnson, C. C. and Shore, M. L., Eds., Publ. (FDA) 77-8010, U.S. Department of Health, Education and Welfare, Rockville, Md., 1977, 377.

378. **Lu, S. T., Lotz, W. G., and Michaelson, S. M.,** Advances in microwave-induced neuroendocrine effects the concept of stress, *Proc. IEEE,* 68, 73, 1980.

379. **Petrov, I. R. and Syngayevskaya, V. A.,** Endocrine glands, in *Influence of Microwave Radiation on the Organism of Man and Animals,* Petrov, I. R., Ed., (NASA TT F-708), Meditsina Press, Leningrad, 1970, 31.

380. **Demokidova, N. K.,** The effects of radiowaves on the growth of animals, in Biological Effects of Radiofrequency Electromagnetic Fields, Gordon, Z. V., Ed., U.S. Joint Publications Research Service No. 63321, Arlington, Va., 1974, 237.

381. **Lenko, J., Dolatowski, A., Gruszecki, L., Klajman, S., and Januszkiewicz, L.,** Effect of 10-cm radar waves on the level of 17-ketosteroids and 17-hydoxycorticosteroids in the urine of rabbits, *Przegl. Lek.,* 22, 296, 1966.

382. **Mikolajczyk, H.,** Hormone reactions and changes in endocrine glands under influence of microwaves, *Med. Lotnicza,* 39, 39, 1972.

383. **Parker, L. N.,** Thyroid suppression and adrenomedullary activation by low-intensity microwave radiation, *Am. J. Physiol.,* 224, 1388, 1973.

384. **Milroy, W. C. and Michaelson, S. M.,** Thyroid pathophysiology of microwave radiation, *Aerosp. Med.,* 43, 1126, 1972.

385. **Baranski, S., Ostrowski, K., and Stodolnik-Braranska, W.,** Functional and morphological studies of the thyroid gland in animals exposed to microwave irradiation, *Acta Physiol. Pol.,* 23, 1029, 1972.

386. **Magin, R. L., Lu, S. T., and Michaelson, S. M.,** Stimulation of dog thyroid by local application of high intensity microwaves, *Am. J. Physiol.,* 233, E363, 1977.

387. **Magin, R. L., Lu, S. T., and Michaelson, S. M.,** Microwave heating effect on the dog thyroid, *IEEE Trans. Biomed. Eng.,* 24, 522, 1977.

388. **Vetter, R. J.,** Neuroendocrine response to microwave irradiation, *Proc. Natl. Electron. Conf.,* 30, 237, 1975.

389. **Lu, S. T., Lebda, N. A., Michaelson, S. M., Pettit, S., and Rivera, D.,** Thermal and endocrinological effects of protracted irradiation of rats by 2450 MHz microwaves, *Radio Sci.,* 12(S), 147, 1977.

390. **Michaelson, S. M., Thomson, R. A. E., and Howland, J. W.,** Physiologic aspects of microwave irradiation of mammals, *Am. J. Physiol.,* 201, 351, 1961.

391. **McLees, B. D., and Finch, E. D.,** Analysis of the Physiologic Effects of Microwave Radiation, U.S. Naval Medical Research Institute, Proj. MF12 24.015-0001B, Rep. No. 3, Bethesda, Md., 1971.

392. **Collins, K. J. and Weiner, J. S.,** Endocrinological aspects of exposure to high environmental temperatures, *Physiol. Rev.,* 48, 785, 1968.

393. **Lu, S.-T., Lebda, N., Pettit, S., and Michaelson, S. M.,** Delineating acute neuroendocrine responses in microwave-exposed rats, *J. Appl. Physiol.: Respir. Environ. Exercise Physiol.,* 48(6), 927, 1980.

394. **Abhold, R. H., Ortner, M. J., Galvin, M. J., and McRee, D. I.,** Studies on acute *in vivo* exposure of rats to 2450-MHz microwave radiation. II. Effects on thyroid and adrenal axes hormones, *Radiat. Res.,* 88, 448, 1981.

395. **Michaelson, S. M.,** Physiologic regulation in electromagnetic fields, *Bioelectromagnetics,* 3, 91, 1982.

396. **Lu, S., Lebda, N., Pettit, S., and Michaelson, S. M.,** Microwave-induced temperature, corticosterone, and thyrotropin interrelationships, *J. Appl. Physiol.: Respir. Environ. Exercise Physiol.,* 50, 399, 1981.

397. **Lotz, W. G. and Michaelson, S. M.,** Effects of hypophysectomy and dexamethasone on rat adrenal response to microwaves, *J. Appl. Physiol.: Respir. Environ. Exercise Physiol.,* 47, 1284, 1979.

398. **Cooper, T., Pinakatt, T., and Richardson, A. W.,** Effect of microwave induced hyperthermia on the cardiac output of the rat, *Physiologist,* 4, 21, 1961.

399. **Cooper, T., Jellinek, M., Pinakatt, T., Richardson, A. W., and Cooper, T.,** Effects of adrenalectomy, vagotomy and ganglionic blockade on the circulatory response to microwave hyperthermia, *Aerosp. Med.,* 33, 794, 1962.

400. **Pinakatt, T., Cooper, T., and Richardson, A. W.,** Effect of ouabain on the circulatory response to microwave hyperthermia in the rat, *Aerosp. Med.,* 34, 497, 1963.

401. **Pinakatt, T., Richardson, A. W., and Cooper, T.,** The effect of digitoxin on the circulatory response of rats to microwave irradiation, *Arch. Int. Pharmacodyn. Ther.,* 156, 151, 1965.

402. **Birenbaum, L., Kaplan, I. T., Metlay, W., Rosenthal, S. W., and Zaret, M. M.,** Microwave and infra-red effects on heart rate, respiration rate and subcutaneous temperature of the rabbit, *J. Microwave Power,* 10, 3, 1975.

403. **Kaplan, I. T., Metlay, W., Zaret, M. M., Birenbaum, L., and Rosenthal, S. W.,** Absence of heart-rate effects in rabbits during low-level microwave irradiation, *IEEE Trans, Microwave Theory Tech.,* 19, 168, 1971.

404. **Paff, G. H., Boucek, R. J., Nieman, R. E., and Deichmann, W. B.,** The embryonic heart subjected to radar, *Anat. Rec.,* 147, 379, 1963.

405. **Hamrick, P. and McRee, D. I.,** The effect of 2450 MHz microwave irradiation on the heart rate of embryonic quail, *Health Phys.,* 38, 261, 1980.

406. **Liu, L. M., Rosenbaum, F. J., and Pickard, W. F.,** The insensitivity of frog heart rate to pulse modulated microwave energy, *J. Microwave Power,* 11, 225, 1976.

407. **Frey, A. H. and Seifert, E.,** Pulse modulated UHF illumination of the heart associated with change in heart rate, *Life Sci.,* 7, 505, 1968.

408. **Chou, C. K., Han, L. F., and Guy, A. W.,** Microwave radiation and heartbeat rate of rabbits, *J. Microwave Power,* 15, 87, 1980.

409. **Tinney, C. E., Lords, J. L., and Durney, C. H.,** Rate effects in isolated turtle hearts induced by microwave irradiation, *IEEE Trans. Microwave Theory Tech.,* 24, 18, 1976.

410. **Olsen, R. G., Lords, J. L., and Durney, C. H.,** Microwave-induced chronotropic effects of the isolated rate heart. *Ann. Biomed. Eng.,* 5, 395, 1977.

411. **Clapman, R. M. and Cain, C. A.,** Absence of heart rate effects in isolated frog heart irradiated with pulsed modulated microwave energy, *J. Microwave Power,* 10, 411, 1975.

412. **Barron, C. I. and Baraff, A. A.,** Medical considerations of exposure to microwaves (radar), *JAMA,* 168, 1194, 1958.

413. **Daily, L.,** A clinical study of the results of exposure of laboratory personnel to radar and high frequency radio, *U.S. Nav. Med. Bull.,* 41, 1052, 1943.

414. **Drogichina, E. A., Sadchikova, M. N., Snegova, M. N., Konchalovskaya, G. V., and Glotova, K. T.,** Autonomic and cardiovascular disorders during chronic exposure to super-high frequency electromagnetic fields, *Gig. Tr. Prof. Zabol. U.S.S.R.,* 10, 13, 1966.

415. **Hyde, A. S. and Friedman, J. J.,** Some effects of acute and chronic microwave irradiation of mice, in *Thermal Problems in Aerospace Medicine,* Hardy, J. D., Ed., Unwin, Surrey, England, 1968, 163.

416. **Budd, R. A., Laskey, J., and Kelly, C.,** Hematological response of fetal rats following 2450 MHz microwave irradiation, in Radiation Bio-effects, Summary Report, Hodge, D. M., Ed., Publ. PHS, BRH/DBE 70-7, U.S. Department of Health, Education and Welfare, 1970, 161.

417. **Spalding, J. F., Freyman, R. W., and Holland, L. M.,** Effects of 800 MHz electromagnetic radiation on body weight, activity, hematopoiesis and life span in mice, *Health Phys.,* 20, 421, 1971.

418. **Sadchikova, M. N. and Orlova, A. A.,** Clinical picture of the chronic effects of electromagnetic microwaves, *Ind. Hyg. Occup. Dis. (U.S.S.R.),* 2, 16, 1958.

419. **Michaelson, S. M., Thomson, R. A. E., Tamami, M. Y. E., Seth, H. S., and Howland, J. W.,** Hematologic effects of microwave exposure, *Aerosp. Med.,* 35, 824, 1964.

420. **Michaelson, S. M., Thomson, R. A. E., and Howland, J. W.,** Comparative studies on 1285 and 2800 mc/sec pulsed microwaves, *Aerosp. Med.,* 36, 1059, 1965.

421. **Michaelson, S. M., Thomson, R. A. E., and Howland, J. W.,** Biological Effects of Microwave Exposure, Griffiss Air Force Base, Rome Air Development Ctr. (ASTIA Doc. No. AD 824-242), Rome, N.Y., 1967.

422. **Kitsovskaya, I. A.,** The effect of centimeter waves of different intensities on the blood and hemopoietic organs of white rats, *Gig. Tr. Prof. Zabol.,* 8, 14, 1964.

423. **Baranski, S. and Czerski, P.,** Investigations of the behavior of corpuscular blood constituents in persons exposed to microwaves, *Lek. Woisk.,* 42, 903, 1966.

424. **Baranski, S.,** Effect of chronic microwave irradiation on the blood forming system of guinea pigs and rabbits, *Aerosp. Med.,* 42, 1196, 1971.

425. **Baranski, S.,** Effect of microwaves on the reactions of the white blood cell system, *Acta Physiol. Pol.,* 23, 685, 1972.

426. **Sacchitelli, F. and Sacchitelli, G.,** Protection of personnel exposed to radar microwaves, *Folia Med. (Naples),* 43, 1219, 1960.

427. **Vacek, D. R. A.,** Effect of high-frequency electromagnetic field upon haemopoietic stem cells in mice, *Folia Biol. (Praha),* 18, 292, 1978.

428. **Czerski, P., Paprocka-Slonka, E., Siekierzynski, M., and Stolarska, A.,** Influence of microwave radiation on the hematopoietic system, in *Biologic Effects and Health Hazards of Microwave Radiation,* Czerski, P. et al., Eds., Polish Medical Publishers, Warsaw, 1974, 67.

429. **Czerski, P., Paprocka-Slonka, E., and Stolarska, A.,** Microwave irradiation and the circadian rhythm of bone marrow cell mitosis, *J. Microwave Power,* 9, 31, 1974.

430. **Ivanov, A. I.,** Changes of phagocytic activity and mobility of neutrophils under the influence of microwave fields, in Summaries of Reports, Questions of the Biological Effect of a SHF-UHF Electromagnetic Field, Kirov Order of Lenin Military Medical Academy, Leningrad, 1962, 24.

431. **Miro, L., Loubiere, R., and Pfister, A.,** Effects of microwaves on the cell metabolism of the reticulo-endothelial system, in *Biologic Effects and Health Hazards of Microwave Radiation,* Czerski, P. et al., Eds., Polish Medical Publishers, Warsaw, 1974, 89.

432. **Serdiuk, A. M.,** Biological effect of low-intensity ultrahigh frequency fields, *Vrach. Delo,* 11, 108, 1969.

433. **Michaelson, S. M.,** Effects of exposure to microwaves: problems and perspectives, *Environ. Health Perspect.,* 8, 133, 1974.

434. **Deichmann, W. B., Miale, J., and Landeen, K.,** Effect of microwave radiation on the hemopoietic system of the rat, *Toxicol. Appl. Pharmacol.,* 6, 71, 1964.

435. **Deichmann, W. B., Bernal, E., Stephens, F., and Landeen, K.,** Effects on dogs of chronic exposure to microwave radiation, *J. Occup. Med.,* 5, 418, 1963.

436. **Djordjevic, Z. and Kolak, A.,** Changes in the periphral blood of the rat exposed to microwave radiation (2400 MHz) in conditions of chronic exposure, *Aerosp. Med.,* 44, 1051, 1973.

437. **Djordjevic, Z., Lazarevic, N., and Djokovic, V.,** Studies on the hematologic effects of long-term, low-dose microwave exposure, *Aviat. Space Environ. Med.,* 48, 516, 1977.

438. **McRee, D. I., Faith, R., McConnell, E. E., and Guy, A. W.,** Long-term 2450-MHz CW microwave irradiation of rabbits: evaluation of hematological and immunological effects, *J. Microwave Power,* 15, 45, 1980.

439. **Rotkovska, D. and Vacek, A.,** The effect of electromagnetic radiation on the hematopoietic stem cells of mice, *Ann. N.Y. Acad. Sci.,* 247, 243, 1975.

440. **Rotkovska, D. and Vacek, A.,** Modification of repair of X-irradiation damage of hemopoietic system of mice by microwaves, *J. Microwave Power,* 12, 119, 1977.

441. **Michaelson, S. M., Thomson, R. A. E., Odland, L. T., and Howland, J. W.,** The influence of microwaves on ionizing radiation exposure, *Aerosp. Med.,* 34, 111, 1963.

442. **Michaelson, S. M., Thomson, R. A. E., El Tamami, M. Y., Seth, H. S., and Howland, J. W.,** The hematologic effects of microwave exposure, *Aerosp. Med.,* 35, 824, 1964.

443. **Lappenbusch, W. L., Gillespie, L. J., Leach, W. M., and Anderson, G. E.,** Effect of 2450-MHz microwaves on the radiation response of X-irradiated Chinese hamsters, *Radiat. Res.,* 54, 294, 1973.

444. **Smialowicz, R. J., Weil, C. M., Kinn, J. B., and Elder, J. A.,** Exposure of rats to 425-MHz (CW) radiofrequency radiation: effects on lymphocytes, *J. Microwave Power,* 17, 211, 1982.

445. **Pazderova-Vejlupkova, J. and Josifko, M.,** Changes in the blood count of growing rats irradiated with a microwave pulse field, *Arch. Environ. Health,* 34, 44, 1979.

446. **Smialowicz, R. J., Riddle, M. M., Brugnolotti, P. L., Sperazza, J. M., and Kinn, J. B.,** Evaluation of lymphocyte function in mice exposed to 2450 MHz (CW) microwaves, in *Proc. 1978 Symp. on Electromagnetic Fields in Biological Systems,* Stuchly, S. S., Ed., Ottawa, Can., 1979, 122.

447. **Liburdy, R. P.,** Effects of radio-frequency radiation on inflammation, *Radio Sci.,* 12(6S), 179, 1977.

448. **Wiktor-Jedrzejczak, W., Ahmed, A., Sell, K. W., Czerski, P., and Leach, W. M.,** Microwaves induce an increase in the frequency of complement receptor-bearing lymphoid spleen cells in mice, *J. Immunol.,* 118, 1499, 1977.

449. **Wiktor-Jedrzejczak, W., Ahmed, A., Czerski, P., Leach, W. M., and Sekk, K. W.,** Immunologic response of mice of 2450 MHz microwave radiation: overview of immunology and empirical studies of lymphoid spleen cells, *Radio Sci.,* 12(6S), 209, 1977.

450. **Smialowicz, R. J.,** The effect of microwaves (2450 MHz) on lymphocyte blast transformation *in vitro,* In Biological Effects of Electromagnetic Waves, Johnson, C. C. and Shore, M. L., Eds., Publ. (FDA) 77-8010, U.S. Department of Health, Education and Welfare, Rockville, Md., 1976, 472.

451. **Stodolnik-Baranska, W.,** The effects of microwaves on human lymphocyte cultures, in *Biologic Effects and Health Hazards of Microwave Radiation,* Czerski, P. et al., Eds., Polish Medical Publishers, Warsaw, 1974, 189.

452. **Schlagel, C. J., Sulek, K., Ho, H. S., Leach, W. M., Ahmed, A., and Woody, J. N.,** Biological effects of microwave exposure. II. Studies on the mechanisms controlling susceptibility to microwave-induced increases in complement receptor positive spleen cells, *Bioelectromagnetics,* 1, 405, 1980.

453. **Smialowicz, R. J., Brugnolotti, P. L., and Riddle, M. M.,** Complement receptor positive spleen cells in microwave (2450 MHz) irradiated mice, *J. Microwave Power,* 16, 73, 1981.

454. **Yang, H. K., Cain, C. A., Lockwood, J., and Tompkins, W. A.,** Effects of microwave exposure on the hamster immune system. I. Natural killer cell activity, *Bioelectromagnetics,* 4, 123, 1983.

455. **Roberts, N. J., Lu, S.-T., and Michaelson, S. M.,** Human leukocyte functions and the U.S. safety standard for exposure to radiofrequency radiation, *Science,* 220, 318, 1983.

456. **Czerski, P.,** Microwave effects on the blood-forming system with particular reference to the lymphocyte, *Ann. N. Y. Acad. Sci.,* 247, 232, 1975.

457. **Gordon, Z. V.,** Biological Effect of Microwaves in Occupational Hygiene, Izd. Med., Leningrad, TT 70-50087, NASA TT F-633, 1970.

458. **Petrov, I. R., Ed.,** *Influence of Microwave Radiation on the Organism of Man and Animals,* (NASA TT F-708), Meditsina Press, Leningrad, 1979.

459. **Phillips, R. D., Hunt, E. L., Castro, R. D., and King, N. W.,** Thermoregulatory, metabolic and cardiovascular response of rats to microwaves, *J. Appl. Physiol.,* 38, 630, 1972.

460. **Liburdy, R. P.,** Effects of radiofrequency radiation on peripheral vascular permeability (Abstr.), in Annu. Meet. Int. Union of Radio Science, Amherst, Mass., 1976.

461. **Czerski, P. and Siekierzynski, M.,** Analysis of occupational exposure to microwave radiation, in *Fundamental and Applied Aspects of Non-Ionizing Radiations,* Michaelson, S. M., Miller, M. W., Magin, R., and Carstenesen, E. L., Eds., Plenum Press, New York, 1975, 367.

462. **Liburdy, R. P.,** Radiofrequency radiation alters the immune system: modulation of T- and B-lymphocyte levels and cell-mediated immunocompetence by hyperthermic radiation, *Radiat. Res.,* 77, 34, 1979.

463. **Liburdy, R. P.,** Suppression of allograft rejection by whole-body microwave hyperthermia, *Fed. Am. Soc. Exp. Biol.,* Fed. Proc.,(Abstr.), 37 1281, 1978.

464. **Atkins, E. and Bodel, P.,** Fever, *N. Engl. J. Med.,* 286, 27, 1972.

465. **Bennett, I. L., Jr. and Nicastri, A.,** Fever as a mechanism of resistance, *Bacteriol. Rev.,* 24, 16, 1960.

466. **Kluger, M. J., Ringler, D. H., and Anver, M. R.,** Fever and survival, *Science,* 188, 166, 1975.

467. **Frenkel, J. K. and Caldwell, S. A.,** Specific immunity and nonspecific resistance to infection: listeria, protozoa, and viruses in mice and hamsters, *J. Infect. Dis.,* 131, 201, 1975.

468. **Mandell, G. L.,** Effect of temperature on phagocytosis by human polymorphonuclear neutrophils, *Infect. Immunity,* 12, 221, 1975.

469. **Roberts, N. J., Jr. and Steigbigel, R. T.,** Hyperthermia and human leukocyte functions: effects on response of lymphocytes to mitogen and antigen and bactericidal capacity of monocytes and neutrophils, *Infect. Immunity,* 18, 673, 1977.

470. **Huang, A. T. and Mold, N. G.,** Immunologic and hematopoietic alterations by 2450-MHz electromagnetic radiation, *Bioelectromagnetics,* 1, 77, 1980.

471. **Wiktor-Jedrzejczak, W., Ahmed, A., Czerski, P., Leach, W. M., and Sell, K. W.,** Increase in the frequency of Fc receptor (FcR) bearing cells in the mouse spleen following a single exposure of mice to 2450 MHz microwaves, *Biomedicine,* 27, 250, 1977.

472. **Sulek, K., Schlagel, C. J., Wiktor-Jedrzecjzak, W., Ho, H. S., Leach, W. M., Ahmed, A., and Woody, J. N.,** Biologic effects of microwave exposure. I. Threshold conditions for the induction of the increase in complement receptor positive (CR$^+$) mouse spleen cells following exposure to 2450-MHz microwaves, *Radiat. Res.,* 83, 127, 1980.

473. **Liburdy, R. P.,** Radiofrequency radiation alters the immune system. II. Modulation of *in vivo* lymphocyte circulation, *Radiat. Res.,* 83, 66, 1980.

474. **Hamrick, P. E., McRee, D. I., Thaxton, P., and Parkhurst, C. R.,** Humoral immunity of Japanese quail subjected to microwave radiation during embryogeny, *Health Phys.,* 33, 23, 1977.

475. **Szmigielski, S., Jeljaszewicz, J., and Wiranowska, M.,** Acute staphylococcal infections in rabbits irradiated with 3-GHz microwaves, *Ann. N.Y., Acad. Sci.,* 247, 305, 1975.

476. **Shah, S. A. and Dickson, J. A.,** Effect of hyperthermia on the immune response of normal rabbits, *Cancer Res.,* 38, 3518, 1978.

477. **Szmigielski, S., Janiak, M., Hryniewicz, W., Jeljaszewicz, J., and Pulverer, G.,** Local microwave hyperthermia (43°C) and stimulation of the macrophage and T-lymphocyte systems in treatment of Guerin epithelioma in rats, *Z. Krebsforsch.,* 91, 35, 1978.

478. **Marmor, J. B., Hahn, N., and Hahn, G. M.,** Tumor cure and cell survival after localized radiofrequency heating, *Cancer Res.,* 37, 879, 1977.

479. **Szmigielski, S., Pulverer, G., Hryniewicz, W., and Janiak, M.,** Inhibition of tumor growth in mice by microwave hyperthermia, streptolysin S and colcemide, *Radio Sci.,* 12(6S), 185, 1977.

480. **Roszkowski, W., Wrembel, J. K., Roszkowski, K., Janiak, M., and Szmigielski, S.,** The search for an influence of whole-body microwave hyperthermia on anti-tumor immunity, *J. Cancer Res. Clin. Oncol.,* 96, 311, 1980.

481. **Shah, S. A. and Dickson, J. A.,** Effect of hyperthermia on the immune response of normal rabbits, *Cancer Res.,* 38, 3518, 1978.

482. **Lin, J. C., Ottenbreit, M. J., Wang, S., Inoue, S., Bollinger, R. O., and Fracassa, M.,** Microwave effects on granulocyte and macrophage precursor cells of mice *in vitro, Radiat. Res.,* 80, 292, 1979b.

483. **Stodolnik-Baranska, W.,** Lymphoblastoid transformation of lymphocytes *in vitro* after microwave irradiation, *Nature,* 214, 102, 1967.

484. **Lin, J. C. and Peterson, W. D.,** Cytological effects of 2450 MHz CW microwave radiation, *J. Bioeng.,* 1, 471, 1977.

485. **Mayers, C. P. and Habeshaw, J. A.,** Depression of phagocytosis: a nonthermal effect of microwave radiation as a potential hazard to health, *Int. J. Radiat. Biol.,* 24, 449, 1973.

486. **Szmigielski, S.,** Effect of 10-cm (3 GHz) electromagnetic radiation (microwaves) on granulocytes *in vitro, Ann. N.Y. Acad. Sci.,* 247, 275, 1975.

487. **Goldblith, S. A. and Wang, D. I. C.,** Effect of microwaves on *Escherichia coli* and *Bacillus subtilus, Appl. Microbiol.,* 15, 1371, 1967.

488. **Moore, H. A., Raymond, R., Fox, M., and Galsky, A. G.,** Low-intensity microwave radiation and the virulence of *Agrobacterium tumefaciens* strain B6, *Appl. Environ. Microbiol.,* 37, 127, 1979.

489. **Frey, A. H.,** Auditory system response to RF energy, *Aerosp. Med.,* 32, 1140, 1961.

490. **Frey, A. H.,** Human auditory system response to modulated electromagnetic energy, *J. Appl. Physiol.,* 17, 689, 1962.

491. **Chou, C. K., Galambos, R., Guy, A. W., and Lovely, R. H.,** Cochlear microphonics generated by microwave pulses, *J. Microwave Power,* 10, 361, 1975.

492. **Chou, C.-K., Guy, A. W., and Galambos, R.,** Auditory perception of radiofrequency electromagnetic fields, *J. Acoust. Soc. Am.,* 71, 1321, 1982.

493. **Foster, K. R. and Finch, E. E.,** Microwave hearing; evidence for thermoacoustical auditory stimulation by pulsed microwaves, *Science,* 185, 256, 1974.

494. **Guy, A. W., Chou, C. K., Lin, J. C., and Christensen, D.,** Microwave induced acoustic effects in mammalian auditory systems and physical materials, *Ann. N.Y. Acad. Sci.,* 247, 194, 1975.

495. **Lin, J. C.,** Microwave auditory effect — a comparison of some possible transduction mechanisms, *J. Microwave Power,* 11, 77, 1976.

496. **Carpenter, R. L. and van Ummersen, C. A.,** The action of microwave radiation on the eye, *J. Microwave Power,* 3, 3, 1968.

497. **Carpenter, R. L., Biddle, D. K., and van Ummersen, C. A.,** Biological effects of microwave radiation with particular reference to the eye, *Proc. Third Int. Conf. Med. Electron. (London),* 3, 401, 1960.

498. **Guy, A. W., Lin, J. C., Kramar, P.O., and Emery, A. F.,** Measurement of absorbed power patterns in the head and eyes of rabbits exposed to typical microwave sources, in Proc. 1974 Conf. on Precision Electromagnetic Measurements, London, 1974, 255.

499. **Guy, A. W., Lin, J. C., Kramar, P. O., and Emery, A. F.,** Quantitation of Microwave Radiation Effects on the Eyes of Rabbits at 2450 MHz and 918 MHz, Scientific Rep. No. 2, University of Washington, Seattle, January 1974.

500. **Guy, A. W., Lin, J. C., Kramar, P. O., and Emery, A. F.,** Effect of 2450 MHz radiation on the rabbit eye, *IEEE Trans. Microwave Theory Tech.,* 23, 492, 1976.

501. **Kramar, P. O., Emery, A. F., Guy, A. W., and Lin, J. C.,** The ocular effects of microwaves on hypothermic rabbits: a study of microwave cataractogenic mechanisms, *Ann. N.Y. Acad. Sci.,* 247, 155, 1975.

502. **Kramar, P. O., Guy, A. W., Emery, A. F., Lin, J. C., and Harris, C. A.,** Quantitation of Microwave Radiation Effects on the Eyes of Rabbits and Primates at 2450 MHz and 918 MHz, University of Washington, Bioelectromagnetics Research Laboratory Scientific Rep. No. 6, Seattle, Wash., 1976.

503. **Appleton, B.,** Comment, *Ann. N. Y. Acad. Sci.,* 247, 133, 1975.

504. **Hirsch, S. E., Appleton, B., Fine, B. S., and Brown, P. V. K.,** Effects of repeated microwave irradiations to the albino rabbit eye, *Invest. Ophthalmol. Vis. Sci.,* 16, 315, 1977.

505. **Kinoshita, J. H., Merola, L. D., Dikmak, E. D., and Carpenter, R. L.,** Biochemical changes in microwave cataracts, *Doc. Ophthalmol.,* 20, 91, 1966.

506. **Merola, L. O. and Kinoshita, J. H.,** Changes in the ascorbic acid content in lenses of rabbit eyes exposed to microwave radiation, in *Biological Effects of Microwave Radiation,* Vol. 1, Peyton, M. F., Ed., Plenum Press, New York, 1961, 285.

507. **Weiter, J. J., Finch, E. D., Schultz, W., and Frattali, V.,** Ascorbic acid changes in cultured rabbit lenses after microwave irradiation, *Ann. N.Y. Acad. Sci.,* 247, 175, 1975.

508. **Taflove, A. and Brodwin, M. E.,** Computation of the electromagnetic fields and induced temperatures within a model of the microwave-irradiated human eye, *IEEE Trans. Microwave Theory Tech.,* 23, 88, 1975.

509. **Paulsson, L. E.,** Measurements of 0.915, 2.45, and 9.0 GHz absorption in the human eye, presented at the 6th European Microwave Conf., Rome, 1976.

510. **Kalant, H.,** Physiologic hazards of microwave radiation, survey of published literature, *Can. Med. Assoc. J.,* 81, 575, 1959.

511. **McAfee, R. D., Longacre, A., Jr., Bishop, R. R., Elder, S. T., May, J. G., Holland, M. G., and Gordon, R.,** Absence of ocular pathology after repeated exposure of unanesthetized monkeys to 9.3 GHz microwaves, *J. Microwave Power,* 14, 41, 1979.

512. **Chou, C.-K., Guy, A. W., McDougall, J. A., and Han, L.-F.,** Effects of continuous and pulsed chronic microwave radiation on rabbits, *Radio Sci.,* 17(5S), 185, 1982.

513. **Guy, A. W., Kramar, P. W., Harris, C. A., and Chou, C. K.,** Long-term 2450-MHz CW microwave irradiation of rabbits: methodology and evaluation of ocular and physiologic effects, *J. Microwave Power,* 15, 37, 1980.

Chapter 5

MODULATED FIELDS AND "WINDOW" EFFECTS

Elliot Postow and Mays L. Swicord

TABLE OF CONTENTS

I. INTRODUCTION

As the body of scientific literature describing the effects of radio frequency (RF) electromagnetic radiation increases, so does the variety and complexity of reported effects. Because much bioelectromagnetics research is motivated by the need to delineate hazardous conditions, most experiments have been designed to determine a threshold, the unique value of an experimental variable above and below which the experimental outcomes are different (or begin to differ). Since it has been generally assumed that for all values below the threshold, the experimental outcome will be the same, protocols were designed in which the variable under study differed significantly (often by as much as a factor of ten) from experimental series to experimental series. Experiments aimed at documenting a threshold are appropriate for the study of effects that depend on field strength or on the rate or amount of energy absorbed. However, they are not adequate to investigate most effects that depend upon specific modulation conditions. Of particular concern are effects that are reported to vary nonmonotonically over very small regions of the independent variable. As may be expected, threshold-establishing experiments did not uncover such effects. However, the results of recent experiments in which both frequency and field strength were sequentially varied by small increments indicate that some biological outcomes are observed at a lower threshold when exposed to modulated fields or occur only when specific values of the independent variable are chosen. In the latter case, other nearby values above or below the response range will produce a very different outcome. The condition in which a specific result is found only when the independent variable is within certain well-characterized narrow ranges of values is described as a "window" effect. Both frequency and field strength "windows" have been reported in the literature and both types will be reviewed below.

"Windows" are not new to biology. The existence of competing processes can result in a "window". A commonplace example of this phenomenon is the dependence of cell growth on temperature. As temperature increases from the minimum compatible with growth to about 35°C, the growth rate increases monotonically, because the biochemical reactions required for growth proceed more rapidly. As the optimal temperature for a given type of cell is approached, the increase in growth with increase in temperature becomes smaller until finally, above the optimal temperature, increases in temperature produce lower growth rates. This is because at higher temperatures competing processes become more important. In most cells these processes are related to the thermal lability of critical chemical constituents of the cell. The competition of two processes can produce a single "window". However, the existence of multiple "windows" over a range of a single variable has been reported. This case cannot be explained as simple competition between two processes and a more complex theoretical explanation is required.

The existence of "window" effects, especially multiple "window" effects, is highly significant for several reasons. Most important among these is the implication for the development of safety standards. Consider the case where a well-documented hazardous condition occurs within a known range of exposure values, but is not observed at points outside the range. In this case it would be difficult to define a "safe region", because it could always be argued that there may be another "window" at other exposure conditions within which the same hazardous situation would be produced. In this case, safety could be demonstrated only when the location of all "windows" could be predicted with a high degree of confidence, or when an exhaustive search for additional "windows" failed to discover any. The former requires a tested theory and the latter a tedious, perhaps impossible, series of experiments. In either case, it is imperative, certainly for standard-setting purposes, to perform experiments over a range of exposure conditions (e.g., frequency, modulation, dose, dose rate, peak power) to determine the existence or nonexistence of effects.

Medical applications provide another reason for studying "windows". Beneficial effects may be produced only by unique conditions of exposure. If these effects are to be of clinical use, the exposure conditions required to produce them must be carefully characterized.

Frequency and amplitude "windows" have been observed in various biological-effects experiments. These experiments, which will be reviewed below, include genetic, immune, hematologic, nervous system, and reproductive responses. "Window" responses have been observed over a wide range of frequencies from ELF to millimeter waves. The majority of the reported "window"-type effects has been observed from exposure to either millimeter-wave radiation or ELF, ELF-modulated RF fields. The millimeter-wave effects will be discussed first, followed by a discussion of the lower frequency effects categorized by biological response.

One of the first, and still important, applications of microwave radiation is radar, which requires that the radiation be pulse modulated. Therefore, many hazard analyses have considered pulsed radiation. Only a very limited, usually just one, set of pulse parameters is considered by an investigator. Therefore, experiments using pulsed irradiation are not considered to be evaluations of "window" effects. However, if the results of an experiment using pulse-modulated irradiation is significantly different from one using continuous-wave exposure a "window" effect may be present, and additional research is required. Comparisons of continuous wave and pulse-modulated irradiation will also be reviewed later in this chapter. Even if the range of pulse parameters for which the effect can be observed is large and use of the term "window" is then inappropriate, this finding may be of great practical importance.

II. "WINDOW" EFFECTS IN THE MILLIMETER-WAVE REGION

The suggestion that radiation in the millimeter-wave region could have unique frequency-specific effects on biological systems was made in the mid-1970s.[1] Fröhlich[2] predicted that strong macromolecular oscillations in the 100- to 1000-GHz frequency range are of biological significance. Early experimental support for Fröhlich's ideas came from Webb and Dodds[3] who showed that irradiation at 136 GHz slowed the growth of *Escherichia coli*. This was followed by the same group's observation[4,5] that, depending on the frequency, the growth rate of microwave-irradiated *E. coli* was either enhanced or retarded, as compared with control bacteria. Within 2 or 3 GHz, the effect could change from a 50% increase to a 50% decrease. Blackman et al.[6] attempted to replicate the dramatic report of Webb and Dodds,[3] but found only a small enhancement of cell growth which was attributed to a slightly elevated temperature of the irradiated cells.

In other experiments, Webb and Booth[4] presented data showing rapid variations with frequency (in the 64- to 75-GHz frequency range) of the absorption properties of water and other solutions containing DNA, RNA, and proteases. The reported fine structure of water is not expected to be responsible for these effects and these results are made further questionable due to the ratio method employed. This method is briefly discussed by Webb in several different publications. Generally, in the ratio method absorption losses of the sample holder are eliminated by first measuring the loss in the empty holder and comparing this with the loss observed in the filled sample holder. Measurements are often made simultaneously with two identical sample holders. In Webb's[7] study, the sample holder was composed of two mica "windows", located in a waveguide, approximately 1 mm apart, thus providing an approximately 1-mm-thick sample. Accurate absorption measurements using a ratio method can only be obtained when the sample thickness is much less than a quarter wavelength. Otherwise, reflections at both back and front surfaces of the sample will cause large standing waves within the system not accounted for by measurements with the empty sample holder, thus making accurate measurements impossible. The 1-mm sample was on the order of a quarter wavelength in the samples at the frequencies used by Webb and Booth.[4]

Gandhi et al.[8] reported experiments in this frequency range using a ratio-absorption technique and found no variations of the type reported by Webb and Booth.[4] It should also be noted that the ratio method employed by both experimental groups may not be sensitive enough to detect small perturbations in a highly absorbing medium such as water. Thus, contributions to absorption by a few large molecules suspended in highly absorbing water would go unnoticed.

Berteaud et al.[9] reported a marked decrease in the growth of bacteria exposed at 70.5 and 73 GHz. They examined several specific frequencies with replicate experiments at 73 GHz and showed both frequency dependence of growth and a high degree of variability among replicates under the same experimental conditions. These results are similar to those of Webb and Dodds,[3] however, Webb and Dodds investigated 136 GHz only and did not report replicate experiments. In a follow-on attempt to generalize these results and investigate a possible mechanism, Berteaud's group used a diploid strain of yeast that is readily mutated (D_5 Zimmermann).[10] No evidence of altered survival, impaired function, or structural injury was observed in cells exposed to either 70.5 or 73 GHz at power densities as high as 600 W/m², leading the authors to conclude that the previously reported effects of millimeter waves on bacteria are not mediated by the direct action of microwaves on cellular DNA.

The effects of microwaves on protein synthesis of mammalian cells (BHK-21/c13) were assessed at 0.1-GHz intervals over the ranges of 38 to 48 and 65 to 75 GHz. The exposure system allowed for simultaneous examination of possible power density "windows", because the microwave power within the plane of the cell culture varied as $\cos^2$, permitting simultaneous investigation of a wide range of power densities.[11] Using the system described by Partlow et al.,[11] Bush et al.[12] did not detect acute (1 hr) effects of millimeter waves at power densities up to 3.5 kW/m².

Webb[13] also reported that the induction of lambda prophage can be effected by millimeter-wave radiation (70.5 GHz) and that the frequency dependence shows a strong resonance. A shift of 500 MHz to either side of the resonant frequency results in a decrease of 10⁵ in phage induction. This effect is also critically dependent on the time during the growth cycle when the cells are irradiated. This interesting study was recently repeated at the Max Planck Institute in Stuttgart with negative results.[13a]

In 1974 a group of abstract papers presented at a meeting held in 1973 at the Lebedev Institute of Physics in the Soviet Union was published.[1] These abstracts are not detailed enough to permit critical analysis. They are especially deficient in discussion of experimental methods employed and the environmental variables controlled. Thus, the possibility of engineering artifacts cannot be ruled out. However, these reports are consistent in demonstrating very sharp resonance effects of millimeter-wave radiation on biological systems. In many cases, the half-width of the resonance is about 100 MHz. The biological systems examined include: survivability of bacteria (41.96 GHz is most harmful);[14] stimulation of cell division in *Rhodotorula rubia* (maximum at 42 GHz);[15] damage of cell membrane, degeneration of protoplasm, and increase in cell size (46 GHz);[16] decreased fertility of *Drosophila* (maximum at 43 GHz) and increased mutations in the second generation (maximum 46 GHz);[17] and lower weight and retarded development of chicks born from irradiated eggs (maximum effect at 42 GHz).[17] It should be noted, however, that in the nearly 10 years following publication of these abstracts, to the best of our knowledge, a more detailed description of any of these experiments has not been published in the open literature.

One of the most interesting of the papers presented at the Lebedev Institute meeting is by Smolyanskaya and Vilenskaya,[18] describing multiple resonances in the colicin synthesizing capability of a specific plasmid-containing strain of *E. coli*. At the resonant frequencies, the increase was as much as 300%. This is one of a small number of papers in which repeated experiments, up to 25 in this case, were conducted under a single set of experimental conditions and where the statistical significance of the results was determined ($p < 0.001$), using a statistical test that was not named.

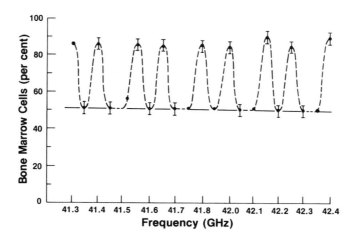

FIGURE 1. Relative number of bone marrow cells as a function of frequency of the microwave radiation used for pretreatment. Cell count in animals exposed to microwaves only represents 100. Animals exposed to X-rays (without the microwave pretreatment) show a 50% reduction in the number of bone marrow cells. Pretreatment with millimeter waves either affords no protection or partial protection from the effect of X-rays, depending on the frequency of the millimeter wave radiation. (Figure redrawn from Sevast'yanova, L. A. and Vilenskaya, R. L.,, *Biol. Nauki*, 6, 48, 1974.)

Smolyanskaya et al.[19] reviewed the earlier reports of millimeter-wave effects on colicinogenesis, however, no further experimental details were provided. They suggested that the resonant effects result from the action of microwaves on rotational states of segments of macromolecules which alters the structure, conformation, and ultimately, the function of the macromolecule. Additional results are presented describing effects on the synthesis of an exogenous enzyme, but at slightly higher frequencies. The results indicate that slight changes in wavelength (± 0.005 mm) result in no effect or inhibition of synthesis. It is also reported that a tenfold decrease in bacterial concentration resulted in a fourfold increase in the magnitude of the effect. Extensive attempts in the U.S. to reproduce the work of Smolyanskaya and Vilenskaya[18] have not been successful.[20] Other U.S. investigators[21,22] also attempted replication of these experiments with inconsistent results. The system of inducible colicin synthesis did not readily lend itself to systematic controlled investigations. This was dramatically demonstrated by unexplainable, highly variable differences in control and sham-irradiated samples. However, the Russian authors do emphasize the high frequency specificity of the effect and claim a capability of measuring wavelength to an accuracy of 0.01%. Whether or not additional frequency components due to an unstable source would cause suppression of the observed response is not known. However, it is possible that the U.S. investigators[20] did not use equipment with the required frequency specificity and stability.

The reports of Sevast'yanova and Vilenskaya[23,24] are also noteworthy because of the presence of several windows between 38 and 45 GHz (6.6 to 7.8 mm) separated by about 0.02 mm and the bimodal distribution of the data (see Figure 1). Depending on the frequency of the millimeter-wave pretreatment, it either afforded protection against the effects of X-rays or it had no protective effect at all. (Protection from the effects of X-ray treatment was assayed by determining the number of undamaged bone marrow cells.) X-ray exposure of mice decreases the number of viable bone marrow cells by 50% from preexposure values. Pretreatment with millimeter waves either has no effect on the number of viable bone marrow cells or affords partial protection from the effects of X-rays depending on the frequency of the millimeter-wave pretreatment (see Figure 1). In the latter case, the number of bone marrow cells is 70% greater than in the case without millimeter-wave pretreatment. Exposures at 100 to 750 W/m² were found to be equally effective, but exposures between 10 and 100 W/m² were not capable of mitigating the effects of X-rays. These reports left a Soviet observer of the field of resonance effects somewhat perplexed.[25] These data have recently been reinterpreted and are said to demonstrate two periodic processes offset by a period of 0.05 mm.[26] The paucity of data (a single data point for each node) does not support the two-processes hypothesis. Furthermore, the fact that over the entire range of frequencies investigated, all data were very close to two values seems unusual.

Irradiation (1.5 W/m²) with 46.31 and 46.38 GHz lowered pencillinase activity of *Staphlococcus aureus,* while similar exposures at 46.22, 26.34, or 46.43 GHz had no detectable effect.[27] Multiple-resonance curves of the effect of x-band and millimeter-wave radiation on growth of the bacteria *Bacillus mesentericus* and *Pseudomonas fluorescens* were reported.[28] Evaluation of a RF device used to rewarm frozen blood led to the finding that when 46-GHz radiation was used for heating, the amount of free hemoglobin was minimal.[29] The amount of free hemoglobin was 30% greater at 41 GHz, but cell injury varied more dramatically at frequencies greater than 46 GHz. Exposures were at 10 W/m².

Three recent Soviet review articles discuss Soviet reports of resonance effects in biological systems irradiated with millimeter-wave radiation.[25,30,31] In general, the Soviet reports of millimeter-wave resonant effects are characterized by (1) sharp resonances (often multiple) where the width is 0.1 to 1% of the mean frequency, (2) insensitivity to power density, and (3) requirement of extended or multiple exposures (for microorganisms this may correspond to irradiation over several cell divisions).[30] Very different results, e.g., beneficial and hazardous effects or stimulation and inhibition, are reported to occur at the same frequency when different test systems are used.

Grundler and Keilmann[32] investigated and extended the work of Devyatkov.[15] They chose to grow yeast (a diploid homozygotic and isogenic wild type strain of *Saccharomyces cerevisiae*) in stirred cell suspensions rather than on agar plates. This allowed for a more homogeneous thermal environment whose temperature was maintained in the neighborhood of 32°C. At a given temperature, the growth rates of controls varied within ± 3% of the mean value, while the growth rates of microwave-exposed samples varied by as much as 20%, depending on the frequency of the radiation (see Figure 2). Multiple resonances between 41.64 and 41.79 GHz with line widths of 10 MHz were reported. (Experimental variability in frequency is estimated as 3% and the absolute value of the frequency was known to 20 MHz.)

More recently, Grundler et al.[33] improved on the dosimetric techniques and temperature monitoring used earlier. The temperature was maintained at 30.7°C in controls, but was allowed to increase slightly by 0.16 to 0.4°C in exposed samples. This slight increase provided an additional calorimetric method of determining absorbed power (0.016°C increase per milliwatt absorbed power per 2.5 mℓ[32]). Total power input to the system was 10 to 25 mW, which over the course of the exposure increased temperature by 0.16 to 0.4°C. Measurements of these growth rates (μ) indicated a weak dependence on temperature ($d\mu/\mu dT \leqslant 0.027°C^{-1}$). This proved to be insignificant compared to the variation of ±4% observed between control and sham-exposed samples. In addition, forward and reflected power were monitored and the reflected power minimized at each frequency with the aid of an x-y tuner. The yeast strain used as well as the frequencies of irradiation were the same as in the earlier study. Again, a very strong frequency dependence of the growth rate was reported. The curves in the two reports cannot be superimposed, because in the earlier study the absolute frequency was known only to an accuracy of ±20 MHz and the resonance width is only about 1 MHz. The report[33] presents data from four separate series of experiments, some of which are replicate experiments done at the same frequency (see Figure 2).

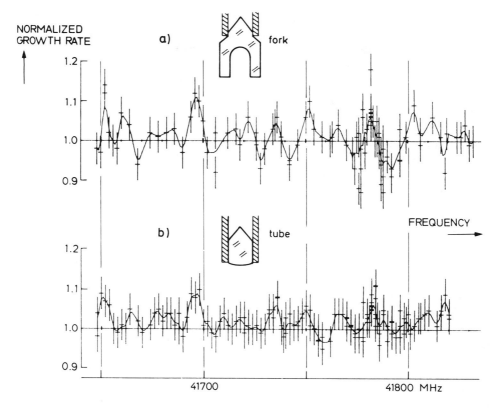

FIGURE 2.   Growth rate of microwave-treated yeast culture relative to the growth rate of untreated culture. In the upper curve (a) a fork-shaped antenna was used, while in the lower curve (b) a tubular antenna was used. (From Grundler, W., Keilmann, F., Putterlik, V., Santo, L., Strube, D., and Zimmermann, I., in *Coherent Excitations in Biological Systems,* Fröhlich, H. and Kremer, F., Eds., Springer-Verlag, Berlin, 1983. With permission.)

Grundler et al.[34] have recently reviewed their past results subjecting them to an extensive statistical analysis. This analysis indicates that the two studies[32,33] yield the same frequency dependence when a +11-MHz correction is applied to the data in the earlier study.[32] Furthermore, their analysis shows that the observed frequency variations are not a mere chance sequence of the frequency-independent microwave effects. Even though the variations in the growth rate are only of the order of ±10%, the authors emphasize the frequency stability of the system. Frequency was stabilized to within ±0.3 MHz.

Raman spectroscopy is an excellent way of determining whether high frequency resonances exist. The presence of anti-Stokes lines of approximately the same intensity as the Stokes lines would substantiate the existence of coherently excited modes. Sharp, distinct lines in the 15- to 300-cm^{-1} region (450 to 9000 GHz) have been reported in Stokes Raman spectra.[35] The peaks are at different frequencies for normal and tumor cells in mammary tissue. However, while the crucial anti-Stokes spectra have not been reported for mammalian cells, this is consistent with the report by Webb and Booth[7] of differences in the millimeter-wave absorption of normal and tumor cells. Webb and Stoneham[36] found narrow Stokes lines, in the region between 100 and 200 cm^{-1} (3000 to 6000 GHz), in a bacterial cell preparation, but only when the cells were actively dividing in synchrony. Webb et al.[37,38] then examined the anti-Stokes spectrum of synchronized, actively metabolizing *E. coli* and noted variations in the spectra with both time and frequency. The fact that the anti-Stokes intensities are similar to the Stokes intensities indicates the existence of coherently excited modes.

Similar results of a preliminary study are reported by Drissler and Santo[39] who provide a more systematic series of observations of variations in the Stokes-Raman spectra during irradiation. Recently, Furia and Gandhi[40] attempted to reproduce the earlier experiment of Webb et al.[38] using *B. megaterium*. However, they did not find Raman lines in the range of 20 to 300 cm^{-1} (600 to 9000 GHz). The work of Cooper and Amer[41] suggests that the time variations observed by Webb et al.[38] in the spectra were due to clumping of the synchronous cells which produced changes in the total light scattered (Mie scattering from clumped cells) and were not due to specific frequency components.

In general, Raman spectroscopy of cell suspensions is difficult. Several effects can produce spurious Raman lines as discussed by O'Sullivan and Santo.[42] At frequencies below 3000 GHz, the probability of experimental artifact in Raman spectroscopy is significant. Grating ghosts and artifactual scattering due to cell division are two possible causes of error.[43,44] Furthermore, when cell suspensions are examined, media that do not fluoresce must be used. Additional problems are encountered, because the cells settle producing time-dependent intensity fluctuation in the scattered light. Gandhi et al.[45] used nonfluorescing media and suspensions of cells in agarose. They did not detect Raman spectra in the region of 15 to 1500 cm^{-1} (450 to 4500 GHz) when examining baby hamster kidney cells (BHK-21/C13), chick fibroblast cells, or *E. coli,* contrary to the reports of others. Although Gandhi et al. could not observe Raman spectra in cell systems, they have not disproved the work of Webb and associates.

Using still another experimental system, Kremer et al.[46] irradiated the salivary glands of the midge *Acricotopus lucidus* with millimeter waves at <50 W/m^2 and observed the constriction of giant chromosome puffs. A specific puff that expresses genes for a secretory protein was reduced in size. While an exhaustive evaluation of the frequency dependence of this effect was not undertaken, swept-frequency irradiation (64.1 to 69.1 GHz) was observed to be less efficient than was irradiation at 67.2 or 68.2 GHz.

The differential millimeter-wave absorption spectra of normal tissue slices and those from tumorous tissues were first reported by Webb and Booth.[7] For a variety of tumors it was reported that tumor cells absorb more strongly at 69, 72, and 75 GHz and normal cells absorb more strongly at 66, 68, and 70 GHz. It was suggested that this difference is the result of differences in the nucleic acid complement of tumorous and normal cells. Stamm et al.[47] investigated effects of irradiation at higher frequencies (77 to 85 GHz) and reported that the absorption spectra of normal tissue and tumors were different and specific to the type of tissue. The use of millimeter-wave absorption in the diagnosis of cancer is an attractive, but yet unproven, application of microwave radiation. These experiments[47] suffer from the use of thick samples, as discussed above, and the results must be substantiated by more appropriate techniques before the diagnostic technique merits clinical acceptance.

As can be seen in Table 1, most reports of frequency-specific effects of millimeter-wave radiation demonstrate several closely spaced "frequency windows". Of the reports reviewed, the most common experiment is an evaluation of the growth of bacteria or yeast, but a wide variety of effects, including biochemical, genetic, and developmental alterations, have been found in these prokaryotic cells. Two results have potential medical application, i.e., protection against the effects of X-radiation and differentiation between tumorous and normal cells. Experiments demonstrating windows were performed in the Soviet Union, Germany, France, Canada, and the U.S., while all the negative reports originate in the U.S.

Very sharp resonances in the millimeter-wave region have been reported by Webb and Booth,[4] Berteaud,[9] Grundler and Keilmann,[32-34] Devyatkov,[15] Smolyanskaya and Vilenskaya,[18] etc. If these sharp resonances are due to effects on DNA, as is likely because of the types of effects observed, they would indicate interaction with specific molecules or molecular systems which, in most of these examples, could result in the removal of a repressor molecule from the DNA, thus allowing transcription to take place. The sharpness

Table 1
SUMMARY OF EXPERIMENTAL RESULTS IN THE
MILLILMETER REGION

Effect	Type of frequency window	Ref.
Decreased bacterial growth	Single	3
Bacterial growth	No window	6
Decreased/enhanced bacterial growth	Multiple	4, 5
Growth of bacteria	Single	14
Stimulation of cell division	Single	15
Decreased cell growth	Single	9
Decreased growth of bacteria	Multiple	28
Stimulation of cell growth	Multiple	32—34
Cell damage	Single	16
Growth of mammalian cells	No window	12
Damage to red blood cells	Multiple	29
Increased bacterial synthesis of protein	Multiple	18
Bacterial synthesis of protein	No window	20
Decreased enzyme activity of bacteria	Multiple	27
DNA, RNA, protein absorption	Multiple	7
DNA, RNA, cell absorption	No window	43
Raman spectroscopy	Multiple	7, 35, 37, 38
Raman spectroscopy	No window	40, 45
Genetic mutations in yeast	No window	10
Induction of lambda prophage	Single	13
Increased mutation of *Drosophila*	Single	17
Chromosomal effects in midge	Probable windows	46
Decreased fertility of *Drosophila*	Single	17
Retarded development of chicks	Single	17
Protection against X-ray damage	Multiple	23
Difference between tumor and normal cells	Multiple	35, 47

of the line widths surprisingly suggest only mildly dampened absorption. This, in turn, indicates some form of isolation of the modes induced in the large molecule from the absorption modes of the surrounding media. Mode dampening would result in considerable line broadening and the absence of rapid variation of absorption with frequency.

It is difficult to draw any overall conclusions concerning the numerous reported effects of electromagnetic fields in the millimeter range. This difficulty stems in part from : (1) the lack of information supplied by most of the investigators, particularly those in the Soviet Union. A reader of this literature is not reassured from the details presented that careful attention was paid to proper engineering practice to assure that the frequency dependence reported, at least in some cases, was not the result of standing waves in the system which would vary rapidly with frequency. It has been the practice in the Soviet Union not to present such details; (2) in some of the early work[4,7,47] insufficient attention was paid to the difficulties of millimeter-wave engineering practice. Thick samples were used in absorption measurements, resulting in frequency variations of the system which would obscure frequency variations in the absorption properties of the sample; (3) the lack of attempts or the inability to replicate or substantiate the reported results by a second investigator. The inability to reproduce the various millimeter-wave experiments reported by Webb is an example. It is unfortunate that more attempts have not been made to replicate other millimeter-wave experiments.

The work of Grundler et al.[34] stands as an example of excellent reporting of experimental details with appropriate attention paid to engineering design. Expense was not spared for equipment or labor resulting in a work of very high quality. This is not to say that no stone was left unturned. This work may be criticized due to the possibility that there may exist a spacial variation of energy distribution with frequency within the sample holder. Points close to the wave guide surface may not be sufficiently stirred and thus selective cells may remain longer in locations that may have different exposure levels at different frequencies. Such criticism, perhaps, can only be properly addressed by performing experiments to either confirm or question the results reported earlier.

III. WINDOW EFFECTS OF ELF AND ELF-MODULATED FIELDS

The subject of frequency-specific and power-density-dependent windows in the effects of electromagnetic radiation has been reviewed recently by Adey,[48-53] Adey and Bawin,[54] and by Myers and Ross.[55] The topic is important for both theoretical and practical reasons and has, therefore, interested a number of scientists. However, there exists today a variety of experimental results and a broad spectrum of opinion on this subject. Research has centered on four areas: electrophysiology, neurochemistry (calcium levels in the brain), immunology, and audition.

A. Electrophysiology

The first reports of dependence of a biological effect of modulation frequency were by Bawin and collegues,[56-58] who showed that weak VHF fields that are amplitude modulated at brain-wave frequencies influence spontaneous and conditioned electroencephalograph (EEG) patterns in the cat. This work has been criticized for the presence (during irradiation) of metallic electrodes which would produce an "antenna effect".[55] Glass or metal electrodes located in the RF field can cause field distortions and hence result in artifacts, due to the difference in electromagentic properties of the electrode material and the solution bathing the nerve. While this artifact may be significant in elucidating the mechanism of interaction, it is unlikely that an electrode artifact would favor specific modulation frequencies from among those in the ELF region that were considered by Bawin et al.[56-58]

In a related experiment, Gavalas et al.[59] showed that the subjective estimate of a monkey of an interval between events was shortened by as much as 0.4 sec when exposed to a 7-Hz electric field at 7 V/m (peak-to-peak outside the head of the animal). Expanding on this finding, they later showed that, for irradiation at 7 Hz, the threshold was lower and the magnitude of the effect larger than at 45, 60, or 75 Hz.[60]

Rabbits exposed to 1 to 10 MHz radiation (amplitude modulated at 15 Hz) for 2 hr/day over a period of several weeks showed changes in their EEG patterns (enhanced low frequency components and decreased higher frequency components). Modulation at lower (4 to 5 Hz) or higher (60 Hz) frequencies was not as effective.[61] This finding is reminiscent of an earlier report by Servantie et al.[62] that a microwave signal illuminating an animal imparted onto the EEG of the animal a component that was the same as the pulsing frequency of the irradiating signal. However, since Servantie et al.[62] only investigated repetition rates of 500 and 600 Hz, their results cannot be used to support or reject the existence of a frequency "window".

Montaigne and Pickard[63] monitored the membrane potential of isolated algeal cells (*Chara braunii* and *Nitella flexilis*) while the cells were exposed to amplitude-modulated 147-MHz radiation at power densities between 2 and 1000 W/m². The importance of modulation frequencies from 4 to 64 Hz, at approximately 90% depth of modulation, was evaluated.

Changes in the membrane potential occurring in phase or in phase quadrature with either the modulation signal or a signal at twice the modulation frequency, using a procedure described by Barsoum and Pickard,[64] were not found under any of the experimental conditions employed.

The absence of reported effects of amplitude-modulated RF radiation on algeal cells (*Chara braunii*,[63] *C. covallina*,[65] and *N. flexilis*[63]) may be due to the fact that these cells use calcium in a different manner than do mammalian cells.

B. Calcium Efflux from Brain

The initial report of frequency-dependent windows of electrophysiological responses[59] was followed a few years later by the report[66] of similar frequency-selective reponses of brain calcium. Cerebral hemispheres from 4- to 8-day-old chicks were incubated for 30 min at 37°C in a physiological solution containing 2.16 mM CaCl$_2$ (the same proportion of calcium as in cerebral spinal fluid), 2.4 mM NaHCO$_3$, and 0.2 µCi radioactive calcium. Following three rinses in radioactivity-free medium the hemispheres were exposed to 147-MHz radiation, sinusoidally amplitude modulated (80 to 90% depth of modulation). Samples from the exposed brain sections exhibited increased concentrations of preloaded calcium in the medium (interpreted as increased calcium efflux) that were dependent on the modulation frequency of the 147-MHz radiation. The 9- to 20-Hz region was effective with maximal efflux being measured at 16 Hz; frequencies both higher and lower were ineffective.[66] Identical results were obtained following incubation with sodium cyanide, indicating that the cytochrome system is not required for this process.[66] From theoretical considerations, Tenforde[67] concluded that a mechanism other than thermal energy transfer was responsible for the observed alterations in calcium binding. The original experimental finding was corroborated and extended by Blackman et al.,[68,69] who reported that the response occurred only within a narrow intensity range (intensity ''window'') around 8.3 W/m². The observations were extended to 450-MHz carrier-wave frequency with similar results.[69a] Again, 16-Hz sinusoidal modulation was effective in producing increased levels of calcium in the medium when exposures were at 1 or 10 W/m², but not at 0.5 or 50 W/m². Later, Blackman et al.[70] extended these experiments to 50 MHz where they found two intensity windows (one between 14.4 and 16.7 W/m² and the second around 36 W/m²). In experiments using ELF-modulated RF fields marked variability was observed in the sham specimens.[69,70] Alterations in the distribution of calcium are of general interest, because of the pivotal role played by calcium in the control of vital processes in the central nervous system (CNS) and other organ systems.

Both frequency and field-strength dependence of calcium efflux in excised chick forebrains[71] and in the cortex of anesthetized cats exposed to ELF fields have been reported.[74] The maximum effect was observed following exposure to 6- to 16-Hz fields and at field strengths of about 10 to 56 V/m in air (producing an electric gradient of the order of 10^{-5} V/m in the tissue). Blackman et al.[72] also suggested the existence of a window at 16 Hz (by examining 1, 16, and 30 Hz) in the efflux of calcium from chick brain tissue. They then proceeded to demonstrate two power density windows for the effect — at 3 to 8 and 40 to 50 V/m (field strengths between 1 and 70 V/m were investigated) (Figure 3). Blackman et al. found that the 16-Hz carrier produces the same type of effect, i.e., increased calcium efflux, as is reported when the 16-Hz signal is used to modulate a RF carrier. This is in contrast to the earlier report[71] that 16-Hz fields effect a reduction in the calcium efflux. Blackman et al.[73] went on to demonstrate a number of ''windows'', both in frequency and in field strength, in the ELF region (see Figure 4). A testable explanation for the difference between the directions of these ELF results has not been proposed. Although very comprehensive, the recent study of ELF ''windows'' on calcium efflux[73] obviously could not evaluate all combinations of frequency and field strength. There may exist very broad windows between 60

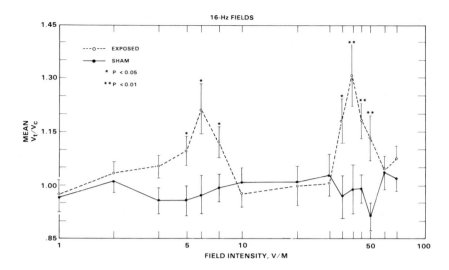

FIGURE 3.   Mean relative quantity of preloaded calcium ions released from chick brain tissue. Treatment, ELF field or sham exposure, is relative to water-bath treatment at 37°C. Electric field strength is peak-to-peak. The bars represent standard error. Probability of sham and ELF-exposed treatments being different is * for $p<0.05$, ** for $p<0.01$. (From Blackman, C. F., Benane, S. G., Kinney, L. S., Joines, W. T., and House, D. E., *Radiat. Res.*, 92, 510, 1982. With permission.)

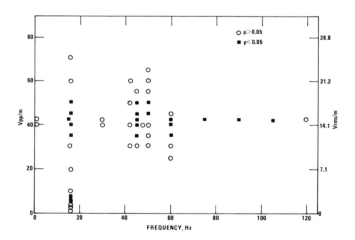

FIGURE 4.   Experimental conditions for which the release of preloaded calcium in chick brains is different from sham exposure. Field strength is in V/m peak-to-peak. Filled squares indicate conditions of frequency and field strength for which the efflux of calcium is significantly greater in the ELF-treated brain tissue ($p<0.05$). (Figure courtesy of C. F. Blackman.)

and 105 Hz. On the other hand, it may be that at any frequency there exists a range of field strengths that will affect calcium levels in the chick brain and other animal tissues. The width and position of this ''field-strength window'' vary with frequency in an as-yet undetermined manner. If this is the case, then it would be inappropriate to describe either ''frequency windows'' or field strength ''windows''. It has recently been suggested that the magnitude of the magnetic field of the earth influences the location of calcium-efflux ''windows''. This evidence is discussed later in this section.

In order to provide additional information helpful in assessing the significance of the experimental results obtained using chick brain sections, studies were conducted using anesthetized cats. Cats immobilized under local anesthesia were surgically prepared in order to allow the introduction of radioactive calcium at the pial surface. Following surgery the cats were exposed to fields similar to those used earlier (450-MHz fields, amplitude modulated at 16 Hz) except at 30 W/m² (specific absorption rate [SAR] = 0.29 W/kg). Efflux of preloaded calcium from the cerebral cortex was greater in exposed cats than in controls. Using a *one-tailed* t test, the difference was significant at $p \leq 0.032$. Different modulation frequencies or field strengths were not evaluated. Therefore, these results by themselves do not demonstrate a window in either frequency or field strength, but in the context of earlier work by this group it must be interpreted as supporting the existence of windows.[74]

By considering the average electric field strength within a homogenous spherical sample, Joines and Blackman[75,76] were able to bring some order to the proliferation of intensity windows. They reported that the intensity windows at 50, 147, and 450 MHz occurred at the same field strength within the specimen as calculated using their homogeneous spherical model of the exposed brain section. This analysis reduced the number of reported intensity windows from six to four. Even though the scaling technique was criticized because of the high degree of uncertainty with which internal field strength could be calculated,[77] it was used successfully to predict the presence of additional intensity windows. More recently,[78] direct measurements of the electric field in chick brains exposed at 50, 147, and 450 MHz were compared with the various theoretical predictions. For a constant external field, the ratio of the measured internal field strength in samples exposed at 50 and 147 MHz was close to that predicted for a layered model.[76] Agreement with predictions for the higher-frequency irradiation or using a simpler model[75] was not good. It must be pointed out that this approach provides no information on how many additional windows might exist and at what field strengths they can be expected.

Motivated by questions of safety, Merritt et al.[79] exposed the brains of anesthetized rats to microwave radiation (1, 2.06, or 2.45 GHz) modulated with 20-msec pulses (at 16 pulses per second) at 5 to 100 W/m². Calcium levels were not altered by the microwave treatment. While these results may be important in a practical sense, because the exposure parameters are similar to those of operating systems, they do not help us to understand the phenomenon reported by others, because the experimental conditions are so different from those reported to produce "window" effects. Significant differences in the experiment performed by Merritt et al. and those performed by Blackman et al. are the value of the internal electric field strength, modulation conditions, and the carrier frequency. Blackman et al. state that the SARs that produce the most significant responses are 1.3 mW/kg at 50 MHz and 1.4 mW/kg at 147 MHz. These absorption rates correspond to an average internal electric field of about 0.05 V/m. Merritt reported SARs of 0.29 to 2.9 W/kg at 1 GHz and 0.3 W/kg at 2.45 GHz. These SARs would correspond to an average internal electric field strength of about 0.4 to 1.6 V/m, a factor of ten or more greater than the reported maximum effective level. Blackman's data indicate that changes of less than a factor of two in field strength can lead to different responses. Thus, it is possible that Merritt used an "ineffective" exposure level to find an effect on calcium efflux. It is also possible that no level of exposure will elicit such responses at carrier frequencies greater than 1 GHz. Thus, Merritt's results are consistent with those of Blackman et al. and Bawin et al. The reported response of the biological system[71] is due to the modulation frequency and not the carrier frequency, therefore, it seems reasonable to assume that the carrier wave is being detected or rectified by some nonlinear component such as the cell membrane. Since the molecular system has momentum and at some high frequency will be unable to keep up with the rapid time-varying field, it also seems reasonable to assume that such a rectification process would relax or cease to respond as the carrier frequency is increased. Unfortunately, no experiments have been performed to date that establish the upper bound on the carrier frequency for this response.

In an effort to investigate the hypothesis that the RF-induced increase in calcium efflux in brain slices was the result of stimulated release from membrane binding sites, synaptosome preparations were studied.[80] The field strength was 43 V/m (in air) in all cases. While the results of experiments using unmodulated or 60-Hz amplitude-modulated 450-MHz fields did not differ from sham irradiation, exposures in which the modulation was 16 Hz produced an increase in calcium efflux. Calcium release was measured as a function of time and was found to increase more rapidly following exposure to 16-Hz modulated fields than following injection of $CaCl_2$ (which releases calcium from the intracellular compartment). Therefore, the authors suggest that intracellular calcium is not involed in the RF effect on calcium movement. Membrane-bound calcium is the probable source of calcium released by electromagnetic radiation. Recently, Dutta et al.[80a] reported an RF-enhanced efflux of calcium ions from human neuroblastoma cells. Monolayer cultures were exposed to 915-MHz fields, with or without sinusoidal amplitude modulation (80%) at frequencies between 3 and 30 Hz. Ten exposure regimens between 0.01 and 5.0 W/kg were used. Significant increase in calcium efflux was observed only at 0.05 and 1.0 W/kg. The increased efflux at 0.05 W/kg was dependent on the 16-Hz modulation but not at 1.0 W/kg.

The mechanism through which RF or ELF fields alter calcium levels in brain tissue cannot be found in the tennets of traditional quantum chemistry. Several suggestions, most commonly involving cooperative phenomena, have been proposed[53,81-83] and are discussed briefly later in this chapter. Following a more classical approach, Spiegel et al.[84] proposed an electromechanical model to explain the results of experiments in which RF radiation alters the levels of calcium. He suggests that an electromagnetically induced force on the surface of the brain causes the brain to vibrate, dislodging calcium ions. Spiegel's model is consistent with the suggestion of Joines and Blackman[75] for extrapolating among carrier frequencies, because that model was based on establishing identical electric fields inside the brain tissue. Two experimental conditions that produced the same electric field would, in Spiegel's model, give the same electric-field-induced surface pressure.[84] Surface pressure becomes the mediating factor.

In another tissue, rat pancreas, an increase in the efflux of calcium ions was also demonstrated[85] following exposure to low levels (SAR estimated to be less than 0.075 W/kg) of 147-MHz radiation amplitude modulated with a 16-Hz sine wave. The magnitude of the effect was similar to that observed in chick brains.[68,71] The efflux of calcium was not accompanied by the release of secretory proteins, as is common for pancreatic tissue. This was shown to be due to the fact that the small volume of medium in which the tissue was incubated did not contain sufficient nutrients to support metabolism. Bawin et al.[66] examined chick skeletal muscle, but did not observe changes in calcium levels following exposure to modulated 147-MHz radiation identical to that which produced changes in chick brain tissue. Although the importance of calcium throughout the body is now generally accepted, the clinical significance of 16-Hz sinusoidal amplitude-modulated fields on these functions remains to be evaluated.

Sanders et al.[86] have examined energy metabolism in the brains of rats during and immediately after exposure to RF radiation. They report that exposure of a surgically uncovered brain to 200- or 591-MHz radiation increases the fluorescence of reduced nicotinamide adenine dinucleotide (NADH) (during exposure) in a dose-dependent manner for exposures between 10 and 100 W/m². No effect was reported subsequent to exposures to 2450 MHz. Postexposure measurement of adenosine triphosphate (ATP) and creatine phosphate (CP) levels in the whole brain showed that ATP levels were decreased by 200- and 591-MHz radiation, but not by 2450-MHz radiation, and that CP levels were affected only by 591-MHz radiation. SAR was comparable for irradiation at all three frequencies. The authors speculate that 200- and 591-MHz radiation inhibits specific, probably divalent-metal-ion-requiring, enzymes in the electron transport chain, but that 2450-MHz radiation is not similarly effective. Although Sanders et al.[86] report an unexpected frequency dependence of RF radiation on alteration of brain bioenergetics under the influence of RF radiation, their results do not indicate the presence of a window. However, a brief description of these results has been included because of their inconsistency with the concept of establishing thresholds on the basis of SAR or field strength. These experiments should be replicated and the frequency-specific nature of the results validated by independent observers.

C. Immunology

Because of its central role in a wide range of medical problems, the immune system has been actively studied in the last decade. Therefore, it is understandable that a significant portion of the bioelectromagnetics literature is concerned with possible effects on the immune system. Several of these studies have considered window effects.

Huang et al.[87] reported that microwave irradiation (2450 MHz, CW) of Chinese hamsters produced an increase in the transformation of circulating lymphocytes into blastic cells. The magnitude of this effect increased with increasing field strength between 50 and 300 W/m². However, the incidence of transformed cells observed following exposure at 450 W/m² was below that found for exposure at 50 W/m². The authors therefore suggest that this effect is another, albeit thermal, microwave "window" effect. Dose-response curves of the lymphocyte transformation-index are commonly of the "inverted-U" shape.[88] Therefore, it is not necessary to postulate a unique type of microwave effect and the use of the term "field strength window" is inappropriate in this case.

Reversible inhibition of allogenic cytotoxicity by murine T lymphocytes (CTLL-1) was reported[89] in the presence of a 450-MHz field that was amplitude modulated at 60 Hz. Modulation frequencies between 3 and 40 Hz (including 16 Hz) as well as 80 and 100 Hz produced inhibitions of cytotoxicity that were not as large as the 20% inhibition observed at 60-Hz modulation. Unmodulated carrier waves did not influence cytotoxicity. Pretreatment with the RF field was also effective when it was modulated at 60 Hz. Both the requirement for calcium in the initial stage of the cytotoxic reaction and reports that ELF-modulated 450-MHz fields affect calcium levels in brain tissue led the authors[89] to suggest the involvement of calcium in the observed effect. More recently, Adey's group extended their immunologic research to investigate the effect of amplitude-modulated RF fields (450 MHz) on protein kinase activity of human tonsil lymphocytes in culture.[90] While the activity of cyclic-AMP-dependent protein kinase was not affected by ELF-modulated fields, noncyclic-AMP-dependent protein kinase was markedly decreased for a brief period of time (returning to control values in 45 min even though RF exposure continued). This effect is described as being "windowed" in both modulation frequency (16 to 60 Hz) and duration of exposure. These results are potentially important because of the role kinase may play in its interaction with cancer-promoting phorbol esters. However, more research must be completed before their meaning and import are put in proper perspective.

Another calcium-dependent response of lymphocytes is the capping of antigen-antibody complexes in which the membrane-bound antigen-antibody complexes migrate to form a polar cap on the surface of the cell. Sultan et al.[91] investigated the effect of 147-MHz fields amplitude-modulated at ELF frequencies (exposure parameters commonly used in experiments showing window effects on brain calcium levels) on the capping phenomenon in mouse B lymphocytes. No effect was observed with RF fields modulated at 9, 16, or 60 Hz at a variety of power densities between 1 and 480 W/m². As expected, a temperature of dependence of capping was observed. The capping of antigen-antibody complexes is dependent on intracellular calcium. These findings are, therefore, not inconsistent with those experimental results in which the efflux of calcium was altered by ELF-modulated RF fields, for in the latter experiments it is the membrane-bound calcium that is affected.

Ragan and Phillips were thought[52] to have demonstrated a power density window in bone marrow cellularity (number of nucleated cells in a femur) of mice exposed to 2.88-GHz radiation (50 or 100 W/m²), but the authors make no such claim.[92] A more detailed report of this work[93] indicated that a comprehensive parametric study was not undertaken; therefore, a window effect could not have been demonstrated.

D. Auditory System

During the late 1940s and 1950s, engineers and technicians working with microwave systems were familiar with the anecdotal reports of "hearing radar", but it was not until 1961 that this phenomenon began to be investigated in a serious manner[94] and not until 1974 was the physical explanation for the effect described.[95] Hearing individual pulses of microwaves as distinct clicks (hisses or buzzes depending on modulation characteristics) is such an unexpected phenomenon that, even today, more experienced workers delight in describing and demonstrating it to new colleagues.

Frey[94] demonstrated that a subject with otosclerosis which caused a 50-dB loss in air-conduction hearing had normal hearing of RF pulses, but that subjects with hearing loss above 5 kHz (bone-conduction dependent) could not hear microwaves. He extended this work examining various frequencies, pulse durations, and power densities determining thresholds in terms of peak power density. These thresholds vary with frequency, depending on penetration of the radiation into the head.[96,97] After considering several possible mechanisms, Frey[96] suggested that a direct interaction of RF radiation with neurons was the mechanism by which individuals could hear microwaves. Sommer and Von Gierke[98] suggested that the mechanism was radiation pressure, but this was disputed because of errors in computation.[99,100] Additional experiments[101,102] established that for pulses shorter than 30 μsec the threshold is 0.4 J/m² per pulse regardless of pulse width or peak power. (This produces a peak-specific absorption of 16 mJ/kg, elevating the temperature in the head by 5×10^{-6} °C.) However, Frey and Messenger[103] reported that perceived loudness of 1245-MHz pulses is peak-power dependent for 10- to 70-μsec pulses.

Behavioral experiments used to determine if animals sense pulsed RF radiation often measure the movement of the animal from an environment that includes irradiation to an otherwise identical habitat that is free of radiation. It has been reported that rats prefer the absence of pulsed microwaves[104,105] and that rats[106] and cats[100] will appropriately respond to pulsed microwaves when used as a conditioned stimulus. However, these experiments do not prove auditory mediation of the behavioral outcome. Food-deprived rats trained to respond to an acoustic signal hesitantly respond to microwave pulses at the same repetition rate as the acoustic signal.[107] This is the strongest behavioral evidence that rats are sensitive to modulation frequency and that they interpret these stimuli similar to the manner in which they would sound. Hearing of pulsed microwave radiation may be responsible for some reported effects on behavior, especially when small decrements in the performance of the animal are observed during or immediately following irradiation with pulsed microwaves.[62,108]

The response of an animal to auditory stimulation can be documented via electrophysiologic measurements including action potentials of the auditory cortex, thalamus, and auditory nerve as well as cochlear microphonics. Electrophysiologic measurements, therefore, can be used to determine whether pulsed microwaves produce an auditory sensation in laboratory animals. Furthermore, if measurements at all sites along the auditory pathway, for both microwave-induced and ordinary sound, give similar results, the microwave-induced auditory response must to be initiated prior to the initial neurologic event. Therefore, several groups have attempted to measure electrophysiological responses to microwave pulses. Frey[109] measured microwave-induced neural responses of the auditory system of guinea pigs and cats, but could not detect the cochlear microphonic response. Using a more advanced electrode system, Guy et al.[101] were also unable to measure a microwave-induced cochelar microphonic. However, they showed that bilateral ablation of the cochlear eliminated previously observed microwave-induced potentials in the auditory system. Using a circularly polarized waveguide which more efficiently couples microwave energy into the guinea pig head, Chou et al.[110,111] detected a microwave-induced cochlear microphonic response at the round window. Microwave-induced cochlear microphonics were also recorded from cats.[112] For both cats and guinea pigs, the greater the length of the brain cavity the lower the frequency of the microwave-induced cochlear microphonic.[113] This relationship is consistent with the thermoelastic expansion model of the microwave hearing phenomenon (see below). The time between pressure waves corresponds to the time of propagation of acoustic waves in the medium. If thermoelastic waves are generated by absorption of the microwave pulse in the brain, then larger brain cavities would lead to lower-frequency cochlear microphonics. Extrapolation of data from guinea pigs[110,111] and cats[112] predicts that, in man, the microwave-induced cochlear microphonic will be between 7 and 10 kHz.[113] This is consistent with the inability of individuals to hear microwave pulses if they have neural hearing loss above 5 kHz. (It has been suggested[114] that the cochlear microphonics recorded during microwave irradiation are due to vibration artifacts and that the hearing of microwaves is mediated via direct neural stimulation by the radiation. The fact that the vibrations from piezoelectric crystals and laser pulses produce similar cochlear microphonics[111] strengthens the argument that the observed potential is due to bulk absorption in the brain.)

Other neurophysiologic studies of microwave hearing in cats include recording of extracellular potentials from single eighth-nerve fibers[115,116] and brain stem using coaxial electrodes;[109] medial geniculate body and eighth nerve, using glass microelectrodes;[101,117] inferior colliculus using glass microelectrodes;[102] and cortex using high-resistance carbon electrodes.[100] (The use of metal electrodes during exposure to microwave radiation presents the possibility of field intensification at the tip of the electrode.[118])

Through the use of microwave-transparent carbon-loaded Teflon® electrodes to record auditory brain-stem-evoked responses from the skin of guinea pigs, Chou et al.[111] have shown that microwave-induced brain-stem-evoked responses were similar to those produced by acoustic pulses. Both Cain and Rissmann[102] and Lin et al.[119] reported similar results on the auditory brain-stem-evoked response of cats. Using the auditory brain-stem-evoked response to determine if guinea pigs hear microwave pulses, Chou and Guy[120] determined that: (1) for pulses shorter than 30 μsec, microwave hearing depends on the energy per pulse; (2) for pulses between 50 and 500 μsec, microwave hearing depends on peak power.

Wilson et al.[121] used autoradiographic techniques to show that the microwave hearing phenomenon does not involve the middle ear. This is consistent with the results of studies on humans[94] and auditory brain-stem-evoked response experiments using guinea pigs.[122] With the exception of a single report of the absence of a response,[109] all of the neurophysiologic data are consistent with a mechanical model of the microwave hearing phenomenon.

White[123] predicted that pulsed electromagentic energy could produce pulses of pressure waves in absorbing solid or liquid material. Thermal expansion due to the absorption of electromagnetic energy can produce strains in the absorbing medium, leading to the production of a stress wave that propagates away from the heated region. Later analysis[124] showed that, for single long pulses, the induced stress wave is a function of peak power density; and for shorter pulses, it is a function of the product of peak power density and pulse width (or energy per pulse). Foster and Finch[95] suggested that at thermoelastic wave of the type proposed by White[123] and Gournay[124] was the mechanism by which pulses of microwave radiation elicited auditory sensation. Foster and Finch[95] used a sensitive hydrophone to demonstrate that 2- to 27-μsec pulses of 2450-MHz radiation produced pressure waves in distilled water and that the polarity of the pressure wave changed as the temperature of the water was cooled below 4°C. At 4°C the signal disappeared completely because the coefficient of thermal expansion of water is zero at 4°C. This dependence on the coefficient

Table 2

**COMPARISON OF ACOUSTIC WAVES GENERATED BY A
MICROWAVE PULSE[a] IMPINGING IN A SEMI-INFINITE
DIELECTRIC FOR THREE PHYSICAL MECHANISMS OF
ENERGY PRODUCTION**

Material	Electrostricture/ radiation pressure	Thermoelastic stress/ radiation pressure	Thermoelastic stress/ electrostricture
Brain	10.67	1301	122
Muscle	15.67	1290	82
Water	26.0	1225	47

[a]   10-μsec Pulse of 2450-MHz radiation.

From Lin, J. C., *Microwave Auditory Effects and Applications*, Charles C Thomas, Springfield, Ill., 1978. With permission.

of thermal expansion argues strongly for a microwave-induced thermoelastic mechanism of microwave hearing. Consistent with the earlier reported loudness-pulse width data, they found that, for short pulses, the intensity of the pressure wave recorded within a band width corresponding to audible sound depends on the energy per pulse, and for longer pulses, the significant parameter is peak power.

The absorption of microwave radiation in a dielectric material can produce acoustic energy via three mechanisms: radiation pressure, electrostrictive force (elastic deformation of a dielectric material in an applied electromagentic field), and thermoelastic stress.[101,125,126] Lin[127] analyzed this problem by considering a one-dimensional model in which a plane wave impinges normally on a semiinfinite slab of homogenous dielectric material. The equation of motion of a particle in a medium responding to an applied pressure was solved by D'Almberts' method for the three mechanisms stated above. In all materials, thermoelastic-generated stress exceeds other types of stress by a very wide margin. It was found that "microwave-induced thermoelastic stress exceeds radiation pressure by almost three orders of magnitude; pressure generated by electrostriction is about ten times as high as that produced by the radiation pressure mechanism."[127] Numerical solutions were obtained for 10-μsec pulses of 2450-MHz radiation impinging on different dielectric materials and are provided in Table 2. For a semiinfinite slab, regardless of the mechanism analyzed the peak stress (pressure) is usually at the surface being irradiated.[127]

Since the head is more closely approximated by a sphere than by a semiinfinite slab and calculations of microwave absorption in spheres have been made for homogenous and layered spheres,[118,128-131] Lin[132,133] then considered the thermoelastic propagation in a spherical model of the brain. A plane wave pulse of microwave radiation is considered to produce volume heating when absorbed in the sphere. The thermoelastic equation of motion, without shear stress, is then solved for the sound wave using boundary value techniques and Duhamel's principle. It is found that the frequency of the acoustic wave is independent of both the position within the sphere where the peak absorption of microwave energy occurs and the frequency of the microwave radiation. The acoustic range of frequencies depends only on the size of the sphere (varying inversely with the radius) and the propagation velocity of the acoustic wave in the spherical material. Two conditions were analyzed: (1) the surface of the sphere is rigidly constrained and (2) the surface of the sphere is free to move. Qualitatively similar solutions are obtained for both cases. However, the calculated peak acoustic pressure is higher in the case where the surface of the sphere is constrained. The constrained case gives better agreement with experimental results. Theoretical results also indicate that there is an optimal microwave pulse width for the conversion of microwave to acoustic energy which varies with sphere size. For a 7-cm sphere (model of the human head) the smallest optimal pulse width is 5 μsec.

Microwave hearing is a universally accepted "low-level effect" of microwave radiation. It enjoys this unassailable position because it has been experienced by many microwave workers, has been demonstrated and replicated in laboratory animals, and is completely predicted from classical physics. The theoretical explanation of microwave hearing begins with a thermally induced effect of microwave absorption, yet the hearing effect is modulation dependent and is produced only under specific modulation conditions.

IV. APPLICATIONS OF "WINDOW" EFFECTS

In recent years, pulsed electromagnetic fields have been used in clinical practice to accelerate or facilitate the healing of nonunions of bone as well as the growth of soft tissue (reviewed by Spadaro[134] and by Tenforde in Part II, Chapter 5 of this book). Clinical empirical observations have not led to an understanding of the basic mechanism whereby pulsed electromagentic fields aid in such healing processes. Several recent studies, however, have examined the effect of various pulse characteristics in an attempt to maximize the biological response and to gain an understanding of the mechanism. These studies indicate that a window effect does occur, i.e., there are certain pulse characteristics that seem to promote healing of separated bones and others that do not. In one such study,[135] rats which had the inner bone of their forepaw surgically separated were exposed to pulsed ELF fields generated by two parallel Helmholtz coils. Bursts of 5-msec-long pulses repeated at a repetition rate of 5, 11, 15, and 20 Hz were used. The bursts were composed of 200-μsec pulses of one polarity followed by a 20-μsec pulse of the opposite polarity. When the degree of repair was compared to controls, little or no effect was observed at 5 Hz and a maximum effect was noted at 15 Hz. Interestingly, this "window" response in the burst repetition rate is similar to the calcium efflux responses discussed earlier in which 16-Hz sinusoidal modulation elicited a peak response. However, the 15-Hz repetition rate is not the only important parameter. By changing the pulse burst from 5 to 50 msec, Christel et al.[135] found that a 2-Hz repetition rate was as effective as the 5-msec burst with a 15-Hz repetition rate previously used. However, the peak value of the field was greater for the 2-Hz repetition rate. Thus, the "window" in this case may not depend solely on the repetition frequency, but may depend on peak-induced field strength (which would depend on the rise time and amplitude of the pulse), as well as the frequency content of the signal. Spectral analysis was performed[135] on the various wave forms using the methods of Laplace transforms. It is most difficult to draw any conclusions concerning the importance of any given frequency component due to the limited data. However, one can conclude that it is not merely a low-frequency response, since the 5-Hz burst repetition rate proved to be ineffective.

Although magnetic fields are externally applied in the Christel et al.[135] experiment as well as numerous other in vivo, in vitro, and clinical studies, it is generally believed to be the internally induced electric field that is the important parameter. This is explained by the work of Pilla et al.[136] in which current characteristics are calculated and measured for in vitro studies. The induced electric fields, measured in media or saline solution, were on the order of 0.1 to 1 V/m with corresponding induced current densities of the order of 0.01 to 1 A/m². Induced electric field and current density varied with the conductivity of the solutions and with the location of measurement (radial distance from center) in the container.

Pulse characteristics similar to those used by Cristel et al.[135] have been used to investigate several cellular functions. One example is the work of Goodman et al.[137] in which two pulse forms that are in clinical use, a single pulse repeated at 72 Hz and the 15-Hz repetitive pulse train discussed previously, were employed. Effects on RNA transcription patterns in salivary gland chromosomes of a dipteran, *Sciara coprophila,* were studied in three ways: (1) nascent RNA chains were identified by conventional autoradiography; (2) transcriptively active regions of chromosomes were identified using deoxyribonuclease I; and (3) the size classes of synthetized RNAs were examined. It was shown that each type of pulse activates a different set of genes, resulting in a different RNA synthesis pattern. These differ from that observed in controls and from that found following gene activation by heat. The responses to different periods of exposure were different for the two types of pulses. Further studies by the same group[138] confirmed these observations demonstrating that, as expected, changes in transcription patterns induce changes in protein synthesis. These results indicate that a frequency or pulse-shape window does exist, but its exact nature is not understood. They also have implications for both possible beneficial applications as well as hazard evaluation.

Stimulation of bone growth has not been confined to pulsed electromagnetic fields. In fact, early work in this area employed direct current stimulation applied through implanted electrodes.[139] There is an apparent current window for maximum stimulation of healing and promotion of growth using this method. This subject was studied and reviewed by Spadaro and Becker[140] who noted a sharp upper limit on the effective current density of 3 to 5 A/m^2 at the electrode. These values correspond to total currents of about 50 to 100 μA. The lower limit of effective treatment was ambiguous and no systematic study was made to determine the optimal parameters. Some degree of stimulation was reported as low as 0.04 A/m^2 by the reviewers. This response is not, strictly speaking, an electrical ''window'', since chemical toxicity begins to occur above 3 to 5 A/m^2, resulting in a necrosis rather than a bone growth stimulation.

An electrical window of effective stimulation was recently observed by Brighton et al.[141] when 60-kHz sinusoidal fields were capacitively coupled to the fractured forearm. Growth plates of rabbits were stimulated by placing a pair of 1.8-cm stainless steel capacitive plates in parallel over the right proximal growth plate in contact with the skin. Four different experimental groups were treated by applying 2.5, 5, 10, or 20 V (peak to peak) to the parallel plates. The curent density induced in the growth plate was not measured, but was calculated to be 3.3 A/m^2. A statistically significant increase in bone and cartilage was observed in the animals stimulated with 5- and 10-V (peak to peak) 60-kHz signals when compared to controls with a 9.2% increase in the 5-V group and 7.8% increase in the 10-V group. The 2.5-V group showed a 4.3% increase (not statistically significant) and the 20-V group showed a 3.6% decrease (not statistically significant).

Delgado and colleagues[142] have performed numerous experiments to show that weak magnetic fields at extremely low frequencies affect various biological systems, depending both on the amplitude of the field and on the frequency. For example, they suggest that embryological changes can be induced in developing embryos when exposed to low-level magnetic fields. They exposed chick embryos to 10, 100, and 1000 Hz at fields of 0.12, 1.2, and 12 μT. A peak response was found for exposure to 100 Hz at 1.2 μT, which was characterized as a ''consistent and powerful inhibitory effect on embryogenesis.'' Again, a field-strength window is suggested.

Further investigations by the same group,[143] indicated that pulse shape was the important factor. They found effects on embryogenesis that depended on the rise and the fall times of the pulse. These variations in response with the rise time of the pulse again suggest a biological response that is dependent on the frequency and the amplitude of the electric field induced within the cells. It has been suggested that the orientation of the egg with respect to the direction of the magnetic field of the earth may also be a significant factor.[143a]

A ''window'' of magnetic field strengths that alters the growth rate of bacteria was reported by Aarholt et al.[144] During log-growth phase bacteria were exposed to 16.66- or 50-Hz magnetic fields. At 50 Hz, a decrease in the mean growth rate was found with a threshold at 0.48 mT; an exposure of 0.8 mT yielded a mean growth rate that was the same as for controls, and at higher field strengths growth was again depressed by the magnetic field. Exposure to a 16.66-Hz magentic field at 0.8 mT also caused a decrease in growth rate. Again the growth rate decreased or increased depending on the field strength. From these results Aarholt concludes, "The phenomenon is observed at very low levels of field strength. This fact, together with the periodicity of the effect, indicates that where biological measurements are to be made at higher field strengths the applied magnetic field must be specified with considerable accuracy — to within approximately 0.1 mT. Furthermore, the larger effects are not necessarily to be expected at higher field strengths contrary to most expectations."

Moore[145] also found that bacteria and yeast were either stimulated or inhibited by magnetic fields and that the effect depended on both field strength and perhaps frequency, in the case of pulsed fields. However, the 30% increase in growth at 20 mT is well documented in at least four different bacteria. At 10 mT the effect has disappeared and at 30 mT it has reversed producing a 30% decrease in growth.

Aarholt[144] also reported a power "window" in the effect of 50-Hz magnetic fields on the lac operon system. A sharp decrease in the rate of β-galactosidase synthesis was found at 0.27 mT, rapidly returning to control values as field strength increased. At 0.51 mT synthesis begins to increase reaching a value twice that of controls in bacteria exposed at 0.54 mT. Control values of the enzyme are again found when exposure is increased to 0.56 mT. Thus, increased or decreased rates of β-galactosidase synthesis were effected by magnetic fields that differed only slightly in field strength.

It is tempting to compare the results from this section on application of window effects by simply calculating induced internal field strength and relating the results to the magnitude of the induced currents. However, this type of analysis does not provide simple explanations. For example, claims of effectiveness or ineffectiveness of osteogenesis are made for small changes in low-frequency repetition rates with little effect on average current density. In addition, claims[141] of effecting osteogenesis are made for exposures at much higher frequencies (60 kHz) using continuous wave exposures. The mechanisms are probably electrochemical in effects on membrane receptor sites and it is likely that more than one mechanism is involved in these processes. It is also possible that more than one type of receptor site is being affected by the complex wave forms. Thus, two mechanistic processes which compete against or assist one another may be operating simultaneously. The exact nature of these windows, in terms of both frequency and amplitude, may be obscured by the presence of two or more interaction sites. Effects on two different receptor-biochemical processes could explain why the pulse burst (15 bursts per second of 20-μsec assymmetric pulses) which was most effective in osteogenesis was less effective than was a single 72-Hz repeating pulse in the effects on RNA synthesis reported by Goodman et al.[137]

V. THEORETICAL CONSIDERATIONS

Mechanisms by which the observed "window effects" could be produced have been proposed.[82,83,146-148] These mechanisms have been reviewed by Taylor.[149] Some proposed mechanisms suggest interaction with complex membrane structures requiring cooperative phenomena and/or Bose-Einstein-type condensation of energy into a single mode. Others propose interaction with large molecular systems through vibrational modes or nonlinear interactions that result in local disturbances.

Grodsky[82] proposed a cooperative model to explain the release of bound calcium from excitable membranes that are exposed to ELF or ELF-modulated RF signals. A simple example of a cooperative model or process is as follows. Consider a two-dimensional array of small magnets that is located in the X-Y plane, with all of the north poles pointing in the positive Z direction. In this configuration, all magnets are aligned in the same direction, and the system is in the lowest energy state. To flip one magnet, i.e., point its north pole in the negative Z direction, requires energy. However, once a magnet has been flipped, perhaps by some random event, then less energy is required to flip neighboring magents. The process continues with each new neighbor requiring less energy to flip. This example and the Grodsky model, to a certain degree, parallel that of the Ising model originally formulated for ferromagnetism, but which has also been applied to biological problems.[150] Grodsky envisions two two-dimensional parallel sheets; the first is the phospholipid sheet in the plasma membrane and the second is a lattice of "calcium sites". The phospholipid layer is a lattice of "dipole-like sites" where at each site an entity (pointing either up or down) combines with its nearest neighbor to control the occupation (binding) of the calcium sites in the parallel sheet. The "calcium site" sheet is more sparsely populated than the phospholipid-dipole sheet. Changing the state of a single dipole-like entity would thus not immediately affect the occupation of the calcium site, but could potentially lead to such an event. The energy required to change N neighboring dipole-like sites is not simply N times the energy required to change one site. The changing of dipole-like states would closely follow the cooperative model discussed previously in which the second and following site(s) are more easily reoriented (require less energy).

Four states of the Grodsky model exist which can be depicted by giving a direction to both the phospholipid dipole-like sites and the calcium binding sites. Each (dipoles and binding sites) can point either up or down. Those directions, for example, could represent conditions of attraction or repulsion. The four states of the two sheets would then be (1) both pointing up, (2) both pointing down, (3) pointing toward each other, and (4) pointing away from one another. Each configuration is characterized by an energy state. Shifts between these four bands, or energy states, thus produce resonant conditions. Grodsky suggests that the lowest resonant frequency state will possess the highest average amplitude, be the most likely to be excited, and will fall in the 1- to 30-Hz frequency range. This cooperative model suggests four bands of frequencies which may or may not overlap. The lowest frequency of the lowest band is the most susceptible to resonance stimulation. Thus, small perturbations from an ELF signal may "resonate" with the structure resulting in a number of altered states and, consequently, the release of bound calcium. Similar cooperative models could be applied to other membrane properties or to interactions with any large macromolecular system or subsystem.

The limit-cycle theory[151] was expanded by Kaiser,[152] who proposed that free internal oscillations can be entrained by an appropriate external field. Competition between slow external and fast internal oscillations leads to entrainment through a series of quasi-periodic states. This process is both frequency- and field-strength-dependent and is proposed to account for the "windows" that have been observed in the millimeter region of the spectrum.

Fröhlich[2,146,153] explained that order or organization exists in many different forms in a system. Order is not confined to spatial order and order of motion can exist in thermal equilibrium. These "other" forms of order provide a basis for physical theories of frequency-selective responses (windows) that occur in a random 37°C thermal environment. One such system of order stems from membrane structure. A membrane, which is about 10^{-8} m thick, maintains a potential on the order of 10 to 100 mV. It is further noted that any oscillation of the membrane perpendicular to its surfaces will generate an oscillating electric dipole. The frequency of these oscillations will be the natural resonances of the system which will be determined by the speed of sound (accoustic velocity) in the membrane. Fröhlich assumed the sound velocity to be about 10^3 m/sec. The width of the membrane, 10^{-8} m, determines the wave length and will be equal to an odd multiple of half wave lengths (resonance for a violin string). Thus, the lowest natural vibrational mode will have a frequency on the order of 50 GHz. Fröhlich's calculations agree with these predictions. He suggests that similar types of resonant structure will exist in large molecular systems such as protein and DNA.

It is further assumed that damping is not unduly high in these systems. Numerous oscillating polar systems of this type will interact with one another generating a series of normal modes of the system. Fröhlich suggests that the energy for the maintenance of these oscillations is supplied by metabolic processes. The model implies that very little energy is required to excite the various modes. Thus, coherent electric vibrations can exist in the biological system, provided that the energy supplied to systems of normal modes exceeds some critical value, so that a Bose condensation occurs, coherently exciting a single mode, presumably, the lowest. (Bose condensation, or Bose-Einstein condensation, refers to a system of indistinguishable particles capable of oscillating in many different modes; all particles will tend to occupy the lowest energy level, i.e., the mode with the lowest frequency.) Fröhlich offers the highly frequency-dependent biological effects of millimeter waves discussed in this chapter as supportive of not only the presence of these modes, but as evidence of the excitation of these modes by external sources.

The reported genetic effects of ELF[137] or millimeter-wave[33] exposure that exhibit window or resonant-frequency responses could be the result of direct interaction of the electromagnetic field with the membrane, inhibiting or enhancing membrane control function; or such effects may be due to the direct interaction of the applied field with the chromosome or DNA molecule. Calculations have been made that indicate that the DNA molecule may show strong resonant aborption at microwave frequencies.[148] These calculations considered longitudinal and transverse vibrational modes (accoustic or optical) in dehydrated double-stranded DNA of varying chain lengths (number of base pairs). The calculated resonant modes were chain-length dependent, with the longitudinal acoustic vibrational modes being the most likely in the microwave frequency range. Van Zandt et al.[154] recently extended this work to consider critically dampened absorption of DNA in aqueous solutions.

The theoretical work of Lu et al.[155] on DNA absorption is not confined to the microwave region, but extends well into the IR region where the theoretical prediction has been experimentally demonstrated.[156] Swicord et al.[157] have reported considerably enhanced absorption in the x-band region from highly concentrated aqueous solutions containing DNA with a wide distribution of chain lengths. The same group[158] reported resonant absorption in solutions containing DNA strands of a uniform length. The observed frequency of the resonant mode agrees with the predictive theory.[148] The electric component of the microwave field causes the charged DNA molecule to vibrate at the microwave frequency. If this frequency corresponds to a resonant frequency of the DNA, then a condition similar to a mechanical standing wave is established in the molecule. For linearized strands, fundamental and harmonic resonant frequencies are observed which are related to odd multiples of a half wavelength of the excited sound wave. For circular DNA, resonant frequencies are related to multiples of the wavelength of the excited sound wave. Resonances are not observed below a few hundred megahertz. At the lower frequencies the resonant length of the molecule is too long and the mode relaxes before energy can travel from one end to the other. Thus, a standing wave is not established. The Prohofsky model[148] suggests a means of strongly coupling energy into a large molecule. Even though these modes may have a short lifetime, the continuous, coherent excitation of the mode would, in effect, maintain the molecule in an excited state. Whether, over a long period of time, this pumping and continual excitation will result in disruption of function has not been determined. Long-term (100 psec or longer) excitation of a molecule implies isolation from the thermal environment. Prohofsky[160a] has suggested that this model of absorption can explain the observations of Grundler et al.[34] The length of a nucleosome in the yeast chromosome is such that resonant absorption will occur at about 40 GHz. Small variations in frequency could result in rapid shifts into and out of resonance for the same or different nucleosomes.

A pulse-type solution for certain nonlinear partial differential equations is called "soliton" and has been used recently in solid-state physics, hydrodynamics, and elementary particle theory. "Solitons" are localized disturbances that are able to propagate through the nonlinear medium unattenuated. They are not wave trains, thus dispersion does not occur. The soliton (solitary wave) was first described by Scott-Russell[159] more than a century ago when he observed the unusual behavior of a wave generated by, separating from, and proceeding in front of a barge in a shallow canal.

Several applications of soliton dynamics have been made to biological systems. The most notable example is the application to the α-helical protein first considered by Davydov[160] and then reviewed and expanded by Scott.[161]

The Davydov model relies on the nonlinearity of hydrogen bonding which potentially couples the basic structure of the α-helical protein with the three "spines". This excitation of amide-I (or CO stretching) vibrations, which under linear analysis would rapidly disperse, is coupled to longitudinal sound waves and is propagated as a soliton. However, the amide-I and nonlinear sound wave must be strongly coupled and the amide-I vibrations must exceed a threshold value in order to generate the local disturbance or soliton. Davydov points out that the probability of a soliton causing a photon emission, or vice versa, is very small. It seems reasonable to argue that excitation of the molecule by a photon would result in nonlocalized (e.g., vibrational) modes that would not directly cause a localized disturbance. However, one could postulate that coherent excitation of vibrational modes, as modeled by Prohofsky and Van Zandt, could lead to large amplitude excitations even in a highly dampened system. Such large excitations could then possibly result in localized excitations, disturbances, or solitons. Furthermore, the Frölich and soliton models are complementary explanations of the storage and transport of energy in biological systems.[161a]

Lawrence and Adey[83] have suggested that solitons provide an explanation of the way in which signals are conducted through membranes, e.g., signals that would be created by the binding of a molecule to a receptor and consist of the subsequent chain of events leading to some manifestation of an event inside the cell. A chemical event, the arrival of a protein molecule, takes place at the surface of the receptor or channel site. A soliton is created in proteins associated with membrane channels and propagates along the protein and through the membrane. The soliton then supplies energy for the second chemical event on the inside of the cell. It is further suggested that these processes may be affected by the presence of external fields, thus causing a bioelectromagnetic response.

Although most theories of interaction deal with induced or permanent dipole responses to electric fields, recent evidence has been presented suggesting that magnetic fields, through nuclear spin coupling, affect the dielectric properties (including dielectric loss) of cells. Jafary-Asl et al.[162] reported resonant absorption peaks in the dielectric spectra of yeast cultures in solution at frequencies corresponding to the proton NMR frequencies. A very sharp resonance (about 1-Hz bandwidth) was observed in the dielectric spectra a little above 2 kHz when a magentic field of 50 μT was applied. The resonant frequency increased proportionally with the applied field from 100 to 200 μT. Jafary-Asl et al.[162] first observed a resonant relationship of cell properties and magnetic fields in studies of dielectrophoresis (for a discussion of dielectrophoresis see Part II, Chapter 1 of this book and the book by Pohl[163]). In these studies a nonuniform oscillating electric field is applied to a cell in solution. The field both induces a dipole moment on the cell and applies unequal forces on the separated charges, due to the nonuniformity of the field. The resulting net force produces motion of the cell towards an electrode. The dielectrophoretic yield is defined[163] as the number of particles collected at either electrode per unit time. This yield can be observed as a function of frequency. The spectral response varies for different types of cells and for live and dead cells. Jafary-Asl et al. observed a pronounced drop in the dielectrophoretic yield of live yeast cells at about 2 kHz. This corresponds to the proton nuclear magnetic resonance frequency in the magnetic field of the earth. This result is both interesting and perplexing. It is difficult to explain how effects on the state of the nucleus of the hydrogen atom are magnified to affect the macroscopic behavior of cells. The possible existence of such a relationship does, however, provide the basis for a theoretical explanation of a frequency "window" in biological responses. More recently, it has been suggested that DC magnetic fields may interact with biological systems to produce resonance-type effects in the ELF frequency region.[163a,163b] Blackman et al.[163a] have reported that the radiation-induced change in calcium levels in samples from chick brain[66,71-73] is dependent on the magnitude of the local geomagnetic field. A 16-Hz signal produced changes in calcium efflux when the local geomagnetic field is 38 μT but became ineffective when the local geomagnetic field was reduced to 19 μT. Furthermore, the ineffective (at 38 μT) 30-Hz signal altered calcium efflux when the local geomagnetic field is changed to ± 25.3 or ± 76 μT. These results led Blackman et al. to postulate a resonance-like relationship for the alteration of calcium efflux requiring that the frequency of the impressed electromagnetic field be proportional to the product of the local geomagnetic field density and an index (2n + 1, where n = 0, 1 . . .). The relation between the frequency of the impressed electromagnetic field and the local geomagnetic field required to affect calcium efflux was not suggested. Liboff[163b] suggests that the results of Blackman et al.,[163a] and of his own preliminary studies on behavior, can be explained by cyclotron-type resonance of specific molecules in the biological system. For cyclotron resonance, the relation between the DC magnetic field (B) and the frequency of the impressed oscillating field is given by f = (1/2π) (e/m) B; where e/m is the charge to mass ratio of the affected ion. Liboff points out that some of Blackman's results are consistent with charge to mass ratios of calcium and potassium. However, some of Blackman's results do not correspond to cyclotron resonance for any common ions.

The theories described in this section indicate that frequency or amplitude-specific effects are conceptually possible, although none of them has been shown to explain, in an unambiguous manner, the experimentally observed effects in biological systems.

VI. PULSE-MODULATION-DEPENDENT EFFECTS OF RF RADIATION

Even in the early years of research into the potential hazards of microwave radiation, several scientists considered the possibility that pulsed radiation could produce effects other than those elicited by continuous-wave radiation at the same average power. However, because of the limited availability of pulse and continuous-wave equipment that operate at the same frequency, only a few investigators explicitly searched for such differences. Where the comparison was made, the pulsed equipment was most commonly a radar transmitter. Therefore, the pulse width generally ranged from 1 to 10 μsec and the pulse repetition rate from 100 to 1500 pps. For comparison purposes, the average power was the same in both microwave-exposed groups.

The earliest suggestion that continuous-wave and pulse irradiation might elicit different effects on animals was made by Boysen,[164] who reported that pulsed radiation was absorbed more strongly than was continuous-wave radiation. However, this is an anecdotal report not amenable to statistical analysis.

Although the total amount of energy deposited in the absorber will be the same for continous-wave and pulsed irradiation at the same frequency and with the same time-averaged power, the time course of energy generation and dissipation will be different. Depending on the pulse parameters and the size, shape, thermal environment, etc. of the absorbing material, this difference may be significant and sufficient to alter a physiological response in a way that depends on the conditions of pulse modulation.

Several early Soviet experiments in which the effects of pulsed and continuous-wave irradiation were compared were reviewed by Gordon.[165] In one experiment, the mean survival time of mice was more than tenfold greater following pulsed VHF irradiation than following continuous-wave irradiation at the same average power density. However, when 3-GHz radiation was used, lethality was higher and more rapid with pulsed waves. Exposure to continuous-wave 3-GHz radiation, at 100 W/m², produced higher core temperatures in rats compared with exposure to pulsed radiation at the same frequency and the same average power. Both pulsed and continuous-wave 3-GHz radiation produced similar degrees of hypotension in rats following prolonged exposure (26 weeks). On the whole, these early Soviet studies do not suggest that pulse-modulated microwaves are more potent in their biological effect than continuous-wave radiation.

Young adult New Zealand rabbits were exposed (2 hr/day for 3 months) to either continuous-wave or pulsed (10-μsec pulses, 100 pps) 2450-MHz radiation at 15 W/m² (1.6 W/kg).[166] A variety of clinical variables were monitored throughout the exposure period, and following the exposure a complete pathological examination was completed. Because the sample size was small (six), the data show large variances, and differences between experimental groups are not significant.

Chronic exposure (3 hr/day for 3 months) of guinea pigs and rabbits to microwaves at 35 W/m² was associated with increased absolute lymphocyte count in peripheral blood, abnormalities in nuclear structure, and mitosis in erythroblastic cells and in lymphoid cells.[167] In general, the effects of pulsed and continuous-wave radiation, at the same average power density, were the same. The author suggests the nonthermal nature of the effect. However, the results of this experiment do not address this hypothesis.

It is reported that rabbits exposed to 2950-MHz radiation showed a significant decrease in erythrocyte production and incorporation of iron into erythrocytes.[168,169] Animals exposed under pulse-modulated conditions (reported variously as 1-μsec pulses at 1200 pps[168] and 1-sec pulses at 400 pps[169]) showed a greater effect of the RF radiation. In fact, the effect was comparable to continuous-wave exposures of more than twice the duration. This experiment was designed as a direct comparison of the effects of continuous-wave and pulse-modulated microwaves. Therefore, these results are among the most important in evaluating differences in the effects of the two exposure conditions.

At the same average power, pulsed and continuous-wave 2.8-GHz radiation were shown to have the same effect on respiration rate, heart rate, and subcutaneous temperature of rabbits.[170] Similarly, in a study of the heart rate of embryonic quail, no differences between continuous-wave and pulsed radiation were reported.[171]

The effects of continuous-wave and pulsed (10-μsec pulse width) 2.45-GHz radiation (at 250 and 100 W/m²) on serum components in the rabbit were compared.[172] The mean colonic temperature was higher in rabbits exposed to continuous-wave radiation at 250 W/m², but the authors suggest that this is due to differences in exposure geometry leading to significant differences in SAR. Results of the analyses of serum chemistry are consistent with nonspecific thermal stress and do not differ for the two irradiation conditions.

A number of brain-chemistry experiments in which the effects of continuous-wave and pulsed microwave radiation, at the same average power density, were compared were described by Baranski[173] and Baranski and Czerski.[169] Guinea pigs and rabbits were irradiated and the levels of several enzymes (acetylcholinesterase and succinic acid dehydrogenase) were measured. Although exposure to pulsed and continuous-wave radiation often produced statistically different results, the data were highly variable. In some instances the effect was an increase or decrease depending on the choice of an experimental parameter. Because of the high degree of variability in these data, the prudent observer awaits independent confirmation before using these results in any practical manner. McLees et al.[174] examined the effects of pulsed and continuous-wave radiation, at the same average power density, at 13.12 MHz on the mitotic activity and chromosomal aberrations in regenerating rat liver and found no difference between treatment and control values for either exposure condition.

The threshold of microwave exposure required to alter rabbit EEGs is about the same for continuous-wave or pulsed irradiation.[175] While preliminary investigation of the effect of microwaves on the functioning of *Aplysia* neurons[176] did not demonstrate differences between pulsed and continuous-wave irradiation, further study[177] revealed that pulsed 1.5- and 2.45-GHz radiation induced rapid changes in firing rate more readily than did continuous-wave irradiation at the same specific absorption rate. This conclusion is based on a small amount of somewhat subjective data and not amenable to statistical analysis. Isolated frog sciatic nerves exposed to pulsed (10 μsec, 50 pps) 2.45-GHz radiation exhibited the same reduction in survival time as did cells exposed to continuous-wave radiation.[178] Synchronizing the microwave pulse with various segments of the firing cycle of the nerve was of no consequence. In all cases, the loss in vitality of the exposed nerves was highly significant ($p < 0.01$) when compared to that of control cells.

Kholodov[175] states that, under identical experimental conditions, pulsed radiation is more effective than is continuous-wave radiation in altering behavior. This conclusion is based on a series of experiments of which only one is discussed. In this experiment, a conditioned active avoidance reaction was developed in white mice. Following this they were exposed to microwave (pulsed or continuous wave) 2 hr/day for 8 days and then retested. In general, mice exposed to the pulsed or continuous-wave field required a greater number of trials to attain training criterion than did control mice. The difference in behavior between animals exposed to pulsed fields and those exposed to continuous-wave fields was most marked for exposure at the highest level, i.e., 10 W/m². Statistical analysis of the data was not provided.

Frey et al.[179] reported that rats avoided pulse-modulated (0.5-msec pulses, 1000 pps, sawtooth wave shape) 1.2-GHz radiation, but not continuous-wave radiation, even though the average power density of the pulsed exposure was one tenth that of the continuous-wave exposure. In a preliminary comparison of the effects of pulsed (1.3-μsec pulses, 770 pps) and continuous-wave 3-GHz radiation, Roberti et al.[180] found no difference in the spontaneous motor activity of irradiated rats. Low levels of continuous and pulsed S-band microwaves yield different effects on the behavior of trained rats.[181] However, the differences were not consistent. In addition, two different frequencies were used (2.45 GHz for continuous-wave irradiation and 2.88 GHz for pulsed irradiation), further complicating the interpretation of these results.

Recently, Thomas et al.[108] reported that alterations in temporal discrimination were more pronounced when rats were exposed to pulse-modulated fields compared with instances of continuous-wave exposure at the same average power density. Radar-type pulses (2-μsec pulses, 500 pps) of carrier frequency 2.8 GHz were used. Although the differences reported are small they are among the best documented. In interpreting the results of these experiments and all other behavioral experiments in which animals are exposed to pulsed radiation, one must first consider the possibility that the animals were responding to the auditory sensation produced by the pulsed microwaves (discussed in Section III). The principle of Occam's razor requires that the hypothesis that the animals are responding to microwave-induced audio cues must be ruled out before more complex explanations are entertained. This remains to be done.

Greater quantities of i.v.-injected fluorescein were observed in the brains of rats exposed to pulse-modulated microwaves (1.2 GHz, 0.5-msec pulses, 1-kHz sawtooth wave shape) than in continuous-wave exposed rats.[179] Following up on these results, Oscar and Hawkins[182] found that specific modulation conditions (10-μsec pulses, 5 pps) were more effective than continuous-wave radiation in compromising the blood-brain barrier. The interpretation of these results is complicated by a later study[183] in which it was demonstrated that low-power pulsed microwave (2-μsec pulses, 500 pps) radiation markedly increases the local cerebral blood flow in rats. In this experiment, the effect is due to increased temperature of the brain. An attempted replication of the critical aspects of the Oscar and Hawkins[182] experiment has not led to confirmation of the difference between continuous-wave and pulsed irradiation.[184]

The cataractogenic potential of pulse-modulated microwaves has been shown to be the

Table 3
COMPARISON OF THE EFFECTS OF CONTINUOUS-WAVE (CW) AND PULSED MICROWAVE (PW) RADIATION AT THE SAME AVERAGE POWER DENSITY

Effect	Maximum with PW/CW	Ref.
Survival of mice	PW/CW	165
Hypotension in rats	—	165
Erythrocyte production in rabbits	—	168
Iron incorporation into rabbit erythrocyte	PW	168
Heart and respiration rates in rabbits	—	170
Heart rate of embryonic quail	—	171
Rabbit blood chemistry	—	172
Acetylcholinesterase and succinic dehydrogenase levels in rabbit and guinea pigs	PW	173
Chromosal aberrations and mitotic activity in rat liver (13 MHz)	—	174
Survival of frog sciatic nerves	—	178
EEG in rabbits	—	175
Firing rates of *Aplysia* neurons	PW	170, 171
Avoidance by rats	PW	179
Spontaneous motor activity of rats	—	180
Trained behavior in rats	PW	102, 181
Conditioned avoidance in mice	PW	175
Blood-brain barrier alteration in rats	PW	177, 282
	—	184
Cataractogenesis in rabbits	—	185, 186
Rat lens in tissue culture	PW	187, 188

same as that of continuous-wave microwaves at the same average power density and frequency.[185,186] However, irradiation of in vitro preparations of rat lenses with 915 MHz at 120 to 1200 W/kg pulsed irradiation (0.1 to 0.5-μsec pulses) was more effective than continuous-wave radiation, at the same average power density, in altering lens morphology; capsular and epithelial defects were observed under a scanning electron microscope.[187,188] It is proposed that the increased potency of pulsed microwaves is due to the additional mechanical stress produced by thermoacoustic expansion.

Although several reports indicate that pulsed microwaves are more likely to produce biological effects than are continuous-wave radiation at the same average incident power density (see Table 3), a careful reading of the results leads to a far more qualified conclusion. The differences between the changes observed following continuous-wave vs. pulsed irradiation are small and often occur with measures of some variables, but not of others. Statistical analysis of the results is most often omitted and independent corroboration is lacking.

We conclude that there is no compelling evidence that pulsed microwaves, of the type produced by radar transmitters, cause biological effects not found following exposure under conditions of continuous-wave radiation at the same average power density. There are, however, several reports that suggest a larger effect or lower threshold for pulsed irradiation. These reports should be corroborated and perhaps repeated in a more systematic manner. The availability of equipment should not be the deciding factor of experimental design. One must determine a threshold of response for both continuous and pulsed-wave radiation under equivalent exposure conditions before differences or similarities of response can be substantiated. At higher levels of irradiation thermoacoustic expansion is a logical mechanism by which pulsed radiation may alter or even disrupt cellular architecture and thereby produce a biological effect. (Even at low levels thermoacoustic expansion may be the mechanism by which RF radiation produces a biological effect; see discussion of microwave hearing effect.) In general, any mechanism depending on a threshold value of field strength or power would more readily produce a response to pulsed radiation than to continuous-wave radiation at the same average power density. It is expected that, for the biological response to be manifested, it will be necessary that some threshold pulse duration be exceeded. Furthermore, the threshold of pulse duration may depend on pulse intensity in a complex manner.[189] Therefore, it is not unreasonable that, for some effects, lower thresholds will be found for conditions of pulsed irradiation.

VII. CONCLUSIONS

Numerous biological responses to electric and magnetic field exposure that involve frequency or amplitude "windows" have been reported. The majority of these responses have been observed when the biological system is exposed to ELF signals, ELF-modulated RF signals, or millimeter-wave signals. There is some consistency in the ELF and ELF-modulated RF data; i.e., plausible arguments can be advanced to explain the similarity or differences of reported results. However, not enough data exist to be used as the basis of an explanation of the observed responses, or in some cases, firmly establish the nature of the response. There is no consistent repeatable pattern of millimeter-wave responses. Very few experiments have been repeated, duplicated, or confirmed by a second investigator. In some cases where duplication has been attempted, the original results were not confirmed. These experiments do, however, require a high degree of precision in the engineering aspects of experimental design, and lack of such precision could be a reason for nonreproducibility. For pulse-modulated effects, no consistent pattern of results has evolved (except for those effects involving a microwave-induced thermoelastic expansion of a biological tissue). This is most likely due to the limited experimental protocols that treat too few experimental parameters and are dictated by equipment available to a particular investigator.

The existence of highly specific nonmonotonic responses to electromagnetic radiation, at microwave and lower frequencies, is both unexpected and jarring to our sense of simplicity and order. This shock, and the appearance of a few controversial reports, have combined to produce a skepticism on the part of some observers of the field. However, skepticism can be constructive if it leads to the formulation of critical tests. At this time the body of experimental data is not sufficient to unambiguously verify existing theoretical models or even to indicate whether some of these models are useful guides to further experimentation. Often model-building efforts have not been detailed enough to predict decisive experimental outcomes. New measurement approaches and theoretical models are needed to improve our understanding of how electric and magnetic fields interact with living systems.

ACKNOWLEDGMENT

We thank Kenneth Foster, Michael Marron, and Charles Polk for their helpful suggestions and Mattia Ruyter for secretarial assistance.

REFERENCES

1. **Devyatkov, N. D., Ed.,** Scientific Session of the Division of General Physics and Astronomy, U.S.S.R. Academy of Sciences (17—18 January 1973), *Upekhi Fizicheskikh Nauk,* 110(3), 452, 1973; *Sov. Phys. Usp.,* 16, 568, 1974.
2. **Fröhlich, H.,** Long-range coherence and energy storage in biological systems, *Int. J. Quantum Chem.,* 2, 64, 1968.
3. **Webb, S. J. and Dodds, D. E.,** Inhibition of bacterial cell growth by 136-GHz microwaves, *Nature,* 218, 374, 1968.
4. **Webb, S. J. and Booth, A. D.,** Absorption of microwaves by microorganisms, *Nature,* 222, 1199, 1969.
5. **Webb, S. J.,** *Time and Motion in Metabolism and Genetics,* Charles C Thomas, Springfield, Ill., 1976.
6. **Blackman, C. F., Beane, S. G., Weil, C. M., and Ali, J. S.,** Effect of nonionizing electromagnetic radiation on single-cell biological systems, *Ann. N.Y. Acad. Sci.,* 247, 352, 1975.
7. **Webb, S. J. and Booth, A. D.,** Microwave absorption by normal and tumor cells, *Science,* 174, 72, 1971.
8. **Gandhi, O. P., Hagmann, M. J., Hill, D. V., Partlow, L. M., and Bush, L.,** Millimeter-wave absorption spectra of biological samples, *Bioelectromagentics,* 1, 285, 1980.
9. **Berteaud, A. J., Dardalhon, M., Rebeyrotte, N., and Averbeck, D.,** Action d'un rayonnement electromagnetique a onde millemtrique sur le croissance bacterienne, *C. R. Acad. Sci. Paris,* 281D, 843, 1975.
10. **Dardalhon, M., Averbeck, D., and Berteaud, A. J.,** Determination of a thermal equivalent of millimeter microwaves in living cells, *J. Microwave Power,* 14, 307, 1979.
11. **Partlow, L. M., Bush, L. G., Stensaas, L. J., Hill, D. W., Riazi, A., and Gandhi, O. P.,** Effects of millimeter-wave radiation on monolayer cell cultures. I. Design and validation of a novel exposure system, *Bioelectromagnetics,* 2, 123, 1981.
12. **Bush, L. G., Hill, D. W., Riazi, A., Stensaas, L. J., Partlow, L. M., and Gandhi, O. P.,** Effects of millimeter-wave radiation on monolayer cell cultures. III. A search for frequency-specific athermal biological effects on protein synthesis, *Bioelectromagentics,* 2, 151, 1981.
13. **Webb, S. J.,** Factors affecting the induction of lambda prophages by millimeter microwaves, *Phys. Lett.,* A73, 145, 1979.
13a. **Kremer, F.,** personal communication, 1984.
14. **Kondrat'eva, V. F., Chistyakova, E. N., Shmakova, I. F., Ianova, N. B., and Treskunov, A. A.,** Effects of millimeter-band radiowaves on certain properties of bacteria, *Sov. Phys. Usp.,* 16, 572, 1974.
15. **Devyatkov, N. D.,** Influence of millimeter-band electromagentic radiation on biological objects, *Sov. Phys. Usp.,* 16, 568, 1974.
16. **Kesselev, R. I. and Zalyubovskaya, N. P.,** Millimeter-band electromagentic waves in the cell and certain structural elements of the cell, *Sov. Phys. Usp.,* 16, 576, 1974.
17. **Zalyubovskaya, N. P.,** Reactions of living organisms to exposure to millimeter-band electromagnetic waves, *Sov. Phys. Usp.,* 16, 574, 1974.
18. **Smolyanskaya, A. Z. and Vilenskaya, R. L.,** Effects of millimeter-band electromagnetic radiation of the functional activity of certain genetic elements of bacterial cells, *Sov. Phys. Usp.,* 16, 571, 1974.
19. **Smolyanskaya, A. Z.,** Action of millimeter electromagnetic waves on fat cells, in *Nonthermal Effects of Millimeter Radiation,* Devyatkov, N. D., Ed., U.S.S.R. Academy of Sciences, Moscow, 1981, chap. 7.
20. **Motzkin, S. M., Benes, L., Block, N., Israel, B., May, N., Kuriyel, J., Birenhaum, L., Rosenthal, S., and Han, Q.,** Effects of low-level millimeter waves on cellular and subcellular systems, in *Coherent Excitations in Biological Systems,* Fröhlich, H. and Kremer, F., Eds., Springer-Verlag, Berlin, 1983, 47.
21. **Swicord, M. L., Athey, T. W., Buchta, F. L., and Krop, B. A.,** Colicin induction by exposure to millimeter-wave radiation (Abstr.), in Open Symp. on Biological Effects of Electromagnetic Waves. URSI 19th General Assembly, Helsinki, 1978.
22. **Athey, T. W. and Krop, B. A.,** Millimeter-wave radiation fails to induce lamda phage expression (Abstr.), in 1st Annual Meeting of Bioelectromagnetic Society, 1979.
23. **Sevast'yanova, L. A. and Vilenskaya, R. L.,** Mouse bone marrow reaction to altered UHF millimeter irradiation, parametric variation, *Biol. Nauki,* 6, 48, 1974.
24. **Sevast'yanova, L. A. and Vilenskaya, R. L.,** A study of the effects of millilmeter-band microwaves on the bone marrow of mice, *Sov. Phys. Usp.,* 16, 570, 1974.
25. **Smolyanskaya, A. Z., Gel'vich, E. A., Golant, M. B., and Makhov, A. M.,** Resonance phenomena in the action of millimeter waves on biological objects, *Usp. Sovrem. Biol.,* 87(3), 381, 1979.
26. **Sevast'yanova, L. A.,** Specific action of millimeter radiowaves on biological systems, in *Nonthermal Effects of Millimeter Radiation,* Devyatkov, N. D., Ed., U.S.S.R. Academy of Sciences, Moscow, 1981, chap. 5.
27. **Smolyanskaya, A. Z., Makhov, A. M., Gel'vich, E. A., and Golant, M. B.,** Influence of electromagnetic waves in the millimeter band on the induction of synthesis of penicillinase by *Staphlococcus aureus, Biol. Nauki,* 5, 24, 1981.

28. **Ostapenkov, A. M., Matiso, V. A., Kaptreva, I. V., Belovolov, A. V., and Laurova, V. L.,** Investigation of the effects of low intensity UHF electromagnetic waves on *Bacillus mesentericus* and *Pseudomonas fluorescens, Biol. Nauki,* 6, 47, 1976.

29. **Zalyubovskaya, N. P., Gordiyenko, O. I., and Kisselev, P. I.,** Action of electromagentic fields of superhigh frequency on erythrocytes preserved at low temperature, *Probl. Gemalol. Perelivaniya Krovi,* 20(4), 31, 1974.

30. **Devyatkov, N. D., Betskiy, O. V., Gel'vich, E. A., Golant, M. B., Mahkov, A. M., Rebrova, T. B., Sevast'yanova, L. A., and Smolyanskaya, A. Z.,** Effects on biological systems of electromagnetic oscillations in the millimeter range of wavelengths, *Radiobiologica,* 21(2), 163, 1981.

31. **Sevast'yanova, L. A.,** Distinctive features of biological effects of millimeter-range radiowaves and possible uses thereof in medicine, *Vestn. Akad. Med. Nauk. U.S.S.R.,* 2, 65, 1979.

32. **Grundler, W. and Keilmann, F.,** Nonthermal effects of millimeter microwaves on yeast growth, *Z. Naturforsch.,* 33c, 15, 1978; **Grundler, W., Keilman, F., and Fröhlich, H.,** Resonant growth rate response of yeast cells irradiated by weak microwaves, *Phys. Lett.,* 62A, 463, 1977.

33. **Grundler, W., Keilmann, F., and Strube, D.,** Resonant-like dependence of yeast growth rate on microwave frequencies, *Br. J. Cancer.,* 45 (Suppl. 5), 206, 1982.

34. **Grundler, W., Keilmann, F., Putterlik, V., Santo, L., Strube, D., and Zimmermann, I.,** Non-thermal resonant effects of 42-GHz microwaves on the growth of yeast cultures, in *Coherent Excitations in Biological Systems,* Fröhlich, H. and Kremer, F., Eds., Springer-Verlag, Berlin, 1983, 21.

35. **Webb, S. J., Lee, R., and Stoneham, M.,** Possible viral envolvement in human carcinomas, *Int. J. Quantum Chem.,* Suppl. 4, 277, 1977.

36. **Webb, S. J. and Stoneham, M.,** Resonance between 10^{11} and 10^{12} Hz in active bacterial cells as seen by laser Raman spectroscopy, *Phys. Lett.,* 60A, 267, 1977.

37. **Webb, S. J., Stoneham, M., and Fröhlich, H.,** Evidence for nonthermal excitation of energy levels in active biological systems, *Phys. Lett.,* 63A, 407, 1977.

38. **Webb, S. J.,** Coherent excitations in biological systems, *Phys. Rep.,* 60, 201, 1980.

39. **Drissler, F. and Santo, L.,** Coherent excitations and Raman effects, in *Coherent Excitations in Biological Systems,* Fröhlich, H. and Kremer, F., Eds., Springer-Verlag, Berlin, 1983, 6.

40. **Furia, L. and Gandhi, O. P.,** Absence of biologically related Raman lines in cultures of *Bacillus megaterium, Phys. Lett.,* 102A, 380, 1984.

41. **Cooper, M. S. and Amer, N. M.,** The absence of coherent vibrations in the Raman spectra of living cells, *Phys. Lett.,* 98A, 138, 1983.

42. **O'Sullivan, R. A. and Santo, J.,** Experimental aspects of Raman spectroscopy of microorganisms, *Can. J. Spectrosc.,* 26, 143, 1981.

43. **Illinger, K. H.,** Electromagnetic-field interaction with biological systems in the microwave and far-infrared region: physical basis, in *Biological Effects of Nonionizing Radiation,* Illinger, K. H., Ed., American Chemical Society, Washington, D.C., 1981, 1.

44. **Illinger, K. H.,** Spectroscopic properties of in vivo biological systems: boson radiative equilibrium with steady state non-equilibrium molecular systems, *Bioelectromagentics,* 3, 9, 1982.

45. **Gandhi, O. P., Hill, D. W., Riazi, A., Wahid, P., Wang, C. H., and Iskander, M. F.,** *Biological Effects of Millimeter Wave Irradiation,* Final report USAFSAM-TR-82-49, Brooks Air Force Base, Tex., 1982.

46. **Kremer, F., Koschnitzke, C., Santo, L., Quick, P., and Poglitsch, A.,** The non-thermal effect of millimeter wave radiation on the puffing of giant chromosomes, in *Coherent Excitations in Biological Systems,* Fröhlich, H. and Kremer, F., Eds., Springer-Verlag, Berlin, 1983, 10.

47. **Stamm, M. E., Winters, W. D., Morton, D. J., and Warren, S. L.,** Microwave characteristics of human tumor cells, *Oncology,* 29, 294, 1974.

48. **Adey, W. R.,** Models of membranes of cerebral cells as substrates for information storage, *Biosystems,* 8, 163, 1976.

49. **Adey, W. R.,** Neurobiological effects of radiofrequency and microwave radiation, *Ann. N.Y. Acad. Med.,* 55, 1079, 1979.

50. **Adey, W. R.,** Long-range electromagnetic field interaction at brain cell surfaces, in *Magnetic Field Effects in Biological Systems,* Tenforde, T. S., Ed., Plenum Press, New York, 1979, 56.

51. **Adey, W. R.,** Frequency and power windowing in tissue interactions with weak electromagnetic fields, *Proc. IEEE,* 68, 119, 1980.

52. **Adey, W. R.,** Tissue interactions with nonionizing electromagnetic fields, *Physiol. Rev.,* 61, 435, 1981.

53. **Adey, W. R.,** Ionic nonequilibirum phenomena in tissue interactions with electromagnetic fields, in *Biological Effects of Nonionizing Radiation,* Illinger, K. H., Ed., American Chemical Society, Washington, D.C., 1981, 271.

54. **Adey, W. R. and Bawin, S. M.,** Binding and release of brain calcium by low level electromagnetic fields: a review, *Radio. Sci.,* 17(5S), 149, 1982.

55. **Myers, R. D. and Ross, D. H.,** Radiation and brain calcium: a critique, *Neurosci. Neurobehav. Rev.,* 5, 503, 1981.
56. **Bawin, S. M.,** Cat EEG and Behavior in Very High Frequency Electric Fields Amplitude-Modulated at Brain Wave Frequencies, Ph.D. dissertation, University of California, Los Angeles, 1972.
57. **Bawin, S. M., Gavalas-Medici, R. J., and Adey, W. R.,** Effect of modulated very high frequency fields on specific brain rhythms in cats, *Brain Res.,* 58, 365, 1973.
58. **Bawin, S. M., Gavalas-Medici, R. J., and Adey, W. R.,** Reinforcement of transient brain biorhythms by amplitude-modulated VHF fields, in *Biological and Clinical Effects of Low Frequency Magnetic and Electric Fields,* Llaurado, J. G., Sances, A., Jr., and Battocletti, J. H., Eds., Charles C Thomas, Springfield, Ill., 1975, 172.
59. **Gavalas, R. J., Walter, D. O., Hammer, J., and Adey, W. R.,** Effect of low-level low-frequency electric fields on EEG and behavior in *Macaca nemestrina, Brain Res.,* 18, 491, 1970.
60. **Gavalas-Medici, R. J. and Day-Magdaleno, S. R.,** Extremely low frequency, weak electric fields affect schedule-controlled behavior of monkeys, *Nature,* 261, 256, 1976.
61. **Takashima, S., Oronal, B., and Schwan, H. P.,** Effects of modulated RF energy on EEG of mammalian brains, *Radiat. Environ. Biophys.,* 16, 15, 1979.
62. **Servantie, B., Servantie, A. M., and Etienne, J.,** Synchronization of cortical neurons by a pulsed microwave field as evidenced by spectral analysis of electrocorticograms from the white rat, *Ann. N.Y. Acad. Sci.,* 247, 82, 1975.
63. **Montaigne, K. and Pickard, W. F.,** The membrane potential of charcean cells exposed to amplitude-modulated low power 147-MHz radiation, *Bioelectromagnetics,* 5, 353, 1984.
64. **Barsoum, Y. A. and Pickard, W. F.,** The vacuolar potential of characean cells exposed to electromagnetic radiation in the range 200—8200 MHz, *Bioelectromagnetics,* 3, 393, 1982.
65. **Liu, L.-M., Garber, F., and Cleary, S. F.,** Investigation of the effects of continuous-wave, pulse- and amplitude-modulated microwaves on single excitable cells of *Chara corallina, Bioelectromagnetics,* 3, 203, 1982.
66. **Bawin, S. M., Kaczmarek, K. L., and Adey, W. R.,** Effects of modulated VHF fields on the central nervous system, *Ann. N.Y. Acad. Sci.,* 247, 74, 1975.
67. **Tenforde, T. S.,** Thermal aspects of electromagnetic field interactions with bound calcium at the nerve cell surface *J. Theor. Biol.,* 83, 517, 1980.
68. **Blackman, C. F., Elder, J. A., Weil, C. M., Benane, S. G., Eichinger, D. C., and House, D. E.,** Induction of calcium ion efflux from brain tissue by radiofrequency radiation: effects of modulation, frequency and field strength, *Radio Sci.,* 14(6S), 93, 1979.
69. **Blackman, C. F., Benane, S. G., Elder, J. A., House, D. E., Lampe, J. A., and Faulk, J. M.,** Induction of calcium ion efflux from brain tissue by radiofrequency radiation: effect of sample number and modulation frequency on the power density window, *Bioelectromagnetics,* 1, 35, 1980.
69a. **Sheppard, A. R., Bawin, S. M., and Adey, W. R.,** Models of long-range order in cerebral macromolecules: effect of sub-ELF and of modulated VHF and UHF fields, *Radio Sci.,* 14(6S), 141, 1979.
70. **Blackman, C. F., Benane, S. G., Joines, W. T., Hollis, M. A., and House, D. E.,** Calcium ion efflux from brain tissue: power density versus internal field intensity dependencies at 50 MHz RF radiation, *Bioelectromagnetics,* 1, 277, 1980.
71. **Bawin, S. M. and Adey, W. R.,** Sensitivity of calcium binding in cerebral tissue to weak environmental electric fields oscillating at low frequencies, *Proc. Natl. Acad. Sci. U.S.A.,* 73, 1999, 1976.
72. **Blackman, C. F., Benane, S. G., Kinney, L. S., Joines, W. T., and House, D. E.,** Effect of ELF on calcium-ion efflux from brain tissue in vivo, *Radiat. Res.,* 92, 510, 1982.
73. **Blackman, C. F., Benane, S. G., House, D. E., and Joines, W. T.,** Effects of ELF (1—120 Hz) and modulated (50 Hz) Rf fields on the efflux of calcium ions from brain tissue *in vivo, Bioelectromagnetics,* 6, 1, 1985.
74. **Adey, W. R., Bawin, S. M., and Lawrence, A. F.,** Effects of weak amplitude-modulated microwave fields on calcium efflux from awake cat cerebral cortex, *Bioelectromagnetics,* 3, 295, 1982.
75. **Joines, W. T. and Blackman, C. F.,** Power density, field intensity and carrier frequency as determinants of RF-energy-induced calcium ion efflux from brain tissue, *Bioelectromagnetics,* 1, 271, 1980.
76. **Joines, W. T. and Blackman, C. F.,** Equalizing the electric field intensity within chick brain immersed in buffer solution at different carrier frequencies, *Bioelectromagnetics,* 2, 411, 1981.
77. **Athey, T. W.,** Comparison of RF-induced calcium efflux from chick brain tissue at different frequencies: do the scaled power density windows align?, *Bioelectromagnetics,* 2, 407, 1981.
78. **Weil, C. M., Spiegel, R. S., and Joines, W. T.,** Internal field strength measurements in chick forebrains at 50, 147 and 450 MHz, *Bioelectromagnetics,* 5, 293, 1984.
79. **Merritt, J. G., Shelton, W. S., and Chamnes, A. F.,** Attempts to alter $^{45}Ca^{2+}$ binding to brain tissue with pulse-modulated microwave energy, *Bioelectromagnetics,* 3, 475, 1982.
80. **Lin-Liu, S. and Adey, W. R.,** Low frequency amplitude-modulated microwave fields change calcium efflux rates from synaptosomes, *Bioelectromagnetics,* 3, 309, 1982.

80a. **Dutta, S. K., Subramoniam, A., Ghosh, B., and Parshad, R.,** Microwave radiation-induced calcium ion efflux from human neuroblastoma cells in culture, *Bioelectromagnetics,* 5, 71, 1984.

81. **Adey, W. R. and Bawin, S. M.,** Brain interactions with weak electric and magnetic fields, *Neurosci. Res. Prog. Bull.,* 15(1), 1977.

82. **Grodsky, I. T.,** Possible physical substrates for the interaction of electromagnetic fields with biologic membranes, *Ann. N.Y. Acad. Sci.,* 247, 117, 1975.

83. **Lawrence, A. F. and Adey, W. R.,** Non-linear wave mechanisms in interaction between excitable tissue and electromagnetic fields, *Neurol. Res.,* 4, 115, 1982.

84. **Spiegel, R. J., Joines, W. T. and Blackman, C. F.,** Calcium-induced efflux from isolated brain tissue: is it caused by an electromagnetically-induced pressure wave? (Abstr.), in *4th Annual Meeting Bioelectromagnetics Society,* 1982.

85. **Albert, E., Blackman, C. F., and Slaby, F.,** Calcium dependent secretory protein release and calcium efflux during RF irradiation or rat pancreatic tissue slices, in *Ondes Electromagnetiques et Biologie,* Berteaud, A. J. and Servantie, B., Eds., Paris, 1980, 325.

86. **Sanders, A. P., Joines, W. T., and Allis, J. W.,** The differential effects of 200, 591 and 2450 MHz radiation on rat brain energy metabolism, *Bioelectromagnetics,* 5, 419, 1984.

87. **Huang, A. T., Engle, M. E., Elder, J. A., Kinn, J. B., and Ward, T. B.,** The effect of microwave radiation (2450 MHz) on the morphology and chromosomes of lymphocytes, *Radio Sci.,* 12(6S), 173, 1977.

88. **Janosy, G. and Greaves, M. F.,** Lymphocyte activation. II. Stimulation of lymphocyte subpopulations by phytomitogens and heterologous antilymphocyte sera, *Clin. Exp. Immun.,* 10, 525, 1972.

89. **Lyle, D. B., Schecter, P., Adey, W. R., and Lundak, R. L.,** Suppression of T-lymphocyte cytotoxicity following sinusoidally amplitude-modulated fields, *Bioelectromagnetics,* 4, 281, 1983.

90. **Byus, C. V., Lundak, R. L., Fletcher, R. M., and Adey, W. R.,** Alterations in protein kinase activity following exposure of cultured human lymphocytes to modulated microwave electromagnetic fields, *Bioelectromagnetics,* 5, 341, 1984.

91. **Sultan, M. F., Cain, C. A., and Tompkins, N. A. F.,** Immunological effect of amplitude-modulated radiofrequency radiation: B-lymphocyte capping, *Bioelectromagnetics,* 4, 157, 1983.

92. **Ragan, H. A. and Phillips, R. D.,** Hematologic effects of mice exposed to pulsed and CW microwaves (Abstr.), in *Proc. Int. Union of Radio Sciences Symp. on Biological Effects of Electromagnetic Waves,* 1978, 48.

93. **Ragan, H. A., Phillips, R. D., Buschbom, R. L., Busch, R. H., and Moris, J. E.,** Hematologic and immunologic effects of pulsed microwaves in mice, *Bioelectrmagnetics,* 4, 383, 1983.

94. **Frey, A. H.,** Auditory system response to RF energy, *Aerosp. Med.,* 32, 1140, 1961.

95. **Foster, K. R. and Finch, E. F.,** Microwave hearing: evidence for thermoacoustic auditory stimulation by pulsed microwaves, *Science,* 185, 256, 1974.

96. **Frey, A. H.,** Human auditory system response to modulated electromagnetic energy, *J. Appl. Physiol.,* 17, 689, 1962.

97. **Frey, A. H.,** Some effects on human subjects on ultra-high-frequency radiation, *Am. J. Med. Electron.,* 2, 28, 1963.

98. **Sommer, R. C. and Von Gierke, H. E.,** Hearing sensations in electric fields, *Aerosp. Med.,* 35, 834, 1964.

99. **Ingalls, C. E.,** Sensation of hearing in electromagnetic fields, *N.Y. State J. Med.,* 67, 2992, 1967.

100. **Frey, A. H.,** Biological functions as influenced by low power modulated RF energy, *IEEE Trans. Microwave Theory Tech.,* 19, 153, 1971.

101. **Guy, A. W., Chou, C.-K., Lin, J. C., and Christensen, D.,** Microwave-induced acoustic effects in mammalian auditory systems and physical materials, *Ann. N.Y. Acad. Sci.,* 247, 194, 1975.

102. **Cain, C. A. and Rissman, W. J.,** Mammalian auditory responses to 3.0 GHz microwave pulses, *IEEE Trans. Biomed. Eng.,* 25, 288, 1978.

103. **Frey, A. H. and Messenger, R.,** Human perception of illumination with pulsed ultra high frequency electromagnetic energy, *Science,* 181, 356, 1973.

104. **Frey, A. H. and Feld, S. R.,** Avoidance by rats of illumination with low power nonionizing electromagnetic energy, *J. Comp. Physiol. Psychol.,* 89, 183, 1975.

105. **Hjeresen, D. L., Doctor, S. K. and Shelton, R. L.,** Shuttle box side preference as mediated by pulsed microwaves and conventional auditory cues, in *Proc. Symp. on Electromagnetic Fields in Biological Systems,* International Microwave Power Institute, Edmonton, Alberta, Canada, 1979, 194.

106. **King, N. W., Justesen, D. R., and Clarke, R. L.,** Behavioral sensitivity to microwave radiation, *Science,* 172, 398, 1971.

107. **Johnson, A. B., Myers, D. E., Guy, A. W., Lovely, R. H., and Galambos, R.,** Discriminative Control in Appetitive Behavior by Pulsed Microwaves, HEW Publ. FDA77-8010, U.S. Department of Health, Education, and Welfare, Rockville, Md., 1976, 238.

108. **Thomas, J. R., Schrot, J., and Banvard, R. A.,** Comparative effects of pulsed and continuous wave 2.8-GHz microwaves on temporally defined behavior, *Bioelectromagnetics,* 3, 227, 1982.

109. **Frey, A. H.,** Brain-stem evoked responses associated with low-intensity pulsed VHF energy, *J. Appl. Physiol.,* 23, 984, 1967.

110. **Chou, C.-K., Galambos, R., and Lovely, R. H.,** Cochlear, microphonics generated by microwave pulses, *J. Microwave Power,* 10, 361, 1975.

111. **Chou, C.-K., Guy, A. W., and Galambos, R.,** Microwave induced auditory responses: cochlear microphonics, in *Biological Effects of Electromagentic Waves,* Vol. 1, Johnson, C. C. and Shore, M. I., Eds., Health Education and Welfare, Rockville, Md., 1976, 89.

112. **Chou, C.-K., Guy, A. W., and Galambos, R.,** Microwave-induced cochlear microphonics in cats, *J. Microwave Power,* 11, 171, 1976.

113. **Chou, C.-K. and Guy, A. W.,** Auditory perception of radiofrequency electromagnetic fields, *J. Acoust. Soc. Am.,* 71, 1321, 1982.

114. **Frey, A. H. and Coren, E.,** Holographic assessment of a hypothesized microwave hearing mechanism, *Science,* 206, 232, 1979.

115. **Lebovitz, R. M. and Seaman, R. L.,** Single auditory unit responses to weak pulsed microwave radiation, *Brain. Res.,* 126, 370, 1976.

116. **Lebovitz, R. M. and Seaman, R. L.,** Microwave hearing: the response of single auditory neurons in the cat to pulsed microwave radiation, *Radio Sci.,* 12(6S), 229, 1977.

117. **Taylor, E. M. and Ashleman, B. T.,** Analysis of central nervous system involvement in the microwave auditory effect, *Brain Res.,* 74, 201, 1974.

118. **Johnson, C. C. and Guy, A. W.,** Nonionizing electromagnetic wave effects in biological materials and systems, *Proc. IEEE,* 60, 692, 1972.

119. **Lin, J. C., Meltzer, R. J., and Redding, F. K.,** Microwave evoked brain stem potentials in cats, *J. Microwave Power,* 14, 291, 1979.

120. **Chou, C.-K. and Guy, A. W.,** Microwave-induced auditory responses in guinea pigs: relation of threshold and microwave-pulse duration, *Radio Sci.,* 14(6S), 193, 1979.

121. **Wilson, B. S., Zook, J. M., Joines, W. T., and Cassedy, J. H.,** Alterations in activity at auditory nuclei of the rat induced by exposure to microwave radiation: audiographic evidence using ^{14}C 2-deoxy-d-glucose, *Brain Res.,* 187, 291, 1980.

122. **Chou, C.-K. and Galambos, R.,** Middle ear structures contribute little to auditory perception of microwaves, *J. Microwave Power,* 14, 321, 1979.

123. **White, R. M.,** Generation of elastic waves by surface heating, *J. Appl. Physiol.,* 34, 3559, 1963.

124. **Gournay, L. S.,** Conversion of electromagnetic to acoustic energy by surface heating, *J. Acout. Soc. Am.,* 40, 1322, 1966.

125. **Lin, J. C.,** Microwave auditory effect — a comparison of some possible transduction mechanisms, *J. Microwave Power,* 11, 77, 1976.

126. **Borth, D. E. and Cain, C. A.,** Theoretical analysis of acoustic signal generation in materials generated with microwave energy, *IEEE Trans. Microwave Theory Tech.,* 25, 944, 1977.

127. **Lin, J. C.,** *Microwave Auditory Effects and Applications,* Charles C Thomas, Springfield, Ill., 1978.

128. **Shapiro, A. R., Lutomirski, R. F., and Hura, H. T.,** Induced fields and heating within a cranial structure irradiated by an electromagnetic plane wave, *IEEE Trans. Microwave Theory Tech.,* 19, 187, 1971.

129. **Kritikos, H. and Schwan, H. P.,** Hot spots generated in conducting spheres by E.M.: waves and biological implications, *IEEE Trans. Biomed. Eng.,* 19, 53, 1972.

130. **Kirtikos, H. and Schwan, H. P.,** The distribution of heating potential inside lossy spheres, *IEEE Trans. Biomed. Eng.,* 22, 457, 1973.

131. **Weil, C. M.,** Absorption characteristics of multilayered sphere models exposed to UHF/microwave radiation, *IEEE Trans. Biomed. Eng.,* 22, 468, 1975.

132. **Lin, J. C.,** Microwave hearing sensation, *IEEE Trans. Microwave Theory Tech.,* 25, 605, 1977.

133. **Lin, J. C.,** Further studies on the microwave auditory effect, *IEEE Trans. Microwave Theory Tech.,* 25, 938, 1977.

134. **Spadaro, J. A.,** Bioelectrical stimulation of bone formation: methods, models and mechanisms, *J. Bioelectr.,* 1, 99, 1982.

135. **Christel, P., Ceff, G., and Pilla, A. A.,** Modulation of rat radial osteotomy repair using electromagnetic current induction, in *Mechanisms of Growth and Control,* Becker, R. O., Ed., Charles C Thomas, Springfield, Ill., 1981.

136. **Pilla, A. A., Sechaud, P., and McLeod, B. R.,** Electrochemical and electrical aspects of low frequency electromagentic current induction in biological systems, *J. Biol. Phys.,* 11, 51, 1983.

137. **Goodman, R., Bassett, C. A. L., and Henderson, A. S.,** Pulsing electromagnetic field induces cellular transcription, *Science,* 220, 1283, 1983.

138. **Ryaby, J. T., Goodman, R., Henderson, A. S., and Bassett, C. A. L.,** Electromagnetic field effects on cellular biosynthetic processes (Abstr.), in *Trans. 3rd Annu. Meeting Bioelectrical Repair and Growth Soc.,* 1983, 25.

139. **Brighton, C. T., Black, J., Friedenburg, Z. B., Esterhai, J. L., Day, L. J., and Connolly, J. F.,** A multicenter study of the treatment of non-unions with constant direct current, *J. Bone Jt. Surg.*, 63A, 2, 1981.

140. **Spadaro, J. A. and Becker, R. O.,** Function of implanted cathodes in electrode-induced bone growth, *Med. Biol. Eng. Comput.*, 17, 769, 1979.

141. **Brighton, C. T., Pfeffer, G. B., and Pollack, S. R.,** In vivo growth plate stimulation in various capacitively coupled electric fields, *J. Orthoped. Res.*, 1, 42, 1983.

142. **Delgado, J. M. R., Leal, J., Monteagudo, J. L., and Garcia, M. G.,** Embryological changes induced by weak extremely low frequency electromagnetic fields, *J. Anat.*, 143, 533, 1982.

143. **Ubeda, A., Leal, J., Trillo, M. A., Jimenez, M. A., and Delgado, J. M. R.,** Pulse shape of magnetic fields influences chick embryogenesis, *J. Anat.*, 137, 513, 1983.

143a. **Leal, J.,** personal communication, 1985.

144. **Aarholt, E., Flinn, E. A., and Smith, C. W.,** Effects of low frequency magnetic fields on bacterial growth rate, *Phys. Med. Biol.*, 26, 613, 1981.

145. **Moore, R. L.,** Biological effects of magnetic fields: studies with microorganisms, *Can. J. Microbiol.*, 25, 1149, 1979.

146. **Fröhlich, H.,** The biological effects of microwaves and related questions, *Adv. Electron. Electron Phys.*, 53, 85, 1980.

147. **Davydov, A. S.,** Solitons in molecular systems, *Phys. Scr.*, 20, 387, 1979.

148. **Kohli, M., Mei, W. N., Prohofsky, E., and Van Zandt, L. L.,** Calculated microwave absorption of double helical B-conformation, poly(dg).poly(dc), *Biopolymers*, 20, 853, 1981.

149. **Taylor, L. S.,** The mechanisms of athermal microwave biological effects, *Bioelectromagnetics*, 2, 259, 1981.

150. **Thompson, C. J.,** *Mathematical Statistical Mechanics,* Princeton University Press, Princeton, 1972, chap. 5 to 7.

150. **Scott, A.,** *Active and Nonlinear Wave Propagation in Electronics,* Wiley-Interscience, New York, 1970.

152. **Kaiser, F.,** Coherent oscillations in biological systems: interaction with extremely low frequency fields, *Radio Sci.*, 17(5S), 17, 1982.

153. **Fröhlich, H.,** Coherent electric vibrations in biological systems and the cancer problem, *IEEE Trans. Microwave Theory Tech.*, 26, 613, 1978.

154. **Van Zandt, L. L., Kohli, M., and Prohofsky, E. W.,** Absorption of microwave radiation by DNA in aquo, *Biopolymers*, 21, 1465, 1982.

155. **Lu, K. C., Prohofsky, E, and Van Zandt, L. L.,** Vibrational modes of A-DNA, B-DNA, and A-RNA backbones; an application of a Greens function refinement procedure, *Biopolymers*, 16, 2491, 1977.

156. **Small, E. W., Peticolas, N., and Warner, L.,** Conformational dependence of the Raman scattering intensities from polynucleotides. III. Order-disorder changes in helical structures, *Biopolymers*, 10, 1377, 1971.

157. **Swicord, M. L., Edwards, G. S., Sagripanti, J. L., and Davis, C. C.,** Chain-length-dependent microwave absorption of DNA, *Biopolymers*, 22, 2513, 1983.

158. **Edwards, G. S., Davis, C. C., Saffer, J. P., and Swicord, M. L.,** Resonant absorption of microwave energy by DNA (Abstr.), *Proc. Bioelectromagnetics Society,* 1984.

159. **Scott-Russell, J.,** Report on waves, *Proc. R. Soc. (Edinburgh)*, 319, 1844.

160. **Davydov, A. S.,** *Biology and Quantum Mechanics,* Pergamon Press, Elmsford, N.Y., 1982.

160a. **Prohofsky, E.,** personal communication, 1983.

161. **Scott, A. C.,** Dynamics of Davydov solitons, *Phys. Rev. A*, 26, 578, 1982.

161a. **Scott, A. C.,** Solitons in biological molecules, *Comments Mol. Cell. Biophys.*, 3, 15, 1985.

162. **Jafary-Asl, A. H., Solanki, S. N., Aarholt, E., and Smith, C. W.,** Dielectric measurements on live biological materials under magnetic resonance conditions, *J. Biol. Phys.*, 11, 13, 1982.

163. **Pohl, H. A.,** *Dielectrophoresis, The Behavior of Matter in Nonuniform Electric Fields,* Cambridge University Press, Cambridge, England, 1978.

163a. **Blackman, C. F., Benane, S. G., Rabinowitz, J. R., House, D. E., and Joines, W. T.,** A role for the magnetic field in the radiation-induced efflux of calcium ions from brain tissue in vitro, *Bioelectromagnetics*, 6, 327, 1985.

163b. **Liboff A. R.,** Cyclotron resonance mechanisms for electromagnetic energy transfer to cells, Bioelectromagnetics Soc. 7th Annu. Meet., (Abstr.), 1985.

164. **Boysen, J. E.,** Hyperthermic and pathologic effect of EMR (350 Mc), *AMA Arch. Ind. Hyg.*, 38, 261, 1953.

165. **Gordon, Z. V.,** *Biological Effects of Microwaves in Occupational Hygiene* Kaner, N., Translator, Keter Press, Jerusalem, 1970.

166. **Chou, C.-K., Guy, A. W., McDougall, J. A., and Han, L.-F.,** Effects of continuous and pulsed chronic microwave exposure on rabbits, *Radio Sci.*, 17(5S), 185, 1982.

167. **Baranski, S.,** Effect of chronic microwave irradiation on the blood forming system of guinea pigs and rabbits, *Aerosp. Med.,* 42, 1196, 1971.

168. **Czerski, P., Paprochy-Stanka, E., Srekierzynski, M., and Stolarska, A.,** Influence of microwave radiation on the hematopoietic system, in *Biological Effect and Health Hazards of Microwave Radiation,* Czerski, P., Ostrowski, K., Shore, M. L., Silverman, C., Suess, M. J., and Waldeskog, B., Eds., Polish Medical Publishers, Warsaw, 1974, 67.

169. **Baranski, S. and Czerski, P.,** *Biological Effects of Microwaves,* Dowden, Hutchinson & Ross, Stroudsburg, Pa., 1976.

170. **Birenhaum, L., Kaplan, I. T., Metaly, W., Rosenthal, S. W. and Zaret, M. M.,** Microwave and infrared effects on heart rate and respiration rate and sub-cutaneous temperature of the rabbit, *J. Microwave Power,* 10, 3, 1975.

171. **Hamrick, P. E. and McRee, D. I.,** The effect of 2450 MHz microwave irradiation on the heart rate of embryonic quail, *Health Phys.,* 38, 261, 1980.

172. **Wangemann, R. T. and Cleary, S. F.,** The in vivo effect of 2.45 GHz microwave radiation on rabbit serum components and sleeping times, *Radiat. Environ. Biophys.,* 13, 89, 1976.

173. **Baranski, S.,** Histological and histochemical effect of microwave irradiation on the central nervous system of rabbits and guinea pigs, *Am. J. Phys. Med.,* 54, 182, 1972.

174. **McLees, B. D., Finch, E. D., and Albright, M. L.,** An examination of regenerating hepatic tissue subjected to radiofrequency radiation, *J. Appl. Physiol.,* 32, 78, 1972.

175. **Kholodov, Yu. A.,** *Reaction of the Nervous System to Electromagnetic Fields,* Moscow, 1975.

176. **Wachtel, H., Seaman, R., and Joines, W. T.,** Effects of low intensity microwave on isolated neurons, *Ann. N.Y. Acad. Sci.,* 247, 46, 1975.

177. **Seaman, R. and Wachtel, H.,** Slow and rapid response to CW and pulsed microwave radiation by individual *Aplysia* pacemakers, *J. Microwave Power,* 13, 77, 1978.

178. **McRee, D. I. and Wachtel, H.,** Pulse microwave effects on nerve vitality, *Radiat. Res.,* 91, 212, 1982.

179. **Frey, A. H., Feld, S. R., and Frey, B.,** Neuronal function and behavior: defining the relationship, *Ann. N.Y. Acad. Sci.,* 247, 433, 1975.

180. **Roberti, B., Heebels, G. H., Hendricx, J. C. M., deGreef, A. H. A. M., and Wolthuis, O. L.,** Preliminary investigation of the effect of low-level microwave radiation on spontaneous motor activity in rats, *Ann. N.Y. Acad. Sci.,* 247, 417, 1975.

181. **Thomas, J. R., Finch, E. D., Fulk, D. W., and Burch, L. S.,** Effects of low level microwave radiation on behavioral baselines, *Ann. N.Y. Acad. Sci.,* 247, 425, 1975.

182. **Oscar, K. J. and Hawkins, T. D.,** Microwave alterations on the blood-brain-barrier of rats, *Brain Res.,* 126, 281, 1977.

183. **Oscar, K. J., Grueneau, S. P., Folker, M. T., and Rapoport, S. I.,** Local cerebral blood flow after microwave exposure, *Brain Res.,* 204, 220, 1981.

184. **Merritt, J. H., Chamnes, A. F., and Allen, S. J.,** Studies on blood-brain-barrier permeability after microwave irradiation, *Radiat. Environ. Biophys.,* 15, 367, 1978.

185. **Birenbaum, L., Kaplan, I. T., Metaly, W., Rosenthal, S. W., Schmidt, H., and Zaret, M. M.,** Effect of microwaves on the rabbit eye, *J. Microwave Power,* 4, 232, 1969.

186. **Birenbaum, L., Grosof, G. M., Rosenthal, S. W., and Zaret, M. M.,** Effect of microwaves on the eye, *IEEE Trans. Biomed. Eng.,* 16, 7, 1969.

187. **Stewart-DeHaan, P. J., Creighton, M. D., Larsen, L. E., Jacobi, J. H., Ross, W. M., and Trevithick, J. R.,** Microwave and temperature effect on murine occular lenses in vitro, *1980 IEEE MTT-S Int. Microwave Symp. Dig.,* 341, 1980.

188. **Stewart-DeHaan, P. J., Creighton, M. D., Larsen, L. E., Jacobi, J. H., Ross, J. H., Sanwall, W. M., Guo, T. C., Guo, W. W., and Trevithick, J. R.,** In vitro studies of microwave-induced cataract: separation of field and heating effects, *J. Exp. Eye Res.,* 36, 75, 1983.

189. **Wachtel, H.,** A model for predicting minimum RF threshold energy pulse widths for evoking fast neuronal effects (Abstr.), Bioelectromagnetics Society 3rd Annu. Conf., 1981, 27.

Appendixes

Appendix 1

IMPORTANT CONSTANTS AND FREQUENTLY USED UNITS OF MEASUREMENT

A. Constants

Boltzman constant	k	$1.380662 \ (10^{-23})$ J/K
		$0.861735 \ (10^{-4})$ eV/K
Planck constant	h	$6.626176 \ (10^{-34})$ J/H
Elementary (electron) charge	e	$1.6021892 \ (10^{-19})$ C
Speed of light in vacuum	c	$2.99792458 \ (10^{8})$ m/sec
Magnetic permeability of vacuum	μ_0	$4\pi \ (10^{-7})$ H/m
Dielectric permittivity of vacuum	ϵ_0	$= \dfrac{1}{\mu_0 c^2} = 8.85418782 \ (10^{-12})$ F/m
		$\approx \dfrac{1}{36\pi} \ 10^{-9}$ F/m
Avogadro's number	N_A	$6.022045 \ (10^{23})$ Molecules/g mol

B. Abbreviations for Prefixes

p	pico	$= 10^{-12}$
n	nano	$= 10^{-9}$
μ	micro	$= 10^{-6}$
m	milli	$= 10^{-3}$
c (as in cm)	centi	$= 10^{-2}$
k (as in kg)	kilo	$= 10^{3}$
M	mega	$= 10^{6}$
G	giga	$= 10^{9}$

C. Units of Measurement

	Standard SI unit		Commonly used fractional or multiple units
Length	meter	m	$1 \ m = 100 \ cm = 1000 \ mm = 10^{10}$ Angstrom $= 10^{10}$ Å
Area	square meter	m^2	$1 \ m^2 = 10^4 \ cm^2$
Mass	kilogram	kg	$1 \ kg = 10^3 \ g$
Energy.	Joule	J	$1 \ J = 10^7 \ erg = 2.389 \ (10^{-4})$ kcal
			1 electron volt $= 1 \ ev \approx 1.6 \ (10^{-19})$ J
Power = time rate of change of energy	Watt	W	

$1 \ W = 1$ Joule per second $= 1$ J/s

Power per unit mass		1 W/kg	1 W/kg = 1 mW/g
Power per unit area		W/m²	1 W/m² = 0.1 mW/cm²
Frequency (symbol f)	Hertz	Hz	1 Hz = 1 cycle per second
Radian frequency ω = $2\pi f$	Radians per second	rad/sec	
Electric resistance	Ohm	Ω	
Electric conductance	Siemens	S	
Electric resistivity	Ohm-meter	Ωm	1 Ωm = 10² Ωcm
Electric conductivity (symbol g or σ)	Siemens per meter	S/m	1 S/m = 10⁻² S/cm
			0.1 S/m = 1 mS/cm
			Note: in older nomenclature 1 S = 1 Mho = 1℧ and 1 ℧/m = 10⁻² ℧/cm; 0.1 ℧/m = 1 m℧/cm
Electric capacitance	Farad	F	
Relative dielectric permittivity $\epsilon_r = \epsilon' - j\epsilon''$	Dimensionless ratio		
$\epsilon'' = \dfrac{\sigma}{\omega\epsilon_o}$	Dimensionless ratio		
Dielectric permittivity of any material $\epsilon = \epsilon_0\epsilon_r$	Farad/meter	F/m	
Loss tangent = $\dfrac{\text{conduction current}}{\text{displacement current}}$			
$= \dfrac{\epsilon''}{\epsilon'} = \dfrac{\sigma}{\omega\epsilon_o\epsilon'}$	Dimensionless ratio		
Electric potential difference	Volt	V	
Electric current	Ampere	A	
Electric current density	Ampere/m²	A/m²	1 A/m² = 10⁻⁴ A/cm² = 0.1 mA/cm²
Electric field intensity	Volt per meter	V/m	1 V/m = 10⁻² V/cm = 10 mV/cm
Magnetic field intensity H	Ampere per meter	A/m	
Magnetic flux density or magnetic induction B = μH (where μ = magnetic permeability of material; in vacuum μ = μ_0) (For a more detailed discussion of magnetic units see Appendix to Part II, Chapter 4.)	Telsa	T	1 T = 10⁴ Gauss = 10⁴ G Note: if $\mu = \mu_0$ then H = 1 A/m corresponds to B = $4\pi(10^{-7})$ T = 4π mG 1 gamma = 10⁻⁵ G = 10⁻⁹ T = 1 nT; 1 milligamma = 1 pT
Electric dipole moment	Coulomb m	Cm	1 Debye = 3.33(10⁻³⁰) Cm
Coefficient of viscosity (shearing stress/rate of shearing strain)	Newton second/m²		1 poise = 10⁻¹ N s m⁻²

Appendix 2

SAFETY STANDARDS

The standards, proposed standards, and recommendations summarized in the tables of this appendix represent the concensus of various committees and private or government agencies. Standards promulgated by different agencies and at different times are inconsistent in several instances and are therefore reproduced here only for reference and not because they are endorsed by the editors of this book. For detailed considerations of safety provided by any particular electromagnetic environment the appropriate chapter of this book as well as the current scientific literature should be consulted.

Table 1
EXPOSURE STANDARDS FOR 50/60-HZ ELECTRIC FIELDS

Standard	Population to which it applies	Exposure limit (kV/m)	Exposure duration	Comments
World Health Organization — 1984		10		There is no need to limit access to regions where the field strength is less than 10 kV/m; exposures should be limited to levels as low as can be reasonably achieved
Australia Victoria — 1976	General public	10 2		Under transmission line At edge of right-of-way
New South Wales — 1976	Occupational	10 2		When practical At edge of right-of-way
Japan — 1976 (ordinance of ministry of international trade and industry)	General public	3	Continuous	Under power lines; buildings permitted directly under transmission lines
Poland — 1980	General public	1	Continuous	Unlimited occupancy, residences permitted
		1—10	Less than full day	Recreational use permitted, buildings prohibited
		10		Occupancy prohibited for general public
U.S.S.R. — 1975	Occupational	5 10 15 20 25	Continuous 3 hr/day 1.5 hr/day 10 min/day 5 min/day	If workers are exposed to fields of 10 kV/m or more for the full time allowed, then they must remain in fields less than 5 kV/m for the rest of the day; workers must not be subjected to spark discharges
U.S.S.R. recommendation	General public	1 12 20	Continuous Short durations	Maximum value where transmission line crosses road Maximum possible exposure to the general public in unpopulated areas
U.K. — 1982 (National Radiological Protection Board, proposed)	General public	10 30	Continuous	Considered to be safe Unlikely to be harmful
Minnesota — 1979	General public	8		At edge of right-of-way
Montana — 1983 (proposed)	General public	1 7	Continuous	At edge of right-of-way In populated areas, at road crossings
New Jersey — 1981 (guideline)	General public	3	Continuous	At edge of right-of-way
New York — 1979 (temporary)	General public	1 7 11	Continuous	At edge of right-of-way Over public roads Over private roads
North Dakota	General public	9		Within the right-of-way
Oregon	General public	9		Within the right-of-way

Table 2
EXPOSURE RECOMMENDATIONS FOR DC MAGNETIC FIELDS

Agency	Population to which it applies	Exposure limits (T)	Uniform field (U) or spatial gradient in T/m	Part of body exposed	Exposure duration
Alpen Committee on interim standards for occupational exposure to magnetic fields (July 1979)	Research personnel and others working in high magnetic fields	2	U	Extremities	Less than 10 min
		0.5	U	Whole body	Less than 10 min
		1	U	Extremities	Less than 1 hr
		0.1	U	Whole body	Less than 1 hr
		0.1	U	Extremites	8-hr work day
		0.01	U	Whole body	8-hr work day
U.S.S.R. Safety Standards recommended by Vyalov	Occupationally exposed persons	0.07	0.2	Hands	8-hr work day
		0.03	0.2	Whole body	8-hr work day
Stanford Linear Accelerator	Occupationally exposed persons	2	U	Arms and hands	Minutes
		0.2	U	Whole body or head	Minutes
		0.2	U	Arms and hands	Extended periods
		0.02	U	Whole body or head	Extended periods
National Accelerator Laboratory	Occupationally exposed persons	0.01—0.5	U	Whole body	Work in areas is minimized
		0.5—1.0	U	Whole body	Less than 1 hr

Note: For effects of magnetic fields on persons with *cardiac pacemakers* see Pavlicek, W. et al., *Radiology*, 147, 149, 1983; "Threshold for initiating asynchronous mode of pacemaker is 17 Gauss (= 0.0017 T) . . . time-varying magnetic fields can generate pulse amplitudes and frequencies to mimic cardiac activity."

Table 3

RADIOFREQUENCY RADIATION EXPOSURE STANDARDS AND GUIDELINES

Standard	Population to which it applies	Frequency range	Exposure limit	Exposure duration	Comments
American National Standards Institute — 1982 (recommendation)	General public and occupational	0.3—3 MHz	$4(10^5)$ V²/m² 2.5 A²/m² 1 kW/m²	Continuous	Guidelines for 300 kHz to 100 GHz may be exceeded if specific absorption rate (SAR) does not exceed 0.4 W/kg for whole body and 8 W/kg in any 1 g of tissue; for exposures at more than one frequency the sum of the fractions of the maximum value must not exceed unity; guideline for 300 kHz to 1 GHz may be exceeded if radiofrequency input power to radiating device is 7 W or less; exposures and SARs are to be averaged over a 0.1-hr period; thus, at 2 GHz a 1 min exposure at 300 W/m² is allowed.
		3—30 MHz	$4,000 (900/f^2)$ V²/m² $0.025 (900/f^2)$ A²/m² $9/f^2$ kW/m²	Continuous	
		30—300 MHz	4,000 V²/m² 0.025 A²/m² 10 W/m²	Continuous	
		300—1,500 MHz	$4,000 (f/300)$ V²/m² $0.025 (f/300)$ A²/m² $f/30$ W/m²	Continuous	
		1.5—100 GHz	$2 (10^4)$ V²/m² 0.125 A²/m² 50 W/m²	Continuous	
American Council of Government Industrial Hygienists — 1983	Occupational	0.01—3 MHz	377,000 V²/m² 2.65 A²/m² 1 kW/m²	Work day	Exposures are averaged over a 0.1-hr period; guidelines for frequencies less than 1 GHz may be exceeded if power input into the radiating device is 7 W or less; guidelines may be exceeded if SAR does not exceed 0.4 W/kg averaged over the person and 8 mW/g in any 1 g of tissue; maximum peak electric field intensity should not exceed 10 kV/m; for exposures at several frequencies, the sum of the fractions of the protection guide within each frequency range should not exceed unity
		3—30 MHz	$3,390,000/f^2$ V²/m² $23.9/f^2$ A²/m² $9/f^2$ kW/m²	Work day	
		30—100 MHz	3,770 V²/m² 0.027 A²/m² 10 W/m²	Work day	
		0.1—1 GHz	37.7f V²/m² $f/3,770$ A²/m² $f/10$ W/m²	Work day	
		1—300 GHz	37,700 V²/m² 0.265 A²/m² 100 W/m²	Work day	

Standard	Category	Frequency range	Exposure limit	Duration	Notes
Occupational Safety and Health Administration U.S. (1972, March 1983 revision of 29 CFR 1910)	Occupational	10 MHz—100 GHz	100 W/m²	Work day	Averaged over 0.1-hr period; same exposure limit for whole-body and for partial-body exposures
International Radiation Protection Association (IRPA) — 1983 (guideline)	General public	0.1—1 MHz	87 V/m 0.23 A/m	Continuous 24 hr/day Continuous	For all frequencies, the instantaneous peak value may not exceed 100 times the exposure limit; whole-body exposures for short periods of time at power densities up to ten times the exposure limit should not produce any untoward effect; devices of 7 W or less are excluded; limits are for 0.1-hr averages; the same exposure limits apply for partial-body exposures; when exposures are to a mixture of frequencies, the sum of the fractions of the exposure limit must not exceed unity
		1—10 MHz	87/$f^{1/2}$ V/m 0.23/$f^{1/2}$ A/m	Continuous	
		10—400 MHz	27.5 V/m 0.073 A/m 2 W/m²	Continuous	
		0.4—2 GHz	1.375 $f^{1/2}$ V/m 0.0037 $f^{1/2}$ A/m f/200 W/m²	Continuous	
		2—300 GHz	61 V/m 0.16 A/m 10 W/m²	Continuous	
	Occupational	0.1—1 MHz	194 V/m 0.51 A/m	8 hr/24 hr	Limits are for 0.1-hr averages; devices of 7 W or less excluded; for frequencies greater than 10 MHz, guidelines may be exceeded if the whole-body SAR does not exceed 0.4 W/kg and spatial peak SAR does not exceed 4 W/kg when averaged over any 1 g of tissue as averaged over any 6-min period; for frequencies below 10 MHz guidelines may be exceeded to 615 V/m or 1.6 A/m if necessary precautions are to be taken to prevent RF burns; instantaneous peak values should not exceed 100 times the limits provided for 0.1-hr averages; when exposures are to a mixture of frequencies, the sums of the fractions of the exposure limits must not exceed unity
		1—10 MHz	194/$f^{1/2}$ V/m 0.51/$f^{1/2}$ A/m	8 hr/24 hr	
		10—400 MHz	61 V/m 0.16 A/m 10 W/m²	8 hr/24 hr	
		0.4—2 GHz	3 $f^{1/2}$ V/m 0.008 $f^{1/2}$ A/m f/40 W/m²	8 hr/24 hr	
		2—300 GHz	137 V/m 0.36 A/m 50 W/m²	8 hr/24 hr	
World Health Organization Preliminary — 1981 (recommendation)	Occupational	Radiofrequencies	1—10 W/m²	Work day	Special considerations may be indicated in the case of pregnant women
	General public	Radiofrequencies	Lower level than occupational levels	Continuous	Should be kept as low as readily achievable

Table 3 (continued)

RADIOFREQUENCY RADIATION EXPOSURE STANDARDS AND GUIDELINES

Standard	Population to which it applies	Frequency range	Exposure limit	Exposure duration	Comments
North Atlantic Treaty Organization — 1982	Occupational	10—1000 kHz	1 kV/m 2.6 A/m 2.65 kW/m²	Not specified	Averaged over a 0.1-hr period; for frequencies less than 30 MHz, both electric and magnetic field strengths must be measured; exposure to electromagnetic pulse radiation must be limited to 100 kV/m in the pulse
		1—10 MHz	500 V/m 1.3 A/m 660 W/m²	Not specified	
		0.01—300 GHz	200 V/m 0.5 A/m 100 W/m²	Not specified	
U.K. National Radiological Protection Board — 1983 (proposed)	General public	3—30 kHz	400,00 V²/m² 2.5 A²/m²	Continuous	See ANSI (1982)
	General public	30 kHz—100 GHz 100—300 GHz	Same as ANSI (1982) 20,000 V²/m² 0.125 A²/m² 50 W/m²	Continuous Continuous	
	Occupational	3—10 kHz	377,000 V²/m² 2.65 A²/m²	Continuous	See ACGIH (1983)
		10 kHz—300 GHz	Same as ACGIH	Continuous	
Japan	General public	3 kHz—300 GHz	None		
France	Occupational	3 kHz—300 GHz	None		
	General public	0.03—300 GHz	10 W/m²	Greater than 1 hr	
	Military	0.03—300 GHz	10 W/m² 10—100 W/m²	100/P² hr	Maximum exposure level 100 W/m²
Canada — Health and Welfare Canada — 1979	Occupational	10—1,000 MHz	60 V/m 0.16 A/m 10 W/m²	8 hr	Averaged over 1-hr period; for whole- or partial-body exposures with exception of extremities; short duration exposures are permitted where t ≤10/P with maximum P = 250 W/m²

Standard	Applicability	Frequency range	Field strength	Duration	Remarks
		1—300 GHz	140 V/m 0.36 A/m 50 W/m²	8 hr	Averaged over 1-hr period; for whole- or partial-body exposures with exception of extremities; short duration exposures are permitted where $t \leq 10/P$ with maximum value of $P = 250$ W/m² for $100 < P < 250$ and $t = 50/P$ with maximum $P = 100$ W/m² for $P \leq 100$ W/m² Ceiling values
		0.01—300 GHz	300 V/m 0.8 A/m 250 W/m²		
		0.01—300 GHz	200 V/m 0.5 A/m 100 W/m²	8 hr	Exposure of extremities averaged over 1-hr period; shorter duration exposures permitted if exposures do not exceed $t = 10/P$ Whole or partial body
	General public	0.01—300 GHz	60 V/m 0.16 A/m 10 W/m²	Continuous	
Federal Republic of Germany — 1984	General public and occupational	10—30 kHz	1.5 kV/m 250 A/m	Unlimited	Values are averaged over 0.1-hr periods; in frequency range 10—300 kHz exposures to arms and legs of 5 kA/m are permitted for 5 min; in the frequency range 10—30 kHz, peak field strengths of 2 kV/m and 500 A/m are allowed; for frequencies greater than 30 kHz the exposure limit may be exceeded for exposures of less than 0.1 hr; specific exposure duration/field strength limits are provided in the standard for each frequency range
		30—2,000 kHz	1.5 kV/m 7.5/f A/m	Unlimited	
		2—30 MHz	3/f kV/m 7.5/f A/m	Unlimited	
		30—3,000 MHz	100 V/m 0.25 A/m 25 W/m²	Unlimited	
		3—12 GHz	$100\,(f/3{,}000)^{1/2}$ V/m $0.25\,(f/3{,}000)^{1/2}$ A/m $25\,(f/3{,}000)$ W/m²	Unlimited	
		12—300 GHz	200 V/m 0.5 A/m 100 W/m²	Unlimited	
Sweden — 1976 (national regulation)	Occupational	10—300 MHz 0.3—300 GHz 0.3—300 GHz 0.01—300 GHz	50 W/m² 10 W/m² 250 W/m² >250 W/m²	8 hr 8 hr 10/P Prohibited	Averaged over 0.1-hr-period; 10/P is the fraction of an hour for which exposure is permitted where P is the power density in W/m²
Netherlands	Occupational	0.03—300 GHz	10 W/m² 100 W/m²	Work day less than 6 min	

Table 3 (continued)
RADIOFREQUENCY RADIATION EXPOSURE STANDARDS AND GUIDELINES

Standard	Population to which it applies	Frequency range	Exposure limit	Exposure duration	Comments
Finland — 1983 (proposed)	Occupational	0.3—3 MHz	200 V/m 0.5 A/m 100 W/m²	Work day	Average over 1-hr period; exposures to body extremities may be permitted to ten times the limit; in frequencies between 300 kHz and 500 MHz exposure limits may be exceeded if total radiated power from the device does not exceed 5 W
		3—10 MHz	110 V/m 0.3 A/m 30 W/m² 0.16 A/m 10 W/m²	Work day	
		30—5,000 MHz	43 V/m 0.12 A/m 5 W/m²	Work day	
		5—300 GHz	10 W/m²	Work day	
		0.3—10 MHz	600 V/m 1.6 A/m 1 kW/m²	1 sec	
		10 MHz—300 GHz	300 V/m 0.8 A/m 250 W/m²	1 sec	
	General public	0.3—3 MHz	90 V/m 0.23 A/m 20 W/m²	Continuous	Exposures averaged over a 6-min period; limiting values do not apply to patients being treated during medical therapy; in the range 0.3—500 MHz exposure limits may be exceeded if total radiated power from the device does not exceed 5 W
		3—10 MHz	43 V/m 0.12 A/m 5 W/m²	Continuous	
		10—30 MHz	30 V/m 0.07 A/m 2 W/m²	Continuous	
		30—5,000 MHz	20 V/m 0.05 A/m 1 W/m²	Continuous	

Standard	Type	Frequency	Field strength/power density	Averaging time	Comments
		5—300 GHz	30 V/m 0.07 A/m 2 W/m²	Continuous	
		0.3—10 MHz	270 V/m 0.70 A/m 200 W/m²	1 sec	
		10 MHz—300 GHz	140 V/m 0.36 A/m 50 W/m²	1 sec	
Italy — 1984 (draft)	Occupational	0.3—3 MHz	140 V/m 0.36 A/m 50 W/m² 300 V/m 0.83 A/m 250 W/m²	8-hr average 8-hr average 8-hr average Maximum level Maximum level Maximum level	Averaged over 0.1-hr period; with prior authorization, workers may be exposed at field strengths twice the maximum (four times maximum power density) for a total of 1 hr in each 6 months
		3 MHz—300 GHz	60 V/m 0.17 A/m 10 W/m² 200 V/m 0.50 A/m 100 W/m²	8-hr average 8-hr average 8-hr average Maximum level Maximum level Maximum level	
	General population	0.3—3 MHz	45 V/m 0.11 A/m 5 W/m²	Continuous	
		3 MHz—300 GHz	20 V/m 0.05 A/m 1 W/m²	Continuous	
Australian Standard (AS-2772) 1985	Occupational	0.3—9.5 MHz	37,700 V²/m² 0.265 A²/m² 100 W/m²	8 hr	In area where possibility of RF shocks and burns exists

Table 3 (continued)

RADIOFREQUENCY RADIATION EXPOSURE STANDARDS AND GUIDELINES

Standard	Population to which it applies	Frequency range	Exposure limit	Exposure duration	Comments
		0.3—3 MHz	377,000 V²/m²; 2.65 A²/m²; 1,000 W/m²	8 hr	In areas where the possibility of RF shocks and burns has been eliminated
		3—9.5 MHz	3,390,000/f² V²/m²; 23.9/f² A²/m²; 9,000/f² W/m²	8 hr	
		9.5—30 MHz	3,390,000/f² V²/m²; 23.9/f² A²/m²; 9,000/f² W/m²	8 hr	All values are averages over a 60-sec period. When exposures are to a mixture of frequencies, the sums of the fractions of the exposure limits must not exceed unity. For periods shorter than 30 min, the limits may be increased by a factor 30/t, up to a maximum of 5, where t is the time in minutes. For frequencies in the range 300 kHz to 300 GHz the following peak levels shall not be exceeded for any period: 10 kW/m², 1,940 V/m, 5.15 A/m
	General population	0.3 MHz—300 GHz		Continuous	The mean square electric or magnetic field strengths shall not exceed one-fifth of the occupational exposure level
U.S.S.R. — 1976 (all-Union state standard 12/1/006-76)	Occupational	0.06—3 MHz	50 V/m	Work day	Fixed and rotating antennas
		3—30 MHz	20 V/m	Work day	Fixed and rotating antennas
		30—50 MHz	10 V/m	Work day	Fixed and rotating antennas
		60—1,500 kHz	5 A/m	Work day	Fixed and rotating antennas
		30—50 MHz	0.3 A/m	Work day	Fixed and rotating antennas
		50—300 MHz	5 V/m	Work day	Fixed and rotating antennas
			0.15 A/m		
U.S.S.R. — 1982 (modification 1 to all-Union state standard)	Occupational	0.3—300 GHz	2/t W/m²; 10 W/m²; >10 W/m²	t >0.4 hr; 0.1—0.4 hr; Prohibited	Exposure limits ten times greater permitted for rotating antennas; different standard applies for exposure of military personnel

Standard	Group	Frequency	Field strength	Duration	Comments
U.S.S.R. — 1984	General public	30—300 kHz	25 V/m	24 hr	
		0.3—3 MHz	15 V/m	24 hr	
		3—30 MHz	10 V/m	24 hr	
		30—300 MHz	3 V/m	24 hr	
		0.3—300 GHz	0.1 W/m²	24 hr	
Poland — 1980	General public	0.1—10 MHz	<5 V/m	Unlimited	Residences permitted
			5—20 V/m	Less than full day	Recreational use, buildings prohibited
		10—300 MHz	>20 V/m	Prohibited	
			<2 V/m	Unlimited	Residences permitted
			2—7 V/m	Less than full day	Recreational use, buildings prohibited
			>7 V/m	Prohibited	
		0.3—300 GHz	<0.25 W/m²	Unlimited	Residences permitted
			0.25—1 W/m²	Less than full day	Recreational use, buildings prohibited
			>1 W/m²	Prohibited	
	Occupational	0.1—10 MHz	<20 V/m	Work day	No limitations for workers
			<2 A/m	Work day	No limitations for workers
			20—70 V/m	Work day	Medical monitoring
			2—10 A/m	Work day	Medical monitoring
			70—1,000 V/m	560/E hours	Medical monitoring
			10—250 A/m	80/H hours	Medical monitoring
			>100 V/m	Prohibited	
			>250 A/m	Prohibited	
		10—300 MHz	<7 V/m	Work day	No limitations for workers
			7—20 V/m	Work day	Medical monitoring
			20—300 V/m	3200/E hours	Medical monitoring
			>300 V/m	Prohibited	
		0.3—300 GHz	<0.1 W/m²	Work day	No limitations for exposure of workers to fixed antennas
			<1 W/m²	Work day	No limitations for exposure of workers to rotating antennas
			0.1—2 W/m²	Work day	Medical monitoring of workers exposed to fixed antennas
			1—10 W/m²	Work day	Medical monitoring of workers exposed to rotating antennas

Table 3 (continued)
RADIOFREQUENCY RADIATION EXPOSURE STANDARDS AND GUIDELINES

Standard	Population to which it applies	Frequency range	Exposure limit	Exposure duration	Comments
Czechoslovakia — 1970 (legal standard)	Occupation		2—100 W/m²	32/P² hours	Medical monitoring of workers exposed to rotating antennas
			10—100 W/m²	800/P² hours	Medical monitoring of workers exposed to fixed antennas
			>100 W/m²	Prohibited	All exposure configurations
		0.03—30 MHz	50 V/m	Work day	400/t V/m for shorter periods
		30—300 MHz	10 V/m	Work day	80/t V/m for shorter periods
		0.3—300 GHz	0.25 W/m²	Work day	Continuous wave exposures only, 2/t W/m² for shorter periods
			0.1 W/m²	Work day	Pulsed exposures, 0.8/t W/m² for shorter periods
	General public		16 W/m²	1 hr	Continuous wave exposure only
			6.4 W/m²	1 hr	Pulsed exposures
		0.03—30 MHz	5 V/m	Continuous	120/t V/m for shorter periods
		30—300 MHz	1 V/m	Continuous	24/t V/m for shorter periods
		0.3—300 GHz	25 mW/m²	Continuous	Continuous wave exposure only, 600/t W/m² for shorter periods
			10 mW/m²	Continuous	Pulsed exposures, 240/t W/m² for shorter exposures
Czechoslovakia — 1984 (proposed)	Occupational	0.03—10 MHz	70 V/m	8 hr	
		10—70 MHz	700/f V/m	8 hr	
		70—300 MHz	10 V/m	8 hr	
		0.3—300 GHz	250 mW/m²	8 hr	
	General population	0.03—10 MHz	14 V/m	Continuous	
		10—70 MHz	140/f V/m	Continuous	
		70—300 MHz	2 V/m	Continuous	
		300—500 MHz	(0.1 f − 20) mW/m²	Continuous	
		0.5—300 GHz	30 mW/m²	Continuous	
German Democratic Republic — 1978 (legal standard)	Occupational	0.06—3 MHz	50 V/m	Work day	
		3—30 MHz	20 V/m	Work day	
		30—50 MHz	10 V/m	Work day	

Standard	Population	Frequency	Limit	Time	Notes
Bulgaria — 1979 (legal standard)	Occupational	50—300 MHz	5 V/m	Work day	
		0.3—300 GHz	0.1 W/m²	8 hr	Stationary antenna
			1 W/m²	2 hr	Stationary antenna
			10 W/m²	20 min	Stationary antenna
			1 W/m²	8 hr	Rotating antenna
			10 W/m²	2 hr	Rotating antenna
		0.06—3 MHz	50 V/m	Work day	
		3—30 MHz	20 V/m	Work day	
		30—50 MHz	10 V/m	Work day	
		50—300 MHz	5 V/m	Work day	
		60—1,500 kHz	5 A/m	Work day	
		30—50 MHz	0.3 A/m	Work day	
		0.3—300 GHz	0.1 W/m²	Work day	Stationary antenna
			1 W/m²	2 hr	Stationary antenna, 0.1 W/m² maximum for remainder of work day
			10 W/m²	20 min	Stationary antenna, 0.1 W/m² maximum for remainder of work day, protective goggles required
			1 W/m²	Work day	Rotating antenna
			10 W/m²	2 hr	Rotating antenna, reduce to 1 W/m² if ambient temperature exceeds 28°C or if X-radiation is also present
Hungary — 1982 (draft)	General public	0.03—30 MHz	480/t V/m	t = 1—24 hr	
			≥1 kV/m	Prohibited	
		30—300 MHz	240/t V/m	t = 1—24 hr	
			≥500 V/m	Prohibited	
		0.3—300 GHz	2.4/t W/m²	t = 1—24hr	
			≥100 W/m²	Prohibited	
	Occupational	0.03—30 MHz	1,200/t V/m	t = 1—24 hr	Higher level short-duration occupatinal exposures are allowed if proper protective equipment is used
			≥1 kV/m	Prohibited	
		30—300 MHz	480/t V/m	t = 1—24 hr	
			≥500 V/m	Prohibited	
		0.3—300 GHz	24/t W/m²	t = 1—24 hr	
			≥100 W/m²	Prohibited	
U.S. Air Force — 1984	Occupational	0.01—3 MHz	1 kW/m²		Equipment that radiates at frequencies below 1 GHz and delivers less than 7 W of power to the radiating device is considered nonhazardous
		3—30 MHz	9/f² kW/m²		
		30—100 MHz	10 W/m²		
		0.1—1 GHz	f/10 W/m²		
		1—300 GHz	100 W/m²		

Table 3 (continued)
RADIOFREQUENCY RADIATION EXPOSURE STANDARDS AND GUIDELINES

Standard	Population to which it applies	Frequency range	Exposure limit	Exposure duration	Comments
	General public	0.01—3 MHz 3—30 MHz 30—300 MHz 0.3—300 MHz 1.5—300 GHz	1 kW/m² 9/f² kW/m² 10 W/m² f/30 W/m² 50 W/m²		Permissible exposure limits may be increased provided that: (1) the SAR does not exceed 0.4 W/kg when averaged over the whole body over any 0.1-hr period; (2) the spacial peak SAR does not exceed 8 W/kg averaged over any one 1 g of tissue; (3) the maximum (peak) electric field intensity does not exceed 100 kV/m; (4) personnel are adequately protected from electric shock and burns through the use of appropriate safety techniques (primarily applicable for frequencies below 30 MHz where the primary hazard is from electric shock and burns)
Massachusetts Department of Public Health — 1983	General public	0.3—3 MHz	200 W/m² 80,000 V²/m² 0.5 A²/m²	Continuous	Exempts all sources that have less than 7 W effective radiated power; for mixed frequencies, the sum of the fractions of the exposure guidelines incurred shall not exceed unity
		3—30 MHz	1800/f² W/m² 7.2(10⁵)/f² V²/m² 4.5/f² A²/m²	Continuous	
		30—300 MHz	2 W/m² 800 V²/m² 0.005 A²/m²	Continuous	
		300—1,500 MHz	f/150 W/m² 8.67f V²/m² f/1500 A²/m²	Continuous	
		1.5—100 GHz	10 W/m² 4000 V²/m² 0.025 A²/m²	Continuous	

Portland, Ore.	General public	0.1—30 MHz	5 W/m²	Continuous	This is a regulation of the emission rather than an exposure standard; times refer to emission durations
			10 W/m²	12 hr or less	
		30—300 MHz	1 W/m²	Continuous	
			2 W/m²	6—12 hr	
			4 W/m²	3—6 hr	
			5 W/m²	Less than 3 hr	
		0.3—10 GHz	5 W/m²	Continuous	
			10 W/m²	12 hr or less	
Connecticut — 1984	General public	0.3 MHz—300 GHz	See ANSI (1982)	Continuous	Same as ANSI (1982)
New Jersey — 1983	General public	0.3 MHz—300 GHz	See ANSI (1982)	Continuous	Same as ANSI (1982)
San Diego County, Calif. — 1984	General public	0.3 MHz—300 GHz	See ANSI (1982)	Continuous	Same as ANSI (1982)
Multnomah County, Ore. — 1982	General public	0.1—3 MHz	80,000 V²/m² 0.5 A²/m² 200 W/m²	Continuous	Average over 0.5-hr period; for shorter periods more intense exposures are permitted according to the formula P = Po/t, where Po is the exposure limit for that frequency range as averaged over a 0.5-hr period
		3—30 MHz	720,000/f² V²/m² 4.5/f² A²/m² 1,800/f² W/m²	Continuous	
		30—300 MHz	800 V²/m² 0.005 A²/m² 2 W/m²	Continuous	
		0.3—1.5 GHz	2.7 f V²/m² 1.7×10^{-5} f A²/m² f/150 W/m²	Continuous	
		1.5—300 GHz	4000 V²/m² 0.025 A²/m² 10 W/m²	Continuous	
Johns Hopkins University Applied Physics Laboratory — 1984	Occupational	0.3—3 MHz	200 V/m 0.5 A/m 100 W/m²	8 hr	Averages over 0.1-hr period; maximum peak power density 1 kW/m²; exposures to higher intensities for very brief periods (less than 6 min) is permitted; personnel must ensure that exposures are as low as reasonably achievable
		3—30 MHz	600/f V/m 1.5/f A/m 900/f² W/m²	8 hr	
		30 MHz—100 GHz	20 V/m 0.05 A/m 1 W/m²	8 hr	

Note: f = Frequency in MHz; P = power density in W/m²; t = time in hours; E = electric field strength in V/m; and H = magnetic field strength in A/m.

Index

INDEX

B

C